BIOLOGICAL FIXATION OF
ATMOSPHERIC NITROGEN

D1545423

Biological Fixation of Atmospheric Nitrogen

by

E. N. MISHUSTIN

Corresponding Member of the USSR Academy of Sciences and Honoured Man of Science of the RSFSR.

and

V. K. SHIL'NIKOVA

Translated by Alan Crozy

The Pennsylvania State University Press
University Park and London

International Standard Book Number: 0–271–01110–6
Library of Congress Catalog Card Number: 78–177914

First published 1971

Published by
THE MACMILLAN PRESS LIMITED

London and Basingstoke,
Associated companies in New York,
Toronto, Melbourne, Dublin, Johannesburg
and Madras

БИОЛОГИЧЕСКАЯ ФИКСАЦИЯ АТМОСФЕРНОГО АЗОТА

Мишустин, Е. Н. и Шильникова, В. К.

Издательство "Наука" Москва 1968. 531 pp.

Translated from the Russian in co-operation with
THE DEPARTMENT OF EDUCATION AND SCIENCE
National Lending Library for Science and Technology
Boston Spa, Yorks

Published in the United States of America
by The Pennsylvania State University Press
Printed in Great Britain

CONTENTS

CONTENTS

LIST OF PLATES

1 Electron micrograph of cells of *Rhizobium* with monotrichous flagella. (*a*) From the nodule of *Phaseolus* (after Ziegler); (*b*) from the nodule of *Glycine hispida* (after De Ley and Rassel).

2 Electron micrograph of *Rhizobium* cell with peritrichous flagella from peanut nodule (after De Ley and Rassel).

3 Cell of nodular tissue of *Trifolium subterraneum* filled with bacteroids. The cytoplasm occupies a limited volume of the cell and the mitochondria are pushed against the cell wall. Residues of the infection thread can be seen containing bacteria.

4 Nodules on the roots of *Tribulus cistiodes* (after E. and O. Allen).

5 Fimbria in *Azotobacter* (electron microscopy preparation after Nikitin).

6 *Tolypothrix tenuis.*

7 *Cylindrospermum constricta.*

8 *Anabaena cylindrica.* The free-floating oval bodies are spores and the rounded cells within the algal filament are heterocysts, the sites of nitrogen fixation.

9 *Anabaena flos-aquae* with a heterocyst and gas vacuoles within the other cells. (*a*) Spiral form; (*b*) a straight filament.

10 *Nostoc muscorum.*

11 Open system for culture of algae (after Pinevich).

12 *Mycobacterium azot-absorption* (after L'vov).

NOTE TO READER

Throughout this book the references are cited according to the author's name as it appears on the work. In the case of works by Western authors originally appearing in Cyrillic script publications or quoted from a Russian translation, this will occasionally result in a somewhat unfamiliar form. In such cases the usual form of the name is also cited in the bibliography.

FOREWORD

The biological fixation of molecular nitrogen attracts the attention of bio-chemists, biologists and agricultural workers. From the general biological point of view most interest focusses on the chemical pathways, which can be set up experimentally only with great difficulty, although they are commonplace in living cells. The unravelling of the biochemistry of this process may not only be of great theoretical importance, but may also be useful in revealing new ways to obtain compounds for the binding of atmospheric nitrogen on a commercial scale.

Nitrogen fixation is widespread in the micro-organisms which live in the soil, fresh-water, sea-water and tissues of higher plants and animals. In each of these environments nitrogen-fixing organisms play their own specific roles.

This book looks at the process of biological nitrogen fixation in the soil. It pays special attention to the practical utilization of nitrogen fixation in agricul-ture, since nitrogen assimilated by micro-organisms ('biological' nitrogen) can replenish substantially the nitrogen reserves of the soil, and help to improve soil fertility. Biological nitrogen makes it possible to counteract a shortage of mineral nitrogenous fertilizers and to use these fertilizers more economically.

There is an immense literature concerned with nitrogen fixation by micro-organisms–either symbiotic with plants or living free in the soil. Shil'nikova and I do not intend to give an exhaustive general survey of the world literature, but take as our principal starting point recent investigations, which in our judge-ment are of fundamental interest. Undoubtedly, factors of a subjective nature will be involved.

As well as published material, the book presents the results of our own experi-mental work with various groups of nitrogen-fixing micro-organisms.

The book is concerned chiefly with problems of the biology and ecology of nitrogen-fixing organisms. There is only a brief review of the biochemistry of the process—this problem warrants a special monograph.

E. N. Mishustin. Corresponding
Member of the USSR Academy of
Sciences and Honoured Man of
Science of the RSFSR.

FOREWORD

1 Introduction[*]

Long before the beginning of the present century, Boussingault (1838) established the importance of nitrogen in plant nutrition, and gave pride of place to nitrogenous among other forms of fertilizers in terms of their influence on the harvest.

Usually only a little nitrogen in the soil is accessible to plants, and so in all countries the problem of increasing the yield of farm crops is, in practice, primarily associated with improving their nitrogen nutrition. Of course, the maximum effectiveness of nitrogenous fertilizers can be achieved only by

Table 1.1. World Production of Fertilizers

Year	N tons $\times$ 10^6	P_2O_5 tons $\times$ 10^6	K_2O tons $\times$ 10^6
1913	0·51	2·14	1·27
1938	2·4	3·50	2·50
1962	11·5	10·40	9·30
1964	14·5	12·5	10·6

From *The Food and Agriculture Organization Production Yearbook*, 1965.

'balancing the nutrition of plants' (Donald *et al.*, 1965) with other necessary nutritional elements, and by following many essential agricultural practices. We shall not go into these questions here, but concentrate on the nitrogen nutrition of plants.

Utilization of 'industrial' nitrogen—the nitrogen of mineral fertilizers—is one of the most effective ways of making good the nitrogen deficit of the soil (Koren'kov, 1966). Accordingly, the pace of development of world production of mineral fertilizers, particularly nitrogenous types, is steadily quickening (Peterburgskii, 1965b). In the past fifty years the world production of nitrogenous fertilizers has increased twenty-nine-fold, of potassium fertilizers 8·4-fold and of phosphorus fertilizers 5·8-fold (Table 1.1). The approximate figures for world nitrogen production from 1956/1957 to 1961/1962 are given in Fig. 1.

The use of mineral fertilizers varies in different countries. Naturally the yield

[*] This introduction was written in collaboration with A. V. Peterburgskii.

1

of farm crops in individual countries bears a definite relation to the quantity of mineral fertilizers used. On average, per hectare of cultivated land, the following quantity of mineral nitrogen (measured in kilogrammes) is used annually: Japan 112 (Peterburgskii, 1964), Holland 128 (Peterburgskii, 1965a), German Federal Republic 54 (Sinyagin, 1965), Great Britain 39·3 (Cooke, 1964), France 36·6, the United States 14·6 (Garmen and White, 1964) and the German Democratic Republic 43·5 (Statistical Handbook, 1963).

In other countries less nitrogenous fertilizer is used, and the yield of crops is proportional to this use. In Japan, on average, 4,700 kg of rice per hectare is

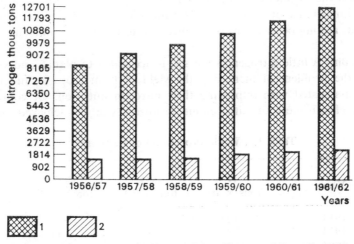

Fig. 1. World nitrogen production (after Sharp and Powell, 1965).
1, Agriculture; 2, industry

harvested, in Holland the average yield of winter wheat is 4,400 kg per hectare. In the German Democratic Republic the figure is 3,180 kg per hectare, while in France the same crop yields 2,390 kg per hectare and in the United States 1,610 kg per hectare.

The various countries fall into a different order when compared with respect to the distribution of mineral fertilizers per head of population rather than per unit of farm land. In this case the use of mineral fertilizers in Holland amounts to 20 kg, in the United States 17, in France 14, in the USSR 10, in Great Britain 9·3, in Japan 9, in the Chinese People's Republic 1 and in India 0·7 (*Bulletin des Engrais*, 1963).

The average quantity of nitrogenous fertilizers used in the USSR per hectare of arable land approaches 11 kg (converted to nitrogen). Of course this average is still very low. This is largely because such vast areas are farmed; the position will, of course, improve. According to the directives of the twenty-third congress of the Communist Party of the Soviet Union, by 1970 the production of mineral fertilizers would be more than double that in 1965.

It should be noted that mineral fertilizers, and especially the nitrogenous types, are usually not used on all agricultural land. Thus, in the USSR commercial crops (cotton, tea, sugar beet, citrus fruits and so on) receive virtually all their principal requirements for nutritive elements (N, P and K) in the form of mineral fertilizers, which, however, are not applied to cereals, everywhere, nor in full measure. For example, in 1965 in the RSFSR, only 15 per cent of sowings of cereals received mineral fertilizers in one form or another.

In many West European countries mineral fertilizers are introduced in large amounts into the soil under perennial grasses and also on pasture land. This ensures the supply of a large amount of food and the maintenance of a large cattle stock. Thus, in Holland, on average 150 kg per hectare of nitrogen is placed under perennial grasses, 125 kg per hectare under potatoes and 60 kg under cereals. With such an outlay of fertilizer it is possible to keep 160 head of cattle, 127 pigs and 193 fowl on 100 hectares of farmland.

In Great Britain the optimum dose of nitrogen for tilled crops and perennial grasses is put at 125 kg per hectare; for cereals it is 75 kg. The number of cattle kept in these conditions is already more than 100 per 100 hectares.

It is interesting to compare the amounts of nitrogenous fertilizers mentioned with the gross utilization of nitrogen by the crop. On average, it can be taken that for a grain yield of 1,000 kg/hectare about 30 kg of nitrogen is removed from the field (with the grain and the chaff). The coefficient of utilization of nitrogenous fertilizers at best is about 75 per cent. Thus, to ensure the yield of each 1,000 kg of grain it is necessary to add about 37·5 kg of nitrogen per hectare in the form of fertilizer. A yield of 3,000 kg per hectare requires about 112 kg of nitrogen per hectare of arable land.

Plants cannot utilize nitrogenous fertilizers completely, and their efficiency depends on several factors, which vary for different forms of nitrogen. The ammoniacal nitrogen in fertilizers is partially fixed in a non-exchangeable form by the minerals of clay soil (Peterburgskii and Korchagina, 1965). There is a fairly large store of firmly bound ammoniacal nitrogen in the soil. Its content in the cultivated layer of different soils ranges from 134 to 344 kg per hectare. Ammonia is fixed to the greatest degree by the clay fraction of the soil. Bound ammoniacal nitrogen is extremely inaccessible to plants.

Nitrate nitrogen, even with an optimum air-water regime in the soil, is partially lost as a result of denitrification. This is not surprising since in the soil, especially on the surfaces of roots and around the root systems, are many denitrifying bacteria of the genus *Pseudomonas*. The amount of nitrogen lost as a result of denitrification depends on the amount of nitrate nitrogen introduced, on the type of soil and and on other conditions. In certain cases these losses reach 10–35 per cent of the nitrogen introduced.

In part, the losses of nitrogenous mineral compounds are a consequence of their being washed out from the arable layer of the soil. This is particularly true of wet districts with light soils.

Our rough calculations of the nitrogen requirements of graminaceous plants have not indicated that the appropriate amounts of this element must necessarily be introduced in the form of mineral fertilizers. It should not be forgotten that part of the nitrogen may be obtained from reserves in the soil. The more fertile the soil the greater its role in supplying plants with nitrogen. This naturally makes possible a more economic use of mineral fertilizers. There have been many analyses of the dynamics of the nitrogen store of the soil (for example Harmsen and van Schreven, 1955; Jansson, 1958; Harmsen, 1961).

As a rule, mineral fertilizers sufficient to almost cover the removal of nitrogen from the soil by the crop are used only in countries with a small area under cultivation and a high population density, for example Holland and Japan. In such countries, particular importance attaches to intensification of farming in order to reduce imports of foodstuffs and concentrated feeds. Since these countries use not only mineral but also organic fertilizers, the nitrogen balance of the soils is positive. In countries with a large farming area such as France, Britain and the United States, mineral nitrogen is used in much smaller amounts, not covering its depletion by plants. This is connected with the profitability of running a farm: mineral nitrogen is relatively expensive and its use is not always paid for by the extra yield of cereals. It is, therefore, applied in sufficiently large quantities only for valuable forage and industrial farm crops.

In some countries, for example Britain, the nitrogen balance of the soil is achieved without a deficit, largely because of the use of manure and leguminous crops—nitrogen accumulators. Thus, according to rough calculations, in Britain mineral fertilizers cover about 27 per cent of the removal of nitrogen by farm plants, manure (taking into account home and imported supply) accounts for about 31 per cent and nitrogen accumulators for about 42 per cent.

In countries with large areas under cultivation the nitrogen balance of the soil is often negative. In particular in the United States as much as 30 per cent of the nitrogen removed is not returned to the soil. This affects the yield of farm crops, which is considerably lower (except for maize and irrigated sugar beet) than in countries where larger quantities of fertilizers are used.

In the USSR the gross production of nitrogenous fertilizers is 2·694 million tons of nitrogen a year—17 per cent of world production. The USSR is now successfully applying chemistry to agriculture, and its production of mineral fertilizers is developing rapidly. By 1970 it is hoped to more than double the 1965 level of production. Because of the huge farming areas the average fertilizer level in the USSR is low. According to A. V. Sokolov (1963) in the period 1954 to 1959 the nitrogen deficit in agriculture in the Soviet Union was about 70 per cent. But since the use of nitrogenous mineral fertilizers is increasing considerably, it must be supposed that the amount of nitrogen returned to the soil will increase in the near future.

According to the rough calculations of Kolar and Greenland (1961) world agricultural production removes about 100–110 million tons of nitrogen from the soil annually. The world chemical industry produced only 14·5 million tons of nitrogen in the form of fertilizer in 1963–1964. At present the gap is largely compensated biologically. Hence the great importance of biological nitrogen, that is nitrogen obtained from micro-organisms.

D. N. Pryanishnikov, the founder of Soviet agrochemistry, in his classic work *Nitrogen in the Life of Plants and Farming in the USSR* (1945) wrote that however high the production of mineral fertilizers one cannot imagine the nitrogen problem being solved solely by the chemical industry. It must be solved to a large extent with the aid of nitrogen-gathering plants and nitrogen-fixing bacteria, that is nitrogen must be fixed biologically. This conclusion still holds for most countries.

In several black earth regions of the USSR quite satisfactory yields are obtained without mineral fertilizers. These soils have been farmed for more than three hundred years and according to calculations ought by now to have lost all their nitrogen. Nevertheless, they remain quite rich in nitrogen and give satisfactory harvests. For this we are indebted to the nitrogen fixers.

It is well known that two groups of micro-organism fix atmospheric nitrogen. One is in symbiosis with higher plants, usually forming nodules on their roots, while the other lives in the soil independently of the plants. The potentialities of the symbiotic and free living nitrogen-fixing bacteria are not equal—something that will be made clear in this book.

Leguminous plants play an unquestionable role in maintaining the fertility of agricultural soils. During the third to first centuries BC this topic was discussed by the Greek philosopher Theophrastus and the Romans Cato and Varro. The latter, in particular, noted that leguminous plants should be sown especially in light soils, not so much for the sake of their yield as for the benefit to succeeding crops.

Later, it was shown that leguminous plants benefit the soil in large measure by enriching it with nitrogen. The first exact experiments were carried out by Boussingault (1838). On the Beshelbron estate in Alsace he carried out field experiments on the rotation of different crops, in which he studied the turnover of nitrogen and other elements. He established beyond doubt that lucerne and clover enrich the soil with nitrogen. Growing these plants increases the yield of the succeeding winter crops without the use of manure. Cereal and root crops turned out to deplete the soil, confirming earlier suspicions. Boussingault drew his conclusions fifty years before Hellriegel and Wilfarth discovered the role of nodule bacteria in the fixation of atmospheric nitrogen (Hellriegel, 1886; Hellriegel and Wilfarth, 1888).

Boussingault deserves the most credit for establishing the role of leguminous plants in the enrichment of soil nitrogen. As is well known, he carried out a series of experiments with the aim of verifying that legumes fix molecular nitrogen. He took young clover plants from the field, carefully washed them free of soil and replanted them in calcined sand. The plants, of course, were inoculated by nodule bacteria in the field, and when grown in the sand they appreciably increased the content of nitrogen.

Later (1851–1853) Boussingault conceived the possibility of the uptake of ammonium carbonate from the air. He grew some plants from seeds sown in calcined sand under a bell jar. In these conditions, which prevented infection of the root system by nodule bacteria and thus the formation of nodules, no nitrogen was fixed. Boussingault, who knew nothing of nodule bacteria, concluded that his first experiments were faulty and abandoned them. As Pryanishnikov said, he threw the baby out with the bath water.

But field experiments provided ever stronger evidence of the value of leguminous plants. Experiments at Rothamsted Experimental Station in England in 1850 and 1860 with different alternating crops showed that legumes were better precursors for wheat than clean fallow ground. In ten years five harvests of wheat and five harvests of legumes yielded 872 kg of nitrogen per hectare, whereas ten harvests of wheat (not alternating with legumes) gave only 262 kg nitrogen per hectare. Schulz (1881) showed that in sandy soils where lupins had previously grown, the potato yield could be increased from 8 to 20 tons per hectare with simultaneous enrichment of the soil with nitrogen. In the United States Atwater (1885) confirmed the accumulation of nitrogen by leguminous crops. During the second half of the nineteenth century many Russian investigators also concluded that leguminous plants were of undoubted value.

The discovery of the causes of the enrichment of soil nitrogen by leguminous plants came in a series of microbiological studies (Lachmann, 1858; Voronin, 1866, 1867; Beijerinck, 1888; Pražmowski, 1890; Kossovich, 1890; Timiryazev, 1892).

Lachmann (1858) discovered that the root nodules of leguminous plants contain vibrio-like micro-organisms. Independently and somewhat later, Voronin (1866, 1867) also established this. In a paper published in *Zapiski Rossiiskoi Akademii Nauk* (1866) he described innumerable rod-shaped bodies in root nodules. He also established that they were motile. Voronin detected inflated cells as well as rod shapes. His drawings leave no doubt that he was describing nodule bacteria.

Beijerinck isolated a pure culture of nodule bacteria (*Rhizobium*) in 1888. He demonstrated that they caused the formation of nodules in which molecular nitrogen was assimilated. Somewhat later a pure culture of nodule bacteria was also isolated by Pražmowski (1890).

The extensive experience now accumulated, points to the importance of leguminous plants in soil fertility. According to Pryanishnikov, after the introduction of field rotation with clover the average yield of cereal crops in Europe increased from 700 to 1,700 kg per hectare. At the Timiryazev Agricultural Academy (Moscow) rotation involving one year of clover gave, in the course of fifty years, a rye harvest of about 1,400 kg per hectare (without mineral fertilizer). The same crop rotation without clover gave only 700 kg per hectare (Egorov, 1961). The introduction of mineral fertilizers, of course, helps to increase the cereal yield still further.

Leguminous plants can also increase yields of cereals in more fertile soils, where agricultural practice is good. In Chayanov's work, for example, the

Voronezh Experimental Station four-plot rotation without leguminous plants and fertilizers gave about 2,000 kg of winter wheat per hectare; with the ground under clover for one year the yield increased to 2,500 kg per hectare, and with two years of clover it was 2,800 kg per hectare. Similar yields persisted during the seventeen years of the experiment.

In view of all these findings it is not surprising that in countries with a highly developed agriculture up to 20–25 per cent of the cultivated land is usually taken up by leguminous plants; this helps both to give valuable fodder and to enrich the soil with nitrogen.

Pryanishnikov concluded that leguminous grasses grown from pure seed are

Table 1.2. Comparison of the Effectiveness of Clover and
Rye Grass as Precursors of Winter Wheat

Quantity of nitrogen under wheat (kg/hectare)	Yield of wheat 100 kg/hectare		Difference with use of clover
	With clover	With rye grass	
0	61·7	44·1	+17·6
63	70·5	67·9	+ 2·6
125	73·0	68·0	+ 5·0

much more valuable in rotation than mixtures with graminaceous grasses. Convincing support for this contention has come from other countries. Thus, in an experiment begun in 1942 on the experimental farm of the University of Iowa, the yield of maize grain in four-plot rotation (maize-maize-oats-grass) was 5,990 kg per hectare when lucerne alone was sown in the grassy field, but when a leguminous-graminaceous mixture was sown in the same field the yield of maize was 3,950 kg per hectare. In an experiment at Rothamsted yields of winter wheat were compared in rotation with one-year clover (not receiving nitrogen fertilizer) and rye grass (fertilized with 151 kg per hectare of nitrogen). As Table 1.2 shows, clover was obviously superior. It not only ensured a higher yield of wheat but helped to save a large amount of nitrogen and fertilizer. At Rothamsted too a lucerne bed was far superior as a precursor to annual grasses (Table 1.3).

It would be wrong, however, to imagine that all types of leguminous plants enrich the soil with nitrogen in equal measure.

Table 1.4 gives some comparative data (Pryanishnikov, 1945; Tyurin, 1957). Clearly the level of fixation possible depends on the type of leguminous plant grown. Of course, the conditions in which the seeds are grown is also important.

It has also been established that the distribution of assimilated nitrogen differs in the organs of perennial and annual legumes. Whereas in perennials no less

Table 1.3. Comparison of the Effectiveness of Lucerne and
Annual Grasses as Precursors of Winter Wheat

Quantity of nitrogen under wheat (kg/hectare)	Yield of wheat 100 kg/hectare		Difference for lucerne
	With lucerne	With grasses	
0	56·9	35·3	+21·6
50	68·8	49·1	+19·7
100·4	70·0	60·5	+ 9·5
150·6	68·8	65·0	+ 3·8

than one third of bound nitrogen is concentrated in the root mass, in annual legumes practically all the nitrogen assimilated from the atmosphere has passed to the aerial parts by the time of ripening.

Economic conditions also encourage the choice of particular legumes in different countries. But as Pryanishnikov has emphasized, with a widespread use of annual leguminous plants the nitrogen they have accumulated will be returned to the soil in the form of manure.

Tyurin and Mikhnovskii (1961) and Mikhnovskii et al. (1965, 1966) measured the levels of nitrogen accumulation by different types of leguminous crops in the Moscow Region, on soddy-podzolic, medium loamy soil. The soil was fertilized with 30 tons of manure per hectare and with phosphorus-potassium fertilizers. The experiments lasted for several years. Table 1.5 shows the results obtained for leguminous plants typical of the central USSR, with reference to the amounts of bound nitrogen left in the soil as root and crop residues after harvesting. This table shows the mean values of the annual change in the store of nitrogen in the soddy-podzolic soil under different types of leguminous crops. Allowing for some approximation, it can be said that in optimum conditions of growth in the southern zone, lucerne is capable of enriching the arable layer of one hectare of soil with 150–200 kg of nitrogen annually. Somewhat lower

Table 1.4. Nitrogen Accumulated Annually by Different Leguminous Crops

Plant	Total quantity of nitrogen bound by plant (kg/hectare)	Loss or gain of nitrogen in post-harvest soil
Lucerne	300 (to 500–600)	+100 (to 150–200)
Clover	150–160 (to 250–300)	+75–100 (to 125–150)
Lupins	to 150	+ about 130
Pulse legumes	50–60	−5 (to −15)

Table 1.5. Nitrogen Accumulated by Different Leguminous Crops

Leguminous plant	Nitrogen enrichment or impoverishment (kg/hectare)
Clover (mixed varieties)	+77
Red clover	+50 −55
Perennial lupin	+90
Annual lupin	+30
Bean–pea mixture	−40

enrichment is given in the central zone by clover and lupin (especially the annual form) while annual pulse legumes at best leave 10–20 kg of nitrogen in the soil, but they may remove some nitrogen from the soil.

Others have reached similar conclusions (for example Zakharchenko and Shilina, 1964; Litvinyuk, 1965; Bazilevich and Dement'eva, 1965). Thus, in agriculture, perennial legumes are of the greatest interest as nitrogen fixers.

In the Argentine, according to Halbinger (1965), symbiotic nitrogen fixation leads to an annual entry of 100 kg of nitrogen per hectare into the soil.

As we shall see, many, but by no means all, types of leguminous plants can fix molecular nitrogen. Many plants not belonging to the Leguminosae can also fix nitrogen. According to Bond (1963), who has studied this group in detail, 200 different species bind atmospheric nitrogen in symbiosis with micro-organisms, usually forming root nodules (sometimes leaf nodules).

Such symbiotic nitrogen fixers are mostly woody and shrub plants, and are therefore more important for forestry than field husbandry. Of course, these plants may also be used as twig fodder for cattle, and their nitrogen is then introduced into the soil in the form of manure. But the contribution of twig fodder to animal husbandry is very small and cannot seriously influence the nitrogen balance in agriculture.

In the soil there is a considerable variety of free-living fixers of molecular nitrogen. Their activity may enrich the soil with nitrogen. This was first demonstrated by Jödin (1862) who established that nutrient solutions in closed vessels containing organic non-nitrogenous substances can be enriched with the bound nitrogen of the atmosphere when micro-organisms develop in them (Mycoderma). Later, Berthelot (1885) confirmed this in relation to the soil. In sterilized soil in the summer, the content of nitrogen did not change, but in non-sterile soil it increased. This suggested that micro-organisms aid in the enrichment of the soil with nitrogen. This became obvious when Vinogradskii (1893) isolated from the soil *Clostridium pasteurianum*—an anaerobic sporulating bacterium that binds gaseous nitrogen. Later, Beijerinck (1901) discovered *Azotobacter*— an aerobic non-sporulating bacterium with the same function. Other free-living nitrogen fixers have been described.

A survey of world experience showed Pryanishnikov that the nitrogen deficit of 13–14 kg per hectare is compensated by the activity of the free-living nitrogen fixers, and by the entry of nitrogen compounds with rain water and so on.

Tyurin (1957) estimated that for yields of 1,000 kg per hectare of cereals saprophytic nitrogen fixers may enrich the arable layer of one hectare of soil with about 5 kg of nitrogen, while for yields of 2,000 kg per hectare the figure for enrichment is 10 kg. We feel that these figures are rather exaggerated, but they rightly indicate the limited possibilities of this group of micro-organisms. Certain amounts of bound nitrogen in the soil may, of course, be contributed by microscopic blue-green algae—photosynthesizing micro-organisms living on atmospheric carbon dioxide. But the entry of nitrogen into the soil, as a consequence of their activity in dry soil, can barely exceed a few kilograms per hectare per year.

The poor effectiveness of free-living, nitrogen-fixing micro-organisms in natural conditions has been confirmed by many years of farming. Thus, in West Europe before the use of leguminous plants in crop rotation, average yields of cereals were 700 kg per hectare. At Rothamsted Experimental Station, in England, wheat sown without rotation between 1843 and 1925 on podzolized, acid, loamy soil with no fertilizer gave an average yield of 880 kg per hectare. A similar result was obtained with cultivated rye without rotation in the soddy-podzolic soil of the Timiryazev Agricultural Academy (Moscow). Without fertilizers the yield of rye was 700 kg per hectare.

Only free-living nitrogen fixers can function when there is no rotation. They stabilize the harvest, but its level is plainly unsatisfactory. Saprophytic nitrogen fixers undoubtedly play a beneficial role in the soil, but cannot ensure good harvests.

In some exceptional cases, for example when large doses of carbon-containing substances, such as straw and green matter, are added to the soil, the level of nitrogen accumulation by free-living nitrogen fixers may increase considerably.

Undoubtedly, blue-green algae are important in rice fields; they assimilate nitrogen and multiply in large numbers. Published findings indicate that nitrogen accumulation through the activity of algae may reach 50 kg (and more) per hectare per year. This is of undoubted practical interest and justifies attempts to use algae in the cultivation of rice. Apparently, good results can be obtained in rice fields by using the water fern *Azolla* living in symbiosis with the blue-green alga *Anabaena*. The proper exploitation in agriculture both of symbiotic and free-living nitrogen fixers is clearly possible only with a full understanding of the factors that determine the assimilation of molecular nitrogen. Ways should be found to intensify this process. This will help to solve several practical problems connected with the agricultural use of microbiological preparations intended to intensify nitrogen accumulation. All these problems come within the scope of the other chapters of this book, which summarize both our own observations and published findings.

References

Atwater, W. O. 1885. On the acquisition of atmospheric nitrogen by plants. *Amer. Chem. J.*, **6**, 365–388.

Bazilevich, N. I. and Dement'eva, T. G. 1965. Uptake and return of ash elements and nitrogen in cotton-lucerne crop rotation. *Agrokhimiya* (9), 28–41.

Beijerinck, M. W. 1888. Die Bakterien der papilionaceen Knollchen. *Bot. Ztg.* **46**, 725–735, 797–803.

Beijerinck, M. W. 1901. Über oligonitrophile Mikroben. *Zbl. Bakteriol. Parasitenkunde, Infektsionskrankh. und. Hyg*, part 2, **7**, 561–582.

Berthelot, M. 1885. Fixation directe de l'azote atmospherique. *C. R. Acad. Sci.*, **101**, 784.

Bond, G. 1963. The root nodules of non-leguminous angiosperms. Symbiotic associations. *Thirteenth Symposium, Soc. Gen. Microbiol.*, 72–92.

Boussingault, J. 1838. Recherches sur la vegetation entreprises dans le but d'examiner si les plantes fixent dans leur organisme l'azote qui est a l'etat gazeux dans l'atmosphere. *C. R. Acad. Sci.*, **6**, 102–112.

Cooke, G. W. 1964. *The Basis of Modern Manuring*. Plymouth.

Donal'd, L., Stendzhel, Kh. Dz. and Pesak, D. T. (Donald, L., Stangel, H. J. and Pasak, D. T.). 1965. Advances in the application of nitrogen fertilizers in the USA since 1950: in *Proizvodstvo i primenenie mineral'nykh udobrenii* (Production and Use of Mineral Fertilizers). (Translated from English), Kolos, Moscow.

Egorov, V. E. 1961. Results of half a century of experience in the study of soils and fertilizers in developing the fertility of podzolic type soils. *Doklady TSkhA* (71), 5–25.

Garmen, W. and White, W. 1964. Maintaining and improving soil productivity. *Agric. Chemicals*, 17.

Gro, A. *Prakticheskoe rukovodstvo po primeneniyu udobrenii*, (Practical Textbook on the Application of Fertilizers), Moscow.

Halbinger, R. E. 1965. The microbiological utilization of atmospheric nitrogen. *Ann. Inst. Pasteur*, **109**, Suppl. 3, 161–166.

Harmsen, G. W. 1961. Einfluss von Witterung, Dungung und Witterung auf den Stickstoffgehalt des Bodens. *Landwirtsch. Forsch.*, **8**, 61–74.

Harmsen, G. W. and van Schreven, D. A. 1955. Mineralization of organic nitrogen in soil. *Advances Agron.*, **7**, 299–398.

Hellriegel, H. 1886. Welche Stickstoffquellen stehen der Pflanze zu Gebote? *Landwirtsch. Versuchs-Stat*, **33**, 464–465.

Hellriegel, H. and Wilfarth, H. 1888. Untersuchungen über die Stickstoff-Nahrung der Gramineen und Leguminosen. *Beilageheft Z. Vers. Rübenzükerind.*, 1–234.

Jansson, S. L. 1958. Tracer studies on nitrogen transformations in soil with special attention to mineralization immobilization relationships. *Kgl. lantbrukshogskolans ann*, **24**, 101–136 (in English).

Jödin, C. E. 1862. Du role physiologique de l'azote faisant suite à un précédant travail presenté à l'Academie dans la séance du 28 avril, 1862. *C. R. Acad. Sci. Paris*, **55**, 612–615.

Katon, Varron, Kolumella, Plinii (Cato, Varro, Columella and Pliny), *O sel'skom khozyaistve* (Farming), OGIZ-Sel'khozgiz Moscow-Leningrad.

Kolar, C. and Greenland, P. 1961. Nitrogen balance of the earth's surface. *Austral. J. Sci.*, **23** (9), 290–297.

Koren'kov, D. A. 1966. *Voprosy teorii i praktiki primeneniy a azotnykh udobrenii* (Aspects of the theory and practice of nitrogen fertilizers). Author's summary of Doctoral thesis, Moscow.

Kossovich, P. S. 1890. Origin of the nitrogen of leguminous plants. *Izvestiya Petrovsk. s-kh. akad.*

Lachmann, J. 1858. Über Knöllchen der Leguminosen. *Landwirtsch. Mitt. Z. Kaiserl. Lehranst. Vers. Sta. Poppeldorf (Bonn)*, **37** (1858). Secondary source: Fred, E. B., Baldwin, J. L., McCoy, E. 1932. *Root Nodule Bacteria and Leguminous Plants*, 52. University of Wisconsin Press, Madison.

Litvinyuk, R. S. 1965. Accumulation of nitrogen in the soil by certain graminaceous-leguminous crops. *Tezisy dokladov nauchnoi konferentsii Khar'kovskogo sel'sko-khozyaistvennogo instituta* (Summaries of speeches of the scientific conference of the Khar'kov Agricultural Institute) part 3, Khar'kov.

Mikhnovskii, V. K. 1966. Role of nitrogen in the fertility of soddy-podzolic soils, in *Balans azota v dernovo-podzolistykh pcohvakh* (*Nitrogen Balance in soddy-podzolic soils*) 3–18, Nauka, Moscow.

Mikhnovskii, V. K., Yartseva, A. K. and Bobritskaya, M. A. 1965. Nitrogen balance in soddy-podzolic soils, *Pochvovedenie*, (7), 72–80.

Mikhnovskii, V. K., Yartseva, A. K. and Morozova, A. V. 1966. Nitrogen balance in soddy-podzolic soils under various farm crops, in *Balans azota v dernovo-podzolistykh pochvakh*, 38–88, Nauka, Moscow.

Peterburgskii, A. V. 1964. World production of mineral fertilizers and their use in certain foreign countries. *Agrokhimiya*, (2).

Peterburgskii, A. V. 1965a. Fertilizers and the harvest. *Izvestiya TSKhA*, part 3.

Peterburgskii, A. V. 1965b. Role of nitrogen in modern farming. *Mezhdunarodnyi sel'sko-khozyaistvennyi zhurnal*, (6), 82–86.

Petersburgskii, A. V. and Korchagina, Yu. I. 1965. Ammonium fixation by certain soils from fertilizers and the accessibility of this form of nitrogen for farm plants. *Agrokhimiya*, (7), 15–25.

Pražmowski, A. 1890. Die Wurzelknöllchen der Erbse. *Landwirtsch. Versuch. Stat.*, **37**, 161–178.

Production Yearbook. 1965, **18**, FAO, Rome.

Pryanishnikov, D. N. 1945. *Azot v zhizni rastenii i v zemledelii SSSR* (*Nitrogen in the life of Plants and Farming in the USSR*), Izd-vo AN SSSR, Moscow-Leningrad.

Schultz, L. 1881. Reinerträge auf leichtem Boden,—ein Wart der Erfahrung zur Abwehr der wirtschaftlichen Noth. *Landwirtsch. Jahrb.*, **10**, 777–848.

Sharp, Dzh. K. and Pauell, R. K. (Sharp, J. K. and Powell, R. K.). 1965. Progress in nitrogen fixation, in *Proivzodstva i primenenie mineral'nykh udobrenii* (*Production and Use of Mineral Fertilizers*), Kolos, Moscow (translated from English).

Sinyagin, I. I. 1965. Use of fertilizers in the German Federal Republic. *Sel'skoe khozyaistvo za rubezhom*, (9).

Sokolov, A. V. 1963. Next problems in the study of soil fertility and means of improving it. *Pochvovedenie*, (1).

Statistical Handbook (*Statisticheskaya karmannaya knizhka*), 1963. German Democratic Republic, Berlin. (In Russian.)

Timiryazev, K. A. 1892. Gas exchange on the root nodules of leguminous plants. *Trudy SPb. ob-va estestvoizpytat.*, **23**.

Tyurin, I. V. 1957. Soil fertility and the problem of nitrogen in pedology and farming, in *Soveshchanie po voprosam effektivnykh sposobov ispol'zovaniya udobrinii* (*Conference on Effective Ways of Using Fertilizers*). Izd-vo Min-va sel'sk. khoz-va SSSR.

Tyurin, I. V. and Mikhnovskii, V. K. 1961. Effect of green fertilizer on humus and nitrogen content of soddy-podzolic soils. *Izvestiya Akad. Nauk SSSR, Ser. Biol.*, (3).

Vinogradskii, S. N. 1893. Sur l'assimilation de l'azote gazeux de l'atmosphère par les microbes. *C. R. Acad. Sci.*, **116**.

Voronin, M. S. (Woronin, M. S.). 1866. Über die bei der Schwartzerle (*Alnus glutinosa*) und der gewöhnlichen Gartenlupine (*Lupinus mutabilis*) auftretenden Wurzelan-schwellungen. *Mem. Acad. Imp. Sci. St. Petersbourgh*, ser. 7, **10**, (6), 1–13.

Voronin, M. S. (Woronin, M. S.). 1867. Observations sur certaines excroissances que présentent les racines de l'aune et du lupin des jardins, *Ann. Sci. natur. Bot*, ser. 5, **7**, 73–86.

Voronin, M. S. 1886. Mycorrhiza. *Zapiski Peterb. Akad. nauk.*

Zakharchenko, J, G. Shilina, L. I. 1964, Method of studying the role of leguminous crops in the accumulation of nitrogen in the soil. *Visnyk s-kh. navki*, (12), 35–38.

2 Assimilation of Molecular Nitrogen by Leguminous Plants in Symbiosis with Bacteria

Nodule Bacteria of Leguminous Plants

Palaeontology has shown that the oldest legumes with tubercles belonged to the Eucaesalpinioideae, although this group also contained species without root nodules. It is now known that among existing legumes (of the order Leguminosae) nodules are not formed by representatives of the families Caesalpiniaceae or Mimosaceae which may be considered phylogenetically more primitive than the nodule-forming representatives of the Papilionaceae. But it should be noted that according to Grobbelar *et al.* (1964) not all representatives of the Caesalpiniaceae are devoid of nodules.

There are many species of leguminous plants (about 10,000), of which only some 200 are exploited agriculturally when their nodules become able to assimilate atmospheric nitrogen. They can also feed on bound nitrogen, in the form of, for example, ammonium salts and nitric acid. According to Castelli (1951) only *Hedysarum coronarium* assimilates molecular nitrogen alone. Consequently this species must be inoculated with nodule bacteria if it is to develop.

In 1888, Beijerinck isolated bacteria from the nodules of leguminou splants (pea vetches, peavine, bean, trefoil, Dakota vetch). He called the bacteria *Bacillus radicicola;* it was small, aerobic and changed in shape during development and when the environment changed. A year later Pražmowski called it *Bacterium radicicola* (1889). At the same time, Frank (1889) proposed changing the generic name of nodule bacteria to *Rhizobium*, which has been used in the literature ever since.

Frank, like his contemporaries, considered that one species of nodule bacteria infects all leguminous species. But this view proved to be false. Several rhizobial species have now been recognized, each capable of infecting a particular leguminous species or group of related species. Without discussing all the morphological and physiological features of *Rhizobium* we shall note the most important. The different species and races of nodule bacteria have certain differences in morphological and physiological properties.

The cells of nodule bacteria contain nuclear material but the structure of the

14

nuclear apparatus has still not been studied adequately (Milovidov, 1935; Uher, 1937; Lewis, 1938; Schaede, 1941; Baylor *et al.*, 1945; Voets, 1949; Bergersen, 1955; Bergersen and Briggs, 1958; Chizhik, 1959). When young, nodule bacteria are usually rod-shaped, measure about $0.5-0.9 \times 1.2-3.0 \mu m$ and are motile. Some are monotrichous, sometimes with a subpolar flagellum (Plate 1); others are peritrichous (Plate 2).

As they age, nodule bacteria lose their motility and become 'belted' rods, so-called because alternate parts of the cell protoplasm stain lightly and heavily with aniline dyes. This happens because the ageing bacterial cell becomes filled with fatty inclusions that do not take stain; young *Rhizobium* cells stain uniformly.

In culture, immobile and mobile coccoid cells may form. According to Nowak and Netzsch-Lehner (1965) nodule bacteria usually exist as cocci in the soil.

Several observations suggest that cultures of *Rhizobium* contain small forms filterable in the Berkefeld candle. Some of these may regenerate in favourable conditions and give normal cells (Almon and Baldwin, 1933; Krasil'nikov, 1941; Izrail'skii, 1953). But some investigators (Barthel and Bjalfve, 1933) have been unable to detect filterable forms of *Rhizobium*.

Thickened, branched, pear-shaped or almost spherical formations also develop in bacteriological media and in nodules. These formations, which Brunchorst (1885) named bacteroids, are considerably larger than the usual bacterial cells in nodules. Each bacteroid cell contains many volutin and lipo-protein granules (O. Allen and E. Allen, 1940; Forsyth *et al.*, 1958) and the glycogen content is higher than in the rod-shaped cells (Avvakumova *et al.*, 1966).

Jordan thought (1962) that the bacteroids may also have a rod shaped form. This is convincingly shown in the nodules of *Sesbania grandiflora* where the bacteria are always in the form of rods and never have branched or curved forms (Harris *et al.*, 1949). The bacteroids are immobile and unable to multiply (Bazarewski, 1927). Fonbrune (1949), Heuman (1954) and Yakovleva (1964), however, showed that in certain conditions bacteroids start to multiply. Almon (1933), who carried out many experiments with single bacteroid cells of different species of *Rhizobium* in different nutrient media, considers that if bacteroids multiply they do so very rarely.

Jordan and Coulter (1965) established that the principal properties of bacteroids formed in bacteriological media and in tubercles are practically identical. Among other things, the nucleic acid content is the same as in the rod-shaped forms but the magnesium content of the membrane is considerably lower, making it less rigid. There are certain differences in the intracellular distribution of polysaccharides and lipids (Avvakumova *et al.*, 1966).

Starting with Pražmowski (1890) who studied the developmental cycle of nodule bacteria both in nutrient media and in nodules, most investigators have considered bacteroids to be inactive involutive forms. In fact, as we have noted, bacteroids do not multiply nor do they grow in ordinary media suitable for *Rhizobium*. Nevertheless, the available evidence compels us to recognize that vigorous nitrogen assimilation in the plant coincides with the period of formation of the bacteroids. Therefore, it must be assumed that the bacteroids, especially those that are young, retain several active physiological functions. There have been several detailed studies of these bacteroids (for example, O. Allen and E. Allen, 1940; Bergersen, 1955; Jordan, 1962).

In experimental conditions it is possible to speed up the transition of typical *Rhizobium* cells to the other forms, including those that are involutive. The cells that form, resemble bacteroids. The appearance of these forms is encouraged in a depleted nutrient medium (Demolon *et al.*, 1950); in cultures containing other species of micro-organisms (Naundorf and Nilsson, 1942, 1943; Heuman 1955), and when there is an excess of thiamine (Nilsson *et al.*, 1938), blood (Heumann 1952c) or alkaloids of the strychnine and caffeine types (Carroll, 1894; Barthel, 1926; Sembrat, 1934; Itano and Matsuura, 1936, 1938) in the nutrient medium. Many of the morphological forms of nodule bacteria that we have described appear at a definite stage in the cycle of development (Thornton and Gangulee, 1926).

Nodule bacteria of the various leguminous species grow at different rates in bacteriological nutrient media. Rapid growth is a feature of those from clover, pea, bean and lucerne or alfalfa; the slow developers come from soy bean, lupin, peanut and cowpea.

On solid media nodule bacteria usually form colourless, transparent mucilaginous colonies, of the type peculiar to S forms. Sometimes the rough colonies peculiar to R forms are found (Izrail'skii and Starygina, 1930).

The mucilage formed by nodule bacteria is a complex compound of a polysaccharide nature incorporating hexoses, pentoses and uronic acids (Kleczkowska and Kleczkowski, 1952; Leizaola and Dedoner, 1955; Humphrey and Vincent, 1959; Dudman, 1964b). The mucilage of the strains that develop slowly consists essentially of polysaccharides that are poorly soluble in water, whereas that of fast growing strains is composed of water soluble polysaccharides (Graham, 1965). Acid hydrolysis gives a compound close to glucose-1,4-glucuronic acid (Harworth and Stacey, 1948).

Nodule bacteria can assimilate different carbohydrates, sometimes including polysaccharides (dextrin, glycogen). When grown in media containing carbohydrates many cultures form acids, although some make the medium alkaline. Norris (1965) considered that the acid-producing races of nodule bacteria are peculiar to leguminous plants growing in neutral and alkaline soils, while those that make the medium alkaline are peculiar to plants of acid soils.

According to Virtanen *et al.* and Wilson (Virtanen and Hausen, 1931; Virtanen *et al.*, 1933, 1947; Wilson, 1940) nodule bacteria convert carbohydrates into lactic acid. The formation of volatile acids and gaseous hydrogen and carbon dioxide is also possible. About 30 per cent of the sugar utilized is transformed into mucilage by nodule bacteria.

Nodule bacteria can assimilate organic acids and higher alcohols. There are indications that the rapidly growing strains of *Rhizobium* oxidize carbohydrates and organic acids with equal vigour, while slowly developing forms oxidize carbohydrates more rapidly. The enzyme complex associated with the oxidation of carbohydrates is constitutive, while that associated with the conversion of organic acids is adaptive (Virtanen and Miettinen, 1963).

Tokhver and Lokk (1966) considered that this low rate of growth of the slowly developing strains in laboratory conditions might be explained by their greater requirements for assimilable nitrogen, growth factors and complex forms of organic matter, compared with the fast growing strains. The latter, they considered, may be much like the prototrophs in their ability to live and multiply using simple mineral substances and the minimum amounts of nitrogen present in the form of impurities in synthetic mineral media.

The growth rates of both slow and fast developing strains can be speeded up by adding to the medium compounds containing bound nitrogen, especially combined with vitamins, or a complex of organic substances in the form of yeast extract. Fast growing strains are more resistant to X-rays than slow growing strains (Tokhver and Lokk, 1966). Rhizobia are aerobic, but they can live in a medium containing insignificant amounts of oxygen—about 0·01 atmospheres (Wilson, 1940).

When considering the nitrogen nutrition of nodule bacteria it must be noted that so far it has not been possible to establish convincingly their ability to assimilate molecular nitrogen in pure cultures. This problem has often been investigated, with contradictory results (for example, Fred *et al.*, 1926; Allison, 1929; Löhnis, 1930; Rabotnova, 1938; Fedorov, 1940; Tove and Wilson, 1948; Burris *et al.*, 1943; Uspenskaya, 1953; Filippova, 1953; Wilson, 1958).

Nodule bacteria can multiply in media containing very little nitrogen, but in such cases they are usually depending on their nitrogenous reserves. Thus, while an initial culture of *Rhizobium* has about 8 to 9 per cent nitrogen (based on dry mass) a culture multiplying in a poor medium has only about 3·5 per cent nitrogen (Burris and Wilson, 1945). In our view, the fact that several investigators could not establish nitrogen fixation in pure cultures of *Rhizobium* does not mean that nodule bacteria do not fix molecular nitrogen in pure cultures. Although it must be assumed that symbiosis grew out of a prolonged interaction between plant and bacteria, conditions have not been found which support the assimilation of molecular nitrogen in the absence of the host plant. In this respect some investigations (Kalininskaya and Il'ina, 1965) with a nitrogen-fixing *Mycobacterium* provide a useful example. Until a complete nutrient medium was found this bacterium utilized molecular nitrogen only when the medium contained an associated micro-organism (not a nitrogen fixer). But when a suitable nutrient medium was used the *Mycobacterium* assimilated molecular nitrogen almost as briskly as did cultures of *Azotobacter*.

Nevertheless, the failure to demonstrate assimilation of molecular nitrogen by nodule bacteria encouraged the view that nitrogen fixation is accomplished not by the bacteria but by the leguminous plants, and that the nodule bacteria penetrating the root only stimulate this process (for example, Frank, 1892;

Stoklasa, 1895; Krasil'nikov, 1954; Yukhimchuk, 1957). Turchin (1959) and Bergersen (1960) expressed similar views, developing these ideas into elaborate schemes. They studied the distribution of labelled nitrogen in the individual structures of the nodule, and found that the bacterial cells contained very small amounts of ^{15}N, while a large part of the bound nitrogen was in the structural elements of the plant cell (cell membrane). We shall consider Bergersen's ideas on the mechanism of nitrogen assimilation in the nodule later in this chapter.

Many amino acids are available as nitrogen sources for nodule bacteria: alanine, glycine, proline, asparagine, cystine, cysteine and aspartic acid (Pohlmann, 1931; Nielsen, 1940; Jordan, 1952; Jordan and San Clemente, 1955a; Beresniewicz, 1959; Proctor, 1963; Fan Yun-Liu and Huang Sing-Fu, 1965; Strijdom and Allen, 1966; Handi, 1966). Kaszubiak's experiments (1965) established that certain amino compounds in the tissues of leguminous plants, for example tingitanine isolated from the seeds of *Lathyrus tingitanus* (Nowacki and Przybylska, 1961) and di-aminobutyric acid and homoserine, have a beneficial influence on various species of nodule bacteria. The exact nature of their influence differed from strain to strain. Some cultures assimilate peptone.

The ammonium-fixing powers of the cultures are very weak and they take up little ammonia from media containing organic nitrogen (Virtanen and Laine, 1936). According to Muller and Stapp (1925) nodule bacteria can assimilate ammonium and nitric acid salts, and according to Jordan and San Clemente (1955b) they can also assimilate purine and pyrimidine bases. Cultures of most nodule bacteria can use sources of biuret nitrogen (Jensen and Koumaran, 1965) and also urea (Jensen and Schröder, 1965) the latter as a result of the ability to produce urease (Benjamin, 1915; Beijerinck, 1923; Hutchinson, 1924).

When nodule bacteria are grown in nutrient media containing a high concentration of nitrogenous substances the bacteria may lose the ability to inoculate a plant and produce nodules. Therefore, in commercial laboratory conditions nodule bacteria are usually grown in plant extracts (legume decoction, hay decoction and so on).

The nodule bacteria can obtain the phosphorus they require from mineral and organic compounds, and they have phosphatase (for example, Żelazna, 1962; Moustafa, 1964; Lorkiewicz et al.; 1965; Zaremba et al., 1966). Potassium, calcium and other elements can be assimilated from inorganic substances. Evidently, nodule bacteria also need certain trace elements such as iron, titanium and molybdenum.

Nodule bacteria also require certain vitamins (Rigaud, 1965; Badawy, 1960) but these requirements vary in different cultures, for example, some do not develop without biotin while others can do without it (J. Wilson and P. Wilson, 1942). Apparently in the latter case the nodule bacteria synthesize their own biotin. According to Murphy and Elkan (1963) and Badawy (1966) biotin even suppresses the development of certain strains of *Rh. japonicum*. The inhibitory effect of biotin decreases somewhat in the presence of calcium pantothenate.

There are indications that nodule bacteria also need large amounts of thiamine and pantothenic acid (E. Allen and O. Allen, 1950). Several investigations have shown that nicotinic acid is not important for nodule bacteria (Lilly and Leonian, 1945; Graham, 1963b). It is evidently harmful to biotin-independent cultures, for it interferes with the biosynthesis of biotin (Abdel-Ghaffar and Jensen, 1966). In general, nodule bacteria can synthesize several vitamins (thiamine, vitamin B_{12} and riboflavine) and growth substances, in particular β-indolylacetic acid and its derivatives, gibberellins and so on (for example, Oplištilova and Vančura, 1963; Bulard et al., 1963 Rigaud, Bulard, 1965 Chailakyan et al., 1965, Naumoua, 1966). The synthesizing capacity of the different species and strains of nodule bacteria differs. Some cultures, for example, Rh. lupini do not seem to produce heteroauxin at all (Novikova and Irtuganova) which is possibly connected with the culture conditions.

For most cultures of Rhizobium the optimum pH is in the range 6·5–7·5, and growth ceases at pH 4·5–5·0 and 8·0. Results obtained in our laboratory show that cultures of nodule bacteria develop well between 24°C and 26°C, while at temperatures of about 0°C and 37°C growth practically stops. The cell of a nodule bacterium contains many enzymes including intracellular deoxyribonuclease (Kaszubiak, 1964). We shall not describe the cultural properties of nodule bacteria, for they are given in all classifications of bacteria.

Kluyver and van Niel (1936) considered Rhizobium to be systematically close to the Pseudomonadaceae. Jensen (1958) agreed with this, but in addition noted a similarity between nodule bacteria and Agrobacterium radiobacter. He regarded Agrobacterium as a possible original form of Rhizobium which in the process of evolution acquired the ability to penetrate the roots of leguminous plants. Earlier than this, Izrail'skii (1933) suggested the kinship of Rhizobium with Agrobacterium tumefaciens, which was also noted by Voronkevich (1966).

The closeness of Rhizobium and Agrobacterium has been confirmed by several DNA transformation experiments. Kern (1965) induced the formation of tumours in Datura, Solanum and other plant genera by injecting them with cultures of nodule bacteria and DNA from Agrobacterium tumefaciens. The DNA alone did not induce tumours. Results were not always positive and Klein and Klein (1953) were successful only with a definite combination of bacterial cultures and DNA.

But different views have been expressed on the origin of nodule bacteria. For example, Bisset (1952) postulated a connection between Rhizobium and Bacillus polymyxa. This has been challenged by several investigators (for example, Graham et al., 1963). Jacobs (1949) and Jensen (1952) thought that the family Corynebacteriaceae was also close to the family Rhizobiaceae, and according to Nowak (1966) so is the family Chromobacteriaceae.

Derx (1953) assumed that nodule bacteria originated from non-symbiotic, nitrogen fixers of the Beijerinckia type. This view has raised a justifiable objection (Jensen, 1958). It is difficult to believe that Beijerinckia acquired virulence and at the same time lost the ability to fix molecular nitrogen in pure culture.

Nodule bacteria are now divided into species according to their infective powers in relation to the host organism. Bergey's classification (1957) contains six species of *Rhizobium—Rh. leguminosarum, Rh. phaseoli, Rh. trifolii, Rh. lupini, Rh. japonicum* and *Rh. meliloti.*

Dorosinskii (1965a), using as the basis of his classification cross-infection and several morphological and cultural properties of nodule bacteria, distinguished eleven groups of the genus *Rhizobium* (Table 2.1).

From the practical standpoint such a scheme for the classification of nodule bacteria must be considered convenient. But as far as systematics is concerned it would be desirable to enlarge certain groups and not break them up as this scheme does, for example, with slow growing bacteria. Several investigators

Table 2.1. The Genus *Rhizobium*

Plant infected

Rh. leguminosarum	Pea, vetch, broad bean, peavine and lentil
Rh. phaseoli	Runner bean
Rh. japonicum	Soy bean
Rh. vigna	Cowpea, Oregon pea, peanut
Rh. cicer	Chick pea
Rh. lupini	Lupin
Rh. trifolii	Clover
Rh. meliloti	Lucerne, sweet clover, trigonella
Rh. simplex	Sainfoin
Rh. lotus	Dakota vetch
Rh. robinii	Acacia

have tended to enlarge the groups (for example, Manninger, 1962; Tešić and Todorović, 1963; Drożańska, 1963; Graham, 1964).

There are some other groupings of nodule bacteria, also based on their ability to infect leguminous plants (for example, Fred *et al.*, 1932; Krasilnikov, 1949; E. Allen and O. Allen, 1958; Galli, 1959). Later we shall give some possible principles for a classification of nodule bacteria which is now being developed.

Tešić and Todorović (1963) based their classification of nodule bacteria on the type of flagella they possessed (Table 2.2). By this criterion the genus *Rhizobium* divides into two species which in turn subdivide into 'special forms'.

The scheme just considered is similar to that proposed by Graham (1964). Biochemically it has been confirmed by the investigations of De Ley and Rassel (1965), who established that in monotrichous nodule bacteria the DNA content of guanine and cytosine usually fluctuates from 62·8 to 65·6 per cent. Peritrichous nodule bacteria have a somewhat lower content of guanine and cytosine, approximately equal to 58·6–63·1 per cent.

Work on the purine and pyrimidine composition of the nucleic acids, especially DNA, of the individual bacterial species including representatives of the genus *Rhizobium*, is now developing on quite a large scale. Differences have been found in the ratio $\dfrac{\text{guanine}}{\text{adenine}} + \dfrac{\text{cytosine}}{\text{thymine}}$ between the fast and slow growing forms of nodule bacteria and also for the bacteria of the related genus *Agrobacterium*. These differences can be used in a classification of the nodule bacteria (Wagenbreth, 1961*a*).

Microbiologists have done a great deal of work on the serological properties of nodule bacteria and their use in classification (for example Stevens, 1923; Wright, 1925; Vincent, 1942; Purchase *et al.*, 1951*b*; Koontz and Faber, 1961;

Table 2.2. Scheme for a Classification of Nodule Bacteria

Flagellar arrangement	Common characteristic signs	Species	Special forms
Monotrichous	Slow growth, slightly branched bacteroids, resistance to acid reaction of medium, formation of bright ring in milk, bacteriophage with slow lysogenic activity	*Rhizobiomonas leguminosarum*	*lupini japonica*
Peritrichous	Fast growth, branched bacteroids, low resistance to acid reaction of medium, absence of ring in milk, bacteriophage with rapid lysogenic activity	*Rhizobacterium leguminsarum*	*leguminosarum, melilloti, trifolii, phaseoli*

This scheme is after Tešić and Todorović (1963).

Vintikova *et al.*, 1961; Drożańska, 1963, 1965; Johnson *et al.*, 1963, 1964; Ziemiecka, 1963; Graham, 1963*a*; Date and Decker, 1965; Škrdleta, 1965*a*).

Some have proposed schemes based on serological properties, but this work cannot be considered to be complete. Many investigators have noted a high degree of serological heterogeneity even among individual strains of nodule bacteria, and this adds to the problems involved. Nevertheless we are interested in the recommendations of Date and Decker (1965) for the use of group and polyvalent antisera in the diagnosis of nodule bacteria.

It has been proposed that the resistance of individual species of nodule bacteria to antibiotics could be used as a criterion for classification. But results have been contradictory. Thus, Graham (1963*c*) claimed that slow growing species of nodule bacteria are less sensitive to antibiotics, but according to Kecskés and Manninger (1962) no such relationship can be established. Manninger (1962) proposed that *Rhizobium* be classified on the basis of the biochemical formation of gas from carbohydrates, acids and so on. Knösel (1962) considered using as a systematic criterion, the ability of nodule bacteria to form stellate configurations in different media, especially carrot juice.

Graham (1964) made an interesting attempt to establish a relationship between representatives of *Rhizobium* and the forms claimed in the literature to be similar to it (*Agrobacterium, Beijerinckia* and so on) on the basis of the similarity of about a hundred features, morphological, cultural, physiological, and biochemical, as well as virulence and sensitivity to antibiotics. The information was fed into a computer. These hundred features had previously been grouped (Graham and Parker, 1964) on the basis of the data obtained. Graham considers it desirable to: (1) bring into one species *Rh. leguminosarum, Rh. trifolii* and *Rh. phaseoli*; (2) leave as a special species *Rh. meliloti*; (3) bring into one species *Agrobacterium tumefaciens* and *Ag. radiobacter* and include them in the genus *Rhizobium* as *Rh. radiobacter*; and (4) assign the slow growing nodule bacteria to the genus *Phytomyxa* with the species name *Ph. japonicum*. In our view these proposals, based on formal signs, are likely to be suitable for practical purposes.

From the point of view of the taxonomy of nodule bacteria there is considerable interest in investigations of the bacterial variation that can be induced by many external factors, and may also be a consequence of transduction and transformation. Several investigators, whose work we shall discuss later, have shown that under the influence of bacteriophage, nodule bacteria often form mutants with new properties.

In the past few years, work has been published on the transformation of nodule bacteria under the influence of DNA. Among these investigations we would particularly mention the work of Balassa (1954–1963) and Kleczkowska (1965). Investigations of a similar nature have been carried out by other microbiologists, but not always successfully (Wagenbreth, 1965). This is partly due to methodological differences, for the success of the DNA transformation treatment depends on several factors (Ellis *et al.*, 1962; Żelazna, 1963, 1964 *a* and *b*). Using DNA preparations Balassa transmitted to nodule bacteria various properties such as a change in their antigenic structure, resistance to antibiotics, acquisition of prototrophicity (mutant requiring cysteine) and the ability to form turbercles in another host plant. Using DNA obtained from streptomycin-resistant mutants she was also able to achieve intraspecies and interspecies transformation in nodule bacteria of the lupin. The mechanism of genetic control of the streptomycin-resistance marker was established. (Much of Balassa's work was carried out with Gabor, 1964, 1965.)

In 1965, Kleczkowska found that *Rh. trifolii* lost its ability to fix nitrogen in symbiosis when treated with the DNA from an ineffective strain. Similar attempts to transfer effectiveness to the ineffective culture failed. In her view effectiveness in nitrogen fixation is determined both by the genes of the bacteria and by the genes of the plant (their compatibility). Therefore, nodule bacteria become effective only when the host plants have certain genetic properties, and a change in the hereditary properties of the bacteria alone may be insufficient to bring about nitrogen fixation.

Further study of variation in nodule bacteria may be rewarding not only for solving theoretical problems, but also for providing cultures of micro-organisms that are of practical value.

Specificity, Virulence and Activity of Nodule Bacteria

Beijerinck (1888), working with isolated nodule bacteria, suggested that there was a single species that can be divided into two. In each of the two groups he isolated varieties associated with particular plant species.

Later work by many other investigators suggested that individual cultures of nodule bacteria can infect only a definite group of leguminous plants. Sometimes it is a large group, sometimes small. The selectivity of nodule bacteria in relation to the host plant was termed specificity, and was taken as the basis of the classification which we have already considered.

Some cultures of nodule bacteria infect with equal success each of a group of species or several different varieties of leguminous plants. For example, the nodule bacteria of the pea may infect vetch, the peavine and broad beans. Nodule bacteria of the *Vigna* group (cowpea) can inoculate many leguminous plants belonging to different subfamilies (Norris, 1965). According to Wróbel and Allen (1961) the American varieties of alfalfa are efficiently inoculated by nodule bacteria of both Polish and American origin.

Other cultures of nodule bacteria have a narrow adaptability, even to a particular variety. For example, *Rhizobium* cultures form good tubercles only on certain species or varieties. The tetraploid and diploid forms of the same species of plant are infected equally well (for example Nilsson and Rydin, 1954; Buśko, 1959; Lange, 1961; Norris and Mannetje, 1964; Rubenchik et al., 1966).

Nodule bacteria have been noted to have a high specificity for certain tropical legumes, for example *Lotonius bainesii* (Norris, 1958a) and *Centrocema pubescens* (Bowen, 1959; Bowen and Kennedy, 1961). The existence of species and sometimes varietal specificity in the nodule bacteria of alfalfa has been established by several investigators (Manil and Bonnier, 1951; Schwendimann, 1955; Wikén, 1956; Hamatová-Hlaváčková, 1963, 1965; Krasil'nikov and Melkumova, 1963), lupin, (Federov and Svitych, 1959; Apltauer, 1963), pea (Avvakumova, 1956) and clover (Loos and Louw, 1964).

The nodule bacteria of the clover group are considered to possess a high specificity, for they do not form tubercles on other plants. For this reason the clover group is called monolithic. For a long time it was thought that nodule bacteria of the clovers also have strict species specificity. Recently, however, this has been disproved (Fig. 2). Individual strains of bacteria of this group can form tubercles on many, though not all, species of clover (Nutman, 1956; Burton, 1965). Individual strains of nodule bacteria of this group, such as those of *Trifolium alexandrium*, infect *Trifolium pratense* poorly. Only a few cultures

2

have this property. Nutman (1956) considered that in terms of specificity the nodule bacteria of the clover group can be divided into at least three sub-groups.

As for varietal specificity, Hamatová-Hlaváčková (1965) feels that it is weakly marked in the nodule bacteria of the pea, sainfoin, clover and lucerne, but quite definite in those of soy bean, lupin and beans.

The specificity of nodule bacteria often seems to be upset, even in relation to leguminous species. Many cultures of nodule bacteria give 'cross' infection, that is they infect other, sometimes not very closely related leguminous species (for example Fred *et al.*, 1932; J. Wilson, 1939*a* and *b*; Hauke–Pacewiczowa, 1952; Erdman and Means, 1956; Norris, 1959*b* and *c*; Lange, 1961; Lo Ming-Tien, 1961; Razumovskaya, 1963, 1965; Watanabe and Tuzimura, 1962; Rangaswami

Species of clover	Strain of rhizobium								
	B_1	B_2	B_3	B_4	K_2	K_4	K_{10}	P_{28}	P_{29}
T. nigrescens	○	●	◉	●	○	○	○	◉	○
T. alexandrinum	●	●	◉	●	●	●	●	●	◉
T. incarnatum	◉	○	○	●	●	●	●	○	○
T. pratense	○	○	○	○	●	○	○	●	●

●1 ◉2 ○3

Fig. 2. Inoculation of various clover species by strains of nodule bacteria (after Burton). Nitrogen fixation: (1) active; (2) moderate; (3) weak.

and Oblisami, 1962; Parker and Oakley, 1963; Bjalfve, 1963; Hoffmann, 1964; Jensen, 1964; Trinick, 1965; Zaremba and Malyns'ka, 1966). Thus, for example, some species of clover may be infected by the bacteria of vetch, and lupin species may be infected by the bacteria of lucerne and so on. There are many such examples but we shall confine ourselves to a description of the experiments of Vyas and Prasad (1959), Norris (1959*b*) and Dorosinskii and Lazareva (1966).

Vyas and Prasad isolated nodule bacteria from several plants growing in India and used them to infect seeds of various leguminous species. As a result, nodules were found on the roots of plant species which are not usually cross infected. Thus, the nodule bacteria of the pea caused nodule formation on *Vigna* and, in part, on beans (*Dolichos*). *Vigna* was infected by the nodule bacteria isolated from the nodules of *Sespania aegyptica*. However, the *Dolichos* bacteria did not cause nodules to form on *Vigna*, in spite of the fact that these plants are considered to belong to the same group.

Norris infected three species of clover with five cultures of nodule bacteria isolated from plants of the vetch group (*Vicia sativa*, *Pisum arvense* and *Lathyrus ochrus*). In each case nodules formed, although some were very small. We should recall that in terms of bacterial infectivity and specificity the clover group is considered to be unique. When vetches were infected with bacteria of clover, nodules formed only in five out of twelve cases.

After cross infecting several leguminous plants with nodule bacteria, Dorosinskii and Lazareva established that the peanut, for example, is readily infected by strains of bacteria isolated from the nodules of various species. On the other

Table 2.3. Specificity of Nodule Bacteria of Different Species of Leguminous Plants

Plant	Rhizobial culture used	Weight of green matter (g)	Infectivity of roots	Activity	Description of nodules
Arachis	*Arachis*	75·9	+++	+++	Large pink
	Vigna	41·3	+++	—	Small yellow
	Lupin	41·1	++	—	
	Oregon pea	39·4	+++	—	
	Soy	44·2	++	—	
Lupin	*Arachis*	34·3	+	—	Small
	Vigna	47·5	+++	—	Small
	Lupin	112·0	+++	+++	Large pink
	Oregon pea	82·6	+++	++	Small
	Soy	101·1	+++	++	Small
Soy	*Arachis*	71·2	—	—	None
	Vigna	73·7	—	—	—
	Lupin	75·2	—	—	—
	Oregon pea	75·7	—	—	—
	Soy	110·0	+++	+++	Large

+++, Considerable infection and activity; ++, moderate infection and activity; —, infection absent.

This table is after Dorosinskii and Lazareva.

hand, only specific races isolated from nodules of *Arachis* proved to be effective for cross infection. Lupins were also infected by different species of nodule bacteria, but specific cultures were the most active. Soy beans have proved to be the most specialized in relation to nodule bacteria. They can be infected only by their own nodule bacteria (Table 2.3).

It should be noted, however, that on cross infection nodule bacteria, foreign to a given species of legume, give rise to but few nodules and nitrogen fixation is relatively weak. Thus, maximum nitrogen accumulation is ensured by races of bacteria specific to a given leguminous species (or related species).

In Jensen's view (1964) the ability to inoculate a comparatively wide range of plants may promote the maintenance and survival of *Rhizobium* in the soil in the absence of its specific host plant.

There is some evidence (Thorne and Brown, 1937; Chailakhyan and Megra-byan, 1958*b*; Drozdowicz, 1959) that when the roots of leguminous plants are surrounded by nodule bacteria they rarely encounter, they produce substances to impede penetration. These compounds are possibly related to antibiotics. Thompson (1960) found in the seed coats of many species of perennial legumes a heat stable, bacteriostatic substance that depresses the growth of nodule bacteria. Fottrell *et al.* (1964) isolated from the seeds of *Trifolium repens* an antibiotic toxic to *Rh. leguminosarum*. He identified it as myricetin, which together with tannin and quercetin was found by Masterson (1965) in the seeds

Table 2.4. Intraspecific Symbiotic Adaption in *Trifolium pratense*

Line of host plant	Strains of *Rhizobium trifolii*				
	A	Mutants A			F
		A211	Af12	A11	
Unselected	E	E	I	—	E
Homozygous for i_t	I	E	I	—	E
Homozygous for i_e	I	I	I	—	E + I

E, effective; I, ineffective symbioses; —, no nodules.

This table is from Nutman (1959*a*).

of various species of clover. Particularly high concentrations of these substances were found in seeds of *Trifolium repens* and *T. fragiferum*.

So far it has not been possible to relate the specificity of nodule bacteria to their antigenic structure or morphological or physiological properties (Kleczkowski and Thornton, 1944; Purchase *et al.*, 1951*b*; Johnson and Means, 1963; Kalnin'sh *et al.*, 1966). Of course, this refers only to those properties that have been studied—by no means all the properties of nodule bacteria. Certain properties of these micro-organisms must be correlated with specificity. There is confirmation for this in the work of Balassa (1960, 1963) and Yamane and Higashi (1963), who showed that the factor responsible for specificity is the DNA of the bacterial cultures. It is possible to transform the specificity of nodule bacteria using DNA as donor strains.

The specificity of nodule bacteria can be changed. There are indications that the specific properties of nodule bacteria may be influenced by factors operating on the host plant (Hoffman, 1927). Using different lines of red clover and mutant strains of *Rhizobium trifolii* (Table 2.4) Nutman (1959*a*) showed that on the one hand, the host may determine the nature of the interrelations between the symbionts and, on the other, the influence of the host largely depends on the strain of *Rhizobium* involved. (In this case the host was i_e.) In Nutman's experiments strain Af12 represented a loss of effectiveness while strain A11 represented a loss of virulence. Another demonstration of genetic control of the relationship between the symbionts is the loss of the ability to form nodules in some lines of soy bean (Lynch and Sears, 1952; Williams and Lynch, 1954; Weber, 1966).

When the root system of a legume is being infected by nodule bacteria their virulence is very important. While the specificity of the nodule bacteria determines their activity spectrum, their virulence characterizes their activity within a given spectrum. (Virulence is the ability to penetrate root tissues, multiply within them and form tubercles.)

Frank first suggested that there are natural strains of nodule bacteria that differ in virulence (1889). Later Hiltner and Störmer (1903) confirmed his assumption and related it to the concepts of plant immunity. According to them bacteria which penetrate a plant induce an immunity that prevents further infection of the root system. The non-uniform distribution of nodules over the root systems of legumes is thought to favour this idea. The old tissue of the root, especially the upper part of the main root, is densely covered, while in the younger, lateral roots there are no nodules or only small ones.

The assumption that the plant acquires immunity through primary infection has been confirmed by several authoritative investigators (for example Fred *et al.*, 1932; Izrail'skii *et al.*, 1933; Virtanen, 1945). Virtanen and Lincola (1947) established that the first *Rhizobium* culture to penetrate the root system holds up but does not suppress infection by another culture. Thus when a plant is infected first with an inactive *Rhizobium* culture (see below) and with an active culture after 1 to 2 weeks, the second culture forms tubercles after approximately 1·5 months. A serological method (Thornton and Kleczkowski, 1950) can be used to analyze cultures penetrating the roots of leguminous plants.

Nutman (1963) considered that inhibition of *de novo* formation of tubercles by existing nodules is explained by the depressant property of the meristem of the nodular tissue.

Vigorous nitrogen assimilation may occur when the root system of the leguminous plant is infected with a virulent and active race of nodule bacteria. Several investigators have felt, however, that it is not always the more virulent culture that is first to inoculate the root system in natural conditions. Sometimes the culture is not sufficiently competitive with other micro-organisms in the soil and is depressed by them. The virulent culture must be competitive (Nicol and Thornton, 1941; Burton and Allen, 1950; Harris, 1954; Baird, 1955; Kalnin'sh, 1961; Abel and Erdman, 1964; Petrović, 1963; Ireland, 1966).

Using bacteria labelled with phosphorus-32, Kerpely *et al.* (1963) established that a culture of nodule bacteria introduced into the soil usually competes with the bacteria already there, no matter how many there are. But for successful inoculation, there must be a definite (relatively large) number of virulent bacteria. Often, even the presence of virulent and competitive nodule bacteria in the soil does not ensure inoculation, for infection of the root system depends largely on the number of virulent bacteria introduced or present in the rhizosphere (for example Porter *et al.*, 1962; Lim, 1963). This will be discussed in greater detail in an analysis of the infectious process (page 40).

In certain conditions nodule bacteria may lose some or all of their virulence, for example when cultured in media containing glycine and alanine (for example Wolf and Baldwin, 1940; Longley *et al.*, 1957; Gostkowska, 1963). According to Izrail'skii *et al.* (1964) loss of virulence is accompanied by loss of motility and flagella H antigen. Bacteria with weak virulence lost their agglutinability in relation to specific serum but maintained the precipitation reaction.

The factors which determine the virulence of nodule bacteria have not been studied sufficiently. Some investigators attach significance to the ability of nodule bacteria to produce indolylacetic acid, promoting the process of invasion (Thornton, 1930; Thimann, 1936; Link and Klein, 1951; Kefford *et al.*, 1960; Oplištilová, 1963; Tanner and Anderson, 1963). But the effect of pure indolylacetic acid on the root hairs differs somewhat from the effect of cultures of nodule bacteria. Evidently auxin is only one factor influencing the process of inoculation (Manil, 1958; Stenz, 1962). The role of indolylacetic acid in the infectious process will be considered later. Substances that Rolitski (1966) called bacteriocins (melilothicin, trifolicin, phaseolicin and so on) may be involved in infection. They form in large quantities in cultures of nodule bacteria especially under the influence of ultraviolet radiation.

The properties of the legumes themselves are also of fundamental importance in the infectious process (Nutman, 1956, 1963). Certain substances that promote inoculation are thought to accumulate in the cotyledons and the root system (M. Raggio and N. Raggio, 1956; Valera *et al.* (1956). Their nature has still to be elucidated.

A change in virulence can be brought about artificially, for example by passaging a culture of nodule bacteria through a plant (Izrail'skii and Artem'eva, 1937; Sarić, 1963). This was noted earlier by Hiltner and Störmer (1903) who considered that repeated passaging of nodule bacteria through the plant so enhances their virulence that they may even change to a parasitic way of life. According to Hamatová-Hlaváčková (1963; 1965) one passage of a bacterial culture of *Rh. japonicum* through the plant enhances virulence to a greater degree than do two passages. But a single passage has no such effect on certain cultures of nodule bacteria of lucerne. Evidently, the effect of passaging may change with the character of the bacterial cultures and the leguminous species. Some investigators doubt that the properties of nodule bacteria can be altered by passaging (Virtanen and Miettinen, 1963). Virulence can be increased by exposure to radium and thorium (Krasil'nikov *et al.*, 1955), colchicine (Schiel *et al.*, 1963), γ-rays (Hamatová-Hlaváčková, 1964), introduction of DNA from virulent strains (Ljunggren, 1961; Lange and Alexander, 1961), exposure to a weak electric current (Kravtsov, 1965) and so on.

Factors that promote an increase in the number of root nodules in leguminous plants include γ-rays and colchicine (Bonnier, 1954; Migahid, El Nady and Abd El Rahman, 1959), vitamins (Wagenbreth, 1961b; Weir, 1964) and growth substances (for example Chailakhyan et al., 1961, 1963; Tagiyev, 1965, 1966). Lie (1964) observed an increase in the number of nodules produced in beans which were germinated on a medium containing root secretions from the pea and the bean, especially when they were inoculated with an active strain of nodule bacteria. No effect was observed if root secretions of uninoculated plants were used. Nodule formation was suppressed when the root secretions used were from old plants. The active substance of the root secretions was initially identified as kinin.

Manil (1960) attempted unsuccessfully to transmit the virulence factor to an inactive race of nodule bacteria using DNA obtained from an effective culture.

An important property of nodule bacteria is their activity (effectiveness)— the ability to assimilate molecular nitrogen when in symbiosis with a plant. Some that form tubercles actively fix nitrogen, while others do so only to a limited extent or have lost the capacity to do so. Depending on the extent to which the nodule bacteria promote an increase in the yield of the leguminous crop, it is common to denote them active (effective) or inactive (ineffective). An intermediate, weakly effective group can also be distinguished. The presence in nature of cultures of nodule bacteria with differing activities was first noted by Süchting (1904) and later confirmed by many other investigators.

Bonnier (1962) considered the activity of nodule bacteria to be related to their specificity. Some investigators believe that loss of the capacity to fix nitrogen leads to loss of virulence in nodule bacteria (Izrail'skii et al., 1964). There are indications, however, that inactive cultures are often fairly virulent (Berezova and Dorosinskii, 1961). Evidently, activity and virulence are not very intimately connected (Rubenchik et al., 1966).

The activity of nodule bacteria is not constant and may change under the influence of many different factors. Thus, cultures often lose their activity in the laboratory (for example E. Allen and O. Allen, 1950; Lopatina, 1960). Either the whole culture loses its activity or individual cells with low activity appear. Certain antibiotics added to the culture medium can markedly change the activity of nodule bacteria (Gupta and Kleczkowska, 1962). Schwinghamer (1964) found that inactive forms of nodule bacteria developed under the influence of viomycin and neomycin. Amino acids have a similar effect, in particular glycine and alanine (Holding et al., 1960), methionine (Hamdy, 1966) and glutamine (Badawy, 1966). According to Strijdom and Allen (1966) the D-isomers of amino acids have a stronger effect than the L-isomers. The properties of cultures of nodule bacteria change in response to phage (Kleczkowska, 1950; Gupta and Kleczkowska, 1962), electric current (Kravtsov, 1965), ultraviolet radiation (Zagor'e, 1966) and X-rays (Jordon, 1952; Emtsev, 1962; Dýgdala, 1963; Hamatová-Hlaváčková, 1964). Treating a culture of nodule bacteria with 45,000 r. of X-rays, Jordan (1952) obtained highly active mutants from an inactive strain of lucerne bacteria. Emtsev (1962) and Hamatová Hlaváčková (1965) obtained quite similar results with a series of cultures. Smaller doses of X-rays may also stimulate nitrogen fixation, but this effect is transient (Sokurova, 1956). An increase in the activity of cultures of nodule bacteria treated with colchicine has also been observed (Schiel et al., 1963).

Many investigators have been able to achieve a considerable increase in the activity of nodule bacteria by passaging them through the plant (Allen and Baldwin, 1931; Bonnier, 1962; Hamatová-Hlaváčková, 1965).

It is well worth noting the work of Balassa (1960) who showed that the DNA of an active strain may transmit high activity to another strain of nodule bacteria. When so treated the recipient strain may even be more active than the donor strains.

The signs that distinguish active from inactive strains of nodule bacteria are most important. As a rule, the differences are manifest chiefly in a different distribution of the nodules within the root system. Usually, when plants are inoculated with active strains numerous nodules form on the main root, with a few on the lateral roots. According to Radulović (1966) this is connected with the growth rate of the root system, in particular of the root hairs on the main root, through which the bacteria can penetrate. The time of growth of the root hairs decreases as the root develops, so reducing the possibility of nodule formation on the main root. Ineffective strains of nodule bacteria usually form small nodules scattered over the whole root system.

Sarić (1963b) thought that the siting of the nodules on the roots changed significantly with ecogeographical conditions. Thus, according to her findings, nodules on lupin roots in Yugoslavia have a different pattern of distribution from those in tropical conditions. The pattern of distribution and the shape and size of the nodules are primarily determined by the leguminous species.

Many investigators have confirmed Virtanen's finding (1965) that the nodules formed by active strains of bacteria are pink, as had been noted in the nineteenth and early twentieth centuries (Pražmowski, 1888, 1889; Kossowitsch, 1892; Krasheninnikov, 1916). The chemical composition of the red pigment of the nodules, however, was not known until 1939—the first attempt to establish it was unsuccessful (Pietz, 1938). The pigment Pietz isolated was incorrectly identified as 5,6-quinone-2,3-dihydroxyindole-2-carboxylic acid—the intermediate product of oxidation of dihydroxyphenylalanine. The haeme nature of the pigment was established by Kubo (1939), and later confirmed (Burris and Haas, 1944; Keilin and Wang, 1945). Kubo obtained a pure preparation of the pigment (leghaemoglobin) from soy nodules. Leghaemoglobin is present in the cytoplasm and vacuoles of the plant cell and is readily extracted with water. There is no pigment in the cells of the bacteria and the outer cortical parenchyma (Smith and Jordon, 1949).

Smith and Jordan (1949) used a microspectroscopic method to demonstrate that the pigment was in the nodules. Smith (1949) found in different leguminous species between 1·09 and 3·25 mg of haemoglobin per gram of nodule. Leghaemoglobin can be precipitated from aqueous solution using ammonium sulphate (Kubo, 1939) and can be separated electrophoretically into several fractions. The heterogeneity of leghaemoglobin is due to differences in the protein part of the molecule (Peive and Zhiznevskaya, 1966; Peive, 1967). Detailed study of two fractions (Thorogood, 1957; Ellfolk, 1959, 1960) has shown that one has an isoelectric point at about pH 4·4 and a molecular weight of 16,800 while the other has its isoelectric point at pH 4·7 and a molecular weight of 15,400 and moves less rapidly when subjected to electrophoresis.

The haeme group of leghaemoglobin is similar to that in blood haemoglobin, but the protein components differ and evidently have a more complex structure (Pieve et al., 1966), as indicated by the molecular weight, amino acid composition and isoelectric point, which for leghaemoglobin is the lowest for any haemoglobin. The nodules formed by active races of Rhizobium contain as much as 4 per cent leghaemoglobin (dry matter). Jordon and Garrard (1951) and Berg (1965) have examined the content of leghaemoglobin in many species of leguminous plants and established its behaviour during the growth of the legume.

In annual legumes, at the end of the growing period when the process of nitrogen fixation is terminated, the red pigment turns green. The change in colour begins at the base of the nodule and later the apex greens. In perennial legumes greening of the nodules does not occur or is observed only at the base.

The change of red into green nodules is irreversible. The green pigment can easily be isolated from a solution of ammonium sulphate and contains 0·29 per cent iron. It has specific absorption curves (Fig. 3). The transition from red to green pigment is accomplished at different rates and with different intensities in different species of leguminous plants. All available evidence indicates that the green pigment is formed by oxidation of the red pigment; one of the methene bonds is ruptured in the porphyrin ring and oxygen attached to it, closing the ring. Pietz (1938) assumed that the red pigment maintains a definite redox level, also promoted by the presence of vitamin C and the activity of catalase. When greening occurs, the vitamin C content of the nodules decreases and catalase activity diminishes. There is also a change in phenoloxidase activity. Iron is in the ferrous form in the red pigment but when the nodules are destroyed it is oxidized.

Leghaemoglobin apparently catalyses nitrogen assimilation (Virtanen and Laine, 1945). This has been confirmed, in particular by experiments carried out in Wilson's laboratory (1958), where it was shown that the reduced form of leghaemoglobin is oxidized by nitrogen. Appleby and Bergersen (1958) established that Rhizobium bacteroids in the absence of oxygen may reduce leghaemoglobin. Perhaps leghaemoglobin is reduced by bacteria and oxidized during nitrogen fixation, promoting one of the intermediate reactions.

It is worth noting that leghaemoglobin almost quantitatively reduces hydroxylamine to ammonia (Colter and Quastel, 1950). It reduces hydrazine to ammonia rather more slowly, though actively (Virtanen and Miettinen, 1950). All this

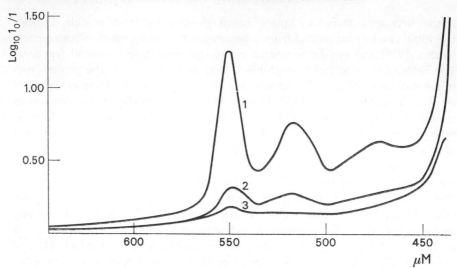

Fig. 3. Absorption spectra of the pyridine extract of nodules of the pea (after Virtanen and Miettinen)

1, Red nodules; 2, partially green nodules; 3, green nodules.

evidence suggests a relation between the presence of leghaemoglobin and the process of nitrogen fixation in leguminous plants, although the role of leghaemoglobin is still not clear.

Several workers have noted a connection between the formation of this red pigment and the efficiency of symbiosis (for example Virtanen *et al.*, 1947;

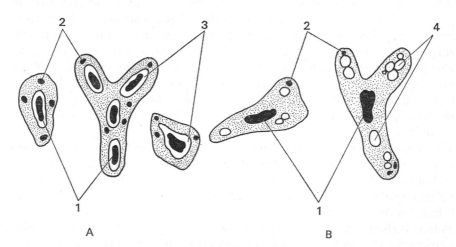

Fig. 4. Cytological differences between active and inactive bacteroids (after Bergersen).
A, Active bacteroids; *B*, inactive bacteroids; 1, nuclei; 2, analogues of mitochondria; 3, perinuclear zone; 4, glycogen.

Smith and Jordan, 1949; Garrard and Jordon, 1951; Materassi, 1956; Razumov-skaya and Fan Yun-Liu, 1961; Atwall and Sidhu, 1964; Sen, 1966). But there is not always any reliable pointer to a relationship, for there are cases in which the activity of the nodule bacteria does not depend on the accumulation of pigment in the nodules (for example Verbolovich, 1949; Nowotny-Mieczynska, 1952; Egle and Munding, 1953; E. A. Fedorov, 1955; M. F. Fedorov and Egorova, 1957).

Falk *et al.* (1959) felt that a guide to the ineffectiveness of nodule bacteria may be the lack of ability to form certain porphyrins (in particular, coproporphyrin and compounds of the uroporphyrin type) from α-aminolaevulinic acid and porphobilinogen.

Leghaemoglobin is also found in the nodules of plants other than the Leguminosae, but not in cultures of micro-organisms assimilating molecular nitrogen, nor in cultures of *Rhizobium* (Egle and Munding, 1953).

Cytochrome *a* is absent from effective nodules, but is detected in ineffective nodules and in the cells of *Rhizobium*. It is possibly destroyed in nodules that are vigorously binding nitrogen because of the low redox potential (Appleby and Bergersen, 1958). Cytochromes *b* and *c* are present in both effective and ineffective nodules.

Lucerne nodules formed by active strains of *Rhizobium* contain more cobalt and vitamin B_{12} than do ineffective nodules (Bertrand and Wolff, 1954). There are indications that the active strains of nodule bacteria form bacteroids more rapidly in nodular tissue. There is not always a positive link to be found between the number of bacteroids and nitrogen fixation. Thus, Spicher (1954) found such a link in species of *Lupinus* and *Ornithopus* but not in *Trifolium pratense*.

Glycogen does not accumulate in the bacteroids of active strains but is usually present in those of inactive strains. In the nucleus of bacteroids in active cultures it is possible to detect the process of division, with the nucleus surrounded by a 'perinuclear' zone (Fig. 10). There is no 'perinuclear' zone around the nucleus of the bacteroids of inactive strains (for example Chen and Thornton, 1940; Bergersen, 1957, 1958). According to Chizhik (1959) the nuclei of bacteroids of active cultures segregate when the plant is in flower.

According to Petrosyan and Avvakumova (1963, 1964), starch (glycogen) is absent but glucose is present in nodules formed by effective cultures of bacteria when the host plants are budding. On the other hand, in ineffective cultures, at this time when nitrogen fixation is highest, there is always glycogen but no glucose. Gołębiowska and Sypniewska (1962b) found glycogen in the bacteroids of the nodules formed by weakly effective strains, and also in plants grown in poor light or low temperatures.

Löhnis (1930) considered that nodules formed by active cultures contain starch only in the bacteroid tissue, whereas nodules formed by ineffective bacteria contain starch throughout.

Experiments carried out by Federov and Laslo (1956) and Nitse (1958) suggested that the more active strains of nodule bacteria were characterized by more vigorous utilization of molecular oxygen during respiration. We have not been able to confirm this (Mishustin and Shil'nikova, 1966) using suspensions of pure cultures and slices of nodular tissue. As Fig. 5 shows, the uptake of

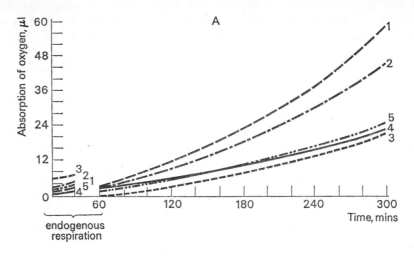

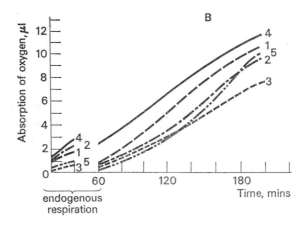

Fig. 5. Uptake of oxygen by nodule bacteria of broad beans with differing degrees of activity (increase per unit of cell nitrogen)

A, Suspensions: 1, 5, inactive, 2–4, active strains; *B*, nodular tissue: 1, 2, inactive; 3–5, active strains.

oxygen depends primarily on the characteristics of the strains in the cultures.

Numerous attempts have been made to determine the dehydrogenase activity of the nodular and cell tissues in bacterial cultures that differ in their activity. Results have been extremely contradictory (Jordan, 1955; Katznelson and Zagallo, 1957; Manil and Wernimont, 1961; Shmidt, 1964; Shil'nikova *et al.*,

1965, 1966; Dorosinskii *et al.*, 1966), and there is still no evidence of a correlation between the activity of individual cultures of bacteria and the activity of this enzyme.

Combined with a biochemical investigation of dehydrogenase activity we have made a cyto-morphological comparison of the cells of active and inactive cultures of nodule bacteria, in order to make a more detailed study of the structures involved in energy metabolism. The respiratory apparatus of bacterial cells is known to be represented by mesosomes—structures functionally identical to mitochondria. As with mitochondria, definite cytochemical activity is associated with mesosomes (Meisel' *et al.*, 1964; Yakovlev and Levchenko, 1964). The ability of mesosomes specifically to accumulate tetracycline antibiotics means that they can be studied under the luminescent microscope (Du Buy and Showackre, 1961; Poglazova, 1964).

We treated nodule bacteria with a luminescent tetracycline in a dilution of 1:50,000 for 12–14 hours according to the method of Du Buy and Showackre (1961) and also with triphenyltetrazolium chloride, and then inspected them under luminescent and light microscopes (Shil'nikova *et al.*, 1966). The tetracycline was selectively accumulated by the mitochondroid structures, and caused these structures to luminesce when excited by visible blue-violet light.

The mitochondroid structures thus detected corresponded to those found after treatment with triphenyltetrazolium chloride (by phase contrast). No differences were found in the number of mitochondroid structures per cell, or in their shape and size corresponding to the effectiveness of the strain in pure cultures or in a symbiotic relationship.

Each cell usually has a luminescing granule at one pole. The glow of the mesosomes was always brightest in the first minutes of an investigation and then the intensity decreased, irrespective of the biological features of the cultures. Thus no differences were demonstrated in the structure, size or intensity of the glow of mitochondroid bodies in active and inactive cultures.

In the long run a study of the enzymes of pure cultures of nodule bacteria and the tissues of the nodules formed by them (Fottrell, 1966; Fottrell and Graims, 1966) will probably give interesting results, which will be used to establish criteria of nitrogen fixation. So far, however, in spite of the work already described and several other investigations of the physiological and cyto-morphological features of the nodules (for example Deherain and Demousy, 1900; Spicher, 1954; Razumovskaya and Vasil'eva, 1956b; Wróbel and Gołębiowska, 1956; Melkumova, 1957; Bowen, 1959; Radulović, 1959–1960; Rangaswami and Oblisami, 1962; Holding and King, 1963; Sarić, 1963b; Hoffman, 1964; Gelin and Blixt, 1964; Döbereiner, 1966; Cheremisov, 1966) no definite conclusions can be drawn.

We feel that interesting results could be gained from a study of the isoelectric point of the tissues of the nodules. In the conditions of symbiosis this property is extremely labile.

In young nodules the isoelectric point is about 3·0–4·0, with ageing it shifts to pH 6·0–7·0 (Table 2.5). When the isoelectric point is in the acid zone, nitrogen assimilation is most active (Table 2.6). At this time the bacteria contain much DNA (Yakovleva, 1959, 1961).

Table 2.5. Change in the Isoelectric Point of Nodule Bacteria with the Age of the Plant

Plant	Developmental phase	Zone of nodule		
		Apical (pH)	Middle (pH)	Base (pH)
Lucerne in first year	Before budding	3·4	3·6	4·2
	On ripening	4·2	5·0	7·8
Pea	Before budding	3·0	3·0	3·0
	On ripening	5·8	5·8	7·2

This table is taken from Yakovleva (1960).

The isoelectric point of nodules infected by weakly active cultures of bacteria is more alkaline than that of nodules formed by active cultures (Tittsler and Lisse, 1936; Yakovleva, 1959, 1961; Emtsev et al., 1963, 1964; Ivanova and

Table 2.6. Isoelectric Point and Vigour of Nitrogen Assimilation by Nodule Bacteria of Lucerne

Zone of nodule	Isoelectric point (pH) of nodules formed	
	Spontaneously	By active strain
Apical	6·8	5·2
Middle	7·2	6·3
Base	7·6	7·6
Vigour of nitrogen assimilation, mg of N_2 per nodule		
Undifferentiated nodular tissue	0·7	2·4

This table is after Yakoleva.

Sakharnova, 1965; Dinchev, 1965; Til'ba and Golodyaev, 1966). But these differences are difficult to make use of for diagnostic purposes. They are observed only in nodules formed by cultures with sharply marked differences in effectiveness (Table 2.7). Moreover, this indicator changes with the age of the plant, the character of the substrates in which it is grown and so on. Changes in

Table 2.7. Isoelectric Point of Nodules of Broad Beans from Budding to the Beginning of Flowering

Strain	Level of nitrogen supply to plants		
	0·3 of norm*	0·8 of norm	1·3 of norm
Inactive No. T1	4·2–4·4	4·8–5·2	—
Inactive No. T15	4·4–4·6	4·6–5·0	5·6–5·8
Active No. T20	3·8–4·0	4·6–5·0	5·4

* A norm is equal to 0·3 g nitrogen in the form of NH_4NO_3 per kg of substrate.

the isoelectric point of the nodules of broad beans were found to be as in Table 2.8.

Wróbel (1956) obtained interesting results when he established that the pH and rH_2 of the root sap of inoculated plants were lower than in uninoculated plants. These values changed with the degree of effectiveness of symbiosis.

Wieringa and Bakhuis (1957) noted that the activity of nodule bacteria affects the level of the free amino acids in the bleeding sap of leguminous plants. Many

Table 2.8. Isoelectric point of Nodules of Broad Bean Plant with Ageing

Age (days)	Inactive strain T16	Active strain T20
30	4·6–5·0	3·8 4·0
50	5·6	5·6
75	6·0–6·4	6·0–6·4

earlier and later studies yielded similar results, but they are not adequate for an estimate of the effectiveness of nodule bacteria entering into symbiosis with leguminous plants (Orcutt, 1937; Hunt, 1951; Mityushova, 1955; Butler and Bathurst, 1958; Krasil'nikov and Aseeva, 1959; Ebertova, 1960; Dorosinskii et al., 1962; Garkavenko, 1962, 1962c; Dinchev, 1965; Shilnikova, 1962; Nita, 1963; Wang Chin T'ung, 1963; Nyval'ko and Sokorenko, 1966). Moreover, the determination of effectiveness of nodule bacteria by such a method is no less laborious than the use of the essential criterion of effectiveness—the yield and accumulation of nitrogen in the plant. It is to be hoped that further investigations will provide simpler and more reliable criteria of the effectiveness of nodule bacteria. It is worth noting the work of Ziemiecka (1963) who showed that the activity of bacteria actively forming nodules influences the intensity of respiration of the underground tissue of leguminous seedlings. There have been many investigations of the morphological and cultural properties of nodule bacteria that differ in activity, but all have failed to reveal any characteristic differences associated with the effectiveness of the cultures (Konokotina, 1934; Kobus, 1952; Peterson, 1954; Fedorov and Laslo, 1956; Nitse, 1958; Modrić, 1963; Gupta and Sen, 1964a).

Investigations carried out by Gupta and Sen (1964b; Sen, 1965) are particularly interesting. These two established a distinct correlation between the effectiveness of cultures of nodule bacteria and their ability to utilize glucose in pure cultures. Their observations, however, have not been followed up. Nor has it been possible to establish differences in the antigenic properties of cultures of nodule bacteria differing in activity (for example Izrail'skii, 1928; Shtern, 1953; Vincent, 1953; Read, 1953; Filippova, 1958; Kalnin'sh, 1964). The serological differences which are found do not always correlate with activity (Loos and Louw, 1964).

Different views have been expressed about the resistance to antibiotics of nodule bacteria that differ in their effectiveness (Lopatina, 1960; Gamatoua-Glavachkova, 1961; Schwinghamer, 1964). Individual cultures of nodule bacteria seem to differ in their resistance to antibiotics, but this property cannot be reliably related to their effectiveness.

There is no direct connection between the activity of nodule bacteria and their ability to synthesize vitamins (Okuda and Yamaguchi, 1960). As we have said (page 17), in laboratory conditions in artificial nutrient media nodule bacteria either do not fix molecular nitrogen at all or assimilate insignificant amounts. Therefore it is unfortunately not possible to use the nitrogen-fixing power as a criterion of the activity of nodule bacteria in pure culture.

Thus, the information we have outlined shows that no indirect criteria of the activity of nodule bacteria are yet of practical value. This means that when selecting and checking cultures of nodule bacteria for commercial experiments, it is necessary to grow the plants in sterile or semi-sterile conditions, inoculating them with the test bacterial cultures.

For practical purposes it is important to know the frequency of inactive nodule bacteria in nature. Several investigations have shown that a saprophytic existence in the soil frequently leads to inactivation of nodule bacteria (for example Purchase et al., 1951; Kalnin'sh, 1951; Thornton, 1952; Egorova, 1955; Nutman, 1956). The rapidity of inactivation is influenced by the properties of the soil, acidity being particularly unfavourable. Thus, Egorova isolated considerably more active cultures of nodule bacteria of clover from limed soddy-podsolic soils of the Moscow Timiryazev Agricultural Academy than from unlimed soils.

The relatively rapid loss of useful bacterial properties in an acid soil is demonstrated in Table 2.9 (after Egorova). Similar findings were obtained previously by Izrail'skii and Artemeva (1937). Federov and Glavachkova (1956) established a decline in nitrogen assimilation by nodule bacteria depending on the time when farm crops were used. They isolated more active cultures of nodule bacteria from the nodules of leguminous plants of the first year than from the nodules of plants in the second year of exploitation.

Thus, one cause of inadequate symbiosis between nodule bacteria and leguminous plants is a weakening of the activity of the bacteria in natural conditions. A second cause stems from the fact, noted at the beginning of this section,

Table 2.9. Activity of Nodule Bacteria in Limed and Unlimed soils

Culture of farm plant	Non-limed soil	Limed soil
Clover (first year)	53	100
Potato	12	112
Oats	9	111

100 is taken as the activity of a culture isolated from limed soil planted with clover of the first year.

that the nodule bacteria of one group of plants can often infect plants of another group. In most such cases the nodules formed accumulate nitrogen only weakly. As a result, nodule bacteria are active for some plants and inactive for others (for example Nutman, 1956; Norris, 1959b; Vyas and Prasad, 1959; Bonnier, 1962a; Bjalfve, 1963; Jensen, 1964).

In field conditions, for various reasons, the root system of leguminous plants is not always properly inoculated by active cultures of nodule bacteria, and so artificial infection of seed is recommended, using tested pure cultures of the bacteria (page 87).

Nodulation of Leguminous Plants and Nitrogen Fixation

Nodules form on the roots of leguminous plants as a result of infection by nodule bacteria. The infectious process has been extensively investigated, and there is now a onsiderable amount of information available about it (for example Beijerinck, 1888; Pražmowski, 1890; Němec, 1915; Milovidov, 1926, 1928; Thornton and Gangulee, 1926; Thornton, 1930, 1936; Fähraeus, 1957; Purchase, 1958; Bürgin-Wolff, 1959; Bergersen, 1958; Nutman, 1959, 1963; Vincent, 1962; Rothschild, 1963; Grilli, 1963; Jordan et al., 1963, 1965; Dart and Mercer, 1963a, b and c, 1964b; Mosse, 1964; Goodchild and Bergersen, 1966; Higashi, 1966). Nodule bacteria usually penetrate the roots through the root hairs. Sometimes they enter through damaged epidermal and cortical cells, especially at sites of branching of the lateral roots. In aquatic legumes, for example Neptunia oleracea, which have no root hairs, the nodule bacteria penetrate through the undamaged cells of the cortex or epidermis.

We shall consider a normal case of infection with nodule bacteria entering through the root hair. When the root of a leguminous plant develops in the soil it accumulates an abundant microflora peculiar to the rhizosphere of the plant to which it belongs. Here there is a sharp stimulation of multiplication in nodule bacteria which is not observed in the rhizosphere of non-leguminous plants (for example grasses). Purchase and Nutman (1957) showed clearly that any species of leguminous plant gives rise to conditions especially favourable to the multiplication of its own particular nodule bacteria. The reasons for this are still not clear. Purchase and Nutman infected clover seedlings with a mixture of nodule bacteria of clover and lucerne. After three weeks the titre of the former had risen by millions while the latter had hardly multiplied at all. According to Rovira (1957) the stimulating effect of the root affects the nodule bacteria even at a distance of 20–30 mm.

From a theoretical point of view it might seem that the infection of the root system of a leguminous plant would require but a single nodule bacterium. But, experiments have shown that effective inoculation is ensured only with a fairly large number of bacterial cells, which varies considerably according to different sets of data. According to Allen (1966) inoculation of plants which produce small seeds requires 500–1,000 cells per seed, whereas inoculation of those that produce large seeds requires not less than 70,000 cells per seed. Vincent (1966a and b) thought that at the moment of inoculation there must be at least several hundred viable and active nodule bacteria per seed. Lasting and Küüts (1966) found in field conditions that the success of inoculation depends on the number of nodule bacteria. The yield of white sweet clover in their experiments was maximum when seed was provided with not less than 40,000 cells of nodule bacteria. Vincent (1966a), Cloonan (1965), Cloonan and Vincent (1966) and Ireland (1966) found that if ineffective rhizobia predominate in the soil, the number of nodule bacteria introduced into the seeds must be increased to 100,000–1,000,000. It is possible that large numbers of nodule bacteria in the rhizosphere are necessary to ensure completion of the processes attendant on infection.

Among the products of exosmosis from the root system (sugars, organic acids, amino acids, nucleotides, vitamins, enzymes and so on) are small amounts of the amino acid, tryptophan. Under the influence of nodule bacteria tryptophan is converted to indolyl-3-acetic acid (IAA), which causes a distinctive change in the shape of the root hairs, bending them in the shape of an umbrella handle. (Thornton, 1929b, 1948; O. Allen and E. Allen, 1954; Kefford, 1960; Nutman, 1962). Curling of root hairs during inoculation was first noticed by Ward (1887, 1889).

In 1964, Haak questioned the importance of IAA in the bending of the root hairs, knowing that its effect is unspecific and can be exerted on the root hairs of non-leguminous plants. But nodule bacteria bring about changes only in leguminous species and, moreover, sometimes they have a certain selectivity. Thus, for example, *Rh. japonicum* does not affect the hairs of *Ornithopus sativum*.

If the bending were due only to IAA, there would be no such specificity. Moreover, root hairs respond in rather different ways to nodule bacteria and to IAA.

All this calls for more careful study of the factors responsible for the bending of root hairs, which is undoubtedly a very important part of the process of infection. It should be noted, however, that in individual cases infected roots can be uncurled (Bürgin-Wolff, 1959). Observation has shown that in lucerne and the pea 60–70 per cent of the root hairs are bent and curled, while in clover the figure is about 50 per cent (McCoy, 1932; Razumovskaya *et al.*, 1952). In some species of clover (*Trifolium parviflorum* and *Tr. glomeratum*) Nutman (1959*b*) recorded bending in not more than a quarter of infected hairs. The condition of the root hair is evidently important in the bending reaction. Razumovskaya *et al.* (1952) noted that root hairs are most sensitive to the action of substances formed by the bacteria, while growing.

Tagiev (1965) working in our laboratory, found that 0·0005 per cent IAA considerably enhanced the process of inoculation and development in lucerne growing in monobacterial culture. In control plants an average of thirteen nodules formed on each plant, but when the IAA was given they increased to twenty-one. More or less IAA had slightly less effect. Clearly IAA is not inert during inoculation.

Pate (1958) demonstrated chromatographically three types of auxin in the nodules of clover; β-indolylacetic acid predominated. There were also other auxin-like substances.

Nodule bacteria also have a different kind of influence on the root system of the leguminous plant. They form an extracellular mucilage of a polysaccharide nature (Anderson, 1933; Schluchterer and Stacey, 1945; Dudman, 1964*a*), and this steps up the plant's production of the enzyme polygalacturonase which is thought to act on the wall of the root hair, making it more plastic and permeable to bacteria. When nodule bacteria of a definite strain are unable to induce infection, they also fail to stimulate the formation of polygalacturonase of the leguminous plants in question (Ljunggren and Fähraeus, 1959). Thus even in the pre-infectious process there is interaction between nodule bacteria and the root system of legumes.

It is convenient to observe changes in the root hairs and their penetration by nodule bacteria in monobacterial culture, using the technique developed by Fähraeus (1957). Sterile seeds are grown between vertical glass slides, the lower parts of which are immersed in a sterile solution of mineral salts serving as nutrient. Nodule bacteria can be introduced into the solution, and the whole apparatus is placed in a test tube. As the root system develops between the slides it can be inspected under the microscope. The upper part of the plant extends beyond the slides and develops in the tube. Excised roots can also be cultured for study of the process of inoculation (McGonagle, 1944; Nutman, 1952; Raggio *et al.*, 1959; Avilov, 1961; Bunting and Horrocks, 1964).

We now consider the mechanism by which nodule bacteria penetrate the root hair. The investigations of Dart and Mercer (1964b) suggest that on the surface of the root there is a granulated mass of a substance enclosed in a membrane. Figure 6 shows schematically the structure of the infected root hair. In the presence of nodule bacteria the membrane has the form of a fibrillar (fibrous) matrix with the bacterial cells on its surface. Nodule bacteria may form conglomerations in this area but it is not known whether such an association is necessary for infection. Groups of cells at the tips of the roots were previously noted by Viermann (1922) and Thornton (1936), but Fähraeus (1957) from his own

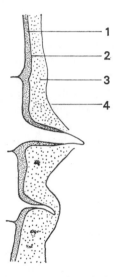

Fig. 6. The structure of the surface of the root hair of a leguminous plant after injection with *Rhizobium* (after Dart and Mercer, 1964b). 1, Plant cell wall; 2, cuticular layer over which there is a layer of granulated substance (3) which includes reuzobial cells; 4, outer layer of (3).

observations on the infectious process using the light microscope, assumed that infection is mediated by individual bacterial cells.

Nodule bacteria penetrate the root hair at certain places where the wall is sufficiently permeable. The wall of the root hair contains cellulose and pectin substances that nodule bacteria cannot decompose. In Nutman's view (1956) bacteria penetrate through the tip of the root hair when it is covered only by the first wall layer of pectin substances, polygalacturonides, galactanes, arabanes, hemicelluloses, waxes and polypeptides. The fibres of cellulose comprise most of the wall, in the form of a matrix. The gaps between the fibres are up to 0·3 μm wide and are not filled with calcium pectate. Therefore, with some stretching of the membrane the bacteria can pass into the hair (Frey–Wyssling, 1952).

Rudakov and Birkel' (1954) consider that penetration is accomplished with the aid of soil bacteria associated with the nodule bacteria. Soil bacteria destroy the pectin of the cell wall, and so aid penetration by the nodule bacteria. Of possible bacterial associates special attention has been paid to *Bacillus polymyxa* and *Achromobacter radiobacter*. It may be that in natural conditions some soil saprophytes promote the inoculation of leguminous plants by nodule bacteria. But inoculation can readily be achieved in a culture medium containing only nodule bacteria.

We think it likely that Ljunggren and Fähraeus (1959, 1961) were right when they considered that the enzyme polygalacturonase is essential to increase the permeability of the wall of the root hair. It is always present in small amounts in the root hair and evidently, by causing partial dissolution of the components of the wall, it enables the cell to stretch. Substances of the polysaccharide type, secreted by the nodule bacteria, greatly promote the formation of polygalacturonase. In turn, this leads to softening of the wall of the root hair, which enables the nodule bacteria to enter the root more easily.

Beijerinck (1888) assumed that the small, motile cells of the nodule bacteria can penetrate the modified wall of the root hair. Izrael'skii (1933) suggested that penetration of nodule bacteria into the root occurs at the filtrable stage of development. Dart and Mercer (1964*b*) using the replica method for electron microscopy found among the microfibrils of the walls of root hairs small coccoid cells (0·1–0·4 μm). These had numerous flagella. The gaps between the fibrils of the wall reached 0·3–0·4 μm wide, that is large enough for the cocci to pass through. (The existence of coccoid nodule bacteria in the soil has been noted on page 15).

In the root hair the nodule bacteria form a so-called 'infection thread'—a hypha-like mucilaginous mass in which are buried the multiplying rod-shaped bacteria. According to Dart and Mercer (1963), the bacteria in the thread have a cytoplasmic membrane and a cell wall 8 nm thick. On the outside the bacteria are surrounded by a microcapsule (Dixon, 1964).

The infection thread moves towards the base of the hair and the cells of the epidermis. This distance, about 100–200 μm, is covered in 1–2 days at about 5–8 μm per hour. The thread may move as a result of the pressure which builds up when the bacteria develop within it (Goodchild and Bergersen, 1966). As a rule, one thread forms for each root hair, although there are cases when several threads are seen in one hair, or several root hairs have a common thread.

By no means all deformed root hairs form an infection thread. The percentage of infected root hairs compared with those that are deformed is 0·6–3·2 and rarely reaches 8 (McCoy, 1932; Fähraeus, 1957; Nutman, 1959). The development of the infection thread is not infrequently arrested. Septa form in the thread, which is destroyed, although sometimes it continues to develop. The number of infection threads which cease to develop may vary from 10 to 80 per cent (Nutman, 1959). In some plants there are more of them, in others fewer.

If the infection thread continues to develop, it passes from the root hair into the external cells of the cortex and, in part, into the parenchyma. It never seems to penetrate either the endoderm or the pericycle.

When it penetrates the plant cells the infection thread is enclosed in a cellulose sheath from which, according to all available evidence, it is formed. The formation of the membrane around the infection thread can be regarded as a defence

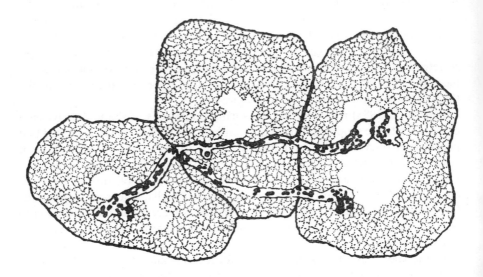

Fig. 7. Diagram of the passage of the infection thread through the cells of the root

reaction of the cytoplasm of the host plant. Only the growing apex of the infection thread is free of a membranous covering. The infection thread only grows actively in the root cell if it enters it close to the nucleus.

Where the threads come into contact with the host cell wall, the wall usually thickens. The mechanism by which the bacteria penetrate the cell walls is not clear, for the pores of the plasmodesmata are too small for the bacteria to pass through them. The infection thread, still without a membrane, crosses the intercellular space. But at this time the wall of the nearest plant cell proliferates and forms a funnel around the thread. Lengthening of this funnel produces a tubular sheath surrounding the infection thread. Duplication of this process in several cells gives the impression that the strand has a continuous hypha-like structure (Jordan *et al.*, 1963; Dart and Mercer, 1964*b*). The passage of the infection thread through the cells of the root tissue is shown in Fig. 7.

According to some investigators nodule bacteria can multiply only in tetra-ploid cells (Wipf and Cooper, 1938–1940; Senn, 1938; Razumovskaya *et al.*, 1952). When the infection thread reaches the tetraploid cells of the cortex, some nodule bacteria pass from the thread into the cytoplasm of these cells and there they begin to multiply. This pattern of infection is also true of polyploid plants. For example, the cells of *Nedigo sativa* are usually tetraploid but rhizobia infect only root cells which are octoploid. Diploid plants form nodules most slowly (Hely, 1957; Evans and Jones, 1966).

When the cytoplasm of the plant cell is infected by nodule bacteria division is induced in the cell concerned and in the adjacent uninfected cells. Division usually begins one to two cell layers from the end of the infection thread. The result of this process is a nodule.

The infection thread branches and spreads further through the tetraploid cells without involving the diploid cells. The cortex and conducting vessels of the nodule are formed from the diploid tissue of the plant. We shall consider later the structure of the nodule formed, and go on now to the next stage of infection—the passage of the bacteria from the infection thread into the host plant cells.

The bacteria may pass out of the infection thread, especially from its apex if this is not covered with a membrane, or after rupture of the vesicle-like (alveo-lar, vesicular) evaginations formed by the infection thread in the plant cells. These evaginations do not usually have a cellulose sheath and therefore the bacterial cells surrounded, as it were, by vacuolar cavities readily branch off into the cytoplasm of the host cells. There are different views about the formation of the membranous envelope of this vacuolar cavity. Dart and Mercer (1963) con-sidered that the membrane was formed *de novo* by bacteria which had emerged from the plant cell. But they did not ignore the possibility that the membrane was produced by the plant cell in response to the invasion of a foreign body. In the opinion of Jordan *et al.* (1963), after nodule bacteria have passed from the filament into the host cytoplasm they seem immediately to enter the small channels and cisternae of the endoplasmic reticulum, through which they pass. By joining one with another, the elements of the endoplasmic reticulum form a membrane around the bacteria. Bergersen and Briggs (1958) thought that the membrane was formed from the membrane of the plant cell, and Dixon (1964*a*) agreed that it was formed from the shell of the infected plasmalemma.

Light microscopy shows that bacteria which have passed into the cytoplasm elongate to rods. In this state they continue to multiply within each cavity. Later these cells are transformed into bacteroids. Their shape, as Beijerinck has indicated (1888) may differ. Bacteroids cannot divide (Almon, 1933) but greatly increase in size. In fast growing plants fewer large bacteroids form in the nodules than in slow growing plants. The bacteroids gradually swell and begin to occupy much of the plant cell (Plate 3), which usually expands considerably.

In some cases the volume of the bacteroid zone reaches a maximum early in the development of the nodule and then remains relatively constant. Therefore, when the nodule grows, the ratio of bacteroid tissue to the total volume of the nodule decreases (Allen *et al.*, 1955). The central bacteroid region, according to Bond (1941), Tuzimura (1950) and Rautanen and Saubert (1955) accounts for 16 to 50 per cent of the dry weight of the nodule.

There are several signs of subsequent changes in the bacteroids. The perinuclear zone becomes distinctly larger, with fragmentation of chromatin and invagination of the cytoplasmic membrane. The cells are vacuolated and volutin, glycogen and lipoproteins accumulate. As we noted in the previous section, bacteroids of active strains of nodule bacteria differ somewhat in their cytochemical properties from the bacteroids of inactive cultures.

The passage of bacteria into the cytoplasm from infection threads formed by inactive strains seems to be retarded. As a rule the formation of bacteroids is associated with the appearance of leghaemoglobin. According to Virtanen (1945) only rod-shaped bacteria are present in effective nodules. If the pink nodule becomes green the bacteroids are rapidly lysed. All this prompts us to accept the view of Nobbe and Hitner (1904) that the process of nitrogen accumulation is in some way related to the formation of the bacteroids.

The bacteria in the cytoplasm, including the bacteroids, are covered with an outer membrane the origin of which, as we have noted, is debatable. The numbers of bacteroids within such a membrane differ and are determined by the leguminous species concerned. Thus in the nodules of *Vigna sinensis*, *Acacia longifolia* and *Viminaria juncea* each membranous envelope contains eight or more bacteroids, while in the nodules of *Lupinus angustifolius* the number of bacteroids within a membranous envelope is one or, less often, two (Dart and Mercer, 1966).

Several observers, (for example, Bergersen and Briggs, 1958; Jordan *et al.*, 1963; Dart and Mercer, 1964*b*; Mosse, 1964) have found that penetration by the infection thread and its bacteria causes some substantial changes in the root cells. For example the nucleus of the plant cell usually has a fibrillar structure and is filled with ellipsoid granules measuring about 15 nm across. And there are cracks in the nuclear membrane. There are also appreciable changes in the endoplasmic reticulum, with increased numbers of ribosomes and mitochondria and a change in the Golgi apparatus. Small electron-dense particles measuring 6 nm across form in the plastids of the cells.

While the bacteroids are forming, the mitochondria and cell plastids move towards the cell membrane and line up along it. All this time leghaemoglobin is found in the nodules; it is believed to take part in the process of nitrogen assimilation. Since the numbers of ribosomes increase just before the formation of haemoglobin, it can be supposed that these particles are responsible for the latter process.

All these changes seen in the infected cell indicate that it is in an excited state, similar to that observed in cells penetrated by parasitic organisms. Multiplication is enhanced in the infected cells and these in turn stimulate adjacent uninfected cells, leading to the formation of an inflated nodule. In the mature nodule the following zones can be distinguished. (1) The cortex consisting of several layers of uninfected parenchymatous cells. These cells are small and differ from the cells of the root cortex. When the nodule forms, the root cortex usually ruptures. Its remains may be maintained on the surface of the nodule. (2) A zone of briskly dividing, uninfected cells located under the nodule cortex. The cells of this zone form the nodule cortex, the vascular cells, the meristem and the cells that are attacked by the bacteria. The various types of proliferation of the meristems account for the different shapes of the nodules: spherical, cylindrical and so on. (3) A zone of infected tissue in which cells infected by bacteria alternate with uninfected cells. This zone is in the deeper layers of the nodules. Cells that contain bacteroids are severely hypertrophied and inflated. They may be eight times as large as the uninfected cells. Bacteroids do not exceed 0·62–4·5 per cent of all bacteria in the nodules of *Ornithopus sativus* and *Lupinus luteus* (Spicher, 1954) but their proportion is greater in the nodules of *Trifolium pratense*. The ratio of cells infected with bacteria and bacteroids to cells without them differs from species to species. Practically all the cells in this zone of the clover plant are infected, whereas in *Caragana* only 50–70 per cent of them are infected. (4) The vascular system, formed early in the development of the nodules, comes from the cells present between the cortex of the nodule and the infected tissue. The vascular bundles of the nodule, which later join up, consist of the tracheids of the xylem and the fibres and sieve tubes of the phloem. They are surrounded by a sheath of parenchymatous cells.

There are between one and twelve vascular bundles in the nodules, in many cases only one. Often the vascular system of the nodule is separated from its cortex on the outside by a layer of suberized cells ('nodular endodermis') linked with the endodermis of the root.

Heuman (1952*a*) made a quantitative estimate of the content of bacteroids in different zones along the nodule. In the apical zone there were none, in the intermediate zone 70 per cent of bacterial cells were bacteroids, in the central zone the proportion was 93 per cent and in the basal zone up to 45 per cent. The percentage of live bacteroids in these zones, however, was 60, 35 and 20 respectively (Heuman, 1952*b*).

Nodules formed by active cultures of nodule bacteria usually have a whitish colour when young. By the time of optimum activity they are pink. Nodules that develop after infection with inactive cultures are green. These nodules are almost indistinguishable in structure from those that develop when active strains are involved, but they disintegrate prematurely. In some cases nodules formed by inactive bacteria have an abnormal structure.

Occasionally black nodules are formed. They have been found on the roots of *Dolichos* and *Centrosema* (Cloonan, 1963; Donald, 1965); presumably the dark coloured compounds they contain are similar to the melanins (Döbereiner, 1965).

An interesting abnormal relationship between nodule bacteria and a leguminous plants was described by Erdman *et al.* (1956). They found that some quite effective cultures of nodule bacteria cause chlorosis in soy-bean leaves. Later work has shown that such strains can produce a phytotoxin—a low-molecular weight compound containing an amino group. Certain lines of soy bean, however, are insensitive to the phytotoxin (Johnson *et al.*, 1959, 1960, 1964; Owens and Wright, 1965; Owens and Thompson, 1966).

Sterile nodules found on the roots of lucerne can be assumed to be the result of abnormal development. Histologically they are similar (in the presence of

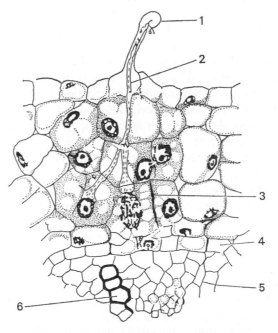

Fig. 8. Infection of lucerne with nodule bacteria (after Thornton, 1930)
1, Infected root hair; 2, infection thread containing bacteria; 3, dividing cells; 4, endodermis; 5, central cylinder of root; 6, xylem.

bacteroid tissue) to normal nodules (Bonnier, 1962). In Bonnier's view the cause of such nodules may be either a disturbance of the auxin equilibrium in the root tissues—the accumulation of excess auxin—or a virus-like agent that might be present under the seed coat.

In many leguminous species nodules are formed as we have described (Figs. 8 and 9), but sometimes the infectious process is rather different. Thus in the peanut, infection takes place not through a root hair but through the growing point of a lateral root. In the lupin no infection thread can be found penetrating the cells of the root cortex, and the distribution of the bacteria through the root is brought about by division of the cells of the infected tissue. In this way the volume of tissue containing bacteroids is increased. Around it are the uninfected

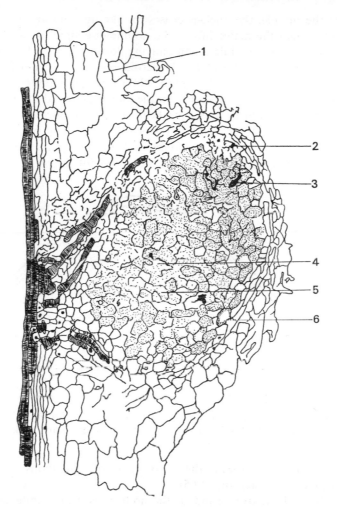

Fig. 9. Longitudinal section 5 mm thick through 15 day old clover nodule (after
Virtanen and Miettinen, 1963)

1, Cortical cells (of the root disrupted by the developing nodule); 2, uninfected cells
of the meristem in the outer layer of cells of the nodule; 3, fragment of infection thread;
4, hypertrophied and deformed nucleus of the plant cell; 5, uninfected cells; 6, hypertrophied plant cell with large vacuole. The vessels of the root are shaded.

cells of the cortex. The cells of the cambium at the base of the nodule are transformed into vascular bundles which link the vascular system of the central root
to the nodule.

In time the nodule degenerates and dies away. Suberization of the cells of the
vascular system is important in this respect for it retards the exchange of nutrient
substances between the host plant and the nodular tissues. Vacuoles appear in

the cells of the nodule, the nucleus ceases to take up stain and the bacteroids undergo lysis when the active life of the nodule is ending, usually coinciding with the necrosis of the nodule accompanying flowering in the host plant. This necrosis begins at the centre of the bacteroid region (Harris *et al.*, 1949) or close to the base of the nodule (Thornton, 1947) and progresses to the periphery. Nodule bacteria which have passed from the infection thread to the cells of the tissues of the nodule during this time begin to decompose the cell contents. Leghaemoglobin is destroyed and is replaced by a yellow pigment. Various species of saphrophytic bacteria are found in degenerating, and sometimes also in active nodules (Subba Rao and Vasantha, 1965*b*). Microscopic fungi are

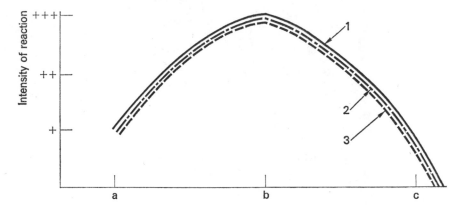

Fig. 10. Activity of the nodules in different phases of development
1, Cytochrome oxidases; 2, amino-acids; 3, sulphydryl reaction. (*a*) Young nodules (green); (*b*) mature nodules (pink); (*c*) senescent nodules (green and dark if destroyed).

also found in large quantities in the rhizospheres of leguminous plants (Taylor and Parkinson, 1965; Jackson, 1965).

The length of the active period of the nodules is greatly influenced by the state of the host plant. For example, by castrating or removing the flowers it is possible to prolong the vegetative life of the plant and extend the time of activity of the nodule bacteria (Yukhimchuk, 1957). Nodules are short-lived in annual plants, but in perennials they may function for several years. By the end of the season their bacteroid tissue degenerates but the whole nodule does not die off. In the following year it begins to function again. It should be noted that the cortical cells of the nodules in annual plants do not usually have inclusions. In perennial woody plants the walls are often suberized or include resins and tannins which evidently have a protective function. The biological activity of the nodule changes with the different phases of plant development. Figure 10 is a qualitative characterization of certain indicators of the activity of nodular tissue (results of Shemakhanova).

As we have indicated, the formation of the nodule has now been studied in considerable detail. Several important details have been elucidated largely thanks to the electron microscope. Callao and Olivares have suggested that the fluroescent antibody method be used for examining the bacteria in the nodules (1964).

Figures 11–13 give diagrammatic representations of the cell of a nodule bacterium undergoing a series of consecutive changes in the process of infecting a root tissue, a scheme for the structure of the infected cell of the nodule and a diagram of the nodule so formed.

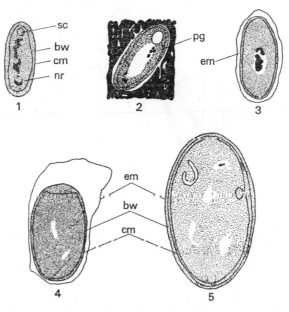

Fig. 11. Changes in the structure of the bacterial cell as the nodule develops (after Mosse, 1964).

Bacterium (1) in the rhizosphere; (2) in the infection thread; (3) passing into the cell of the plant and still dividing; (4) final stage of formation of bacteroid. bw, Bacterial wall; cm, cytoplasmic membrane; em, enclosing membrane of plant origin; nr, nuclear region; sc, storage carbohydrate.

The number of nodules on leguminous plants is always more or less limited. This point has been explained well by Nutman (1949) using morphogenetic data. He considers that the nodules develop only in parts of the root where there is meristematic activity, that is at areas of potential growth of lateral roots. The number of such areas is predetermined, thus restricting the number of nodules on the roots. In legumes that develop in a short time the ability to form nodules is more restricted than in species with a long growing period. Clearly the structure of the root system may have a profound influence on the numbers of nodules. Nutman's view has been confirmed by several experiments.

As we have noted, active strains of nodule bacteria give rise to fewer nodules than do ineffective strains. In Nutman's view the *de novo* formation of nodules is

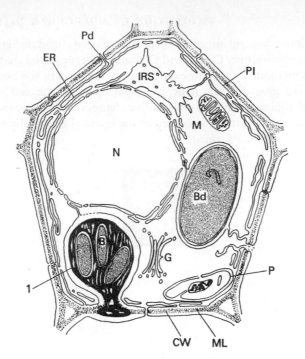

Fig. 12. Schematic structure of the infected cell of the nodule (after Mosse, 1964).
Bd, Bacteroid; N, nucleus; B, bacteria; IT, infection thread; G, Golgi apparatus; P, proplastids; M, mitochondria; IRS, intrareticular space; ER, endoplasmic reticulum; PD, plasmodesma; CW, cell wall; ML, middle lamella; PI, plasmalemma.

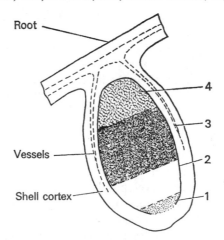

Fig. 13. Scheme of the structure of the nodule of clover during nitrogen fixation (after Mosse, 1964).
I, Meristematic zone with dividing cells and young infection threads; II, zone containing cells with proliferating infection threads and rod-shaped nodule bacteria; III, pink-coloured zone with vacuolated cells filled with bacteroids; IV, greenish-coloured zone containing collapsing plant cells and disintegrating bacteroids.

inhibited by the lateral meristem of the nodules and the root. In active nodules the meristem remains active for a long time and therefore delays the formation of new nodules. Ineffective nodules are incomplete and are not capable of exerting a significant influence on the formation of new nodules.

Other factors that may influence the numbers of nodules and their activity have now been studied quite closely. We shall consider them in the following section.

Several early investigations showed that atmospheric nitrogen is fixed in the nodules (for example, Voronin, 1867, 1886; Hiltner, 1896; Timiryazev, 1893; Krasheninnikov, 1916). Krasheninnikov (1916) kept isolated nodules for 25 to 94 hours in a gas medium containing molecular nitrogen and found that they fixed nitrogen. Ten grams of nodules fixed 0·37–1·2 mg of free nitrogen. Later Virtanen and Laine (1937), using the gasometric method and adding oxaloacetic acid to the nodules, obtained quite convincing evidence that the nodules bind molecular nitrogen. Crushed nodules did not have this function.

Burris et al. (1943) and Aprison and Burris (1952) used nitrogen to demonstrate the nitrogen-fixing activity of the nodules. After a period of failure they established assimilation of molecular nitrogen by slices of nodules. This, however, required the use of a particular experimental technique. Unlike Virtanen these investigators found that oxaloacetic acid and reducing substances decreased the level of nitrogen assimilation by the nodules. Only sucrose had a positive effect. As the nodules aged their nitrogen-fixing capacity diminished. When the nodules were left in the dark for a long time leghaemoglobin was oxidized and less molecular nitrogen was fixed. The optimum oxygen concentration was 0·5 atmospheres.

There have been many hypotheses concerning the chemistry of nitrogen fixation by leguminous plants (for example Vinogradskii, 1936; Wilson, 1939, 1940; Virtanen, 1945; Federov, 1943; Turchin, 1959; Bergersen, 1960a, b; 1963b). It can be supposed that the process that occurs in the nodules of legumes is approximately identical to that in the cells of free living nitrogen fixers, which has been the theme of a great deal of work reviewed in various publications (for example Nicholas, 1963a, b; Virtanen and Miettinen, 1963; Kretovich and Lyubimov, 1964). We shall consider the theory of nitrogen fixation later, and here we merely note that the first product of synthesis is evidently hydroxylamine, which is readily converted to ammonium and then very rapidly transformed into amino acids and then into proteins. In the nodules there are large amounts of protein nitrogen (0·6–1·0 per cent by fresh weight) and less non-protein nitrogen (0·13–0·23 per cent) and no ammoniacal nitrogen (Korsakova and Lopatina, 1934).

The nodules contain far more nitrogen than the rest of the plant. This is indirect confirmation that nitrogen assimilation occurs in the nodules. From here bound nitrogen passes into the aerial parts of the plant. Initially it was assumed that the bacterial cells were lysed and release nitrogen-containing substances. But this was refuted by subsequent investigations which showed that during nitrogen fixation the cells of the nodules are filled with bacteroids. And Grijns established that there is no rhizobial phage in the nodules (1927). All this indicates that bound nitrogen is transferred from the tissues of the nodules at a time when the bacteroids are in a viable state. This confirms the view first advanced by Nobbe and Hiltner (1893).

Virtanen and Hausen (1931), growing pea plants in sterile sand culture, established that part of the nitrogen assimilated by the plant is released into the external environment in the form of organic compounds. The level of exosmosis in individual cases varied from 10 to 30 per cent (sometimes it was 50 per cent) of the assimilated nitrogen. Bond (1936) showed that aspartic acid makes up a considerable proportion of the substances released by exosmosis. Chromatography has shown that the products of exosmosis contain other organic compounds (Miettinen, 1955). In particular when the excreted substances are hydrolysed with 6 N HCl, a compound closely related to γ-aminobutyric acid is found in large quantities (Butler and Bathurst, 1958). The composition of the products of exosmosis possibly differs somewhat in different species.

The composition of the secretion of the roots of leguminous plants seems to be similar to that of bleeding sap. The nitrogen-containing compounds found in it include aspartic acid, asparagine, glutamine, a little hydroxyproline, threonine, alanine, homoserine and certain other substances. The bulk of the nitrogenous substances is accounted for by aspartic acid (for example Wieringa and Bakhuis, 1957; Nirtanen and Santaoja, 1959).

It is interesting that the bleeding sap of leguminous plants infected with effective races of nodule bacteria contains larger amounts of nitrogenous compounds than the bleeding sap of other plants.

The most intense exosmosis in leguminous root systems occurs at an early age. Later (in sterile culture) a considerable proportion of these compounds is utilized by the plant. The pea, for example, may assimilate L-asparagine and L-glutamic acid. When the plants are grown in the soil, of course, the compounds released are rapidly mineralized. All this data on exosmosis is evidence that nitrogen-containing substances are transported from the nodules to the plant principally in the form of amino acids.

Several authors have not been able to establish any considerable exosmosis of nitrogenous compounds from leguminous plants. Some have observed exosmosis only when the nodules are removed from the roots (Wilson and Wyss, 1937; Wilson and Burton, 1938; Weichsel, 1961; Hoffman, 1961). There is no reason to doubt the thoroughness of the experiments carried out by any of these investigators. It must be supposed that the discrepancies in the results are determined by individual experimental conditions. This can be settled only by further investigation. If we assume that nitrogenous substances can be released by the root systems of leguminous plants, then we can suppose that other plant species growing near these legumes will develop better than those cultivated separately.

A similar phenomenon has been observed by several investigators (Lipman, 1912; Lyon and Bizell, 1911; Evans, 1916; Vartiovaara, 1933; Thornton, 1939; Jonsson, 1939; Elsukov, 1961; Evans and Jones, 1966; Jones and Evans, 1966). In passing, we note that according to Demolon (1961) leguminous plants in mixed sowings with grasses fix nitrogen more briskly than those planted alone. This, in his view, is determined by the decrease in the amount of bound nitrogen in the soil as a result of its utilization by the grasses. At the same time, in other experiments, when legumes and other plants were cultivated together the development of the latter was virtually no better than in single stands (for example, Romashev, 1936; Wilson and Wyss, 1937; Bond and Boyes, 1939; Madhok, 1940; Gulyakin et al. 1963a). This problem seems to require fuller investigation.

The identity of the structural elements of the infected cells of the nodule which fix molecular nitrogen has been investigated. Using nitrogen 15 and differential centrifugation of crushed nodules, Bergersen and Wilson (1959) found that fixed nitrogen accumulates in the plant cell membrane and not in the bacteria. Turchin (1956) obtained similar results. This led Bergersen (1962a, b) to develop his scheme for this fixation of atmospheric nitrogen (Fig. 137). Using nitrogen-15, Yakovleva (1963) showed that nitrogen fixation proceeds most vigorously in the bacteroid zone of the nodule tissue, with a lower isoelectric point than the rest of the nodule.

The bacteroids are evidently active in the plant cells when conditions are quite poor, for the plastids and mitochondria are located along the cell wall and utilize the oxygen entering the cell.

Factors Determining Symbiotic Relationships of Leguminous Plants and Bacteria

The relationships between leguminous plants and nodule bacteria that ensure healthy development of the plants depend on a definite set of conditions.

First, soil moisture is important; its importance in inoculation has been emphasized several times (for example Prucha, 1915; Wilson, 1917; Machavariani, 1951; Ukrainskii, 1954; Shchepkina, 1959; Tuzimura and Watanabe, 1960; Kishinevskii, 1966). Nodules usually form when the soil moisture content is 40 to 80 per cent of the total water capacity. The optimum is 60 to 70 per cent. Federov and Pod'yapol'skaya (1951) showed that surplus moisture does not suppress the normal functioning of nodules once they are formed. Moreover, Masefield (1961) observed that wetting of inoculated leguminous plants growing in a humid climate intensifies the accumulation of nitrogen. Our own observations indicate that a lack of moisture in the soil leads to the death of formed nodules. It is known that in regions where water is in short supply many leguminous plants develop without forming nodules, even if inoculated with nodule bacteria.

For practical purposes it is important to note that plants differ in their critical

3

moisture thresholds. For example, sainfoin forms nodules quite well even when the soil is only slightly wet, while lucerne is very sensitive to a lack of moisture (Kornilov and Verteletskaya, 1952). This should be borne in mind when choosing leguminous plants for growing in different soil and climatic conditions. The critical moisture level for nodulation has been insufficiently studied for individual leguminous species. It is also not clear how a lack or excess of water influences the relationship between nodule bacteria and leguminous plants. We can only assume that with very high moisture contents the supply of oxygen to the root system decreases (van Schreven et al., 1954). Beneficial aeration promotes the development of nodules and fixation by them of molecular nitrogen (Allison and Hoover, 1934; Lilly and Leonian, 1945; Bond, 1950; van Schreven, 1958).

As for the cells of nodule bacteria, early investigations (Giltner and Langworthy, 1916; Vandecaveye, 1927) established that when the moisture content of the soil is reduced the bacteria do not multiply, but remain inactive for a long time. Evidently, the minimum soil moisture for nodule bacteria to develop is about 16 per cent of the total water capacity (for example Brockwell, 1954; Bonnier, 1955). But because nodule bacteria do not multiply in the absence of moisture Millington (1955) recommends deep placement of inoculated seeds when the soil is likely to be dry for a long time. It is interesting to note that nodule bacteria are more resistant to drought in soils in a dry climate than in a wet climate (Saubert von Hausen and Van Gylswyk, 1957). Masefield (1958) considered that the factor restricting the spread of nodule bacteria in a dry climate was moisture rather than temperature.

We have noted the favourable effect of aeration on symbiotic nitrogen fixation, which was established by Virtanen et al. (1931, 1936) who showed that when they are weakly aerated even effective rhizobial races form only small nodules. van Schreven (1958) found that the part of the root system of the pea which was aerated developed more briskly and formed nodules more abundantly than the unaerated part.

When the entry of oxygen to the root system is restricted the content of leghaemoglobin in the nodules decreases (Virtanen et al., 1946, 1947; Egle and Munding, 1953; Fedorov, 1955) and fewer nitrogenous compounds leave the root by exosmosis (Virtanen and Torniainen, 1940).

The degree of aeration affects the distribution of the nodules at different levels in the soil. When soil moisture is normal the best formation of nodules is found in the near-surface soil layer. Shevchuka's findings (1964) are a good example; he investigated the distribution of nodules in the root system of the broad bean in relation to the depth of the root system (Table 2.10).

Undoubtedly, the number of nodules is influenced not only by the depth of the root system, but also by the state of the soil (in particular, its moisture content) and the age of the plant. Species specificity is also involved, for oxygen requirements differ from species to species of plants and nodule bacteria (Fedorov and Laslo, 1956).

Nodule bacteria are aerobic and develop most successfully when they have access to air (van Schreven *et al.*, 1953, 1954). When they are cultivated under low partial oxygen pressure there is a tendency for the cellular cytochrome *a* to decrease and cytochrome *c* to increase. The content of cytochrome *b* in rhizobial cells remains more or less constant whatever the conditions of cultivation. The cytochrome balance of bacteria isolated from the nodules resembles that of bacteria growing in conditions of oxygen deficit (Appleby and Bergerson, 1958).

Insufficient aeration also affects other properties of cultures of nodule bacteria, in particular their antigenic properties (Tuzimura and Watanabe, 1964; Dudman, 1964*b*) and respiratory activity (Konishi, 1936; Bond, 1950). According to Cook and Quadling (1962), in a culture of *Rh. meliloti* deficient in oxygen an

Table 2.10. Distribution of Nodules of Broad Bean in the Soil

Depth of roots (cm)	No. of nodules on roots of one plant
0–5	21
5–10	16
10–15	7
15–20	3
20–25	2
25–30	0

excess of hydrogen peroxide forms which suppresses the vital activity of the nodule bacteria. A similar phenomenon may occur in a pure culture of nodule bacteria or within the nodules. In the latter case it may be the cause of ineffective symbiosis.

The redox potential of the tissues of nodules is considerably decreased just before flowering (Rabotnova, 1936, 1957; Kłosowska, 1952). This is quite understandable, for oxygen can be an acceptor of reductants of molecular nitrogen, and so a high degree of aeration will suppress the assimilation of nitrogen. Thus, a certain level of oxygen is useful for inoculation and nodulation but the redox potential must not be high in the nodular tissue.

Temperature is particularly important in the relationship between nodule bacteria and the leguminous plant. Nitrogen fixation occurs most actively only within a certain range of temperature. The vigour of the process diminishes sharply as the temperature falls or rises. According to Gukova (1945) if the temperature goes below the optimum, nitrogen fixation is suppressed less than for an equivalent rise in temperature. The optimum temperatures for the development of leguminous plants, for nodulation and for nitrogen assimilation do not coincide, and the temperatures that promote the formation of nodules on the primary and secondary roots also differ.

The formation of nodules in natural conditions can be observed at temperatures a little above freezing. In such conditions there is virtually no nitrogen assimilation. It is possible that only Arctic symbiotic leguminous plants bind nitrogen at very low temperatures (Allen, *et al.*, 1964). Usually nitrogen fixation can be observed clearly only at temperatures of about 10°C and warmer (Stalder, 1952; Maeda, 1960; Hely and Williams, 1964). According to Pate (1961) maximum nitrogen fixation is observed in vetch and lucerne at about 24°C. The position of the optimum point may vary with the plant species, and even the variety, and also with features of the bacterial culture (Gibson, 1965, 1966). The temperature optimum for the development of leguminous plants may vary according to the form of nitrogen available. At colder temperatures they grow better on molecular than on bound nitrogen (Gukova, 1962).

According to Mes (1959) tropical legumes fix nitrogen symbiotically at temperatures higher than do legumes growing in temperate climates. Thus, an increase in the night time temperature from 10° to 21°C, or the day time temperature from 18° to 27°C, has a negative effect on the level of nitrogen fixation in vetch, pea and lupin. These plants thrive at a lower temperature. On the other hand, in tropical plants such as (*Stizolobium deeringianum, Arachis hypogoea, Glycine max* Merr) a night time temperature colder than 18°C with a constant day time temperature of 33°C has a markedly negative influence on nitrogen accumulation.

According to Possingham *et al.* (1964) in *Trifolium subterraneum* nodules form best at about 30°C, but for vigorous nitrogen accumulation a lower temperature is necessary. Jaffe *et al.* (1961) also noted the harmful effect of a high temperature on symbiotic nitrogen fixation and so did Fan Yun-Luy (1965). At temperatures above 30°C the leghaemoglobin content of the nodules decreases and dark pigment accumulates.

Weak fixation of molecular nitrogen at high temperatures cannot be attributed to a deficit of carbohydrates in the root system of the plant as a result of intensified respiration. As Meyer and Anderson (1959) showed, the addition of sucrose to the medium at 30°C does not increase the effectiveness of symbiosis.

Many leguminous species at a temperature of about 30°C bind molecular nitrogen quite weakly although their bacterial symbiont can multiply actively in these conditions. Nodules form but nitrogen does not accumulate (Meyer and Anderson, 1959; Virtanen and Miettinen, 1963).

Vartiovaara made a very interesting observation (1937) in establishing the ecological adaptability of nodule bacteria to temperature. In his experiments cultures of clover nodule bacteria isolated from Egyptian soils at 6°–13°C formed nodules more poorly and fixed nitrogen more weakly than bacteria isolated from Swedish and Finnish soils. At 19°–20°C these differences levelled out. Petrosyan (1959) also recorded definite differences in relation to temperature between ecological races of nodule bacteria. Bacteria from mountainous regions of Armenia had a lower temperature maximum than those from the warmer valley area of the same republic. According to Bowen and Kennedy

(1959), however, nodule bacteria isolated from tropical soils often develop with a lower temperature maximum than those isolated from the soils of Rothamsted Experimental Station. According to Dart and Mercer (1965a) leghaemoglobin forms best in the nodules at 24°C.

All this evidence indicates that at low temperatures (below 10°C) nodules form but nitrogen is not assimilated. Therefore, if legumes are to be sown early it is desirable to treat the soil with small amounts of nitrogenous fertilizers (Shevchuk, 1964; Korovin and Vorob'ev, 1965). In this connection it is interesting to note the findings of Krylova et al. (1963) indicating that early sowing of legumes creates favourable conditions for the establishment of symbiosis and for its effectiveness.

Dried nodule bacteria tolerate quite well an increase in temperature (50°–70°C) (Parker and Sanderson: quoted by Vincent, 1966a). The degree of this resistance varies with individual cultures. In a medium containing colloids, particularly in the soil, the bacteria more readily tolerate warmer temperatures (Vass, 1919; Marshall and Roberts, 1963; Marshall, 1964; Hely, 1964).

Several investigators have noted resistance to low temperatures by nodule bacteria living free in the soil (for example Dorosinskii and Lampovshchikov, 1948; Lopatina and Lazareva, 1957). Nodule bacteria in wintering nodules are equally resistant to the cold (Pate, 1958; Bergersen et al., 1963).

The acidity of the soil greatly influences the vital activity of nodule bacteria and nodulation. The range of suitable pH values differs somewhat for different species and even different strains (Bryan, 1923). Thus, for example, nodule bacteria of clover are more resistant to low pH than are those of lucerne. Evidently, the adaptation of the micro-organisms to the habitat plays a part here— clover grows on more acid soils than does lucerne. Seradella grows on even more acid soils. Confirmation is provided by the experiments of Mulder et al. (1966). According to their findings the relationship between nodule bacteria and pH also differs for different varieties of most plants.

On the basis of soil acidity, Pyarsim (1966) divided species of nodule bacteria into two. The first group contains bacteria that form nodules on Vicia, Lathyrus, Trifolium and Lotus and develop at pH_{KCl} 4·6–7·1; the second group, infecting Medicago, Ononis and Anthyllis, develop at pH_{KCl} 6·8–7·4.

The effect of the pH of the medium on the vital activity and competence of nodule bacteria has been investigated often (Doolas, 1930; Lampovshchikov, 1951; Kedrov-Zikhman, 1951; Kalnin'sh, 1951, 1958; Egorova, 1955; Vincent, 1958; Loneragan and Douling 1958; Norris, 1958, 1959a; D'yakova and Borodulina, 1960; Mulder and van Veen, 1960b; Holding and King, 1963; Obaton and Blachère, 1963; Virtanen and Miettinen, 1963; Döbereiner et al., 1965; Vikulina and Krilova, 1966). These workers all established an unfavourable effect of low pH on the properties of nodule bacteria. In particular, they noted that nodule bacteria with weak nitrogen-fixing powers are very often isolated from acid soils (Table 2.11).

In response to both acid and alkaline conditions in the medium nodule

Table 2.11. Effectiveness of Nodule Bacteria of Vetch Isolated from Different Soils

Soil	pH of soil (aqueous) used for isolation of bacteria	Dry weight of 100 plants (g)	N in plants (%)
Soddy-podzolic:			
Virgin	5·7–6·0	12–14	2·6–2·7
Cultivated	5·7–7·2	37·9–70·3	3·1–3·9
Black earth	6·8–7·9	66·3–82·5	3·8–4·0

This table is after Izrail'skii and Artem'eva (1937).

bacteria become vacuolated or assume other forms (coccoid, L-forms) which are unable to infect a plant (Sniezcko, 1928; Cabezas de Herrera, 1956). These changes, however, are reversible.

The pH limits for the growth of legumes are usually wider than the pH range suitable for nodulation (Bryan, 1923; Virtanen, 1928; Fletcher, 1958). Thus, Bryan found that soy beans can grow in a pH of 3·9 to 9·6, while nodules are seen to form only between pH 4·6 and 8·0. Nodules form best in a nearly neutral medium. The effect of the pH of the medium on the formation of nodules by soy beans is shown in Table 2.12.

Table 2.12. Effect of pH of Medium on Nodulation in Soy bean

pH of medium	No. of nodules per plant
4·0	0·0
4·5	1·4
5·0	17·0
6·0	30·0
8·3	4·0

The findings of Doolas (1930) and others show that nodulation may occur in leguminous plants when the surrounding pH is somewhat greater than 4·0. Jensen (1943) observed nodules in some species of clover between pH 4·2 and 4·5.

Nodule bacteria in the soil are more sensitive to an acid medium than those living within a root system (Jensen 1961). Possibly in the root tissues the pH is higher. Thus according to Tsyurupa (1937) the pH of root extracts of leguminous plants is approximately at the optimum for the development of nodule bacteria. At any rate, the nitrogen-fixing function of formed nodules is maintained even when the medium is too acid for the formation of new nodules.

Mulder *et al.* (1966) felt that the penetration of nodule bacteria into the roots of a legume is impeded in an acid medium, and also by disturbances in the surface structure of the root which destroy possible sites of entry of infection. When, however, the bacteria do penetrate the root hair at a low pH, the bacteroid tissue develops abnormally and is rapidly destroyed (Gołębiowska and Sypniewska, 1962*a*).

In acid soils aluminium and manganese salts pass into the soil water and influence unfavourably the development of root systems and the process of nitrogen assimilation (for example Schmehl *et al.*, 1950; Rorison *et al.*, 1958; Peterburgskii, 1964; Foy and Brown, 1964; McLeod and Jackson, 1965). And the content of carbon dioxide decreases (Mulder *et al.*, 1966). As a result, there is a sharp decrease in the levels of root secretion by leguminous plants (Mulder *et al.*, 1966).

This *a priori* conclusion is apparently correct, for when cross pollination is exploited to give mutant legumes that develop well in acid soils and, consequently, release sufficient metabolites into their surroundings, nodulation is normal.

The best way of counteracting an unfavourable pH in the soil and ensuring good inoculation is by liming (for example Taranovskaya, 1931; Bernard, 1953; Zaremba, 1953; Borodulina, 1953; Vincent and Waters, 1954; Egorova, 1955; Roizin, 1959; Peterburgskii, 1964; Loos and Louw, 1965). To improve the formation of nodules in legumes planted in acid soils it is recommended that the seeds be granulated with a paste of lime or soda. With some plants, for example clover and lucerne, this treatment will give good results while for others (lupin and seradella) the effect is negative (Parker and Oakley, 1965).

In the presence of bound nitrogen the sensitivity of different leguminous species and varieties to pH decreases considerably, and so Mulder *et al.* (1966) concluded that the relationship between a symbiotic system and the surrounding pH is largely determined by the reaction of the micro-organism. This suggests that the isolation or selection of strains of nodule bacteria resistant to low pH will ensure successful inoculation of leguminous crops in acid soils. Findings suggest that this will prove to be correct (Mulder *et al.*, 1966).

Soils saturated with bases are more favourable for the vital activities of nodule bacteria (Holding and King, 1963). Alkaline soils, according to Norris (1964), support mostly the multiplication of nodule bacteria that are vigorously producing acids. Apparently this is a form of protection against an unfavourable soil pH.

The possibility of an indirect effect from an unfavourable medium cannot be excluded. Thus, Samtsevich (1939) considered that at a low pH nodule bacteria are heavily adsorbed by the soil and that this retards the inoculation process.

Dinchev (1959*a*) established that nodule bacteria are strongly adsorbed by the colloid fraction of the soil, especially by particles measuring less than 0·001 mm across, but according to his findings the pH of the soil does not materially influence the rate of adsorption. Nevertheless, Nutman (1952) has shown that the introduction of adsorbents into the soil releases nodule bacteria from the adsorbed state and improves nodulation. The adsorbents seem to be more active at extreme pH values (5 and 9) than in the neutral zone (Table 2.13).

Table 2.13. Effect of Adsorbents on the Formation of Nodules in Red Clover

Experimental variant	pH of soil				
	5	6	7	8	9
Controls	1·3	30·8	18·9	22·4	15·9
Activated charcoal introduced	21·9	42·5	50·3	55·4	57·4
Bentonite introduced	14·1	31·1	30·1	51·2	47·2

The adsorbents were introduced in a concentration of 1 per cent of soil weight.

Among the mineral nutrients, mobile forms of nitrogenous compounds have a particularly important effect. It was established long ago that the presence in the soil of nitrogen in forms accessible to the plant suppresses the formation of nodules (for example Thornton, 1936; Wilson, 1940). This is demonstrated by the data in Table 2.14, taken from the work of Nicol (1934) using lucerne in a

Table 2.14. Effect of Nitrate Nitrogen on the Formation and Size of Lucerne Nodules

Amount of NO_3–N added	Number of nodules		Length of nodules (mm)	Volume of bacterial tissue of nodules (mm³)
	Per ten plants	Per g of dry weight of roots		
0	496 ± 27	178	2·22 + 0·057	0·240
165	508 ± 67	145	1·45 ± 0·093	0·0100
330	333 ± 49	152	1·03 ± 0·13	0·0055
990	204 ± 27	100	0·71 ± 0·039	0·0025
1650	69 ± 41	42	0·55 ± 0·057	0·0020
2000	68 ± 22	29	0·58 ± 0·076	0·0020

sand culture in 12 kg vessels. As this table shows, the introduction of nitrate nitrogen into the medium inhibited the formation of nodules. It can be taken as established that nitrates inhibit the formation of nodules more strongly than ammonium salts (for example Fred and Graul, 1916; Richardson et al., 1957).

The material we have accumulated shows convincingly that in some conditions bound nitrogen can promote both the development of leguminous plants and the process of nitrogen assimilation. Thus, in the initial period of development, leguminous plants require small amounts of nitrogen in the soil because stores in the seed are insufficient for the plant to develop to the stage of nodule formation, at which time it can begin to make use of molecular nitrogen. This 'starter dose' of nitrogen, exerting a plainly beneficial effect on plant growth, nodulation and nitrogen assimilation, differs somewhat in various plant species (for example Giöbel, 1926; Diener, 1950; Richardson et al., 1957; McConnel and Bond, 1957a; Ionescu, 1958; Ezedinma, 1964a; Korovin and Vorobev, 1965; Dorosinskii, 1965a, b; Marnauza, 1966).

According to Koslov (1962) and Kolosova (1965) even quite large doses of mineral nitrogen fertilizers have a favourable effect on the symbiosis between legumes and nodule bacteria. However, when Koslov increased the dose of nitrogen to more than 0·6 of the Hellreigel standard, nodule formation was suppressed. In this case plants developed quite well but changed completely to using bound nitrogen. Others have made similar findings (for example Bazyrina

Table 2.15. Effect of Nitrogenous Fertilizers on Accumulation of Nitrogen by Lucerne

Amount of NO₃–N introduced, (kg/hectare)		Amount of nitrogen fixed (kg/hectare)	
Per year	In four years	In six years	Per year (mean)
19·7	78·8	346·2	57·7
39·4	157·6	328·2	54·7
78·8	315·2	203·4	33·9
128·2	512·8	171·0	28·5
157·6	630·4	110·4	18·4

The mean values are for six years.

et al., 1949; Gulyakin et al., 1963a, b). The introduction of large doses of mineral nitrogen in small portions given every two days for two to two and a half months does not suppress the formation of nodules and has a favourable effect on symbiosis (Uziakowa, 1959).

Several groups have studied the effect of various doses of nitrogen fertilizers on nitrogen fixation in the soil by leguminous plants. In our view one of the most thorough studies was made by Giöbel (1926) who carried out several experiments, one lasting six years, which we shall describe.

Giöbel introduced varying doses of sodium nitrate into sand fertilized with mineral salts (without nitrogen) and calcium carbonate for four years in succession. The test plant was lucerne, which was inoculated with nodule bacteria before it was sown. Table 2.15 shows the results obtained. Sodium nitrate added for four years in doses of 80 to 160 kg per hectare (20–40 kg nitrogen per year) greatly promoted nitrogen fixation. Large doses of nitrate restricted but did not completely suppress it.

Thus a limited nitrogen supply to leguminous plants may increase the level of nitrogen fixation. Such a conclusion was also reached by Allos and Bartholomew (1959), Darbyshire (1963) and Marnauza (1966) working with other leguminous species. But the doses of nitrogen fertilizers that stimulate nitrogen fixation differ for different species of plants.

A change in the nitrogen content of the surroundings, within narrow limits (0·05–0·2 of the norm), does not influence the fraction of fixed nitrogen in the total nitrogen assimilated by lupins from the air and the medium (Trepachev and Khabarova, 1966). According to Hopkins (1910) and Pieters (1927) in field

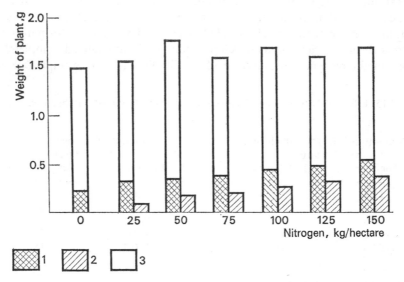

Fig. 14. Ratio of nitrogen assimilated by soybeans from the soil and air with different forms of fertilization of the soil (after Giöbel, 1926)

1, Nitrogen from the soil; 2, nitrogen from NaNO₃; 3, nitrogen from the atmosphere.

conditions leguminous plants absorb one-third of their nitrogen from the soil and assimilate two-thirds from the air.

According to Loginov (1966), who used nitrogen-15 to investigate the level of nitrogen fixation, the amount of nitrogen fixed by the lupin in seventy-seven days of cultivation (forty-eight days in a natural atmosphere and twenty-nine in a gaseous medium with $^{15}N_2$) was 53 per cent of the total nitrogen in the plant. Forty-seven per cent was taken by the plant from the soil, with 66 per cent passing from the atmosphere to the aerial parts and 34 per cent to the roots.

But Sokolov (1957) considered that leguminous plants developing well and giving a high yield fixed considerably more nitrogen from the air than from the soil. This is in good agreement with Giöbel's findings (Fig. 14). Evidently when there is a high level of nitrogen fixation the uptake of nitrogen from the soil may be reduced to below 15 per cent (Kolosova, 1965; Manauza, 1966).

We shall deal briefly with methods for calculating the quantity of atmospheric nitrogen fixed by symbiotic systems of leguminous plants and nodule bacteria. The calculation is usually carried out as follows. Let a be the content of nitrogen in the plant (fixed from the air and utilized from the substrate in the form of bound compounds); b_0 the initial content of bound nitrogen in the substrate; b_1, the final content of bound nitrogen in the substrate (remaining at the end of the experiment); b, the amount of nitrogen utilized by the plant from the soil, and x, the amount of nitrogen fixed by the plant from the air. Then, $a = x + b$; and so $x = a - b$ where $b = b_0 - b_1$.

This method of calculation is used most often (for example Tyurin and Mikhnovskii, 1966; Meshkov, 1966; Trepachev and Khabarova, 1966). An example of such calculation is given in Table 2.16.

Table 2.16. Nitrogen Balance in an Experiment with Lupin

Dose of nitrogen as percentage of Helbriegel norm	Introduction of nitrogen (mg)			Weight of dry plants (g per vessel)	Nitrogen utilized by plants (mg per vessel)	Remaining in sand (mg per vessel)	Balance	Amount of nitrogen fixed	
	With fertilizer	With seeds	Total					mg per vessel	% of total amount
0·05	16·8	39	55·9	5·8	55·5	Traces	−0·4	—	—
0·2	67·2	39	106·3	7·5	83·0	Traces	−23	—	—
				Bacterized plants					
0·05	16·8	39	55·9	15·9	326·6	Traces	270·7	271	83
0·2	67·2	39	106·3	19·5	464·9	Traces	358·6	381·9	82

This table is after Trepachev and Khabarova (1966).

Shevchuk (1965) proposed a different scheme for calculating the amount of nitrogen fixed. The yield of the various parts (leaves, stems, roots, seeds) of inoculated and uninoculated plants is calculated (a) in the leaves; (b) in the seeds; (c) in the roots and (d) in the stems. This is the amount of nitrogen taken in by the plant from the air and the substrate.

The same calculations are done for uninoculated plants. In this case all the nitrogen in the plant is taken from the substrate. Since inoculated plants have a greater weight of aerial parts and roots than uninoculated plants, and thus a greater assimilating capacity in the root system, it is assumed that more nitrogen will be taken from the substrate than is the case with uninoculated plants.

To determine the amount of nitrogen taken from the substrate, the weight of each part of the inoculated plant is multiplied by the percentage content of the corresponding organ in the uninoculated plant (a_1, in the leaves; b_1, in the seeds; c_1, in the roots; d_1, in the stems). This content of nitrogen corresponds, according to Shevchuk, to the amount of nitrogen taken from the substrate by the inoculated plant.

The difference $a - a_1$, $b - b_1$, $c - c_1$, and $d - d_1$ is also the quantity of fixed nitrogen.

This calculation does not require determination of the amounts of initial nitrogen or of those remaining in the soil. If such a method is to be acceptable it is necessary to demonstrate that the specific effectiveness of the root system is the same in providing both the inoculated and uninoculated plants with bound nitrogen. (Specific effectiveness is the fraction of bound nitrogen, that is nitrogen taken from the substrate in the total nitrogen in the plants referring to both inoculated and uninoculated plants.) It seems that the specific effectiveness of well developed and weakly developed plants is different and must depend on the activity of the nodule bacteria.

Therefore, it must be concluded that Schevchuk's proposal for estimating nitrogen fixation, although of some interest, requires additional verification. His calculations of the level of nitrogen fixation by different leguminous species (1965) also require confirmation.

Special attention must be paid to the problem of establishing the cause of the depressant effect of high doses of mineral nitrogen on the formation of nodules and on nitrogen fixation by symbiotic leguminous plants. Several different views have been expressed on this point and we shall discuss the most interesting.

As long ago as 1898, Maze suggested that abundant use of fertilizers containing nitrogen results in an enhanced uptake of carbohydrates for protein synthesis in the aerial parts of plants, and this considerably worsens the conditions for the bacteria that are penetrating the root system. This view was later supported by several others (for example Weber, 1930; Allison and Ludwig, 1934; Wilson, 1940). It is a view supported by the fact that the activity of the nodule bacteria is depressed not only when fertilizers containing nitrogen are present in the soil (Weber, 1966), but also when the leaves are sprayed with them. Urea is such a fertilizer (van Schreven, 1958; Cartwright and Snow, 1962). It is also significant that when a carbon deficit is made good in a plant abundantly supplied with mineral nitrogen (brought about by intensification of photosynthesis, by exposure to light or the addition of carbon-containing substances to the medium) forms in the normal way (Weber, 1930; Georgi, 1933; Demidenko and Timofeeva, 1937).

This idea first suggested by Maze has been confirmed by experiments carried out by Chailakhyan and Megrabyan (1945). They cultivated their plants with the upper part of the root system in sand and the lower part in a solution containing a considerable amount of mineral nitrogen. They found that the formation of nodules was suppressed in both parts of the root system.

The literature contains conflicting findings on this topic, obtained by submitting different parts of roots to different treatments. Some results show that nitrates utilized by one part of the root, which suppress the formation of nodules there, do not influence nodulation in another part of the root which is developing in a nitrogen-free medium (for example Gäumann et al., 1945; Federov and Kozlov, 1954). Experiments carried out by Raggio et al. (1957, 1965), who cultivated excised roots of leguminous plants, have also shown that nitrates suppress the formation of nodules when added to the medium surrounding the roots. However, they found that when introduced through the base of the excised root, nitrates had no harmful effect on nodulation.

These observations (and also several others) compel us to recognize the inadequacy of Maze's theory. Richardson et al. (1957) showed that low doses of ammonium ions (0·5–12 parts per million) stimulate nodulation of leguminous plants, while corresponding doses of nitrate ions suppress it. In considerably higher concentrations both ammonium and nitrate salts prevent nodulation. Thus, not only is the dose of fundamental importance but also the form in which the nitrogenous compound is introduced (Uziak, 1964).

Tanner and Anderson (1964) consider that indolylacetic acid produced in the roots is very important during the formation of nodules. They assume that the process of infection by nodule bacteria proceeds in two stages, the first connected with the conversion of tryptophan to indolylacetic acid (differing with the strain) and the second with the curling of the root hairs under the influence of this product. In the absence of bound nitrogen both stages proceed normally— at different rates depending on the strain of nodule bacteria and the species of plant involved. In the presence of ammonium salts the transformation of tryptophan to indolylacetic acid can be inhibited; thus the first stage is upset. The greater the concentration of ammoniacal nitrogen (urea, glycine and so on) the smaller the amount of the acid formed. In the presence of nitrogen of nitrate, nitrogen-less indolylacetic acid is produced (the second stage is disturbed) because the acid is destroyed. The breakdown of indolylacetic acid is mediated by nitrites formed from the nitrates by the nodule bacteria. The reaction of the nitrite with indolylacetic acid results in the formation of polymerized indolyl-3-aldehyde which is biologically inactive.

Cheniae and Evans (1957) established the ability of nodule bacteria to reduce nitrates, and the interaction of nitrites with indolylacetic acid was established by Tonhazy and Pelczar (1954). Some cultures of nodule bacteria do not reduce nitrates and are incapable of destroying indolylacetic acid. The presence of this acid in the root system of leguminous plants was established by Kefford et al. (1960).

The suppressant effect of the nitrates can be completely or partially reversed by introducing an indolylacetic acid preparation into the substrate (Valera and Alexander, 1965b). The mechanism of suppression of formation of nodules by nitrogen-containing compounds (Tanner and Anderson, 1964) can be represented by the scheme outlined in Fig. 15.

We feel that Tanner and Anderson's hypothesis should not be set against that of Maze. It is most likely that the mechanism of the effect of nitrogen-containing compounds on the plants, and the relationships of the plants with the bacteria, are more complex than they may seem. According to Virtanen (1953) for example, the nitrates in the plant are reduced to nitrites which then react with leghaemoglobin to form a NO–leghaemoglobin complex. As a result nitrogen assimilation is suppressed. Other causes of the suppression of nitrogen fixation may be brought to light by treating leguminous plants with nitrogen-containing fertilizers.

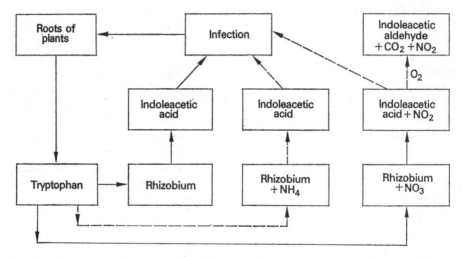

Fig. 15. Mechanism of suppression of infection in the leguminous plant by different forms of nitrogen (after Tanner and Anderson, 1964)

–, Normal reaction; – – –, reaction inhibited, depending on concentrations of bound nitrogen.

There are indications in the literature that the infectious process is stimulated not only by indolylacetic acid but also by other substances such as kinetin and pectinase preparations (Darbyshire, 1963).

Finally, according to Bogacheva (1961) the nodules of leguminous plants usually contain no nitrates and only a few amides. There is protein nitrogen and the nitrogen present in free amino acids (Butler and Bathurst, 1958; Shil'nikova, 1963; Aseeva et al., 1960). In the presence of nitrate fertilizers the roots of leguminous plants contain salts of nitric acid and ammonia. The non-protein fraction of the nitrogen-containing compounds consists mostly of free amino acids and, in part, amides (Table 2.17).

Nitrogen is apparently transported from the nodules to the aerial parts of the plant in the form of amino acids (Ebertova, 1960; Shil'nikova, 1962, 1963). According to Uziak (1964) leguminous plants can develop when amino acids are their sole source of nitrogen. The degree of utilization of amino acids

Table 2.17. Form of Nitrogen-Containing Compounds in the Roots of *Phaseolus*

Phase of plant development	0·3 of nitrogen norm				Twice the nitrogen norm			
	Protein nitrogen	NH₃	Amides	Amines	Protein nitrogen	NH₃	Amides	Amines
Before blooming	13·5	0·6	0·6	00	23·8	0·7	0·3	0·8
Flowering	11·7	0·5	0	7·5	17·7	0·6	2·1	2·8
Ripening	10·3	0·6	0	8·1	17·8	0	0·6	5·6

Plants were grown in Hellriegel's mixture. Measurements are in mg/g.

depends on the species of plant and not on infection by bacteria. Only in the lupin is a negative reaction to amino acids noted when the roots are injected by bacteria.

From their experimental work Dorosinskii *et al.* (1962) concluded that active races of nodule bacteria in normal symbiotic relationships can provide the plant with all its nitrogen. Leguminous plants utilizing 'biological' nitrogen flower and bear fruit earlier than those depending on mineral nitrogen nutrition. Of course, the nodule bacteria can fulfil this function only when all conditions for their development and for the growth of the plant involved are satisfied.

Phosphorus nutrition is important in the activation of nitrogen assimilation by leguminous plants (for example Thornton and Gangulee, 1926; Whyte *et al.*, 1953; Uyas and Desay, 1953; Dinchev, 1959*b*; Sankaram *et al.*, 1963; Pyarsim, 1966). When the supply of phosphorus is low, bacteria penetrate into the root but nodulation is not observed (Diener, 1950). Individual cultures of nodule bacteria which infect the same plant species require different amounts of phosphorus. Normal inoculation requires a definite level of phosphorus compounds, not only in the medium but also within the root system of the host plant (Ash, 1951; Kamata, 1962). Phosphorus enhances the positive effect of molybdenum on nitrogen assimilation. The introduction of molybdenum without an accompanying supply of phosphorus to the plant may even reduce the number of nodules (Anderson and Spencer, 1948; Mulder, 1954; Hewitt, 1958).

Some investigators, particularly Dinchev (1959*b*, 1961*a*), consider that the favourable effect of phosphorus on the activity of nodule bacteria makes it possible to dispense with inoculation if phosphorus fertilizers are introduced into the soil. This only applies of course if active nodule bacteria are present in the soil.

The harvesting of leguminous plants depletes the soil of considerably more potassium than is the case with other farm crops (Gukova and Bogomolova, 1963). And so potassium and, in particular, phosphorus–potassium fertilizers would be expected significantly to increase the quantity of nitrogen accumulated by leguminous plants, as has been found (Il'in, 1939; Roberts and Olson,

1942; Lynch and Sears, 1950; Fedorov and Pod"yapol'skaya, 1951; Nowotny-Mieczyńska and Zinkiewicz, 1959; Dinchev, 1961a; Last and Nour, 1961; Raicheva, 1962; Masefield, 1965; Gulyakin et al., 1965a; Til'ba and Golodyaev, 1966).

Increased doses of potassium seem to have a positive effect on nitrogen fixation only in the presence of definite amounts of nitrogen-containing mineral compounds (Nowotny-Miecsyńska and Zinkiewicz, 1959; Marnauza, 1966). According to Pronin (1966) the highest nitrogen-fixing activity, and as a result, the highest yield of leguminous plants, is noted when the ratio of nutritional elements is 0·5 N:1P:2K (of the Hellriegel norm). There are some indications that potassium is not specifically important in nitrogen assimilation. Its beneficial effect is apparently connected only with an improvement in plant growth (for example Diener, 1950).

We noted the value of calcium when discussing liming of soils, but its importance is not confined to the elimination of soil acidity. Calcium has a specific function and cannot be replaced by other elements. A lack of calcium has an adverse effect on the physiological properties and multiplication of nodule bacteria (for example Vincent and Waters, 1954; Bergersen, 1961; Rovira, 1961; Radulović, 1963; Radcliffe, 1964). In media containing sufficient calcium it is possible to maintain cultures of nodule bacteria in an active state for a long time (Jensen, 1962; Halbinger, 1965).

Calcium deficiency reduces the rigidity of the bacterial cell wall and alters its permeability. The cells become vacuolated and inflated. Calcium concentrates chiefly in the cell wall (Vincent and Colburn, 1961; Humphrey and Vincent, 1962). The antigenic make-up of strains of nodule bacteria cultivated in the presence or absence of calcium is qualitatively identical. Usually, however, in a suspension of bacteria growing in the absence of calcium a heat-stable antigen is found which readily diffuses into the medium. In bacteria grown in the presence of calcium this antigen is found only after the cells have been treated in some way. Mechanically or chemically this indicates that there is some disturbance in the structural integrity of the cells of Rhizobium when calcium is deficient (Humphrey and Vincent, 1965).

The calcium requirements of Rhizobium may in part be compensated by strontium but not by magnesium or barium (Humphrey and Vincent, 1966). Strontium also concentrates chiefly in the cell wall. Nodule bacteria require extremely small amounts of calcium, which can be considered a trace element in this case (Nicholas, 1958).

There are indications that the nodule bacteria of tropical plants do not require calcium (Norris, 1958b). In this connection it is interesting to recall the variant of Rh. trifolii which Vincent and Jancey (1962) described with growth sharply suppressed in the presence of calcium.

But as a rule calcium has an unquestionably favourable effect on nodule bacteria, and obviously (Fig. 16) plays an important role in ensuring normal symbiosis of the bacteria with the leguminous plants (for example Izrail'skii et *al.*, 1933; Holding and King, 1963; Tewari, 1963). But the absolute requirements for calcium during symbiosis differ, for certain plants require high concentrations while others require less. According to Badawy and Allen (1963) nodule bacteria of clover require considerably more calcium than those that infect the soy bean.

Norris (1959) and Vincent (1962) have shown that nodule bacteria also require magnesium, in greater quantities than calcium. A magnesium deficit upsets the multiplication of nodule bacteria and depresses their metabolism.

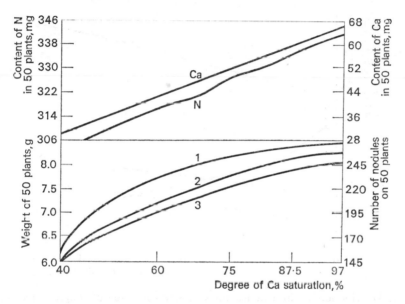

Fig. 16. Effect of the degree of saturation of clay by calcium on the growth and nodulation of soy bean.

1, Height; 2, nodules; 3, weight.

Moore (1905) and Michael (1941) found that magnesium salts stimulate nodulation. And Trepachev and Atrashkova (1965) found that nitrogen fixation is intensified and lasts longer in lupins infected with nodule bacteria in the presence of magnesium. Other elements are also essential for symbiotic nitrogen fixation, in particular sulphur and iron. Their possible significance is considered in reviews by van Schreven (1958), Hallsworth (1958) and in the monograph by (Pochon and de Barjac) Poshon and de Barzhak (1960). Here, we merely note that sulphur compounds act favourably on nodulation (Ivanoff, 1948) and on the synthesis of haemoglobin (Hilder and Spencer, 1954).

Iron has a positive effect on the symbiotic activity of nodule bacteria and leguminous plants. Iron-containing protein particles have been found in the microsomes of pea seedlings. These are analogues of the ferritin in animal tissue

(Hyde *et al.*, 1962). Bergersen (1963*a*), using the electron microscope, found granules similar to ferritin in the young nodules of the soy bean. When the bacteroids and leghaemoglobin appeared, ferritin disappeared. In Bergersen's view, ferritin is a reservoir from which iron is mobilized for the synthesis of various compounds, particularly leghaemoglobin (Fig. 17).

Trace elements too have a significant effect on symbiotic nitrogen fixation. Their optimum quantities do not usually exceed 15–40 g active principle per hectare sown. Molybdenum has often been studied; its positive effect on

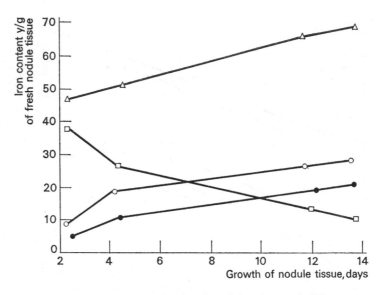

Fig. 17. Distribution of iron in fractions of nodular tissue of different ages (after Bergersen, 1963*a*)

1, Total quantity of iron; 2, non-haem iron forming part of ferritin; 3, soluble and weakly bound iron after treatment with trichloracetic acid at 0°C; 4, haem iron.

symbiotic nitrogen fixation was first noted by Bortels (1930) and confirmed by a series of investigations (for example Anderson and Spencer, 1948, 1949; Vinogradova, 1952; Mulder, 1954; Aizupiete, 1956; Peive, 1965; Kurkaev, 1965; Shumilin and Muravin, 1965). Molybdenum requirements are much lower when the leguminous plants are supplied with sources of bound nitrogen (Mulder, 1954).

When there is a molybdenum deficiency, nodule formation is poor (Bobko, 1963; Atwal and Sidhu, 1964), the synthesis of free amino acids is disturbed (Mulder *et al.*, 1959) and that of leghaemoglobin suppressed (Mulder, 1948*a*). There are indications (Gladkii and Panovskii, 1965) that enrichment of the soil with molybdenum partially compensates for poor illumination.

Several experiments have indicated that the introduction of molybdenum into the soil enhances the multiplication of nodule bacteria (Obraztsova *et al.*, 1937; Klintsare and Kreslinya, 1963; Gordeev, 1964), speeds up inoculation and activates nodule development (Korovin and Vorob'ev, 1965; Efendiev and Khalilov, 1965). Often, however, even fewer nodules are formed than in plants cultivated in soils containing insufficient molybdenum (Anderson and Spencer, 1948; Mulder, 1954; Hewitt, 1958).

A considerable volume of experimental evidence (for example Shkol'nik and Bozhenko, 1956; Sobachkin and Muravin, 1963) suggests that molybdenum has a more favourable effect on plants in acid (pH of salt extract 5·3–5·4) than in neutral soils. To a certain extent this is determined by species. The presence of free aluminium in the soil may also be a decisive factor—molybdenum is ineffective in acid soils with a high aluminium content (Peterburghskii, 1964). Neutralization of the soil (in particular by liming) therefore enhances the positive effect of molybdenum (Mulder, 1954; Bulavskaya, 1959; Sabo, 1964; Peterburgskii, 1964).

Molybdenum is localized chiefly in the cell sap of the nodule tissue (Turchin, 1959) while there is little in the bacteria—about four times less (about 0·005 per cent of dry matter). Its positive role is quite understandable because together with other elements with variable valency (Fe, Co, Cu and so on), molybdenum is a mediator of electron transfer in redox enzyme reactions (Peive, 1964). According to the results of Ratner and Akimochkina (1964) molybdenum has an effect on the plant similar to that of vitamins of the B complex; they are exchangeable in symbiotic fixation of molecular nitrogen.

Vanadium exerts a considerable effect on nitrogen assimilation by leguminous plants. According to Peterburgskii (1963, 1965) this element not only supplements and potentiates the effect of molybdenum but itself acts on nitrogen fixation. Nicholas (1958a) established that V^V and V^{IV} can partially replace molybdenum in nitrogen assimilation, but V^{II} is ineffective.

Boron is also essential. Brenchley and Thornton (1925) demonstrated that when there is a boron deficiency vascular bundles do not form in the nodules, and consequently normal development of the bacteroid tissue is disturbed and nitrogen fixation is weak. A boron deficit sharply narrows the range of pH in which nodules form. Others have noted the necessity of boron for the normal development of leguminous plants and their symbiosis with nodule bacteria (Loustalot and Tefford, 1948; Mulder, 1948a and b; Filippova, 1964; A. and J. Klintsare, 1965; Aleksandrov and Martovitskaya, 1965; Glyan'ko and Shevchuk, 1965).

Cobalt is another requirement for the nitrogen-fixing activity of symbiotic leguminous plants (for example Ahmed and Evans, 1959; Hallsworth *et al.*, 1960; Reisenauer, 1960; Lowe *et al.*, 1960, 1962a; Nicholas, 1962; Powrie, 1964; Wilson and Hallsworth, 1965). When cobalt is added to the soil the yield of leguminous plants increases considerably (Yagodin, 1964).

As Wilson *et al.* (quoted by Jordan, 1962) showed, a considerable part of the cobalt (^{60}Co) (more than 85 per cent) accumulates in the non-protein fraction of the nodules formed by an effective strain of nodule bacteria. According to Nowotny-Mieczynska (1952) the haemoglobin of nodules is not destroyed if the substrate contains cobalt (or ascorbic acid). The requirement for cobalt is also connected with the synthesis of vitamin B_{12}, by both pure cultures of nodule bacteria and those in nodules (Malinska and Pedziwilk, 1958; Kliewer and Evans, 1962, 1963a and b). Active races of nodule bacteria form vitamin B_{12} in considerable quantities (Levin et al., 1954). A cobalt deficit leads to disturbances in the activity of several enzymes (de Hertogh, 1964).

The high cobalt requirements of nodule bacteria enabled Kliewer et al. (1964) to devise a method for determining cobalt using Rh. meliloti. It is a very sensitive method, facilitating the determination of as little as 0.005γ per litre.

In the absence of copper the synthesis of leghaemoglobin is suppressed (Erkama, 1947; Allen, 1956; Parle, 1958) and so is that of amino acids (Pieve et al., 1965) and carbohydrates, while carbohydrate metabolism is also upset (Hewitt, 1954; van Schreven, 1958; Yates and Hallsworth, 1963; Stul'neva, 1965). When copper is in short supply small nodules usually form, scattered over the root system. Apparently, the plants require copper particularly in the spring, the time of most active synthesis of leghaemoglobin. Copper deficiency is usually noted in the soil in the spring because the metal is in the bound state. It is only released in hot weather when the microbiological activity of the soil is intensified (Hallsworth et al., 1958, 1960).

Manganese also affects the nitrogen-fixing activity of the nodules of leguminous plants; a deficiency disturbs the carbohydrate exchange of the plants (Gerretson, 1937; Pirson, 1937). But the exclusion of zinc from the nutrient medium makes little difference to nitrogen fixation (Hewitt et al., 1954). In Mulder's experiments (1948b) when there was a zinc deficiency leguminous plants formed small nodules in abundance but the assimilation of molecular nitrogen was normal. High concentrations of zinc salts are toxic to leguminous plants (Jenkins et al., 1954).

Aluminium salts are toxic for the Leguminosae among other plants (Burgess and Pember, 1923; Ouellette and Desseureaux, 1958; Kliewer, 1961; Foy and Brown, 1964; McLeod and Jackson, 1965; Gromyko, 1966). Their harmful effect is offset to some extent against a background of bound nitrogen (Rorison et al., 1958; Hallsworth, 1958). A drastic method of detoxifying aluminium is liming, for in neutral soils the salt is removed from the soil solution.

Several other elements, including chlorine are also required in small amounts by both leguminous plants and nodule bacteria (Ukrainskii, 1954; Johnson et al., 1957). The problem of radioactive substances has been insufficiently clarified. Some findings indicated that small doses promoted symbiotic nitrogen accumulation (Drobkov, 1945; Krasil'nikov et al., 1955) while others suggest that this is not so (Lepapei and Trannoy, 1934; Tsyurupa, 1937).

Carbohydrate metabolism is of great importance to the formation of nodules. It is determined by several factors—photosynthesis, the presence of carbon dioxide, characteristics of the plant and so on.

In this respect it is interesting to note that Shulyndin (1953) showed that the infectivity of different soy varieties by nodule bacteria largely depends on the accumulation of carbohydrates in the plant tissues. Soy varieties with a high content of carbohydrates form better and more numerous nodules. Results obtained by Kvasnikov and Dolgikh (1955) are in agreement, indicating that the higher the content of monosaccharides in the roots of leguminous plants, the more nodules form in the root system. This problem was also investigated by Fedorova (1966) with different varieties of pea.

The disturbed formation of vascular bundles that Brenchley and Thornton (1925) established when boron was lacking also indicates that in such conditions nodules are absent because there the bacteria have an insufficient supply of carbohydrates.

When solutions of sugars are introduced into the soil or sprayed on to leaves, and when other measures are taken to increase the content of carbohydrates in the root system, inoculation and nitrogen accumulation are affected favourably (Georgi et al., 1933; Demidenko and Timofeeva, 1937; van Schreven, 1959). But there must be a definite optimum concentration of carbohydrates in the root system of leguminous plants (Allen and Baldwin, 1954). When there are high ratios of carbon to nitrogen in the nodules poly-β-hydroxybutyric acid accumulates in the nodules in large quantities. Its role has still not been established (Schlegel, 1962).

From the practical point of view it is very interesting to use straw and culmi-ferous fresh manures as fertilizer for leguminous plants. They rapidly decompose in the soil and greatly improve the carbohydrate nutrition of the plants by providing carbon dioxide. Straw fertilizers appreciably increase the yield of leguminous plants and the amount of nitrogen they accumulate. Several studies have demonstrated this (for example Thornton, 1929a; Sabinin and Vyalovskaya, 1931; Pink et al., 1946; Spencer, 1954; Mishustin and Erofeev, 1965; Meshkov and Khodakova, 1966). Manure and other organic fertilizers also have a favourable effect (Hedlin and Newton, 1948; Ukrainskii, 1954; Klintsare, 1959b; Dinchev, 1961b; Kurbatov et al., 1966).

An increased carbon dioxide content in the soil and in the air next to the soil enhances photosynthesis. Mulder and van Veen (1960a and b) and Lowe and Evans (1962b) showed that enrichment of the nutrient substrates with carbon dioxide sharply stimulates the formation of nodules and the assimilation of molecular nitrogen.

According to Van Veen, and Lowe and Evans, increased atmospheric carbon dioxide greatly stimulates the multiplication of pure cultures of nodule bacteria. Carbon dioxide can be supposed to activate several enzyme systems in the microbial cell (phosphoenolpyruvate carboxylase, propionyl-Co-A-carboxylase and so on). And straw fertilizers contain many organic substances that stimulate the multiplication of nodule bacteria. Many of the breakdown products of straw are also active. A similar effect is obtained by introducing chaff into the soil, especially combined with phosphates (Fig. 18).

In many grain-storing countries (or regions) there are large straw surpluses. These can be put to extremely good use as fertilizers under leguminous crops, and may substantially replenish the nitrogen reserves of the soil.

Our investigations (Mishustin and Erofeev, 1966; Golod, 1966) have revealed that the adverse effect of straw on grasses in their first year in the soil is connected with the biological fixation of nitrogen and with the formation of toxic

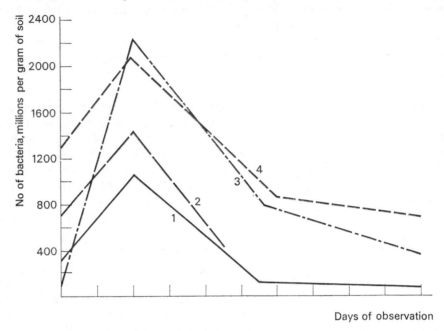

Days of observation

Fig. 18. Development of lucerne nodule bacteria in differently fertilized soils (after Thornton, 1929a)

1, Controls; 2, with phosphates; 3, with chaff; 4, with chaff + phosphate.

compounds. The latter also have a harmful effect on certain species of leguminous plants. Harmful soil substances accumulate in large quantities deeper down in the soil and so it is desirable even in the case of leguminous crops, to place the straw in the upper layer of the soil. This has been borne out by our experience with the lupin. Plants were cultivated either in soddy-podzolic soil or on sand treated with $P_2O_5 + K_2O$. Crushed straw of winter wheat was introduced before sowing (40 g per 8 kg in each vessel). The results are given in Table 2.18.

As experiments have shown, especially those carried out on sand in which nodule bacteria were the sole source of nitrogen, the most favourable results are obtained when straw fertilizers are placed in the surface layer of the soil. Certain species of leguminous plants, for example the pea, are insensitive to the toxic breakdown products of straw. The depth of the straw is not important for them.

Table 2.18. Effect of Straw on Lupin Yield

Experimental variant	Yield (g per vessel)		Change in content of N in vessel with sand (mg)
	Soil	Sand	
Controls, without straw	8·55 ± 0·10	4·87 ± 0·34	+21
Straw introduced into 0–10 cm layer	10·75 ± 0·95	10·43 ± 0·74	+87
Straw introduced into 0–30 cm layer	7·12 ± 0·76	7·12 ± 0·43	+22

Because plants require sunlight to synthesize carbohydrates from carbon dioxide, it is not surprising that several investigators have pointed to a connection between the degree of luminosity and the process of nodule formation (for example Leonard, 1926; Eaton, 1931). Orcutt and Fred (1935) and Marnauza (1966) thought that the formation of nodules in the roots of leguminous plants depended on the ratio of carbohydrates and nitrogenous compounds in the roots. Both lack and excess of carbohydrates have a negative effect.

In poor light nodule formation is improved when the atmosphere is enriched with carbon dioxide (Wilson et al., 1933), when carbohydrates are added to the medium (Georgi et al., 1933) or when leaves are sprayed with sugar solutions (van Schreven, 1958). When plants are cultivated in the dark, carbohydrate supplements in the nutrient solutions do not stimulate the formation of nodules (McGonagle, 1949; M. and N. Raggio, 1956). If root systems of leguminous plants are inoculated by nodule bacteria in the dark, nodules will not form. In poor light, leghaemoglobin is destroyed in the nodules, which turn green (Orcutt and Fred, 1935; Virtanen, 1945). The rate of destruction of leghaemoglobin varies for different plants. In soybean nodules, for example, it is destroyed in the dark more slowly than in pea nodules.

Excessive sunlight causes synthesis of excess carbohydrates which also suppresses nodulation. This effect can be eliminated by adding bound nitrogen to the substrate or by a short period in the shade (P. Wilson, 1939). Individual leguminous species require different degrees of sunlight to ensure normal symbiosis with nodule bacteria (van Schreven, 1959).

Several investigators have been concerned with the connection between the formation of nodules and photoperiodism (Chailakhyan and Megrabyan, 1945; Fedorov and Uspenskaya, 1955; Borodulina, 1951; Sironval, 1958). The results are contradictory, but in the view of most investigators the length of daylight has an important effect on nodule formation (Sironval, 1958; Tewari, 1966). For example, in Sironval's experiments (1958) twelve times as many nodules formed on the roots of soybeans in a fifteen hour day than in eight hours of illumination. Red light has a more favourable effect on nodule formation than blue (Lie, 1964).

Biological factors are definitely important in the symbiosis of nodule bacteria and leguminous plants. In the roots there is an abundant microflora among which are found both inhibitors and activators of nodule bacteria. In certain conditions these micro-organisms may influence the effectiveness of nodule bacteria (for example Konishi *et al.*, 1935, 1936; Krasil'nikov and Korenyako, 1944; Fogle and O. Allen, 1948; Virtanen and Linkola, 1948; Mikhaleva, 1952; Afrikyan and Tumanyan, 1958; Dinchev, 1962*a*; Hattingh and Louw, 1964; Visona and Tavdieux 1964; Uidal and Uisana, 1965; Holland and Parker, 1966; Verona *et al.*, 1966; Shtina *et al.*, 1966; Damirgi and Johnson, 1966). The degree of this positive or negative effect depends on environmental factors and the properties of the cultures of nodule bacteria involved.

Several investigators attach fundamental importance to the part played by microbial antagonism in the inactivation and even death of the nodule bacteria in the soil (Hely *et al.*, 1957; Marshall and Roberts, 1963; Wieringa, 1963; van Schreven, 1964).

Other microbiologists feel that in such a heterogeneous medium as the soil, where micro-organisms and the products of their activity are localized in definite zones, the phenomenon of antagonism and the action of antibiotics are weakened, and cannot be of fundamental importance (for example, Visona and Tardieux, 1964). And this antagonism towards nodule bacteria in the soil is greatly reduced by many soil micro-organisms. Thus, for example, in Stolp's experiments (1952) with sterilized substrates (sand and soil) *Penicillium expansum* suppressed the nodulation of leguminous plants, but in soil with a normal microflora this fungus had no appreciable influence on nodule formation.

Harris (1953) considers that biologically active substances (stimulators) activate only strains of nodule bacteria which are weakly virulent and which, moreover, have reduced powers of nitrogen fixation. According to Hamatová-Hlávčkóva (1961) the resistance of nodule bacteria to antibiotics is not always related to their effectiveness.

When virulence and nitrogen-fixing function are inhibited by biologically active substances, the elimination of the latter from the medium, for example by introducing adsorbents, often restores a culture of nodule bacteria to its original state (Vantsis and Bond, 1950; Turner, 1955; Hely *et al.*, 1957; Emtsev *et al.*, 1963).

Much attention has been paid to the phage of nodule bacteria, which have been the theme of numerous studies, many of them collated in reviews (E. Allen and O. Allen, 1950; Kalnin'sh, 1952, 1966). Several important factors associated with the characterization of the phage and their influence on nodule bacteria have now been elucidated (A. Kleczkowski and J. Kleczkowski, 1951; J. Kleczkowski and A. Kleczkowski, 1952–1954, 1956, 1959).

Phage differ in their cross-lytic effect in relation to various species of nodule bacteria (Staniewski *et al.*, 1962, 1963; Kowalskii *et al.*, 1963; Ziemiecka, 1963), serologically, and in their different reactions to osmotic shock (Kowalskii, *et al* 1963*a*). Individual phage differ in respect of negative colonies, heat tolerance and rate of reproduction.

Most phage of nodule bacteria can lyse the different species of *Rhizobium*, but there are also more specific nodule bacteria (Dorosinskii, 1941; Staniewski *et al.*, 1962; Kowalski, *et al.*, 1963*a* and *b*; Knyazeva, 1966).

Ultraviolet rays at wavelengths below 240 nm irreversibly destroy the nucleic acids of nodule bacteriophage. The effect of longer wave radiations on cells (280–290 nm) is reversible (J. Kleczkowski and A. Kleczkowski, 1965). By cultivating phage on strains of nodule bacteria of different sensitivity it is possible to induce mutations in the phage (Schwinghamer, 1965).

Many investigations have been concerned with the penetration of phage into nodule bacteria. Adsorption of the phage on to the cells was studied by Staniewski *et al.* (1963) and it has been established that the process is negatively influenced by the capsule mucilage (Lorkiewicz and Dusiński, 1963).

Active phage cause lysis of the cell whereas in the presence of temperate phage lysogeny is observed (Marshall, 1956; Davis, 1958; Szende and Ordögh, 1960; Takahaski and Quadling, 1961; Schwinghamer and Reinhardt, 1963; Staniewski and Kowalski, 1965).

The penetration of phage into the nodule bacteria may bring about formation of new variants of nodule bacteria with new properties—resistance to phage, virulence and activity (Kleczkowska, 1950, 1957). A loss of efficacy is not connected with lysogenicity but has a direct link with the acquisition of resistance to viomycin (Schwinghamer and Reinhardt, 1963).

Particular factors affecting the temperature reaction of strains of nodule bacteria to phage are ultraviolet light and the presence of antibiotics (Schwinghamer, 1966). By taking advantage of the phenomenon of transformation it is possible to obtain cells of nodule bacteria that are resistant to phage (Szende *et al.*, 1961).

Opinions differ on the importance of phage in symbiotic nitrogen fixation. Some (Razumovskaya, 1932; Demolon and Dunez, 1934) point out that the accumulation of phage in the soil reduces the yield. Others (Grijns, 1927; Laird, 1932; Dorosinskii, 1941; Kleczkowska, 1957) do not think that phage is important in the soil. These differences arose partly because these investigators worked with phage of varying activity. Krasil'nikov and Melkumova (1963) established that in the presence of phage the efficiency of many phage-sensitive strains of nodule bacteria is reduced, whereas phage-resistant strains are more efficient. This pattern was not noted, however, in several cultures. Hofer (1954) noted that nodules degenerate when they contain active phage.

Extensive experimental results compiled, in particular, by J. Kleczkowska (1945, 1957) and Kalnin'sh, have shown that the phage of nodule bacteria are widespread in the soil wherever the bacteria are to be found. In the soil the lytic effect of the phage is limited because they are adsorbed by the colloids

(Katznelson, 1939; Kleczkowska, 1957). And also in the soil there are always lysogenic cultures of nodule bacteria, ensuring normal formation of nodules even in the presence of phage.

None of this evidence suggests that phage of nodule bacteria reduces the yield of leguminous plants. But in several cases it is difficult to deny the possibility of their having an adverse effect (Dinchev, 1962; Rautenshtein, 1965). Therefore, if inoculation is to be effected artificially it is desirable to use active phage-resistant strains of nodule bacteria.

It is also necessary to note the part played by the plant in symbiotic nitrogen fixation. As Nutman has pointed out (1948), the ability of leguminous plants to form nodules is connected with their ability to form lateral roots. Factors that depress the formation of lateral roots usually suppress the formation of new nodules. Shulyndin (1953) concluded that the intensity of nodulation in different soy bean varieties was connected with differing degrees of accumulation of carbohydrates. Mishustin and Naumova (1955) established that the rate of formation of nodules on lucerne is influenced significantly by alkaloids that accumulate in the roots. When lucerne ages the alkaloids concentrate in its roots and the nodules die off. Alkaloids do not enter the aerial parts of this plant.

Evidently other components of plant tissues (Thompson, 1960), especially those secreted by exosmosis, can influence the formation of nodules in leguminous plants (Zycha, 1932; Nutman, 1957; Rovira, 1961; Miller and Schmidt, 1965). For example, when Elkan (1962) investigated two genetically related lines of soy bean he found that one formed nodules well (parent), but the other (mutant) poorly. It turned out that this was not connected with the penetration of nodule bacteria into the roots. In both cases the nodule bacteria penetrated the root, but nodules formed only in one line. It seemed that in the mutant line the tissues contained certain factors (not identified) that hampered the reproduction of the bacteria in the root. There could of course be a different reason for this phenomenon. We shall summarize briefly some further evidence.

Clark (quoted by Elkan, 1965) selected a few strains of nodule bacteria capable of infecting both soy bean lines. In anaerobic conditions none of the strains reduced nitrates to nitrites; the strains which were able to form nodules only in the parent soy bean line had de-nitrifying powers (Elkan, 1965).

The degree to which plants can be injected with nodule bacteria is undoubtedly determined to some extent genetically. This was first assumed by J. Wilson (1939c) when he examined the effect of cross and self pollination on the susceptibility of plants to infection with nodule bacteria. Later Razumovskaya (1944), Aughtry (1948), and Nutman (1949) confirmed this view experimentally. Nutman was able to obtain a red clover resistant to infection by nodule bacteria. Experimental crosses established that the marker of resistance to nodule bacteria is inherited, and is transmitted by one recessive gene. This gene is independent of the alleles of sterility and of other hereditary factors determining the character of nodulation. Grafting the apex of a plant not susceptible to nodule bacteria on to the roots of a sensitive plant, or *vice versa*, did not transmit resistance and sensitivity from graft to scion or from scion to graft. Several studies (for example Bergersen and Nutman, 1957; Gershon, 1961; Gelin and Blixt, 1964) have provided grounds for the belief that the ability of a plant to enter into symbiotic relationships with nodule bacteria is controlled by many genes located on different chromosomes. Further information about the genetic basis of the inoculation of legumes may be valuable for practical agriculture. Particularly interesting are experiments with mutagenic substances that may enhance the process of nodulation. For example, Bonnier (1954) has shown that colchicine greatly increases the number of nodules on lucerne and clover.

Plants growing in groups form nodules more poorly than those growing individually (Nutman, 1946, 1953). If all the plants in a community belong to one species this depression in growth is much more effective. As experiments have shown, the depression cannot be explained by a deficit of nutrients in the medium. Therefore the suggestion is that the root systems of leguminous plants secrete substances that inhibit the formation of nodules. Associations of different species of leguminous plants have been described in which the development of nodules is stimulated. It must be assumed that the root system of one plant species sometimes produces compounds, still of an unknown nature, that improve root growth in another species.

The idea that the root systems of leguminous plants secrete biologically active substances that influence the formation of nodules has been supported by Turner (1955) who showed that the addition of charcoal (or another adsorbent) to the medium stimulates the development of nodules and shortens the latent period in their formation. This might be explained by the adsorption of a depressant substance secreted by the plant.

When considering compounds produced by exosmosis from the leguminous root system we should remember that they include not only inhibitors, but also substances which enable nodule bacteria to multiply vigorously in the rhizosphere before inoculation. Incidentally, Nutman (1966) noted that when there were few nodules on the roots of leguminous plants (for various reasons) their volume may have been considerably greater than normal.

Finally, we must mention the harm which can be done to nodule bacteria by certain insects and nematodes. From various species of insects harmful to nodule bacteria Mulder (1948) and van Schreven (1958) singled out the striped bean and pea weevil *Sitonema lineatus* (*Sitona lineatus*), the larvae of which destroy the root nodules of many species of leguminous plants, chiefly annuals. Both *Sitona lineatus* and *Sitona crinitus* (the bristly sitona) are widespread in the USSR (Bryantsev, 1966). In the central belt of the European part of the USSR they inflict enormous harm on sowings of legumes.

In early spring female bean and pea weevils lay between 10 and 100 eggs. Within 10 to 15 days small (up to 5·5 mm), worm-shaped, legless, curved white larvae with a light brown head develop from the eggs and feed chiefly on the nodules and root hairs. Newly hatched larvae may penetrate a nodule and eat up its contents. Older larvae destroy the nodules from the outside. In 30 to 40 days of life each larva consumes from two to six nodules. They are particularly harmful in hot, dry weather when the development of the plants is slowed down.

Recommended treatment for combatting the bean and pea weevil is to dust seedlings with 12 per cent hexachlorane (10–15 kg per hectare), 5·5 per cent DDT (10–15 kg per hectare) or calcium arsenite (10 kg per hectare).

Nodules of lucerne and certain other legumes are also damaged by *Otiorrhynchus ligustici*—the large lucerne weevil (Bryantsev, 1966). The females lay as many as 400 eggs from which the larvae develop, which do most harm to leguminous sowings. The larvae, 10·5–14 mm long, are legless, arched and yellowish-white with a brown head and are covered with brown bristles. The developmental cycle of the large lucerne weevil takes two years. Control measures involve dusting the plants in the early stages of development (up to flowering) with 12 per cent hexachlorane (20 kg per hectare) or calcium arsenite (15 kg per hectare).

Many investigators have found nematodes in the root zone of various species of leguminous plants (Tulaganov, 1954; Kir'yanova, 1961; Atakhanov, 1964; Atakhanov and D'yakova, 1964). Around the roots of the pea, for example, forty-seven species of nematodes were found (Atakhanov and D'yakova, 1964). Of these, twenty-five were parasitic, and predominantly to be found in the root system of *Aphelenchus parietinus* and *Pratulenchus pratensis*.

The roots of young plants of *Phaseolus*, lupin and clover may be parasitized by *Ratylenchus pratense* (germ nematode) which is very widespread not only in the Soviet Union (Uzbekhistan) but in Germany, Denmark, the United States, Argentina, Mexico, Brazil and Holland. The females feeding on the roots of plants lay eggs in the tissues of the plant. The complete life cycle of the nematode which develops from the eggs usually takes place within the tissues. Rotation of crops may to a certain extent serve as a control measure against germ nematodes.

On the Russian Steppes, *Ratylenchus multientus* (Steppe nematode) is widespread in the roots of lucerne, clover and the soy bean. The females penetrate the root, where they deposit from twelve to twenty eggs. The larvae developing in the roots pass through three stages, disturbing the functions of the roots and nodules.

In the western belt of the USSR *Aphelenchus avenae* is found in the roots of lupin, *Phaseolus*, soy bean and vetch (Kir'yanova, 1961). *Aphelenchoides kiihni* and *Aphelenchoides parietinus* have been found in the various continents in the root zone of lucerne and other legumes.

Most of the work carried out in connection with this problem has involved either isolated, mostly random, observations, or observations not intended to establish the effect of pests on the relationship between nodule bacteria and leguminous plants. Although it is obviously the case there is still no convincing experimental evidence that these organisms have a pernicious effect on the symbiotic system of legume and nodule bacteria.

Distribution of Nodule Bacteria and Natural Inoculation

The distribution of nodule bacteria in the soil is of crucial practical importance; their dissemination determines the natural inoculation of leguminous plants. In the absence of the appropriate species of nodule bacteria in the soil leguminous plants develop without nodules, in which case they do not fulfil their function as accumulators of nitrogen, and become consumers of soil nitrogen.

As symbiotic organisms, nodule bacteria are distributed in the soil around certain species of leguminous plants. After the destruction of the nodules the bacteria enter the soil and begin to live on various organic substances.

Nodule bacteria are not only found under leguminous plants, but also in soil where no legumes have grown for several years. This was established as long ago as 1890 by Beijerinck and by Kellerman and Leonard (1913), Lipmann and Fowler (1915) and many others. The counts of nodule bacteria that they mentioned do not, however, inspire confidence, for at that time there were no exact methods for counting these bacteria in the soil. Bacteria of the genus *Rhizobium* were counted by inoculating solid nutrient media and their colonies were diagnosed according to a series of external features. Mistakes are easily made with such a technique. The method recommended by Budinov (1907) and later refined by Allen and Baldwin (1931) was applied less often. It is based on chemotaxis. A pipette with a capillary tip, and filled with nutrient medium is introduced into a sample of wetted soil. The soil in the tip of the pipette becomes saturated with nutrient substances and creates conditions conducive to chemotaxis by motile soil micro-organisms. The nodule bacteria, together with other motile soil micro-organisms, enter the capillary. Their presence is established by inoculating the contents of the capillary on to solid nutrient medium (soil agar). It is desirable to add crystal violet to the medium, in a concentration of 1 in 50,000 to 1 in 100,000, which does not affect the growth of nodule bacteria but suppresses the development of many saprophytic bacteria (Konishi, 1931).

Bystryi (1941) modified this method. He took for analysis not soil but a soil suspension (1 in 1,000) into which he placed, for two hours, capillaries containing Ashby's medium plus crystal violet (concentration 1 in 80,000). The contents of the pipettes were transferred to Ashby's agar also containing crystal violet (1 in 100,000).

These techniques made it possible to isolate nodule bacteria from the soil and to determine which species of plant they associate with, but it proved practically impossible to make a quantitative estimate of particular species of nodule bacteria in the soil. This became feasible only after Wilson (1926) had proposed a new method for the quantitative determination of nodule bacteria in the soil. His method is based on the use of a soil suspension to infect aseptically cultivated leguminous plants. It is essentially the titre method. By watching whether nodules form when plants are inoculated with different soil dilutions it is possible to establish both the number of nodule bacteria in the soil and the species they associate with. The accuracy of this method has been checked by many investigators (for example Purchase and Nutman, 1957; Tuzimura and Watanabe, 1961b; Brockwell, 1963).

Wilson's technique or modifications of it (Krasil'nikov and Korenyako, 1940; Robinson, 1957; Date and Vincent, 1962) have often been used in ecological studies of nodule bacteria (for example Petrosyan, 1953; Raicheva, 1957a, 1962; Kalnin'sh, 1958; Moore, 1960; Tuzimura and Watanabe, 1960, 1961c, 1962a and b; Manil and Brakel, 1961, Melkumova, 1961; Hely and Brockwell, 1962; Dinchev, 1961b; Emtsev, 1963, 1964; Ezedinma, 1964b). Some general conclusions can be drawn from the accumulated data.

First, virgin and cultivated soils usually contain quite large numbers of nodule bacteria appropriate to the species of leguminous plants which are present in the wild flora, or which have been cultivated for a long time in a given locality. For example, in our laboratory many podzolic, soddy-podzolic and black-earth soils have been found to harbour *Rhizobium trifolii* and *Rh. leguminosarum*. Soils contain few or no nodule bacteria which infect plants rarely cultivated or not growing in them. Thus we have found that a considerable number of the soddy-podzolic and black-earth zones of the USSR contain no *Rh. phaseolii* or *Rh. japonicum*. But nodule bacteria infecting the broad bean are fairly widespread, possibly because of the presence of species of nodule bacteria which give cross infection. For example, vetch and pea nodule bacteria cross infect broad beans.

Nodule bacteria are encountered quite often in very unfavourable soils, for example those that are saline or alkaline. Evidently, a high content of salts does not have an adverse effect on nodule bacteria, for in such conditions active races are found together with weakly active races (Ukrainskii, 1954; Filippova, 1953; Zhvachkina, 1960; Pillai and Sen, 1966).

Second, the numbers of nodule bacteria in the soil are influenced by the properties of the soil. For example, nodule bacteria multiply better in neutral soils (black-earth, cultivated soddy-podzolic soils and so on) than in acid soils, and active forms are more often encountered. Working of the soil, especially when associated with the introduction of organic fertilizers, improves conditions for the multiplication of nodule bacteria. A favourable effect is also obtained by liming acid soils (for example Raicheva, 1957b, 1962; Kalnin'sh 1958; Ezedinma, 1964b; Jones et al., 1964). The number of nodule bacteria in the soil is also influenced by ecological and geographical factors (Dinchev, 1961b).

The number of nodule bacteria in one field varies substantially with changing weather, and particularly with changing soil moisture (for example J. Wilson, 1930a; Kalnin'sh, 1958; Melkumova, 1961). In favourable conditions the number reaches tens of thousands and more per gram of soil (for example Samtsevich, 1939; Kanin'sh, 1952; Petrushenko and Salakhov, 1956).

Attempts have been made to establish a relationship between soil types and the presence in them of definite serological groups of nodule bacteria, but so far no clear cut results have emerged (Vincent, 1941, 1942; Damirgi, 1964; Kalnin'sh, 1964).

The third fact to be revealed by Wilson's technique is that nodule bacteria can multiply in the rhizosphere of plants. The results of Purchase and Nutman (1957), already noted, have shown that nodule bacteria peculiar to a given leguminous species multiply with particular vigour in the root zone of these plants. Other investigators note that the best conditions for the multiplication of nodule bacteria are to be found in the rhizosphere of legumes, but in this case high selectivity is not found (for example Obaton and Blachère, 1963; Tuzimura and Watanabe, 1961a, 1964; Shevchuk, 1964; Ivanova and Sakharnova, 1965; Petrushenko and Salakhov, 1965).

It would be a mistake to deny the ability of nodule bacteria to multiply in the rhizosphere of non-leguminous plants, for this is supported by several investigations (for example Korenyako, 1942; Mavritskii, 1947; Raicheva, 1957a; Melkumova, 1961; Hely and Brockwell, 1962). Nevertheless, they all agree that the rhizosphere of other species is not so conducive to the multiplication of nodule bacteria, and sometimes there is a loss of virulence in bacteria isolated from the rhizosphere (Dinchev, 1961b; Mczharaupe, 1964). All this suggests that the root secretions of leguminous plants are most favourable to nodule bacteria. It is therefore understandable that in crop rotation, with increasing time after the cultivation of leguminous crops the number of nodule bacteria diminishes.

Fourth, nodule bacteria may exist for a long time as saprophytes in the soil without leguminous plants. Several investigators (for example Lipmann and Fowler, 1915; Albrecht and Turk, 1930; Spencer, 1954) report that nodule bacteria can be found in the soil between thirteen and seventeen years after cultivation of the corresponding species of leguminous plant.

We were able to analyse soils from the Moscow Timiryazev Agricultural Academy which had not been under clover or other species of legumes for fifty years. In some plots of differently fertilized soils, grasses have been cultivated without rotation while the remaining plots in a similar series have been fallow all the time. In all plots occupied by grasses nodule bacteria of clover, *Phaseolus* and the broad bean were found. In fallow plots these had practically disappeared. Thus it must be concluded that a prolonged saprophytic existence of nodule bacteria in the soils is possible only when fresh plant material is introduced. Each year the grasses gave a root mass and crop residues.

Fifth, the soil acidity usually associated with the appearance in the solution of toxic aluminium ions is particularly unfavourable to nodule bacteria living saprophytically in the soil. There is also some evidence that a prolonged saprophytic existence, especially in acid soils, leads to a weakening of the power to assimilate nitrogen.

Sixth, many studies (for example J. Wilson, 1930b; Umbreit, 1944; Kalnin'sh, 1951; Thornton, 1952; Nutman, 1956; McConnel and Bond, 1957b; Holding and King, 1963; Kuznetsov and Natko, 1963; Zaremba and Malyns'ka, 1966) have indicated that ineffective cultures of nodule bacteria, if they do not predominate in the soils, are encountered very frequently. Here we are probably dealing with a complex of adverse properties of the soils, failure to provide them with organic matter and so on. Thus, according to Wróbel and Golebiow-ska (1956) and also Kalnin'sh (1958) inactive races of nodule bacteria are more often encountered in light sandy and sandy-loam soils and active races in clay and loamy soils. Harmsen and Wieringa (1954) isolated ineffective cultures of nodule bacteria chiefly from acid, peat-bog soils. The causes of inactivation of nodule bacteria in the soil have been studied insufficiently so far.

Seventh, soil contains numerous micro-organisms—both antagonists and stimulators of each other's growth (for example Löhnis and Hansen, 1921; Konishi and Fukuchi, 1935; Konokotina, 1936; Robinson, 1945). It has been shown, for example, that when nodule bacteria are cultivated together with *Bac. mycoides*, the former produce more mucilage and are noticeably activated. A favourable effect is also exerted by *Bact. radiobacter* (Konokotina, 1936).

Because microbial coenoses are specific to different soils it can be assumed that the biological factors in different types of soils vary in their effects on the properties of nodule bacteria. This problem awaits investigation.

Finally, appreciation of Wilson's technique has shown that cross infectivity of different species of leguminous plants is often encountered in the wild and on agricultural land. This leads to the appearance in leguminous plants of nodules that are insufficiently active in fixing molecular nitrogen (for example J. Wilson, 1939a and b, 1944; Brockwell and Hely, 1961; Sarić, 1963a and b.

Such infection diminishes the importance of legumes as a factor in soil fertility. The information we have just outlined allows us to conclude that in several cases and evidently quite often, natural inoculation of leguminous plants by highly active nodule bacteria is not ensured. Sometimes, this depends on the absence of the proper species of nodule bacteria in the soil. Such a phenomenon is most often observed when new species of leguminous plants are cultivated. In such cases, either there is inoculation of a cross character which is usually not very effective or the plant develops without nodules.

All this points to the importance of and need for artificial infection when seeds of legumes are sown. This problem will be considered in greater detail in the following section.

Effectiveness of Inoculation

When species of leguminous crops are introduced into a particular area they do not always form nodules and often develop poorly because they are not inoculated. This becomes much more likely if there are no legumes among the wild flora which are systematically close to the new species and which can cross infect them. This was found to be particularly noticeable in the European part of the Soviet Union when the growing of soy beans began to be widespread. In the past few years the same phenomenon has been noted among the expanding growth of *Phasoelus*. In these cases, and when for some reason the soil contains insufficient or only slightly active nodule bacteria, a considerable effect can be obtained by inoculating seeds with a culture of nodule bacteria when they are sown.

It has long been known that if soil in which a fresh species of leguminous crop is to be planted is first infected with soil from places where the crop in question has been growing for a long time, a much better harvest can be obtained. This has not been practised on a large scale because it is so laborious. In each case it would be necessary to transport a great quantity of soil—2–4 tons per hectare.

After nodule bacteria of leguminous plants had been isolated by Beijerinck (1888) the idea of using them for practical purposes became more popular. A commercial preparation of nodule bacteria was first proposed by Nobbe and Hiltner in Germany in 1896. It contained a mixture of cultures of nodule bacteria for the nineteen most widespread species of legumes. The preparation was called 'nitragin'.

The first attempts to use nitragin were not very successful, but later, when the quality of the preparation was improved, its use began to have good results. Thus, for example, in Bavaria in 1903 in 83 per cent of cases the use of nitragin had a positive effect.

The first attempts to inoculate leguminous plants were made in the United States in 1896 by Duggar (1897), and in 1907 Harrison and Barlow proposed the preparation 'nitroculture,' an agar culture of nodule bacteria. In England, in 1906, Bottomley (1910) began to produce the preparation known as 'nitro-bacterin', also containing a culture of nodule bacteria. This preparation, however, was extremely inconvenient since it required preliminary multiplication of the bacterial culture.

In Russia experiments with nodule bacteria for inoculation began at the beginning of the present century (Budinov, 1907; Makrinov, 1915) and were extended after the revolution of 1917. Bacterization was usually carried out with pure cultures of nodule bacteria (nitragin). The use of crushed nodules (Ryumin, 1938) was also recommended for nursery soils (Gel'tser, 1948).

The Soviet agronomist A. G. Doyarenko has written that the use of nodule bacteria in the Soviet Union became the ABC of field husbandry (*Selected Works of Doyarenko*, 1966). Other countries too have carried out, and continued to carry out, intensive work on the effectiveness of inoculation with nodule bacteria. They include France, Belgium, Sweden, the United States, Poland, Czechoslovakia, Hungary and Rumania.

What have been the results of studies of the effectiveness of inoculation of leguminous crops with nodule bacteria? When new crops of leguminous plants are sown inoculation often gives a considerable increase in yield (for example, Erdman and Wilkins, 1928; Voroshilova *et al.*, 1964; Girenko, 1956; Samsonova and Finkel'shtein, 1965; Goss and Shipton, 1965; Kurkaev, 1965; Rougieux, 1966; Cloonan and Vincent, 1966). Girenko's experiments carried out in the northern forest steppes of the Ukraine showed that treating seeds

4

with nodule bacteria substantially increases the soybean harvest in an area where they are newly cultivated (Table 2.18).

Voroshilova *et al.* (1964) also reported favourable results after inoculating soy beans in the Far East (Amur Region). There the gain when a culture of nodule bacteria was used was as great as 40 per cent. When inoculated lucerne seeds were sown in an alluvial plain in Uzbekhistan, where lucerne had not been grown before, there was an increase at the first mowing of 26 per cent and at the second of 33 per cent.

Thus, there can be little doubt of the desirability of inoculating newly introduced leguminous crops and also those planted in reclaimed farming areas. It is much more difficult to solve the problem for old arable, well cultivated soils in which definite leguminous species have long been cultivated. It can be assumed

Table 2.19. Effect of Inoculation of Soy beans on Yield

Experimental variant	Yield (100 kg/hectare)	No. of nodules per plant	Weight of 100 seeds (g)	Content of protein in beans, % in relation to absolutely dry matter
Controls	16·5	0·4	136	35·9
Inoculation with culture:				
638	19·6	39·0	156	41·1
640	20·8	49·0	158	42·2

that in such soils stable microbial coenoses have already formed, involving numerous nodule bacteria of the plants under cultivation. Is inoculation necessary here and is it economically worthwhile? This problem has long interested scientists in several countries. Numerous experiments have been carried out to try and answer it giving, it seems to us, quite definite results.

Analysing the available data we shall deal first with experiences in the USSR. In 1936–1938 in the European part of the USSR, large scale experiments were carried out involving the inoculation of different legumes under the guidance of Mishustin and Bernard. The results obtained, published in 1938, were on the whole quite positive. In the most cases the consequence of inoculation was an appreciably better harvest. The best effect was noted in acid soils where the gain for the pea crop reached 40 per cent. Results were less obvious in neutral soils.

As Table 2.19 shows, inoculation mostly gave positive results. The gains at harvest amply covered the expenditure on the bacterial preparation and its use.

Kalnin'sh (1958) carried out some experiments with clover in the Latvian SSR in soil which had long been in use. The soil contained spontaneous nodule bacteria which might have produced heavy infection in the clover. As Table 2.20 shows, the yield of the green matter of clover increased under the influence of inoculation. According to Bel'skii(1964) and Ezubchik (1965) nitragination of leguminous plants in the Byelorussian SSR ensures good harvests. In the soils of the Perm Region the use of a preparation of nodule bacteria increased the yield of broad beans by 15·5–44·7 per cent (Varaksina, and Pozharinskaya, 1965).

Table 2.20. Effectiveness of Inoculation of Legumes in Field Experiments in the USSR

Crop	Character	No. of tests	Tests with positive results %	Mean harvest in controls 100 kg/ hectare	Increase in harvest 100 kg/ hectare	%
Pea	Green matter	8	75	123·2	29·1	23·5
	Seed	24	83	16·7	2·5	14·9
Broad bean	Green matter	10	90	171·5	29·3	16·6
	Seed	6	66	16·8	3·1	18·0
Lupin	Green matter	14	100	232·9	154·1	66·1
	Seed	9	100	10·5	2·2	20·9
Soybean	Green matter	2	100	79·5	15·5	19·1
	Seed	26	70	12·9	1·6	12·3
Chick pea	Seed	6	100	13·4	2·1	14·9
Lucerne	Hay	4	100	45·7	12·4	26·9
	Seeds	3	100	1·9	1·1	57·9
Sainfoin	Hay	4	100	35·9	11·2	31·1
	Seeds	3	100	6·0	2·8	46·6

Statistical treatment showed the gains to be significant. This table is taken from Dorosinskii (1964). Data are for 1958–1963.

The use of nitragin under vetch in polar farming conditions considerably reduced the costs of the green matter of a mixture of vetch and oats (Roisin and Buleishvili, 1966).

Extensive work on the inoculation of leguminous seeds has been carried out in the Ukrainian SSR. Experiments were run mostly in the black-earth zone and partially in grey forest soils. The available results have been summarized by Samtsevich et al. (1959) and Kudzin et al. (1964). The results must be considered quite positive. Thus, according to Samtsevich and his associates the average gains at harvest were as shown in Table 2.21.

Considerable experience of positive results has been gained by inoculating sowings of leguminous plants in the Armenian SSR (Petrosyan, 1959).Favourable results were also obtained in the test stations of the All-Union Institute of Fertilizers and Agropedology of VASKhNIL (Lenin All-Union Academy of

Agricultural Sciences) where the effectiveness of inoculation was investigated for eight years (Trepachev, 1964, quoted by Dorosinskii, 1965). Consequently we can conclude that in zones of both acid and neutral soil the use of nitragin gives substantial gains in the legume harvest.

It must be noted, however, that in the zones with insufficient water the effectiveness of inoculation diminishes sharply, which is quite understandable since nodules develop poorly when there is a deficit of moisture. Only sainfoin and

Table 2.21. Effects of Inoculation on the Yield of Clover in Field Experiments

Place of experiment	Character	Harvest 100 kg/ harvest		Increase in harvest	
		Controls	Test	100 kg/hectare	%
Ramava Training School Farm of the Latvian Agricultural Academy	Green mass	133·2	157·2	21·1	18·1
Kleieti Experimental Base	Green mass	180·6	224·6	44·0	24·6
October Collective Farm	Hay	40·7	48·8	8·1	19·8
Tsentita Collective Farm	Hay	39·3	46·7	7·4	21·3
	Seeds	4·1	5·5	1·4	34·2
Lenin Collective Farm	Hay	39·4	44·8	5·4	13·7

certain other leguminous species do not react very sharply to a deficiency of soil moisture.

While noting the many findings of a positive effect of inoculation on legumes, we cannot ignore the results which imply the opposite. Thus, for example, Nichiporovich and Perevozchikova (1935), working with the black-earth soils of North Caucasus, did not obtain an increased soybean harvest by means of bacterization. Evidently their results were connected with the use of incomplete preparations of nodule bacteria, for repetition of the experiments gave a positive result (Antropov and Zinevich, 1930; Tsyurupa, 1937).

Reviewing the results of experiments on the effectiveness of nitragin in state crop testing stations, Generalov (1957) concluded that commercial preparations are ineffective. It seems to us that Generalov's data compels us to look seriously at the quality of the nodule bacteria preparations put out by industrial firms.

On the whole the work carried out in France gives a favourable impression of bacterization by cultures of nodule bacteria (Obaton and Blachère, 1963, 1965; Pochon and de Barjac, 1960), and so too does work from Holland

(Harmsen, 1965), Belgium (for example Manil and Bonnier, 1955; Manil and Brakel, 1961; Bonnier *et al.*, 1964; Brakel and Manil, 1965), the United States (for example J. Wilson, 1939) Poland (for example Marszewska-Ziemiecka *et al.*, 1938, 1948; Wróbel and Gołębiowska, 1956; Wróbel 1959), Rumania (for example Negryana *et al.*, 1961; Georgiu *et al.*, 1964), Bulgaria (for example Raicheva, 1959, 1966; Dinchev, 1959), Yugoslavia (Sarić, 1953), Hungary (Segi (Szegi) and Manninger, 1964) and other countries.

In France, the most extensive work on the inoculation of leguminous plants is carried out at the Versailles Testing Station (Obaton and Blachère). Here, the results of treating lucerne with a preparation of nodule bacteria prepared at the testing station (N-germ) have been monitored for several years. The work carried out allowed them to draw the following principal conclusions: (1) inoculation

Table 2.22. Harvest Gains after Treatment with Nodule Bacteria

		100 kg/hectare
Lucerne	Hay	23·7
	Seed	0·7
Pulses	*Phaseolus*	3·2
	Vetch	1·5
	Pea	3·5
Lupin	Hay	51·3
	Seed	2·6

is necessary in the acid soils of Brittany, the Central Massive and the Vosges; (2) it is often beneficial in the decalcined soils of Normandy, the coastal region of the Atlantic Ocean, the Jura mountains and districts bordering the Central Massive; and (3) no appreciable effect is observed in calcium-rich and limed soils, even when they have not previously been sown with lucerne. Successful cases of inoculation gave an increase of 22–113 per cent in the yield of lucerne hay.

In England, preparations of nodule bacteria are used less widely, but the data available indicate that application of the preparations often increased the yield of leguminous plants, although to an extent which varies from 5·3 to 236 per cent.

Nodule fertilizer preparations are also used on a large scale in Sweden. At the Agricultural Institute at Upsala investigations on inoculation have been in progress for more than thirty years (Bjalfve, 1949). The findings are interesting in that they refer entirely to well cultivated soils in conditions very similar to those in the north-west of the European USSR. As Table 2.22 shows inoculation was justified here too.

In Czechoslovakia, Hamatová-Hlaváčková and Marečková (1962, 1964, 1966) concluded from the results of many experiments that, for example, inoculation of sowings increases the yield of lucerne hay on average by 6–13 per cent, of clover hay by 6·31 per cent, of pea by 8–15 per cent and of soybean by 9–13 per cent. Using phosphorus and potassium fertilizers, the effectiveness of bacterization increases still further.

In Poland 10 per cent of leguminous crops sown are inoculated (Juroszek, 1963). Table 2.23, taken from the work of Wróbel (1959) gives the results of field experiments carried out in 1948–1957, involving the seeds of different species of leguminous plants. As the table shows, the treatment did not always have an effect, but, on average, quite an appreciable increase in the harvest was noticed.

Marszowska-Ziemiecka et al. (1938), who bacterized soy beans, found an appreciable gain in several cases (up to 25 per cent) but inoculation sometimes

Table 2.23. Effectiveness of Nodule Fertilizers in Poland

	Lucerne hay	Clover hay
Number of experiments:		
with positive result	307	843
with insignificant gain	7	9
with no effect	3	21
Gain (%):		
minimum	15	13
maximum	900	100
average	90	49 (In southern districts 60%)

had no effect. The same investigator with his colleagues (1948) showed that nitragination of sweet lupin with active strains increased the protein and reduced the alkaloids in the green parts of the plant. In the seeds, the content of protein was found to decrease slightly.

The increased production of preparations of nodule bacteria is a clear indication of the success in the United States. The amount produced in 1960 was more than double that produced in 1940. Preparations of nodule bacteria produced in the United States are used to treat some 650,000 tons of legume seeds each year (Erdman, 1961).

In most cases the use of cultures of nodule bacteria gives positive results and therefore we agree with Pochon and de Barjac (1960) that 'It is necessary to mention the oft expressed and quite unsubstantiated view that infection is unnecessary because legumes have already been cultivated in the soil. In such a case it is impossible to know whether this soil will contain sufficient nodule bacteria of the corresponding race and whether this race will possess maximum effectivity. Properly conducted, infection (guaranteed infecting material, proper methods of application) remains the simplest and most economic method for ensuring a good harvest.'

Bacterization not only increases the yield of leguminous plants but improves their quality. There is much published evidence indicating that leguminous plants inoculated with effective races of nodule bacteria have a considerably higher content of nitrogen (in particular protein). Several sets of relevant data have been published (for example Sakhautdinov, 1939; Razumovskaya and Vasil'eva, 1956a; Bonnier, 1957, Richardson et al., 1957; Sarić, 1959; Samtsevich et al., 1959; Meshkov, 1963, 1966, Ramaswami and Nair, 1965; Glukhovtsev, 1966). We shall confine ourselves to Table 2.24, taken from the work of Doro-sinskii et al. (1962). In this case, all soy beans were grown on sand fertilized with

Table 2.24. Effect of Inoculation on the Yield of Leguminous Plants in Field Experiments in Poland

Crop	Total number of experiments	Experiments with positive result (%)	Harvest	Extreme limits of increase in harvest (%)	Average increase in harvest (%)
Lupin	16	60	Seed	0–220	27
	15	75	Straw	0–40	12
Serradella	5	80	Seed	0–33	14
	11	90	Straw	0–25	15
Pea	2	100	Seed	5–11	8
	4	100	Straw	13–25	19
Lucerne	9	80	Hay	0–230	42
	4	75	Hay	0–29	13
Clover	9	55	Hay	0–18	7
	4	100	Hay	13–45	25
Soy bean	4	100	Seed	11–230	70
	4	100	Straw	14–16	15

different doses of ammonium nitrate. In one series a full dose of this fertilizer was given as recommended by Pryanishnikov, in the second one fifth of the norm was given. As the table shows, in both series of experiments and even with the high dose of ammonium nitrate, inoculation increased the content of total and protein nitrogen in the aerial parts of the plants.

Many investigators have noted that plants with nodules ripen more rapidly than uninfected plants.

It is worth noting that nodule bacteria stimulate the accumulation of vitamins in the host plant. Krasilnikov (1958) established that most nodule bacteria can synthesize vitamin B_1, while Burton and Locchead (1956) noted their ability to accumulate vitamin B_{12}. Ratner's experiments (1965), the results of which we give in Table 2.25, showed that the inoculated plant contained considerably more vitamins of the B complex than the uninoculated plant.

The content of vitamins is considerably greater in the nodules than in the

Table 2.25. Effect of Inoculation on the Content of Nitrogenous Substances in the Aerial Parts of the Soy bean

Dose of nitrogen	Inoculation	Percentage of nitrogen in plant		
		Total	Protein	Non-protein
0·2	+	1·68	1·34	0·34
0·2	−	0·87	0·82	0·05
1·0	+	1·55	1·42	0·13
1·0	−	0·91	0·9	0·01

One dose of nitrogen, according to Pryanishnikov, is equal to 0·24 g of ammonium nitrate/kg of substrate.

roots (Kon, 1955; Shemakhanova and Bun'ko, 1966a and b). The findings of Shemakhanova Lanova and Bun'ko are given in Fig. 19.

Because the positive effect of inoculation extends to the roots as well as the aerial parts of the plants, the crop residues left after the harvest have an effect

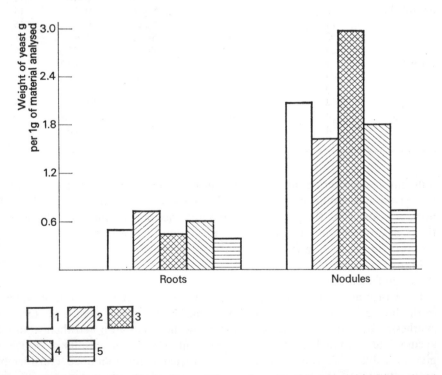

Fig. 19. Content of vitamins in the nodules and roots of legumes (ripening phase). 1, Pantothenic acid; 2, nicotinic acid; 3, biotin; 4, pyridoxine; 5, thiamine.

on the crop which follows them. This has been noted by several Soviet micro-biologists (for example, Shevchuk, 1964; Lasting, 1966; Meshkov and Khoda-kova, 1966).

According to Shevchuk (1966), when the roots of various leguminous species are ploughed into the soil there is a redistribution of the forms of nitrogen, with an increase in the readily hydrolysed fraction. Five to ten per cent of the organic substances of the roots of leguminous plants is converted into humus in the soil. Humic acids from the roots of leguminous plants are distinguished by a higher content of nitrogen than those from other species (Myskow and Ziemiecka, 1966).

Table 2.26. Increase in the Content of Certain Vitamins and Nitrogen in the Pea Flowering after Inoculation

Part of plant investigated	Pyridoxine	Thiamine	Pantothenic acid	Nitrogen
Leaves	116	100	134	161
Stems	141	60	192	129
Roots	488	260	481	120

Values are expressed as percentage of controls.

Table 2.27 shows the results of one of Dorosinskii's experiments (1965a), carried out on light podzolic soil sown with broad beans. The soil in one set of containers was inoculated with nodule bacteria while in the other set there was only natural infection. In the following year, after harvest, the root residues of the broad beans were withdrawn from each container separately, ground up and mixed with the soil from the same container. A mixture of soil plus crushed roots was placed in the container. This operation was repeated for the whole set of containers. Then barley was sown. As Table 2.27 shows, inoculation increased the yield not only of broad beans but also of the next crop.

Table 2.27. Effect of Inoculation on the Yield of Broad Beans and Succeeding Crops

Experimental variant	Sowing of broad beans (1963)		Sowing of barley (1964)	
	Weight of green matter (g)	Weight of pods (g)	Weight of ears (g)	Increase in yield (%)
No inoculation	124·8	72·1	5·2	100
With inoculation	106·1	80·3	6·9	132·6
Controls (42 mg of NH$_4$NO$_3$/kg of soil)	—	—	7·4	142·3

The root system of many leguminous species, as Cherepkov showed (1965), is mineralized approximately three times faster than that of graminaceous grasses. This explains the good after-effect on the soil of leguminous plants. It must be remembered, however, that the roots of different leguminous species decompose in the soil at different rates (Myskow and Morrison, 1963).

All this evidence suggests that in a considerable number of cases the use of preparations of nodule bacteria is economically worthwhile. The known cases of ineffectiveness are usually associated with the use of low quality preparations. Sometimes, the absence of a positive result may be due to the high content of active races of nodule bacteria in the soil.

The common practice of disinfecting seeds with various substances toxic to living organisms raises the question of the compatibility of this procedure with bacterization. In examining it we shall analyse (1) the effect of disinfectants on free-living nodule bacteria, and (2) their effect on nodule bacteria in symbiosis with the plant.

The most widespread preparations for dry and wet disinfection of seed are Granosan, containing 2·0–2·5 per cent chloroethyl mercury, Mercuran containing about 2 per cent chloroethyl mercury, formalin, Thiuram (TMTD, tetramethylthiuram disulphide), copper trichlorophenolate and hexachlorane.

Many investigators have felt that concentrations of Granosan, Mercuran, TMTD, formalin and preparation AB (containing as active ingredient 15–16 per cent $CuSO_4.3 Cu(OH)_2$ converted to copper) which are toxic to nodule bacteria in agar media do not have such an effect in the soil (Zaremba, 1958; Khatipova, 1958; Berezova and Dorosinskii, 1961). Their effect is weakened and the bacteria remaining in a viable state rapidly begin to multiply (for example Dorosinskii, 1965a).

Work done by microbiologists suggests that Thiuram, copper trichlorophenolate and hexachlorane are less toxic to pure cultures of nodule bacteria of the pea, bean and clover than are the organic mercury preparations Mercuran and Granosan (Jakubisiak and Gołębiowska, 1963; Pronchenko and Andreeva, 1964; Dorosinskii, 1965a).

According to Dorosinskii (1965a) disinfection of lupin seeds does not exclude the use of nitragin. Although nitragination of the disinfected seeds has less effect with untreated seeds, nitragin does have a very noticeable positive effect in this case as shown in table 2.28.

Whenever possible it is desirable to avoid disinfection of legume seeds. This is quite feasible, for example by using healthy seed material. But if disinfection is necessary, it should be done in good time and nitragination must be carried out before sowing (Jacks, 1956; Lampovshchikov, 1956; Zaremba, 1959; Klintsare, 1961; Dobrokhleb and Tymchenko, 1963; Dorosinskii 1965a; Avvakumova, 1966).

Disinfection of leguminous seeds with a standard supply of 1–2 kg per ton of Granosan, Mercuran or TMTD 2–4 weeks before sowing, followed by nitragination on the day of sowing, has almost no adverse effect on symbiosis.

The toxic effect is minimal if soil or peat preparations of nitragin are used undiluted. In such a case the seeds are wetted slightly (Dorosinskii, 1965a). It should be noted that some investigators (Izrail'skii and Minyaeva, 1953; Gołębiowska, 1965) have considered it desirable to carry out disinfection and bacterial treatment simultaneously before sowing.

Table 2.28. Effects of Various Treatments of Disinfection and Nitragination on Nodulation

Experimental variant	No. of nodule bacteria per gram of soil (thousands) on sowing seeds, after:	
	5 days	10 days
Seed bacterized, not disinfected	18,000	72,300
Seeds disinfected 1 month before bacterization	14,400	76,800
Seeds disinfected 3 days before bacterization	311	9,000

N.B. Bacterization was carried out before sowing

Type of treatment of seeds	Weight of green parts of 100 plants	
	(g)	(%)
Nitragination without disinfection treatment:	3,945	100
With Granosan 1 month before sowing	1,635	41·5
With Granosan 1 month before sowing and nitragination before sowing	3,144	80·0
With TMTD 1 month before sowing	1,865	47·2
With TMTD 1 month before sowing and nitragination before sowing	3,388	85·9

Much has been published about the effect of various disinfectants—fungicides, bactericides, insecticides—on nodule formation (for example Appelman and Sears, 1946a; Payne and Fults, 1947; Braithwaite et al., 1958; Brakel, 1963; Dorosinskii and Avrov, 1964; McKell and Whalley, 1964; Kováčiková and Ujević, 1965; Gołębiowska and Kaszubiak, 1965; Sidorenko and Fedorova, 1965; Gołębiowska et al., 1966; Vincent, 1966a). Much literature is also devoted to the elucidation of the effect of other chemical poisons—the herbicides, 2,4-dichlorophenoxyacetic acid (2,4-D), 2,4-dichlorophenoxybutyric acid, urea derivatives, symmetrical triazines and others—on nodule bacteria both in pure cultures and in symbiosis.

According to Kaszubiak (1966) strains that grow rapidly are more resistant to high doses of herbicides than those that grow slowly. Thus phenoxy compounds and the pyridazines suppress the growth of the fast growing strains when applied in a concentration of 0·1 per cent, while the development of slow growing strains almost stops after treatment with herbicide in concentrations of 0·01–0·05 per cent. But Carlyle and Thorpe (1947) and Jensen and Petersen (1952), studying the effect of the phenoxy compounds, in particular 2,4-D, on slow and fast growing strains, found no such pattern.

Carlyle and Thorpe (1947) studied the sensitivity of several species of nodule bacteria to 2,4-D. On the basis of their results they divided all the species of nodule bacteria they had investigated into two groups. The first group included the less sensitive bacteria which withstood concentrations of 0·03–0·04 per cent of 2,4-D. They were *Rh. trifolii* and *Rh. leguminosarum*. The second group, which withstood concentrations of 0·3 per cent, comprised *Rh. phaseoli, Rh. lupini, Rh. japonicum* and *Rh. meliloti*. In Worth and McCabe's experiments (1948) and those of Petersen (1952) the bacteria most sensitive to 2,4-D were *Rh. phaseoli* and *Rh. japonicum* and not *Rh. trifolii*. The latter cultures withstood a concentration of up to 2 per cent.

Nilsson (1957) showed that a culture of *Rh. meliloti* can counteract the toxic effect of solutions containing 10 parts per million or 2,4-D of 2-methyl-4-chlorophenoxyacetic acid in 14 days.

Carlyle and Thorpe thought that the doses of herbicides commonly used in agriculture would have no significant influence in soil on nodule bacteria in soil conditions, and the same view was taken by Fletcher and Smith (1964). Several investigations have been concerned with the resistance of nodule bacteria in the soil to 2,4-D and other herbicides (for example Jensen and Petersen, 1952; Fletcher, 1956; Kozlova and Dikareva, 1963). Using much the same experimental techniques, individual investigators have established differences in the resistance of nodule bacteria to chemical poisons. As some have rightly pointed out (Škrdleta, 1964; Kaszubiak, 1966; Bilodub–Pantera, 1966) this is determined by specific features of individual strains of nodule bacteria, including age (Fletcher *et al.*, 1956), and also by the conditions of the medium, in particular pH (Kazsubiak, 1966).

It has already been noted that it takes much greater concentrations of chemical compounds to suppress the development of micro-organisms in the soil than in bacteriological media (Nutman *et al.*, 1945; Fletcher and Alcorn, 1958; Avrov, 1966). This may be a consequence of several factors associated with the state of the compounds introduced into the soil, involving in particular the effect on them of micro-organisms and chemical processes. Toxic substances gradually decompose in the soil and are detoxified. The rate of this process depends on the properties of the substances used, the number of times they are used and also on soil and climatic conditions (Nutman *et al.*, 1945; Norman, 1950; Petersen, 1950; Audus, 1951; Jensen and Petersen, 1952; Kaufman, 1964).

Butyric acid compounds (2,4-dichlorophenoxybutyric acid and 2-methyl-4-chlorophenoxybutyric acid) are less toxic to nodule bacteria than the acetic acid compounds: 2,4-D and 2,4,5-trichlorophenoxyacetic acid (Fletcher, 1956). The most toxic of the acetic acid compounds is 2,4-D; 2-methyl-4 chlorophenoxy-acetic acid is less harmful and 2,4,5-trichlorophenoxy acetic acid is least toxic (Fletcher and Alcorn, 1958). The ammonium salt of 2,4-D is more toxic to nodule bacteria than the sodium salt (Carlyle and Thorpe, 1947). According to Kaszubiak (1966) the halogen derivatives of aliphatic fatty acids (for example sodium salts of trichloracetic acid or 2,4-D) are not at all toxic to nodule bacteria in concentrations up to 2 per cent.

According to Avrov (1966) the butyl ester and sodium salt of 2,4-D and also sodium 2,4-dichlorophenoxyethyl sulphate, even in large doses ($4 \cdot 15$ mg/cm^2), do not arrest the development of nodule bacteria of the pea, whereas 2,4-dichlorophenoxyethyl phosphate is toxic even in a concentration of $0 \cdot 14$ mg/cm^2.

The symmetrical triazines, atrazine and atratone, propazine, prometrine and simazine, now widely used, do not depress nodule bacteria when introduced in small doses consistent with field conditions, nor even in considerably larger doses: according to Avrov up to $4 \cdot 15$ mg/cm^2 (converted to area) and according to Kaszubiak (1966) up to 2 per cent. At the same time the derivatives of urea—diurone, neburone, phenurone (an exception is monurone)—are not, according to Avrov, toxic for nodule bacteria ($4 \cdot 15$ mg/cm^2) but, according to Kaszubiak (1966), urea derivatives such as aphalone, carmex and carbamate urea are toxic even in a concentration of $0 \cdot 0001$ per cent.

The effect of various herbicides on the formation of nodules in legumes has often been studied (for example Carlyle and Thorpe, 1947; Payne and Fults, 1947; Worth and McCabe, 1948; Jensen and Petersen, 1952; Braithwaite et al., 1958; McKell and Whalley, 1964; Zhodan, 1966). Different results were obtained in most cases. In individual cases different concentrations of chemical compounds suppressed the symbiosis between the bacteria and the leguminous plant.

Fletcher et al. (1956) and Nilsson (1957) consider that the reduced formation of nodules in the presence of toxic substances results from an inhibition of root growth. It has been established, however, that the formation of nodules in, for example, Phaseolus vulgaris in the soil is suppressed under the influence of 2,4-D (about 0·04 parts per million). Root growth is not affected by this concentration of 2,4-D (Payne and Fults, 1947; Fletcher et al., 1956, 1958). A similar pattern has been established for white clover—2-methyl 4-chlorophen-oxyacetic acid in a concentration of 0·05 parts per million does not supress growth but inhibits the formation of nodules (Fletcher, 1956).

It has been suggested that in certain concentrations, these substances induce the formation of lateral roots with fewer potential sites of nodulation. The herbicides as it were 'compete' with the nodule bacteria. This hypothesis has not been verified experimentally—lateral roots and nodules have not been counted in the presence and absence of herbicides.

When Payne and Fults (1947) sprayed Phaseolus vulgaris with low concentrations of 2,4-D (8 parts per million) the plants formed as many nodules as controls, but the nodular tissues contained many Gram-positive bacteria and a decreased number of nodule bacteria. This led in turn to a reduction in the vigour of nitrogen fixation.

There are indications (Rankov *et al.*, 1966) of an increase in the number and weight of nodules in the presence of A-1803 (2 kg/hectare), arezine (1·5 kg/hectare), aretite (8 kg/hectare) and prometrine (2 kg/hectare) and of a stimulation of nitrogen accumulation with a simultaneous decrease in the number of nodules (Avrov, 1966) in the presence of simazine (0·5 kg/hectare) and to a lesser degree dinitro-orthocresol (2 kg/hectare) and sodium pentachlorophenolate (4 kg/hectare).

In Fletcher and Alcorn's experiments (1958) the development of inoculated white clover in sterile conditions was suppressed most strongly in the presence of 2,4-dichlorophenoxyacetic acid and to a lesser degree by 2-methyl-4-chlorophenoxyacetic acid; least toxic was 2,4-5-trichlorophenoxyacetic acid. In field conditions different results were obtained with the same concentrations; 2,4,5-trichlorophenoxyacetic acid proved to be most toxic with 2-methyl-4-chlorophenoxyacetic acid coming second and 2,4 dichlorophenoxyacetic acid having the least effect.

This brief analysis of the effect of chemical poisons on nodule bacteria indicates that when chemicals are used to treat seeds and sowings the effectiveness of nitragination is usually reduced. But because such treatment is sometimes essential it is necessary to define more closely the conditions necessary for the combined use of chemicals with bacterization. This applies primarily to those factors which, according to various authors, are in dispute.

Production and Application of Preparations of Nodule Bacteria

We shall discuss the production of preparations of nodule bacteria intended for practical use, for they are now widely used in agriculture. Only high quality preparations (nitragin) widely used on farms can be expected to give good bacterization. Several cases have been recorded of poor quality nitragin that has given a patently insufficient increase in the yield of legumes or no benefit at all (for example Maisuryan, 1962; Shevchuk, 1962). Using a commercial preparation of nodule bacteria, Shevchuk obtained a 22 per cent higher yield of lupin. When however, the inoculum was an active pure culture of nodule bacteria or crushed nodules formed by an active strain, the yield increased by 80–120 per cent. Others report similar findings (for example Bernard, 1953); Avvakumova, 1956; Roizin, 1966). Active races of nodule bacteria should be used for the preparations, which when finished must contain a large number of live bacterial cells. The fulfilment of these apparently simple needs, however, requires the solution of several problems. First, there is the choice of a culture of nodule bacteria. Activity and virulence must be checked in a carefully run experiment with aseptically (or semi-aseptically) cultivated plants.

Nodule bacteria are known to infect a definite species or group of leguminous plants (page 23). It is therefore, necessary to prepare not one bacterial preparation but several, with a view to the inoculation of definite leguminous species.

Until now in most countries the preparation has usually been produced without reference to the possible specific adaptation of the individual strains of nodule bacteria. At best a mixture of bacterial cultures is used. But from all we have said, it seems in several cases to be desirable to make up the preparation for particular species of legumes. This applies in particular to *Trifolium*, individual species of which are inoculated to differing degrees by definite strains of nodule bacteria (Vincent, 1945; Burton and Briggeman, 1948; Brockwell and Hely, 1952; Jones, 1963; Nutman, 1963; Burton, 1965).

The problem of the adaptation of strains of nodule bacteria to definite host varieties has still not been solved but is being studied actively. As long ago as 1895 Kirchner observed that different varieties of lupin form nodules with differing intensities. This encouraged the idea of the existence of races of nodule bacteria adapted even to definite varieties of plants (Petrenko, 1953). It seems to us difficult to expect such narrow specialization in nodule bacteria. In fact, as several studies have shown (for example Fedorov, 1952; Melkumova, 1957; Apltauer, 1963) active races of nodule bacteria increase the yield of several varieties of leguminous plants. Table 2.29 shows the data obtained in Fedorov's laboratory (Fedorov and Svitych, 1954, 1959). Cultures of nodule bacteria were isolated from yellow and blue lupins, and experiments showed that race No. 25 isolated from the yellow lupin was active both for the yellow and blue lupin.

We cannot discount the fact that different cultures of nodule bacteria sometimes infect individual varieties of the same species to differing degrees. For example, according to Petrosyan (1959) one of her races of nodule bacteria (No. 3) gave in the 'Oktemberyanskii belyi' variety of the Oregon pea on average thirty nodules per plant against only two in another variety 'Ashtarak-perpa'. Some races of nodule bacteria inoculated both Oregon pea varieties well. Evidently for a fully effective preparation it is necessary to use strains of nodule bacteria with a wide range of activity, or mixed cultures, ensuring inoculation of the most widespread varieties of particular species of legumes.

The problem of geographic races of nodule bacteria deserves close attention but has not been studied much. Yet, according to Petrosyan (1954) the ecological races of these micro-organisms develop at quite significantly different temperatures. Bacteria from the soils of a colder climate have a lower temperature maximum. In the appropriate climate they are more effective. (Petrosyan carried out her experiments on different soils of the vertical belt of Armenia.)

Vartiovaara's observations (1937) are in complete agreement with this. They established that at a low temperature cultures of nodule bacteria from more northerly areas infect leguminous plants better.

There is much evidence that active races of nodule bacteria manifest their activity in quite different climatic conditions. This was the conclusion reached by Lazareva (1953) working with clover and Zhvachkina (1960) and Zavertailo

Table 2.29. Effect of Various Races of Nodule Bacteria on Lupin

Race of bacteria	Dry weight of test plants (% of non-infected controls)	Nitrogen fixed from air (mg)
Yellow lupin		
from blue lupin		
375	146·4	96·17
390	120·8	66·72
from yellow lupin		
25	178·9	137·20
Controls	100	0
Blue lupin		
from blue lupin		
375	210·8	96·36
390	173·3	73·77
from yellow lupin		
25	214·2	105·57
Controls	100	0

The experiment was carried out in sand culture with 0·1 of the Hellriegel nitrogen standard.

(1963) with lucerne and Sen (1963) with chick pea. Table 2.30, taken from the work of Dorosinskii (1965a), shows the results obtained when different cultures of nodule bacteria were tested on pea, soya and vetch in the Leningrad Region. It was not possible to establish a connection between the activity and the origin of the cultures.

Table 2.30. Effects of Various Cultures of Nodule Bacteria on Yields

Origin of culture	Increase in yield (%)	Origin of culture	Increase in yield (%)
Soya		Sweden	119
North Caucasus	813	Controls	100
Far East	287	Pea	
Bulgaria	295	Moscow Region	121
Urals	293	Moscow Region	99
Controls	100	Vinnitsa	
Vetch		(Ukrainian SSR)	134
Moscow Region	130	Urals	123
Armenia	114	Arkhangel Region	115
Yugoslavia	134	Controls	100

The inconsistency of the available information about the geographic, and possibly also ecological, races of nodule bacteria requires that the problem be given more attention. It is undoubtedly of great scientific and practical importance.

Preparations of nodule bacteria are issued in different forms in different countries. These are made in agar, soil, peat and other substrates. They are described in detail in the manual by Vincent (1966b). In sterilized soil and in agar media, nodule bacteria are in an active state.

In some cases a fine peat crumble, for example kieselguhr, is used as a filler, being mixed with the cultures of nodule bacteria. Such preparations contain little moisture and are friable powders convenient to use.

In the Argentine at the suggestion of Halbinger (1965), phosphorite (trade mark Reno) obtained in North Africa has been successfully used as a filler for nitragin. The composition of Reno is as follows (percentage): water (in the bound) form + organic matter, 3·63; P_2O, 30·49; CaO, 49·34; K_2O, 0·11; SO_3, 2·85; CO_2, 5·37; Si, 3·09; Fe_2O_3, 0·70; Al_2O_5, 0·58; MgO, 0·55; Na_2O, 1·36; F_2, 3·13; Cl_2, 0·50; unidentified elements, 0·07 (giving 101.25 per cent through the excess content of O_2 bound with F_2 and Cl_2). The pH of Reno is almost neutral.

The seeds mixed with friable powders have to be inoculated by dusting, whereas when agar and soil preparations are used they have to be suspended in water. In dry preparations the bacteria are, of course, latent.

Nodule bacteria do not form spores and die quite rapidly in a medium without nutrients. Consequently their numbers gradually decrease in commercial preparations. Therefore, preparations are assumed to be active only for a limited time, usually the season in which they are issued.

We shall not discuss the technical details of the production of preparations of nodule bacteria, but note that the need for more stable preparations easily transported and stored is now encouraging work on the production of dry cultures. The best prospects seem to be offered by lyophilized cultures. In this connection investigations have been in progress for some time (for example Appelman and Sears, 1946b; Samsonova, 1959; Wróbel, 1959; Borodulina, 1961; Manninger, 1964; Škrdleta, 1965; Vincent, 1966b; Krongauz et al., 1966).

There are certain difficulties in obtaining cultures, for nodule bacteria stand up poorly to drying (for example Appelman and Sears, 1946b; Izrail'skii, 1951; Nilsson, 1957; Pumpyanskaya, 1959; Norris, 1963; Borodulina, 1964; Alexander and Chamblee, 1965). Nodule bacteria in the 'belted' rods phase are not resistant to drying (Borodulina and Samsonova, 1966). However members of the Institute of Agriculture Microbiology of VASKhNIL have worked out systems for the multiplication of bacteria in reactors, and techniques for their subsequent separation. 'Protective' substances have been found that lower the incidence of death in a dried preparation (Dovgan' and Krongauz, 1965). And conditions have been established for drying the cultures. These and other developments allow us to hope that in the not too distant future farmers will have at their command concentrated, dry but stable preparations of nodule bacteria.

We cannot exclude the possible use of antibiotics for more prolonged maintenance of the purity of preparations of nodule bacteria (Hamatová–Hlaváčková, 1963; Buyanova and Kachan, 1965; Mikhaleva et al., 1965).

We must consider the problem of the survival of nodule bacteria in the soil. Contrary to expectations the surface of leguminous seeds is not a sufficiently good substrate. Here, nodule bacteria survive less well even than on glass beads (Vincent, 1958; Vincent et al., 1962). Thompson (1960), Bowen (1961) and Masterson (1965) established that the seed coat of clover and lucerne contains substances which act as antibiotics against nodule bacteria.

Many point to the desirability of using protective substances when leguminuous seeds are inoculated with nodule bacteria. The bacteria suffer less in peat preparations than in aqueous suspensions. Molasses also has a protective action (Smith et al., 1945; Dobson and Lovvorn, 1949). Here the active factor is the sugar. Vincent (1962) and Vincent et al. (1962) found that various sugars had a favourable effect, especially maltose. Sorbitol and adhesives, $CaCO_3$ and so on, also have a protective function (Burton, 1965). The optimum concentrations of sucrose are between 8 and 25 per cent.

Brockwell and Phillips (1965) studied the course of mortality of nodule bacteria in inoculated seed when a protective substance was used (sucrose or peat) in different thermal conditions (1965). According to Erdman (1961) it is not rare for seeds of leguminous plants to be sold in the United States after bacterization. In such cases, the farmer is free of several worries.

A study has been made of the technique for pre-inoculating seed with nodule bacteria in which it is evidently desirable to use various protective substances (Hofe and Crosier, 1962). Porter et al. (1962) established the need for very intensive infection of seeds for the prolonged maintenance of even a few bacteria on their surface. It is necessary to apply 800–1,600 bacterial cells per seed if all the seeds are still to be inoculated after 245 days. Of course, when protective agents are used the level of primary infection may be reduced.

An attempt has been made to inoculate legume seeds under pressure, but there were no advantages in the method (Brockwell and Hely, 1962). According to Hely (1965) other and more effective suggestions have been made, but their basic features have not been described.

In 1963, Hely proposed the so called triple inoculation of seeds: the seeds are first covered with a layer of adhesive such as gum arabic, methylethylcellulose, methylhydroxypropylcellulose or similar compounds, and then with a layer of lime. The seeds thus prepared are inoculated with nodule bacteria. The technique was rated highly in Australian conditions (Date et al., 1965; Vincent, 1966b) for the treatment with adhesive almost doubled the time of storage of the inoculated seeds.

Bonnier and Lebrun (1963) describe an original preliminary method of inoculation. The seeds of leguminous plants are grown at 20°C in aqueous, very dense suspensions of nodule bacteria containing millions of cells per millilitre. When an adhesive is applied to the seeds, germination is arrested by a current of cooled air. After such treatment the seeds are preserved without loss of germinative powers, and with a good degree of infectivity, for up to five months.

Most of the methods we have mentioned for improving the use of preparations of nodule bacteria are tentative and require refinement. Nevertheless, their practical significance is obvious.

In conclusion, we note that when van Schreven *et al.* (1954) compared many foreign commercial preparations of nodule bacteria, the Swedish preparations came out best.

References

Afrikyan, E. G. and Tumanyan, V. G. 1958. Antagonistic effect of soil micro-organisms on cultures of nodule bacteria. *Izv. Akad. Nauk ArmSSR, Ser. Biol.*, **11**, (2), 37–46.

Abdel-Chaffar, A. S. and Jensen, H. L. 1966. Inhibition of growth of certain root nodule bacteria by nicotinic acid, in *IX Mezhdunarodnyi Mikrobiologicheskii Kongress* (Ninth International Microbiology Congress), Moscow, Symposium V-1, 65–68.

Abel, G. M. and Erdman, L. W. 1964. Response of Lee soybeans to different strains of *Rhizobium japonicum. Agron. J.*, **56** (4), 423–424.

Ahmed, Sh. and Evans, H. I. 1959. Effect of cobalt on the growth of soybeans in the absence of supplied nitrogen. *Biochem. Biophys., Res. Communn*, 7 (5), 271–275.

Aizupiete, I. P. 1956. Application of trace elements under cereals in the conditions of the Latvian SSR. in *Mikroelementy v sel skom khozyaastre i meditsine* (Trace elements in agriculture and medicine), Riga.

Albrecht, W. A. and Turk, L. M. 1930. Legume bacteria with reference to light and longevity. *Missouri Agric. Exp. Bull.*, **132**, 19.

Alekandrov, V. and Martovitskaya, A. 1965. Treatment of pea seeds with bacterial and micro-fertilizers. *Zernobobovye kul'tury*, (4), 15–16.

Alexander, C. W. and Chamblee, D. S. 1965. Effect of sunlight and drying on the inoculation of legumes with *Rhizobium* species. *Agron. J.*, **51**, 550–553.

Allen, E. K. and Allen, O. N. 1950. Biochemical and symbiotic properties of the rhizobia. *Bacteriol. Rev.*, **14**, 273–330.

Allen, E. K. and Allen, O. N. 1958. Biological aspects of symbiotic nitrogen fixation. in *Encyclopedia of Plant Physiology*. Ed. W. Ruhland, **8**, 48–118. Springer, Berlin.

Allen, E. K. and Allen, O. N. and Klebesadel, L. J. 1964. An insight into symbiotic nitrogen-fixing plant association in Alaska. in *Proc. Fourteenth Alaska Sci. Conf., Proc., Alaska Div. Amer. Assoc. Advanced Sci.*, 54–63.

Allen, E. K., Gregory, K. F. and Allen, O. N. 1955. Morphological development of nodules on *Catagana arborescens. Canad. J. Bot.*, **33**, 139–148.

Allen, O. N. 1966. Rhizobial inoculands in United States Agriculture: in *IX Mezhdunarodnyi Mikrobiologicheskii Kongress* (Ninth International Microbiology Congress). Moscow.

Allen, O. N. and Allen, E. K. 1954. Morphogenesis of the leguminous root nodule. *Brookhaven Symp. Biol.*, No. 6, 209–234.

Allen, O. N. and Allen, E. K. 1940. Response of the peanut plant to inoculation with rhizobia, with special reference to morphological development of the nodules. *Bot. Gaz.*, **102**, 121–142.

Allen, O. N. and Baldwin, I. L. 1931. The direct isolation of *Rhizobium* from soil. *J. Amer. Soc. Agron.*, **23**, 28–31.

Allen, O. N. and Baldwin, I. L. 1954. Rhizobia-legume relationships. *Soil Sci.*, **28** (6), 415–427.

Allen, S. H. 1956. The effects of vitamin B_{12} deficiency and copper deficiency on the porphyrin in the erythrocytes of sheep. *Biochem. J.*, **63**, 461.

Allison, F. E. 1929. Can nodule bacteria of leguminous plants fix atmospheric nitrogen in the absence of the host? *J. Agric. Res.*, **39** (12), 893–924.

Allison, F. E. and Ludwig, C. A. 1934. The cause of decreased nodule formation of legumes supplied with abundant combined nitrogen. *Soil Sci.*, **37**, 431–443.

Allison, F. E. and Hoover, S. R. 1934. An accessory factor for legume nodule bacteria. *J. Bacteriol.*, **27**, 561–581.

Allos, H. F. and Bartholomew, W. W. 1959. Replacement of symbiotic fixation by available nitrogen. *Soil Sci.*, **87** (2), 61–66.

Almon, L. 1933. Concerning the reproduction of bacteroids. *Zbl. Bakteriol., Parasitenkunde, Infektionskrankh. und Hyg.*, **2**, 87, 289–297 (in English).

Almon, L. and Baldwin, I. 1933. The stability of cultures of *Rhizobium*. *J. Bacteriol.*, **26**, 229.

Anderson, A. D. 1933. The production of gum by certain species of Rhizobium. *Iowa Agric. Exptl. Bull.*, 158.

Anderson, A. and Spender, D. 1948. Lime in relation to clover nodulation at sites on the southern tablelands of New South Wales. *J. Austral. Inst. Agric. Sci.*, **14**, 39–41.

Anderson, A. and Spencer, D. 1949. Molybdenum and sulphur in symbiotic nitrogen fixation. *Nature*, **164** (4163), 273–274.

Antropov, T. and Zinevich, V. 1930. Infection of soya with nitragin and effect on its yield. *Masloboino-zhirovoe delo*, (6), 59.

Apltauer, J. 1963. Selection of races of *Rhizobium lupini* and their specific application. *Sbor. Českosl. akad. zeměd. věd. Rostl. výroba*, **9** (7–8), 738–740.

Appelman, M. D. and Sears, O. H. 1946a. Effects of DDT on nodulation of legumes. *J. Amer. Soc. Agron.*, **38**, 545–550.

Appelman, M. D. and Sears, O. H. 1946b. Studies of lyophilized cultures, lyophilic storage of cultures of *Rhizobium leguminosarum*. *J. Bacteriol*, **52**, 209–211.

Appleby, C. and Bergersen, F. J. 1958. Cytochromes of *Rhizobium*. *Nature*, **182** (4643), 1174.

Aprison, M. H. and Burris, R. H. 1952. Time course of fixation of N_2 by excised soybean nodules. *Science*, **115**, 264–265.

Ash, C. G. 1951. The efficiency of *Rhizobium meliloti* on alfalfa as influenced by plant food elements. Ph.D. Thesis. Univ. of Wisconsin, Madison.

Aseeva, K. B., Evstigneeva, Z. G. and Kretovich, V. L. 1966. The amino acid composition of the nodules of alder, *Phaseolus* and lupin. *Dokl. Akad. Nauk SSSR*, **169**, (2), 463–465.

Atakhanov, Sh. A. 1964. Parasitic and saprozoic nematodes of certain wild plants and weeds of Uzbekhistan. *Uch. zapis. Karshinsk. ped. in-ta* (jubilee edition), (16), 229–240.

Atakhanov, Sh. A. and D'yakova, A. N. 1964. Nematode fauna of the pea and the rhizosphere in the Karshinsk Region of the Kashkadar'insk oblast'. *Uch. zapis. karshinsk. ped. in-ta*, (jubilee edition), (16), 214–220.

Atwal, A. S. and Sighu, G. S. 1964. Legumes in the nitrogen economy of soil. I. Fixation and excretion of nitrogen by Indian legumes under controlled conditions in sand culture. *Indian J. Agric. Sci.*, **34** (3), 139–145.

Audus, L. J. 1951. The biological detoxication of hormone herbicides in soil. *Plant and Soil*, **3** (2), 170–192.

Aughtry, J. D. 1948. Effect of genetic factors in *Medicago* on symbiosis with *Rhizobium*. *Cornell. Univ. Agric. Exptl. Stat., Mem.*, 280.

Avilov, I. A. 1961. Culture of isolated roots of leguminous plants. *Vestn. Leningrad. UN–TA.*, (9), 135–136.

Avrov, O. E. 1966. Effect of herbicides on the nodule bacteria and nodulation of leguminous plants. *Dokl. VASKhNIL*, (3), 16–19.

Avvakumova, E. N. 1956. Priemy povysheniya effektivnosti nitraginnisatsii gorokha i klovera (Means of improving the effectiveness of nitragination of the pea and clover) (Author's abstract of M.Sc. thesis).

Avvakumova, E. N. 1966. Effect of pre-sowing treatment of legume seeds with Granosan on the activity and virulence of module bacteria: in *Voprosy mikrobiologii* (Problems of microbiology), part 3, Izd-vo Akad.-Nauk. ArmSSR, Erevan, 43–45.

Avvakumova, E. N., Kats, L. N. and Shamtsyan, M. G. 1966. Cytochemical study of nodule bacteria in connection with the intensity of symbiotic nitrogen fixation: in *IX Mezhdunarodnyi mikrobiologicheskii kongress* (Ninth International Microbiology Congress), Moscow, summaries of speeches, section B-2, 294.

Badawy, F. II. 1966. Growth of strains of *Rhizobium japonicum* and their symbiotic relationship with soybean plants (*Glycine max* Merr.). *Diss. Abstr.*, **26** (9), 4965–4966.

Badawy, F. H. and Allen, O. N. 1963. The effect of calcium on the growth and efficiency of rhizobia. *Agron. Abstr.*, **31**.

Baird, K. J. 1955. Clover root-nodule bacteria in the New England region of New South Wales. *Austral. J. Agric. Res.*, **6**, 15–26.

Balassa, R. 1954. Transformations-mechanismen der Rhizobien. *Acta microbiol. Acad. sci. Hung.*, **2** (1–2), 51–78.

Balassa, R. 1955. *In vivo* induzierte Transformationen bei Rhizobien. *Naturwissenschaften*, **42** (14), 422–423.

Balassa, R. 1957. Durch Desocytibonukleinsauren induzierte Veranderungen bei Rhizobien. *Acta microbiol. Acad. sci. Hung.*, **4** (1), 77–84.

Balassa, R. 1960. Transformation of a strain of *Rhizobium lupini*. *Nature*, **188** (4746), 246–247.

Balassa, R. 1963. Genetic transformation of *Rhizobium*. *Bacteriol. Rev.*, **27** (2), 228–241.

Balassa, R. and Gabor, M. 1964–1965. Transformation of streptomycin markers in rough strains of *Rhizobium lupini*. Transformation of streptomycin-dependence. *Acta microbiol. Acad. sci. Hung.*, **11** (4), 329–340.

Baldwin, I. 1929. Strain variation in the root nodule bacteria of clover *Rhizobium trifolii*. *J. Bacteriol.*, **17**, 17–29.

Barthel, C. 1926. Can legume bacteria in pure culture fix atmospheric nitrogen? *Medd. N.* 308. *Centralanstalt försöksväsendet jordbruks. Bakteriol. avd. C.*, (43) (in Swedish).

Barthel, C. and Bjalfve, G. 1933. Investigation of the occurrence of filtrable forms of *B. radicicola. Meddel. Centralaust. försöksv. jordbrun-Ksomradet. Bakteriol. adv.*, (60) (in Swedish).

Baylor, M. B., Appelman, M. D., Sears, O. H. and Clark, G. L. 1945. Some morphological characteristics of nodule bacteria revealed by the electron microscope. *J. Bacteriol.*, **50**, 249–256.

Bazarewski, S. 1927. Investigations of bacteroids. *Roczn. nauk roln.*, **17**, 1–34.

Bazyrina, E. N., Konokitina, A. G. and Kovaleva, V. I. 1949. Development of nodules in connection with nitrogen nutrition of the leguminous plant: in *Trudy Vsesoyuznogo nauchno-issledovatel'skogo instituta sel'skokhozyaistvennoi mikrobiologii za 1941–1945gg* (Proceedings of the All-Union Scientific Research Institute of Agricultural Microbiology for 1941–1945), part 1.

Beijerinck, M. W. 1888. Die Bakterien der Papilionaceen-Knollchen. *Bot. Ztg.*, **46,** 725–735, 741–750, 757–771, 781–790, 797–804.

Beijerinck, M. W. 1890. Kunstliche infektion von *Vicia faba* mit *Bacterium radicicola. Bot. Ztg.*, **48,** 837–843.

Beijerinck, M. W. 1923. Urease as a product of *Bacterium radicicola. Nature,* **112,** 202.

Bel'skii, B. B. 1964. *Rol' khimii v effektivnost. j ispol'zov. torfyanobolotnykh pochv* (Role of chemistry in improving the utilization of peat-bog soils). Minsk.

Benjamin, M. S. 1915. A note on the occurrence of urease in legume root nodules and other plants. *Proc. Roy. Soc. New South Wales,* **49,** 78.

Beresniewicz, K. 1959. Studies on the acceleration of the development of the slow growing strains of *Rhizobium. Acta microbiol. polon.,* **8** (3–4), 333–337 (in English).

Berezova, E. F. and Dorosinskii, L. M. 1961. *Bakterial'nye udobreniya* (Bacterial fertilizers). Leningrad-Moscow. Izv-vo s.-kh. lit-ry, zhurn. i plakatov.

Berg, M. H. 1965. Variations in porphyrin content in root nodules. *J. Minnesota Acad. Sci.,* **33** (1), 15–16.

Bergersen, F. J. 1955. The cytology of bacteroids from root nodules of subterranean clover (*Trifolium subterraneum* L.). *J. Gen. Microbiol.,* **13,** 411–419.

Bergersen, F. J. 1957. The structure of ineffective root nodules of legumes: an unusual new type of ineffectiveness and an appraisal of present knowledge. *Austral. J. Biol. Sci.,* **10,** 233–242.

Bergersen, F. J. 1958. The bacterial component of soybean root nodules; changes in respiratory activity, total dry weight and nucleic content with increasing nodule age. *J. Gen. Microbiol.,* **19,** 312–323.

Bergersen, F. J. 1960*a*. Incorporation of $^{15}N_2$ into various fractions of soybean root nodules. *J. Gen. Microbiol.,* **22,** 671–677.

Bergersen, F. J. 1960*b*. Biochemical pathways in legume root nodule nitrogen fixation. *Bacteriol. Rev.,* **24** (2), 246–251.

Bergersen, F. J. 1961. The growth of *Rhizobium* in synthetic media. *Austral. J. Biol. Sci.,* **14** (3), 349–360.

Bergersen, F. J. 1962*a*. The effects of partial pressure of oxygen on respiration and nitrogen fixation by soybean root nodules. *J. Gen. Microbiol.,* **29,** 113–125.

Bergersen, F. J. 1962*b*. Oxygenation of leghaemoglobin in soybean root nodules in relation to external oxygen tension. *Nature,* **194,** 1059–1062.

Bergersen, F. J. 1963*a*. Iron in the developing soybean nodule. *Austral. J. Biol. Sci.,* **16** (4), 916–919.

Bergersen, F. J. 1963*b*. The relationship between hydrogen evolution, hydrogen exchange, nitrogen fixation, and applied oxygen tension in soybean root nodule. *Austral. J. Biol. Sci.,* **16** (3), 669–680.

Bergersen, F. 1966. Nitrogen fixation in root nodules of legumes: in *IX. Mezhdunarodnyi Mikrobiologicheskii kongress* (Ninth International Microbiology Congress), Symposium V-1, Moscow, 69–72.

Bergersen, F. J. and Briggs, M. J. 1958. Studies on the bacterial component of soya bean root nodules. Cytology and organisation in host tissue. *J. Gen. Microbiol.,* **1,** 482–490.

Bergersen, F. J., Hely, F. W. and Costin, A. B. 1963. Overwintering of clover nodules in Alpine conditions. *Austral. J. Biol. Sci.,* **16** (4), 920–921.

Bergersen, F. J. and Nutman, P. S. 1957. Symbiotic effectiveness in nodulated red clover. 4. *Heredity,* **11,** 175–184.

Bergersen, F. J. and Wilson, P. W. 1959. Spectrophotometric studies of the effects of nitrogen on soybean nodule extracts. *Proc. Nat. Acad. Sci. USA,* **45** (11), 1641–1646.

Bergey's Manual of Determinative Bacteriology, 1957. 7th Ed. Williams and Wilkins Co., Baltimore.

Bernard, V. V. 1953. Effectiveness of nitragin under clover in conditions of high-grade agricultural technology. *Trudy Vses. n.-i in-ta s.-kh. mikrobiologii*, **13**, 120–126.

Bertrand, D. and Wolf, A. 1954. Nickel and cobalt in root nodules of legumes. *Bull. Soc. chim. biol.*, **36**, 905–906.

Bilodub-Pantera, G. 1966. Study of the effect of Alaphon on *Rhizobium lupini*: in *IX. Mezhdunarodnyi Mikrobiologicheskii kongress* (Ninth International Microbiology Congress), summaries of speeches, Moscow, 294–295.

Bisset, K. A. 1952. Complete and reduced life-cycles in *Rhizobium*. *J. Gen. Microbiol.*, **7**, 233–242.

Bjälfve, G. 1949. Inoculation trials of leguminous plants 1914–1948. *Ann. Roy. Agric. Coll. Sweden*, **16**, 609–617 (in English).

Bjälfve, G. 1963. The effectiveness of nodule bacteria. *Plant and Soil*, **18** (1), 70–76.

Bobko, E. V. 1963. *Izbrannye sochineniya* (Selected works), Sel'kozgiz.

Bogacheva, S. D. 1961. Study of nitrogenous fractions in *Phaseolus* roots in water and sand cultures supplied with different amounts of mineral nitrogen. *Dokl. TSKhA*, (70), 195–199.

Bond, G. 1936. Quantitative observations on the fixation and transfer of nitrogen in the soybean with especial reference to the mechanism of transfer of fixed nitrogen from bacillus to host. *Ann. Bot.*, **50**, 559–578.

Bond, G. 1941. Symbiosis of leguminous plants and nodule bacteria. *Ann. Bot.* New Series. **5**, 313–337.

Bond, G. 1950. Symbiosis of leguminous plants and nodule bacteria. 4. The importance of the oxygen factor in nodule formation and function. *Ann. Bot. New Series*. **15**, 95–108.

Bond, G. and Boyes, G. 1939. Excretion of nitrogenous substances from root nodules; observations on various leguminous plants. *Ann. Bot. New Series.* **3** (12), 901–913.

Bonnier, Ch. 1954. Action de la colchicine sur la symbiose Rhizobium, *Medicago sativa* et *Rhizobium trifolium*. *Bull. Inst. agron. et Stat. rech. Gembioux.*, **12** (3–4), 167–168.

Bonnier, Ch. 1955. La conservation du *Rhizobium* en sols steriles (*Rhizobium* specifique des Trifolium). *Bull. Inst. agron. et Stat. rech. Gembloux*, **23** (4), 359–367.

Bonnier, Ch. 1957. Inoculation bacterienne des graines de soja dans les conditions de la practique agricole. *Bull. Agron. Congo Belge*, (2), 87–92.

Bonnier, Ch. 1962a. Relation entre l'efficacite et specificite des souches de *Rhizobium*. *Ann. Inst. Pasteur*, **103** (3), 403–409.

Bonnier, Ch. 1962b. Anatomie comparée des nodosités à *Rhizobium* et des tumenrs d'aspect nodulaire, bacteriologique ment sterilis, des racines de Medicago sativa. *Bull. Inst. Agron. et Stat. rech. Gembloux*, **30** (1–2), 30–40.

Bonnier, Ch., Laloux, R., Lebrun, F. and Weickmans, L. 1964. Inoculation bacterience des graines de Lupin jaune doux (*Lupinus luteus*). *Bull. Inst. agron. Stat. rech. Gembloux*, **32** (2), 163–170.

Bonnier, Ch. and Lebrun, F. 1963. Methode de preinoculation des graines de legumineuses. *Ann. Inst. Pasteur*, **105** (2), 133–142.

Borodulina, Yu. S. 1951. Varietal features of the leguminous plant as a factor determining its infection with nodule bacteria. *Trudy Vses. n-i. in-ta s-kh. mikrobiologii*, **12**, 92–102.

Borodulina, Yu. S. 1953. Changes in the activity of nodule bacteria in crop rotation. *Trudy. Vses. n-i. in s-kh. mikrobiologii*, **13**, 108–113.

Borodulina, Yu. S. Lyophilic cultures of nodule bacteria, their activity and virulence. *Trudy. in-ta mikrobiol. Akad. Nauk SSSR*, (11), 211–214.

Borodulina, Yu. S. 1964. Drying nodule bacteria to obtain dry nitragin. in *Bakterial'nye udobreniya* (Bacterial fertilizers). Kolos, Moscow, 151–156.

Borodulina, Yu. S. and Samsonova, S. P. 1966. Effect of the age of the culture on the resistance of nodule bacteria when dried and stored: in *Ispol'zovanie mikroorganiz-mov dlya povysheniya urozhaya sel'skokhozyaistvennykh kultur* (Use of microorganisms for increasing the yield of farm crops), Kolos, Moscow, 147–151.

Bortels, H. 1930. Molybdän als Katalysator bei der biologischen Stickstoffbindung. *Arch. Mikrobiol.*, **1** (3), 333–342.

Bottomley, W. B. 1910. The assimilation of nitrogen by certain nitrogen-fixing bacteria in the soil. *Proc. Roy. Soc.*, B, **82**, 627.

Bowen, G. D. 1959. Specificity and nitrogen fixation in the *Rhizobium* symbiosis of *Centrosema pubescens*. *Benth. Queensl. J. Agric. Sci.*, **16**, 267–282.

Bowen, G. D. 1961. The toxicity of legume seed diffusates toward rhizobia and other bacteria. *Plant and Soil.*, **15**, 155–165.

Bowen, G. D. and Kennedy, M. M. 1959. *Queensl. J. Agric. Sci.*, **16**, 177 (quoted by Vincent, J. 1966a).

Bowen, G. D. and Kennedy, M. M. 1961. *Queensl. J. Agric. Sci.*, **18**, 161 (quoted by Vincent, J. 1966a).

Braithwaite, B. M., Jane, A. and Swain, F. G. 1958. *J. Austral. Inst. Agric. Sci.*, **24**, 155 (quoted by Vincent, J. 1966a).

Brakel, J. 1963. Action sur le Rhizobium de divers fongicides et insecticides commerciaux. *Ann. Inst. Pasteur*, **105** (2), 143–149.

Brakel, J. and Manil, P. 1965. La fixation symbiotique de l'azote chez le Haricot (*Phaseolus vulgarus* L.). *Bull. Inst. agron. Stat. rech. Gembloux*, **33** (1), 1–25.

Brenchley, W. E. and Thornton, H. G. 1925. The relation between the development, structure and functioning of the nodules on *Vicia faba* as influenced by the presence or absence of boron in the nutrient medium. *Proc. Roy. Soc.*, B, 98.

Brockwell, J. 1954. *J. Austral. Inst. Agric.*, **20**, 243–246 (quoted by van Schreven, D. 1958).

Brockwell, J. 1963. Accuracy of a plant infection technique for counting populations of *Rhizobium trifolii. Appl. Microbiol.*, **11** (5), 377–383.

Brockwell, J. and Hely, F. W. 1961. Symbiotic characteristics of *Rhizobium meliloti* in peat inoculant on lucerne seed. *Austral. J. Sci.*, **27** (11), 332–333; *Rhizobium Newsletter*, 1965, **10** (2), 194–195.

Brunchorst, J. 1885. Über die Knöllchen an den Leguminosenwurzeln. *Ber. Dtsch. bot. Ges.*, **3**, 241–257.

Bryantsev, B. A. 1966. *Sel'skokhozyaistvennaya entomologiya* (Agricultural entomology), Kolos, Moscow.

Bryan, O. C. 1923. Effect of acid soils on nodule-forming bacteria. *Soil Sci.*, **15**, 37–40.

Budinov, L. I. 1907. Nodule bacteria and clover fatigue. *Vestn bakterial'n. agron. stantsii*, (13), 17–109.

Bulard, C., Guichardon, B. and Rigaud, J. 1963. Mise en evidence de substances de nature auxinique synthetisées par *Rhizobium* cultivé en presence de tryptophane. *Ann. Inst. Pasteur*, **105** (2), 150–157.

Bulavskaya, E. S. 1959. Effect of molybdenum on the microflora of leguminous crops. *Trudy Gor'k. s.-kh. opytn. stantsii*, 118–123.

Bunting, A. and Horrocks, J. 1964. An improvement in the Raggio technique for obtaining nodules on excised roots of *Phaseolus vulgaris* in culture. *Ann. Bot.*, **28** (110), 229–237.

Burgess, P. S. and Pember, F. R. 1923. Active aluminium as a factor detrimental to crop production in many acid soils. *J. Agric. Exptl. Stat. Bull.*, **194**.

Bürgin-Wolff, A. 1959. Untersuchungen uber die Infektion von Wurzeln durch Knollchenbakterien. *Ber. Schweiz. bot. Ges.*, **69**, 75–111.

Burris, R. H., Eppling, T. S., Wahlin, H. B. and Wilson, R. W. 1943. Detection of nitrogen fixation with isotopic nitrogen. *J. Biol. Chem.*, **148**, 349–357.

Burris, R. H. and Haas, E. J. 1944. *J. Biol. Chem.*, **195,** 227 (quoted by Keilin, D. and Wang, C. 1945).

Burris, R. H. and Wilson, P. W. 1945. Biological nitrogen fixation. *Ann. Rev. Biochem.*, **14,** 685–708.

Burton, J. C. 1961. Seed pellet containing inoculant. US Patent No. 2995867.

Burton, J. C. 1965. The rhizobium-legume association: in *Microbiology and Soil Fertility*. Ed. by Gilmour, C. M. and Allen, O. N. (Oregon State University Press.

Burton, J. C. and Allen, O. N. 1950. Inoculation of crimson clover (*Trifolium incarnatum* L.) with a mixture of strains of *Rhizobium. Soil Sci. Soc. America Proc.*, **14,** 191–195.

Burton, J. C., Allen, O. N. and Berger, K. C. 1961. Effects of certain mineral nutrients on growth and nitrogen fixation of inoculated bean plants *Phaseolus vulgaris. J. Agric. Food Chem.*, **9** (3), 187–190.

Burton, J. C. and Briggeman, D. S. 1948. Similarity in response of *Trifolium* spp. to strains of *Rhizobium trifolii. Soil Sci. Soc. America Proc.*, **13,** 275–278.

Burton, M. O. and Locchead, A. G. 1952. Production of vitamin B_{12} by *Rhizobium* species. *Plant Physiol.*, **35,** 454–462.

Buśko, J. 1959. Symbiotic effectiveness and some physiological properties of the nodule bacteria of *Lotus* and *Anthylis. Acta microbiol. polon.*, **8** (3 4), 303–308 (in English).

Butler, G. W. and Bathurst, N. O. 1956. Underground transference of nitrogen from clover to associated grass. *Proc. Seventh International Grassland Congress.* Palmerston, New Zealand. Paper No. 14.

Butler, G. W. and Bathurst, N. O. 1958. Free and bound amino-acids in legume root nodules: bound—aminobutyric acid in the genus *Trifolium. Austral. J. Biol. Sci.*, **11** (4), 529–537.

Buyanova, N. D. and Kachan, T. A. 1965. Possible value of antibiotics in the production of dry nitragin: in *Rol' mikroorganizmov v pitanii rastenii i povyshenii effektivnosti udobreniya* (Role of micro-organisms in plant nutrition and in improving the effectiveness of fertilizers), Kolos, Moscow, 123–125.

Bystryi, N. F. 1941. Isolation of nodule bacteria from the soil by chemotaxis, *Mikrobiologiya*, **10,** (2), 247–249.

Cabezas de Herrera, E. 1956. Infection de la alfalfa cen tistintes estirpes de Rhizobium. *An. edafol. y fisiol. veget.*, **15,** 167–184.

Callao, V. and Olivares, J. 1964. Preparacion de anticuerpos fluorescentes anti-*Rhizobium leguminosarum. Microbiol. esp.*, **17** (4), 181–188 (summary in English).

Carlyle, R. and Thorpe, J. D. 1947. Some effects of ammonium and sodium 2,4-dichlorophenoxyacetates on legumes and the *Rhizobium* bacteria. *J. Amer. Soc. Agron.*, **39** (10), 929–936.

Carroll, W. R. 1894. A study of *Rhizobium* species in relation to nodule formation on the roots of Florida legumes. II—*Soil Sci.*, **37,** 227–241.

Cartwright, P. M. and Snow, D. 1962. The influence of foliar applications of urea on the nodulation pattern of certain leguminous species. *Ann. Bot. New Series.* **26** (102), 251–259.

Castelli, T. 1951. Some considerations of the symbiotic microbe of *Hedysarum coronarium* L.—*Contribs. Conf. Improvm. Pasture Fodder Productions in Mediterranean Area*, Rome, 1–8.

Chailakhyan, M. Kh. and Megrabyan, A. A. 1945. Effect of length of day on the formation of nodules in the roots of legumes. *Dokl. Akad. Nauk SSSR*, **18,** 457.

Chailakhyan, M. Kh. and Megrabyan, A. A. 1958a. Effect of root secretions of leguminous plants on the growth of nodule bacteria. *Izv. Akad. Nauk ArmSSR, ser. biol. i s-kh. nauki*, **11,** (8), 3–12.

Chailakhyan, M. Kh. and Megrabyan, A. A. 1958b. Stimulating effect of leguminous plants on the growth of nodule bacteria peculiar to them. *Dokl. Akad. Nauk ArmSSR*, **26**, (2), 103–111.

Chailakhyan, M. Kh., Megrabyan, A. A., Karapetyan, N. A. and Kaladzhyan, N. L. 1961. Effect of gibberellin and heteroauxin on the growth of leguminous plants and the formation of nodules. *Izv. Akad. Nauk ArmSSR*, **14** (12), 25–28.

Chailakhyan, M. Kh., Megrabyan, A. A., Karapetyan, N. A. and Kaladzhyan, N. L. 1963. Effect of growth activators on the formation of nodules and the growth of lucerne. *Dokl. Akad. Nauk ArmSSR*, **36** (3), 189–192.

Chailakhyan, M. Kh., Megrabyan, A. A., Karapetyan, N. A. and Kaladzhyan, N. L. 1965. Growth substances and secretions of nodule bacteria. *Dokl. Akad. Nauk ArmSSR*, **40**, (5), 307–314.

Chen, H. K. and Thornton, H. G. 1940. The structure of infective nodule and its influence on nitrogen fixation. *Proc. Roy. Soc.*, B, **129** (855), 208–229.

Cheniae, G. M. and Evans, H. J. 1957. On the relation between nitrogen fixation and nodule nitrate reductase of soybean root nodules. *Biochim. et biophys. acta*, **26**, 254–255.

Cheremisov, B. M. 1966. Increase in biological fixation by selection of nitrogen fixers. *Vestn. s-kh. nauki*, (9), 57–64.

Cherenkov, N. I. 1965. Accessibility to plants of the nitrogen of the root systems of leguminous and graminaceous grasses. *Agrokhimiya*, (1), 23–27.

Chizhik, G. Ya. 1959. Structure of nodule bacteria. *Mikrobiologiya*, **38** (1), 28–33.

Cloonan, M. J. 1963. Black nodules on *Dolichos*. *Austral. J. Sci.*, **26** (4), 121.

Cloonan, M. J. 1965. Thesis M. Sci. Agric. Univ. Sydney (quoted by Vincent, J. 1966*a*).

Cloonan, M. J. and Vincent, J. M. 1966. The nodulation of annual summer legumes sown on the far north coast of New South Wales. *Rhizobium Newsletter*, **11** (2), 161–167.

Colter, J. A. and Quastel, J. H. 1950. Catalytic decomposition of hydroxylamine by haemoglobin. *Arch. Biochem. and Biophys.*, **27**, 369–389.

Cook, F. D. and Quadling, C. 1962. Peroxide production by *Rhizobium melilotii*. *Canad. J. Microbiol.*, **8** (6), 933–935.

Damirgi, S. M. 1964. Serogroups of *Rhizobium japonicum* in Iowa soils as affected by different soil types, crop sequence and host genotype. *Diss. Abstr.*, **24**, 3907; *Rhizobium Newsletter*, 1965, **10** (1), 76–77.

Damirgi, S. M. and Johnson, H. W. 1966. Effect of soil actinomycetes on strains of *Rhizobium japonicum. Agron. J.*, **58** (2), 223–224.

Darbyshire. 1963. *Rothamsted Experimental Station Report* (quoted by Nutman, 1966).

Darbyshire. 1964. *Ann. Bot.* (quoted by Nutman, 1966).

Dart, P. J. and Mercer, F. V. 1963*a*. Membrane envelopes of legume nodule bacteroids. *J. Bacteriol.*, **85**, 951–952.

Dart, P. J. and Mercer, F. V. 1963*b*. Development of the bacteroid in the root nodule of barrel medic (*Medicago tribuloides* Desr.) and subterraneum clover (*Trifolium subterraneum* L.). *Arch. Mikrobiol.*, **46** (4), 382–401 (in English).

Dart, P. J. and Mercer, F. V. 1963*c*. The intracytoplasmic membrane system of the bacteroids of subterraneum clover nodules (*Trifolium subterraneum* L.). *Arch. Mikrobiol.*, **47** (1), 1–18.

Dart, P. J. and Mercer, F. V. 1964*a*. The legume rhizosphere. *Arch. Mikrobiol.*, **47** (4), 344–378.

Dart, P. J. and Mercer, F. V. 1964*b*. Fine structure changes in the development of the nodules of *Trifolium subterraneum* L. and *Medicago tribuloides* Desr. *Arch. Mikrobiol.*, **49** (3), 209–235.

Dart, P. J. and Mercer, F. V. 1965*a*. The effect of growth temperature, level of ammonium nitrate, and light intensity on the growth and nodulation of cowpea (*Vigna sinensis* Endl ex Hassk). *Austral. J. Agric. Res.*, **76** (3), 321–345.

Dart, P. J. and Mercer, F. V. 1965b. The influence of ammonium nitrate on the fine structure of nodules of *Medicago tribuloides* Desr. and *Trifolium subterraneum* L. *Arch. Mikrobiol.*, **51** (3), 233–257 (in English).

Dart, P. L. and Mercer, F. V. 1966. Fine structure of bacteroids in root nodules of *Vigna sinensis*, *Acacia longifolia*, *Viminaria juncea* and *Lupinus augustifolius*. *J. Bacteriol.*, **91** (3), 1314–1319.

Date, R. A. and Decker, A. M. 1965. Minimal antigenic constitution of 28 strains of *Rhizobium Japonicum*. *Canad. J. Microbiol.*, **11** (1), 1–8.

Date, R. A. and Murguia, J. L. 1965. The use of pellets in the overseeding of pasture legumes. *Rhizobium Newsletter*, **10** (2), 187–190.

Date, R. A. and Vincent, J. M. 1962. Determination of the number of root-nodule bacteria in the presence of other organisms. *Austral. Exptl. Agric. and Animal Husbandry*, 2.

Davis, R. J. 1958. A temperate bacteriophage associated with a strain of root nodule bacteria. *Bacteriol. Proc.*, **10**

Deherain, P. P. and Demoussy, E. 1900. Sur la culture des lupins blancs. *C. R. Acad. sci.*, **130**, 20–24.

Demidenko, T. T. and Timofeeva, E. F. 1937. Straw as a source of carbohydrate for nodule bacteria. *Dokl. Akad. Nauk SSSR*, New Series, **14**, 209–212.

Demolon, A. and Dunez, A. 1934. Le *Bac. radicicola* et son bacteriophage dans le developpement de la luzerne. *C.R. Acad. agric. franc.*, **20**, 659.

Demolon, A. 1961. *Rost i razvitie kul'turnykh. rastenii* (Growth and development of cultivated plants), Sel'khozgiz, Moscow.

Demolon, A., Rozowska, R. and Jacobelli, G. 1950. Observations biochimiques sur le developpement du *Bacterium Radicicola* (*Rhizobium leguminosarum*). *C.R. Acad. sci.*, **230**, 1015–1017.

Derx, H. G. 1953. *VI Congr. Internat. Microbiol. Riassunti Comm.*, 3, 116 (quoted by Jensen, H. L. 1958).

Diener, T. 1950. Phytophathol. Z., **16**, 129–170 (quoted by van Schreven, D. 1958).

Dilz, K. and Mulder, E. G. 1962. The effect of soil-*p*H, stable manure and fertilizer nitrogen on the growth of red clover and of red clover association with perennial ryegrass. *Netherlands J. Agric. Sci.*, **10** (1), 1–22.

Dinchev, D. 1962. Soil actinomycetes antagonistic to lucerne-nodule bacteria. *Izv. tsentral Int. Pochvoznan Agrotekh. Pushkarov*, **2**, 77–89 (in English).

Dinchev, D. 1959a. Nodule bacteria adsorption from the soil. *Nauchni trudove In-ta pochveni izsledovateno 'N. Pushkarov'*, **4**, Zemizdat, Sofia, 123–136.

Dinchev, D. 1959b. Effect of phosphorus fertilizers on preparation of nodule bacteria into the roots of beans. *Nauchni trudove In-ta pochveni izsledovaneto 'N. Pushkarov'*, **5**, Zemizdat, Sofia, 129–140.

Dinchev, D. 1961a. Nitrogen-fixing activity of bean nodule bacteria. *Izvestiya Tsentr. n.-i. in-ta pochvoznanie i agrotekhn.*, **1**, 127–156 (in Bulgarian).

Dinchev, D. 1961b. Distribution and virulence of alfalfa nodule bacteria in the soils of Bulgaria, *Izvestiya Tsentr. n.-i. in-ta pochvoznanie i agrotekhn.*, **1**, 29–54 (in Bulgarian).

Dinchev, D. 1962a. Soil actinomycetes antagonistic to alfalfa nodule bacteria. *Izvestiya Tsentr. n.-i. in-ta pochvoznanie i agrotekhn.*, **2**, 77–89 (in Bulgarian).

Dinchev, D. 1962b. Distribution of the alfalfa nodule bacteriophage in the soils of Bulgaria. *Izvestiya Tsentr. n.-i. in-ta pochvoznanie i agrotekhn.*, **4**, 187–194 (in Bulgarian).

Dinchev, D. 1962c. Characteristics of nitrogen fixed by bean nodule bacteria. *Izvestiya Tsentr. n.-i in-ta pochvoznanie i agrotekhn.*, **4**, 155–165 (in Bulgarian).

Dinchev, D. 1965. Influence of nodule bacteria on the nitrogen nutrition of plants. in *Plant-Microbe Relationships*. Prague, Czechosl. Acad. Sci., 251–255 (in English).

114   BIOLOGICAL FIXATION OF ATMOSPHERIC NITROGEN

Dixon, R. O. D. 1964. The structure of infection threads, bacteria and bacteroids in pea and clover root nodules. *Arch. Mikrobiol.*, **48** (2), 166–178 (in English).

Döbereiner, J. 1965. Black nodules on *Cetrosema pubescens*. *Soil Biol. Internat. News Bull.* No. 2: *Rhizobium Newsletter*, **10** (2), 196–197.

Döbereiner, J. 1966. Evaluation of nitrogen fixation in legumes by the regression of total plant nitrogen with nodule weight. *Nature*, **210**, 5038, 850–851.

Döbereiner, J., Arruda, N. B. and Penteado, A. F. 1965. Problemas da inoculacao de soja em solos acides: in *IX Internat. Grassland Congr. Sao Paulo, Brazil; Rhizobium Newsletter*, **10** (2), 197–198.

Dobrokhleb, I. F. and Tymchenko, L. F. 1963. Effect of seed disinfectants on the nodulation and yield of the edible lupin. *Zashchita rastenii ot vreditelei i boleznei*, (10), 28; *Trudy Bryansk. gos. s-kh. opytn. stantsii*, (2), 60–65.

Dobson, S. H. and Lovvorn, R. L. 1949. Inoculation of legumes. *North Carolina State Univ. Extension Circular*, **309**, 4.

Donald, C. M. 1965. The progress of Australian agriculture and the role of pastures in environmental change. *Austral. J. Sci.*, **27**, 187–198.

Doolas, G. L. 1930. Local variation of soil acidity in relation to soybean inoculation. *Soil Sci.*, **30**, 273–287.

Dorosinskii, L. M. 1941. Influence of bacteriophage on development of clover. *Mikrobiologiya*, **10**, (2), 208–215.

Dorosinskii, L. M. 1962. Bacterial fertilizers. in *Ispol'zovanie mikroorganizmov v sel'skom khozyaistve* (Use of micro-organisms in agriculture), Sel'khozizdat, Moscow, 169–190.

Dorosinskii, L. M. 1965a. *Bakterial'nye udobreniya—dopolnitel'noe sredstvo povysheniya urozhaya* (Bacterial fertilizers—a new way of increasing the harvest). Rossel'khozizdat, Moscow.

Dorosinskii, L. M. 1965b. Nitragin and its application. in *Uspekhi mikrobiologii* (Advances in microbiology). Nauka, Moscow, 145–169.

Dorosinskii, L. M. and Avrov, O. E. 1964. Effect of seed disinfection on the effectiveness of nitragination. *Izv. Akad. Nauk SSSR*, (6), 920–923.

Dorosinskii, L. M., Zagor'e, I. V. and Buziashvili, D. M. 1966. Determination of nitrogen-fixing capacity of nodule bacteria from enzymatic activity. *Mikrobiologiya*, **35**, (2), 319–322.

Dorosinskii, L. M. and Lazareva, N. M. 1956. Specificity of certain species of nodule bacteria: in *IX Mezhdunarodnyi Mikrobiologicheskii kongress* (Ninth International Microbiology Congress), summaries of speeches, Moscow, 290.

Dorosinskii, L. M., Lazareva, N. M. and Emtsev, V. T. 1962. Role of nodule bacteria in the nitrogen nutrition of leguminous plants. *Mikrobiologiya*, **31**, (6), 1062–1066.

Dorosinskii, L. M. and Lampovshchikov, P. K. 1948. Effect of low temperatures on the efficacy of nitrogen: in *Puti povysheniya aktivnosti kluben'kovykh bakterii* (Ways of increasing the activity of nodule bacteria), Sel'khozgiz, Moscow, 95–100.

Doyarenko, A. G. 1966. *Faktory zhizni rastenii. Izbrannye proizvedeniya* (Factors in the life of plants. Selected works). Kolos, Moscow, 210.

Dovgan', T. I. and Krongauz, E. A. 1965. Preliminary experiments in the production of dry nitragin in the Kiev Bacterial Preparations Plant: in *Rol' mikroorganizmov v pitanii rastenii i povyshenii effektivnosti udobrenii* (Role of micro-organisms in plant nutrition and increasing the effectiveness of fertilizers), Kolos, Moscow, 119–122.

Drobkov, A. A. 1945. Importance of radioactive elements in the development of nodule bacteria and their assimilation of atmospheric molecular nitrogen. *Dokl. Akad. Nauk SSSR*, new series, **49**, (3), 227–229.

Drożańska, D. 1963. Antigenic relationships among strains of *Rhizobium*. *Acta microbiol. polon.*, **12** (3), 163–164 (in English).

Drożańska, D. 1965. Studies of antigens of *Rhizobium trifolii. Acta microbiol. polon.*, 14 (3–4), 275–280 (in English).

Drozdowicz, A. 1959. Resistance of plants to infection by an unspecified trace of *Rhizobium. Acta microbiol. polon.*, 8 (3–4), 295–297, 297–298

Du Buy, H. G. and Showackre, J. L. 1961. Selective localization of tetracycline in mitochondria of living cells. *Science*, 133 (3447), 196–197.

Dudman, W. F. 1964*a*. Growth and extracellular polysaccharide production by *Rhizobium meliloti* in defined medium. *J. Bacteriol.*, 88 (3), 640–645.

Dudman, W. F. 1964*b*. Immune diffusion analysis of the extracellular soluble antigens of two strains of *Rhizobium meliloti. J. Bacteriol.*, 88 (3), 782–794.

Duggar, J. F. 1897. Soil inoculation for leguminous plants. *Alabama Agric. Exp. Stat. Bull.*, 87, 459–488.

D'yadkova, V. E. and Borodulina, Yu. S. 1960 *Nauchnye trudy po izvestkovaniyu dernovo-podzolistykh pochu* (Effect of liming in enhancing the activity of nodule bacteria of red clover and lucerne) in Research on Liming of soddy-podzolic soils. *Minsk, Akademiya sel'sko-khozydistvennykh nauk BSSR*, 120–125.

Dygdała, K. 1963. Influence of irradiation on the morphology and cytology of *Rhizobium. Acta microbiol. polon.*, 12, 155–157 (in English).

Dzhordan, G. (Jordan, G.), Garcia, M. and Garrard, E. H. 1966. Role of herbicides in nitrogen fixation and formation of nodules on bird's-foot trefoil: in *IX Mezhdunarodnyi Mikrobiologicheskii kongress (Ninth International Microbiology Congress)*, summaries of speeches, Moscow, 289.

Efendiev, T. A. and Khapilov, G. R. 1965. Effect of molybdenum on the yield and nutrient value of broad beans in grey-earth soils. *Agrokhimiya*, (10), 149–151.

Eaton, E. 1931. Effects of variation in day length and clipping of plants on nodule development growth of soy bean. *Bot. Gaz.*, 91, 113.

Ebertova, J. 1960. Amino acids in the nodules, aerial parts and roots of soya in the course of symbiosis with *Rhizobium japonicum. Sbor. Českosl. akad. zeměd. věd. Rostl. Výroba*, 6 (10), 1371–1382.

Egle, K. and Munding, H. 1953. Haemoglobin in Pflanzen. *Natur and Volk*, 83 (7), 220–226.

Egle, K. and Munding, H. 1954. *Biol. Zbl.*, 73, 577–602 (quoted by van Schreven, D. 1958).

Egorova, S. V. 1955. *Vliyanie prodolzhnitel'nosti prebyvaniya kluben'kovykh bakterii v pochve vne kornevoi sistemy hobovogo rasteniya na ikh fiziologicheskuyu aktivnost'* (Effect of time spent in the rhizosphere on the physiology of nodule bacteria of leguminous plants) (Author's abstract of M.Sc. thesis).

Elkan, G. H. 1962. Comparison of rhizosphere micro-organisms of genetically related nodulating and non-nodulating soybean lines. *Canad. J. Microbiol.*, 8 (1), 79–87.

Ellfolk, N. 1959. Crystalline laeghemoglobin. *Acta chem. scand.*, 13, 596–597 (in English).

Ellfolk, N. 1960. Crystalline laeghemoglobin. II. The molecular weight and shapes of the two main components. *Acta chem. scand.*, 14, 1819–1827 (in English).

Ellis, N. J., Kalts, G. G. and Doncaster, J. J. 1962. Transformation in *Rhizobium trifolii. Canad. J. Microbiol.*, 8, 835.

Elsukov, V. P. 1961. *Boby kormovye* (The broad bean). Izd-vo Min-va sel'sk. kh-va RSFSR.

Emtsev, V. T. 1962. Selection, variation and storage of cultures of micro-organisms used to produce bacterial fertilizers. *Izv. Akad. Nauk SSSR, Ser. Biol*, (6), 732–739.

Emtsev, V. T., Shil'nikova, V. K. and Agadzhanyan, K. G. 1964. Natural infection of broad and runner beans in black-earth soils of the Tula region and the Kamen steppe. *Dokl. TSKhA*, (99), 389–394.

Emtsev, V. T., Shil'nikova, V. K. and Gromyko, E. P. 1963. Natural inoculation of broad and runner beans in soddy-podzolic soils. *Izv. TSKhA*, (4), 55–64.

Erdman, L. W. 1961. The future of preinoculated seed. *Seed World*, **88** (5), 12–17.

Erdman, L. W. 1965. Inoculants at the crossroads. A presentation as a contribution from the SWCKD, ARS, USDA to the Legume Inoculation Conference. O'Hare Holiday Inn, Chicago, Ill. Feb. 18 (quoted by Allen, O. 1966).

Erdman, L. W., Johnson, H. W. and Clark, F. E. 1956. A bacterial-induced chlorosis in the Lee soybean. *Plant Disease Report*, **40**, 646.

Erdman, L. W. and Means, U. 1956. Strains of rhizobia effective on *Trifolium ambiguum*. *Agron. J.*, **48** (8), 341–343.

Erdman, L. W. and Wilkins, F. S. 1928. Soybean inoculation studies. *Iowa Agric. Exptl. Stat. Res. Bull.*, **114**.

Erkama, J. 1947. Uber die Holle von Kulfer und Mangan im Leben der Hoheren Pflanzen. *Ann. Acad. Sci. Fennicae, ser. A*, 2, Chemica, 25–27, 1–105.

Evans, A. M. and Jones, D. G. 1966. The response to inoculation of the three chromosome races of *Trifolium ambiguum* sown with and without a companion grass. I. The effect of inoculation on yield of clover and grass. *J. Agric. Sci.*, **66** (3), 315–320.

Evans, M. W. 1916. Some effects of legumes on associated non-legumes. *J. Amer. Soc. Agron.*, **8**, 348–357.

Ezedinma, F. O. C. 1964a. Effect of inoculation with local isolates of cowpea *Rhizobium* and application of nitrate nitrogen on the development of cowpeas. *Trop. Agric.*, 243–249; *Rhizobium Newsletter*, 1965, **10** (1), 80–81.

Ezedinma, F. O. C. 1964b. Notes on the distribution and effectiveness of cowpea *Rhizobium* in Nigerian soils. *Plant and Soil*, **21** (1), 134–136.

Ezubchik, A. A. 1966. Role of nitragin in increasing the yield of leguminous crops. *Viesci Akad. Navuk BSSR, Sier. s-kh.*, (3), 28–32.

Fähraeus, G. 1957. The infection of clover root hairs by nodule bacteria studied by a simple glass slide technique. *J. Gen. Microbiol.*, **16** (2), 374–381.

Falk, J. E., Appleby, C. A. and Porra, R. J. 1959. The nature, function and biosynthesis of the haem compounds and porphyrins of legume root nodules. *Sympos. Soc. Exptl. Biol.*, **13**, 73–86.

Fan' Yun'-lyu and Khuan Sin-fu (Fan Yün-lu and Huang Xing-fu). 1965. Requirements for growth substances and the effect of amino acids, vitamins and purine and pyrimidine bases of nucleic acids on the growth of *Rhizobium astragali*. *Acta microbiologica sinica*, **11**, (2), 185–194 (in Chinese; Russian summary).

Fan' Yun'-lyu (Fan Yün-lu). 1965. Effect of temperature on the formation of nodules and symbiotic fixation of nitrogen in *Astralagus sinicus L*. *Acta microbiologica sinica*, **2**, (4), 480–517 (in Chinese; Russian summary).

Fedorov, E. A. 1955. *O roli gemoglobina v kluben'kakh kornei bobovykh rastenii* (Role of haemoglobin in the root nodules of leguminous plants). (Author's abstract of M.Sc. thesis). Moscow.

Fedorov, M. V. 1948. *Fiksatsiya azota atmosferi chistymi kul'turami Bacterium radicicola. Puti povysheniya aktivnosti kluben'kovykh bakterii* (Fixation of atmospheric nitrogen by pure cultures of *Bacterium radicicola*. Ways of increasing the activity of nodule bacteria). Sel'khozgiz. Moscow.

Fedorov, M. V. 1952. *Biologicheskaya fiksatsiya azota atmosfery* (Biological fixation of atmospheric nitrogen). Sel'khozgiz. Moscow.

Federov, M. V. and Glavachkova, E. V. 1956. Nitrogen-fixing capacity of lucerne nodule bacteria. *Izv. TSKhA*, (3), 61–78.

Fedorov, M. V. and Egorova, S. V. 1957. Effect of soil conditions on the virulence and nitrogen-fixing activity of clover nodule bacteria. *Izv. TSKhA*, (2), 98–110.

Fedorov, M. V. and Kozlov, I. V. 1954. Effect of bound nitrogen compounds on the nitrogen-fixing activity of nodule bacteria in soya and pea nodules and the relationships between them and the legume. *Mikrobiologiya*, **23**, (5), 534–543.

Fedorov, M. V. and Laslo, D. 1956. Nitrogen fixing activity of pea and vetch nodule bacteria in different phases of development of the nodules of leguminous plants. *Izv. TSKhA*, (2), 61–82.

Fedorov, M. V. and Pod"yapol'skaya, V. P. 1951. Effect of the conditions of cultivation of leguminous plants on the formation of nodules and the yield of plants. *Dokl. Akad. Nauk SSSR*, **77**, (1), 121–124.

Fedorov, M. V. and Svitych, K. A. 1954. Specificity of races of lupin nodule bacteria. *Dokl. Akad. Nauk SSSR*, **97**, (6), 1069–1072.

Fedorov, M. V. and Svitych, K. A. 1959. Effectiveness of various races of lupin nodule bacteria living together with different lupin varieties. *Izv. TSKhA*, (6), 39–44.

Fedorov, M. V. and Uspenskaya, T. A. 1955. Effect of duration of illumination of legumes (pea and soya) on the nitrogen-fixing activity of nodule bacteria in nodules. *Mikrobiologiya*, **24**, (3), 291–302.

Fedorova, L. N. 1966. Effect of varietal features of the pea on the rate of nodule formation. *Agrokhimiya i pochvovedenie*, (4), 31–40.

Filippova, K. F. 1953. Effect of salination on the nitrogen-fixing capacity of nodule bacteria. *Izv. Estestv.-nauchn. in-ta, pri Permskom gos. in-te*, **13**, (7), 661–679.

Filippova, K. F. 1958. Study of certain serological features of lucerne nodule bacteria. *Izv. Estestv.-nauchn. in-te pri Permskom gos in-te*, **14**, (2), 51–56.

Filippova, F. K. 1964. Effect of boron and molybdenum on the biological fixation of nitrogen by lucerne. *Agrokhimiya*, (3), 150–155.

Fletcher, W. W. 1956. Effect of hormone herbicides on the growth of *Rhizobium trifolii*. *Nature*, **177** (4522), 1244.

Fletcher, W. W. 1958. *Phyton*, **10**, 129–134 (quoted by van Schreven, D. 1958).

Fletcher, W. W. and Alcorn, J. W. S. 1958. The effect of translocated herbicides on *Rhizobia* and the nodulation of legumes: in *Nutrition of the Legumes*. Ed. by Hallsworth, E. G. London, 284–288.

Fletcher, W. W., Dickenson, P. B. and Raymond, J. C. 1956. *Phyton*, **7** (2), 121–130 (quoted by Fletcher, W. and Alcorn, J. 1958).

Fletcher, W. W. and Smith, J. 1964. Herbicides and soil microbes. *New Scientist*, **24** (418), 527–528.

Fogle, C. E. and Allen, O. N. 1948. Associative growth of actinomycetes and rhizobia. *Bacteriol. Proc.*, **53**.

Fonbrune, P. de. 1949. Technique de micromanipulation. Masson et Cie, Paris. 204.

Forsyth, W. G. C., Hayward, A. C. and Roberts, J. B. 1958. Occurrence of poly-β-hydroxybutyric acid in aerobic Gram-negative bacteria. *Nature*, **128**, 800–801.

Fottrell, P. F. 1966. Dehydrogenase isoenzymes from legume root nodules. *Nature*, **210** (5032), 198–199.

Fottrell, P. F., O'Connor, S. and Masterson, C. L. 1964. Identification of the flavanol myricetin in legume seeds and its toxicity in nodule bacteria. *Irish J. Agric. Res.*, **3** (2), 247–249.

Fottrell, P. and Graims, E. (Fottrell, P. and Grimes, E.). 1966. Dehydrogenase and amino-transferases in the root nodules of legumes and in *Rhizobium*: in *IX Mezhdunarodnyi Mikrobiologicheskii kongress* (Ninth International Microbiological Congress), Summaries of speeches. Moscow, 293.

Foy, C. D. and Brown, J. C. 1964. Toxic factors in acid soils. 2, Differential aluminium tolerance of plant species. *Soil Sci. Soc., America Proc.*, **28**, 27–32.

Frank, B. 1889. Über die Pilzsymbiose der Leguminosen. *Ber. Dtsch. bot. Ges.*, **7** 332–346.

Frank, B. 1892. Die Assimilation freien Stickstoffs bei den Pflanzen in ihrer Abhangigkeit von Species von Ernahrungsverhaltnissen und Bocenarten. *Landwirtsch. Jahrb.*, **21**, 1–14.

Fred, E. B., Baldwin, I. L. and McCoy, E. 1932. Root Nodule Bacteria and Leguminous Plants. Madison.

Fred, E. B. and Graul, E. J. 1916. The effect of soluble nitrogenous salts on nodule formation. *J. Amer. Soc. Agron.*, **8**, 316–328.

Fred, E. B., Whiting, A. L. and Hastings, E. G. 1926. Root nodule bacteria of leguminosae. *Wisconsin Agric. Exptl. Stat. Res. Bull.*, **72**, 1–43.

Frey-Wyssling, A. 1952. Growth of plant cell walls. *Symposia Soc. Exptl. Biol.*, (4), 320–328.

Galli, F. 1959. Caracteres culturais das bacterias dos nodulos de algumas leguminosas tropicais. *Anais Escola Super. Agric. Luiz. de Queiroz*, **16**, 113–122 (in Portuguese, summary in English).

Garkavenko, A. I. 1962. *Izuchenie aktivnykh i maloaktivnykh shtammov kluben'kovykh bakterii* (Study of active and slightly active strains of nodule bacteria) (Author's abstract of M.Sc. Thesis).

Gäumann, E., Jaag, O. and Roth, S. 1945. Über einen Immunisiernugsversuch mit Wurzelknollchanbakterien bei Leguminosen. *Ber. Schweiz. bot. Ges.*, **55**, 270–277.

Gelin, O. and Blixt, S. 1964. Root nodulation in peas. *Agri. Hort. Genet.*, **22** (1–2), 149–159.

Gel'tser, F. Yu. 1948. New way of obtaining active nodule bacteria (bacterial nurseries). in *Puti povysheniya aktivnosti kluben'kovykh bakterii* (Ways of increasing the activity of nodule bacteria). Sel'khozgiz, Moscow, 42–55.

Generalov, G. F. 1957. Results of quality control of nitragin in state crop testing stations. *Zemledelie*, (2), 50–52.

Georgiu, V., Negranu, V. and Dumitru, I. (Gheorghiu, V., Negreanu, V. and Dumitru, I.). 1964. Production, use and effectiveness of certain biopreparations in the Rumanian People's Republic. in *Bakterial'nye udobreniya* (Bacterial fertilizers), Kolos, Moscow, 66–78.

Georgi, C. E., Orcutt, F. C. and Wilson, P. W. 1933. Further studies on the relation between carbon assimilation and nitrogen fixation in the leguminous plant. *Soil Sci.*, **36**, 375–382.

Gerretsen, F. C. 1937. Manganese deficiency of oats and its relation to soil bacteria. *Ann. Bot.*, **1** (2), 207–230.

Gershon, D. 1961. Genetic studies of effective nodulation in *Lotus* spp. *Canad. J. Microbiol.*, **7** (6), 961–963.

Gibson, A. H. 1963. Physical environment and symbiotic nitrogen fixation. I. The effect of root temperature on recently nodulated *Trifolium subterraneum* L. plants. *Austral. J. Biol. Sci.*, **16** (1), 28–42.

Gibson, A. H. 1965. Physical environment and symbiotic nitrogen fixation. II. Root temperature effects on the relative nitrogen assimilation rate. *Austral. J. Biol. Sci.*, **18** (2), 295–310.

Gibson, A. H. 1966. Physical environment and symbiotic nitrogen fixation. III. Effects of root temperature on shoot and root development and nitrogen distribution in *Trifolium subterraneum*. *Austral. J. Biol. Sci.*, **19** (2), 219–232.

Giltner, W. and Langworthy, H. W. 1916. Some factors influencing the longevity of soil micro-organisms subjected to desiccation, with special reference to soil solution. *J. Agric. Res.*, **5**, 927–942.

Giöbel, G. 1926. The relation of soil nitrogen to nodule development and nitrogen fixation by certain legumes. *New Jersey Agric. Exptl. Stat. Bull.*, **436**.

Girenko, L. 1965. Nitragin for soya. *Zernobobovye kul'tury*, (4), 12–13.

Gladkii, M. F. and Panovskii. 1965. Use of molybdenum fertilizers for growing lucerne on leached black earth. *Khimiya v sel'skom khozyaistve*, (11), 14–15.

Glukhovtsev, V. V. 1966. Nekotorye voprosy biologii i agrotekhniki vozdelyvaniya gorokha i chiny v usloviyakh Kuibyshevskoi oblasti (Some aspects of the biology and agriculture of peas in the Kuibyshev region). (Author's abstract of M.Sc. thesis).

Glyan'ko, A. K. and Shevchuk, V. E. 1965. Effects of trace elements under sugar beat, potato and leguminous crops in the Irkutsk region. *Izv. Irkutsk. s.-kh. in-ta*, 3, (25), 139–147.

Gołębiowska, J. 1965. The influence of fungicides on symbiosis of leguminous plants with bacteria. *Pamiftnik Pulawski*, **18**, 368–382 (in English).

Gołębiowska, J. and Kaszubiak, H. 1965. Sensitivity of *Rhizobium* to the action of thiuram and phenylmercuric acetate. *Ann. Inst. Pasteur*, **109** (3). Suppl. 153–160 (in English).

Gołębiowska, J. and Sypniewska, U. 1962a. The effect of the plant and of ecological conditions on development on symbiosis between lupine and *Rhizobium lupini*. *Acta Microbiol. Polon.*, **11**, 319–328 (in English).

Gołębiowska, J. and Sypniewska, U. 1962b. Studies on the development cycle of *Rhizobium lupini* in root nodules. *Acta Microbiol. Polon.*, **11** (4), 313–318 (in English).

Gołębevska, Yu., Kashubyak, Kh. and Paevska (Gołębiowska, J., Kaszubiak, H. and Pajewska). Adaptation of nodule bacteria to thiuram: in *IX Mezhdunarodnyi Mikrobiologicheskii kongress* (Ninth International Microbiology Congress), summaries of speeches, Moscow, 296.

Golod, B. I. 1966. Effect of various methods of introducing straw on the yield of legumes and the fixation of atmospheric nitrogen by nodule bacteria. *Dokl. TSKhA*, (119), 245–248.

Goodchild, D. J. and Bergersen, F. J. 1966. Electron microscopy of the infection and subsequent development of soybean nodule cells. *J. Bacteriol.*, **92** (1), 204–213.

Gordeev, A. M. 1964. Effect of certain agricultural techniques in increasing the productivity and microflora of eroded soils. *Trud. Vses. n-i in-ta udobrenii i agropochvoved*, (43), 122–129.

Goss, O. M. and Shipton, W. A. 1965. Nodulation of legumes on new light land. 4. Rhizobial strains for pasture establishment. *J. Dept. Agric. West. Austral.*, **6** (11), 663–664, 667–669.

Gostkowska, K. 1963. Effect of amino acids on symbiotical activity of *Rhizobium*. *Acta Microbiol. Polon.*, **12** (3), 158–162.

Graham, P. H. 1963a. Antigenic affinities of the root nodule bacteria of legumes. Antonie van Leeuwenhoek. *J. Microbiol. Serol.*, **29**, 281–291.

Graham, P. H. 1963b. Vitamin requirements of root nodule bacteria. *J. Gen. Microbiol.*, **30**, 245–248.

Graham, P. H. 1963c. Antibiotic sensitivities of the root nodule bacteria. *Austral. J. Biol. Sci.*, **16** (2), 557–559.

Graham, P. H. 1964. The application of computer techniques to the taxonomy of the root-nodule bacteria of legumes. *J. Gen. Microbiol.*, **35**, 511–517.

Graham, P. H. 1965. Extracellular polysaccharides of the genus *Rhizobium*. Antonie van Leeuwenhoek, *J. Microbiol. Serol.*, **31** (4), 349–354.

Graham, P. H. and Parker, C. H. 1964. Diagnostic features in the characterisation of the root-nodule bacteria of legumes. *Plant and Soil*, **20** (3), 383–396.

Graham, P. H., Parker, C. A., Oakley, A. E., Lange, R. T. and Sanderson, I. J. V. 1963. Spore formation and heat resistance in *Rhizobium*. *J. Bacteriol.*, **86** (6), 1353–4354.

Grijns, A. 1927. Clover plants in sterile cultivation do not produce a bacteriophage of *Bact. radicicola*. *Zbl. Bakteriol. Parasitenkunde, Infektionskrankh und Hyg.*, Abt. 2, **71**, 248–251.

Grilli, M. 1963. Osservazioni sui rapporti tra cellule ospiti e rizobi nei tubercoli radicali di pisello (*Pisum sativum*). *Caryologia*, **16** (3), 561–594.

Gritsun, A. T. 1964. Biological nitrogen used in farming in the Primor'e. *Agrokhimiya*, (6).

Grobbelaar, N., Van Beijma, M. C. and Saubert, S. 1964. Addition to the list of nodule bearing legume species. *South Afric. J. Agric. Sci.*, **7**, 265–270.

Gromyko, E. P. 1966. Aluminii kak toksicheskii faktor dlya mikroorganizmov. dernogopodzolistoi pochvy Podmoskov'ya (Aluminium as a toxic factor for microorganisms of soddy-podzolic soils of the Moscow Region. (Author's abstract of M.Sc. thesis).

Gukova, M. M. 1945. Effect of soil temperature on nitrogen fixation by nodule bacteria. *Trudy TSKhA*, (30), 33–42.

Gukova, M. M. 1962. Temperature dependence of symbiotic assimilation of nitrogen by leguminous plants. *Izv. Akad. Nauk SSSR*, (6), 832–839.

Gukova, M. M. and Bogomolova, R. I. 1963. Effect of nutrition on growth and accumulation of nitrogen by broad beans in mixed sowing with maize. *Dokl. Akad. Nauk SSSR*, **149**, (3), 725–727.

Gulyakin, I. V., Gukova, M. M. and Bogomolova, R. I. 1963a. Effect of nutrition on yield and content of nitrogen in maize and broad beans used in mixed sowings. *Izv. TSKhA*, (3), 7–16.

Gulyakin, I. V., Gukova, M. M. and Morugina, M. P. 1965a. Role of potassium in nitrogen assimilation by leguminous crops. *Dokl. TSKhA*, (103), 141–145.

Gulyakin, I. V., Gukova, M. M. and Morugina, M. P. 1965b. Comparative effect of mineral and organic fertilizers on the yield and quality of broad beans. *Dokl. TSKhA*, (103), 191–195.

Gulyakin, I. V., Gukova, M. M. and Sindyashkina, R. I. 1963b. Effect of nitrogen supply to medium on nitrogen fixation by the broad bean. *Dokl. TSKhA*, (94), 5–9.

Gupta, B. M. and Kleczkowska, J. 1962. A study of some mutations in a strain of *Rhizobium trifolii*. *J. Gen. Microbiol.*, **27**, 473–476.

Gupta, K. G. and Sen, A. 1964. Efficiency of isolates of *Rhizobium leguminosarum* (pea) in relation to their size. *Science and Culture*, **30** (7), 348–349.

Gupta, K. G. and Sen, A. 1965a. Relationship between nitrogen-fixing efficiency of *Rhizobium* spp. and their capacity to consume glucose and utilize phosphate. *Indian J. Agric. Sci.*, **35** (1), 39–42.

Gupta, K. G. and Sen, A. 1965b. The relationship between glucose consumption and efficiency of *Rhizobium* spp. from some common cultivated legumes. *Plant and Soil.*, **22** (2), 229–238.

Haak, A. 1964. Über den Einflu der Knöllchenbakterien auf die Wurzelhaare von Leguminosen und Nichtleguminosen. *Z. Bacteriol. Parasitenkunde, Infektionskrankh und. Hyg.*, Abt. 2, **117** (4), 343–366.

Halbinger, R. E. 1965. The microbiological utilization of atmospheric nitrogen. *Ann. Inst. Pasteur*, **109** (3) Suppl., 161–166 (in English).

Hallsworth, E. G. 1958. Nutritional factors affecting nodulation: in *Nutrition of the Legumes*. Ed. by Hallsworth, E. G. London, 183–201.

Hallsworth, E. G., Wilson, S. B. and Greenwood, E. A. N. 1960. Copper and cobalt in nitrogen fixation. *Nature*, **187**, 79–80.

Hamatová-Hlaváčková, (Gamatova-Glavachkova) E. 1961. Effect of penicillin on symbiosis of nodule bacteria with leguminous plants. *Za sots. s-kh. nauku*, **10**, (6), 537–544.

Hamatová-Hlaváčková, (Gamatova-Glavachkova) E. 1962. Investigation of the use of special bacteria for increasing the yield of agricultural crops using nodule bacteria. *Českosl. Akad. zeměd. věd výzkumný ústav rostl. výrobý Praha—Ruzyně*.

Hamatová-Hlaváčková, (Gamatova-Glavachkova) E. 1963. Effect of the host on several physiological properties of *Rhizobium meliloti*. *Publ. Fac. Sci. Univ.*, **30** (448), 432–434.

Hamatová-Hlaváčková (Gamatova-Glavachkova) E. 1963 The specifity of *Rhizobium japonicum* towards various soya varieties. *Sbor. Českosl. akad. zeméd. véd. Rostl, Výroba*, **36,** 728–731 (in English).

Hamatová-Hlaváčková (Gamatova-Glavachkova) E. 1964. Influence of gamma radiation on the nitrogen fixation activity of *Rhizobium japanocum* in symbiosis with two varieties of soybeans. *Acta Microbiol. Polon.*, **13,** (3), 247–254 (in English).

Hamatová-Hlaváčková, (Gamatova-Glavachkova) E., C. Sc. a kol. 1964. Investigation of the use of nodule bacteria for increasing the yield of leguminous plants. *Ústředni výzkumný ústav rostlinné vyroby v Prazei-Ruzyni Sektor vyzivy rostlin.*

Hamatová-Hlaváčková, (Gamatova-Glavachkova) E. 1965. The effectiveness of nitrogen fixation of *Rhizobium meliloti* after several years in contact with different varieties of *Medicago sativa*: in *Plant-Microbes Relationships*. Prague, Czechosl. Acad. Sci., 256–263 (in English).

Hamatová-Hlaváčkova (Gamatova-Glavachkova) E. 1965) Effect of gamma radiation on the fixation activity of *Rhizobium japonicum* in two soya varieties. *Vědecké Práce Ústředniho výz Ústavu Rostl. Výroby v Praze-Ruzyni* 8, Prague, 155–162 (in English).

Hamatová-Hlaváčkova (Gamatova Glavachkova) E. and Marečková, H. 1966. Die Herstellung und Anwendung von Leguminosenimpfstoffen in der Tschechoslowalei. *Hellriegel Symposium*, 4–6 July, 1966.

Hamatová-Hlaváčková, (Gamatova-Glavachkova) E. 1966. The complex of strains of *Rhizobium* used to produce inocula in the Czechoslovak Socialist Republic, and their partial characterization. *Vědecke Práce Ústředniho Výz. Ústavu Rostl. Výr. v Praze-Ruzni,* 10, Prague, 95–109 (in English).

Hamdi, A. M. Y. 1966. Efficiency of strains of *Rhizobium meliloti* as affected by sulfur-containing amino-acids. *Diss. Abstr.*, **26** (9), 4968.

Harmsen, G. W. 1965. Selection of races of *Rhizobium* for improving the inoculation of the Papilionaceae. *Landbouwkund. tijdschr.*, **77** (21), 811–822 (in Dutch, summary in English).

Harmsen, G. W. and Wieringa, K. T. 1954. Attempts to improve the inoculation of legumes. *Landbouwkund. tijdschr.*, **66,** 531–533.

Harris, J. O., Allen, E. K. and Allen, O. N. 1949. Morphological development of nodules on *Sesbania grandiflora* Poir with reference to the origin of nodule rootlets. *Amer. J. Bot.*, **36,** 651–661.

Harris, J. R. 1953. Influence of rhizosphere micro-organisms on the virulence of *Rhizobium trifolii*. *Nature*, **172** (4376), 507–508.

Harris, J. R. 1954. Rhizosphere relationships of subterranean clover. 1. *Austral. J. Agric. Res.*, **5** (2), 247–270.

Harrison, F. and Barlow, B. 1907. The nodule organism of the Leguminosae, its isolation, cultivation, identification and commercial application. *Zbl. Bakteriol., Parasitenkunde, Infektionskrankh. und. Hyg.*, Abt. 2, **19,** 264–272 (in English).

Harworth, N. and Stacey, M. 1948. The chemistry of the immunopolysaccharides. *Annual Rev. Biochem.*, **17,** 97–114.

Hattingh, M. J. and Louw, H. A. 1964. The antagonistic and stimulatory effects of soil micro-organisms on rhizobia of clover. *South Afric. J. Lab. and Clin. Med.*, **10** (1), 32.

Hauke-Pacewiczowa, T. 1952. Inoculants for *Galega officinalis*. *Acta Microbiol. Polon.*, **1** (1), 36–38 (in English).

Hedlin, R. A. and Newton, J. D. 1948. Some factors influencing the growth and survival of rhizobia in humus and soil cultures. *Canad. J. Res.*, sec. C., **26** (2), 174–187.

Hely, F. W. 1957. *Austral. J. Biol. Sci.*, **10,** 1 (quoted by Evans and Jones, 1966).

Hely, F. W. 1963. *CSIRO. Austral. Div. Plant Industr. Field Station Record*, **2**, 89 (quoted by Vincent, J. 1966*a*).

Hely, F. W. 1964. Calcium carbonate as an aid to clover establishment in soils with low winter temperatures. *Commonwealth Sci. Industr. Res. Organisation. Austral. Div. Plant. Ind. Field Stat. Res.*, **2**, 63–68.

Hely, F. W. 1965. Survival studies with *Rhizobium trifolii* on seed of *Trifolium incarnatum* L. inoculated for aerial sowing. *Austral. J. Agric. Res.*, **16** (4), 575–590.

Hely, F. W., Bergersen, F. J. and Brockwell, J. 1957. Microbial antagonism in the rhizosphere as a factor in the failure of inoculation of subterranean clover. *Austral. J. Agric. Res.*, **8** (1), 24–44.

Hely, F. W. and Brockwell, J. 1962. An exploratory survey of the ecology of *Rhizobium meliloti* in inland New South Wales and Queensland. *Austral. J. Agric. Res.*, **13** (5), 864–879.

Hely, F. W. and Williams, J. D. 1964. Influence of root temperature on the early nodulation of two varieties of *Trifolium subterraneum* L. *CSIRO. Austral. Div. Plant. Ind. Field Stat. Res.*, **3**, 45–54.

Hertogh, A. A. de, Mayeux, P. A. and Evans, H. J. 1964. Effect of cobalt on the oxidation of propionate by *Rhizobium meliloti*. *J. Bacteriol.*, **87**, (3), 746–747; *J. Biol. Chem.*, **239**, 2446–2453.

Heuman, W. 1952*a*. Über Wesen and Bedeutung der Bakteroide in den Wurzelknöllchen der Erbse. *Naturwissenschaften*. **39**, 66.

Heuman, W. 1952*b*. Über das Abhängigkeitverhältnis zwischen Hamoglobin-, Stärke-, Bakteroidvorkommen und Stickstoffbindung in den Wurzelknöllchen der Erbse. *Naturwissenschaften*, **39**, 67.

Heuman, W. 1952*c*. Physiologische und morphologische Studien an *Rhizobium leguminosarum* in Knöllchen und auf verschiedenen Nährböden. *Ber. Dtsch. bot. Ges.*, **65**, 229–233.

Heuman, W. 1954. Die Bakteroidbildung von *Rhizobium leguminosarum* in Mischkultur mit einem anderen Bakterium. *Naturwissenschaften*, **41**, 192.

Hewitt, E. 1958. Discussion. in Nutrition of the Legumes. Ed. by Hallsworth, E. G. London, 215.

Hewitt, E. J., Bolle-Jones, E. N. and Miles, P. 1954. Induction of copper, zinc and molybdenum deficiency in crop plants grown in sand culture with special reference to some effects of water supply and seed reserves. *Plant and Soil*, **5** (3), 205–222.

Higashi, S. 1966. Electron microscopic studies on the infection thread developing in the root hair of *Trifolium repens* L. infected with *Rhizobium trifolii*. *J. Gen. and Appl. Microbiol.*, **12** (12), 147–156.

Hilder, E. J. and Spencer, K. 1954. *J. Austral. Inst. Agric. Sci.*, **20**, 171–176 (quoted by van Schreven, D. 1958).

Hiltner, L. 1898. Über Entstehung und physiologische Bedentung der Wurzelknöllchen. *Forstl. naturwiss. Z.*, **7**, 415–423.

Hiltner, L. and Störmer, K. 1903. Neue Untersuchungen über die Wurzelknöllchen der Leguminosen und deren Erreger. *Arb. K. Gesundsamt., Biol. Abt.*, (3), 151–307.

Hofer, A. W. 1954. Description of a virus that destroys the root nodule bacteria of peas. *Soil. Sci.*, **77** (6), 435–436.

Hofer, A. W. and Crosier, W. F. 1962. Preinoculated alfalfa seed. *Agron. J.*, **54** (2), 97–100.

Hoffman, F. W. 1927. Reciprocal effects from grafting. *J. Agric. Res.*, **34** (7).

Hoffmann, G. 1964. Effektivität und Wirtsspezifität der Knollchenbakterien von *Robinia pseudoacacia* L. *Arch. Forstwesen*, **13** (6), 563–576.

Holding, A. J. and King, J. 1963. The effectiveness of indigenous populations of *Rhizobium trifolii* in relation to soil factors. *Plant and Soil*, **18** (2), 191–198.

Holding, A. J., Tilo, S. N. and Allen, O. N. 1960. Modified plant responses induced by *Rhizobia* cultivated on amino-acid media. *Trans. Seventh Internat. Congr. Soil Sci.*, Madison, **21**, 608–616.

Holland, A. A. and Parker, C. A. 1966. Studies on microbial antagonism in the establishment of clover pasture. 2. *Plant and Soil*, **25** (3), 329–341.

Hopkins, C. G. 1904. Nitrogen bacteria and legumes. *Illinois Agric. Stat. Bull.*, **94**, 307–328.

Humphrey, B. A. and Vincent, J. M. 1959. Extracellular polysaccharides of *Rhizobium*. *J. Gen. Microbiol.*, **21**, 477–484.

Humphrey, B. and Vincent, J. M. 1962. Calcium in cell walls of *Rhizobium trifolii*. *J. Gen. Microbiol.*, **29**, 557–581.

Humphrey, B. A. and Vincent, J. M. 1965. Effect of calcium nutrition on the production of diffusible antigens by *Rhizobium trifolii*. *J. Gen. Microbiol.*, **41**, 109–118.

Humphrey, B. A. and Vincent, J. M. 1966. Strontium as a substituted structural element in cell walls of *Rhizobium*. *Nature*, **212** (5058), 212–213.

Hunt, G. E. 1951. A comparative chromatographic survey of the amino acids in five species of legume roots and nodules. *Amer. J. Bot.*, **38**, 452–457.

Hutchinson, C. M. 1924. *Report of the Imperial Agricultural Bacteriologist. 3. Soil Biology.* Sci. Rept. Agric. Res. Inst. Pisa, 1923–1924, 32 (quoted by Jensen, H. and Schroder, M. 1965).

Hyde, B. B., Hodge, A. H. and Birnsteil, M. L. 1962. Phytoferritin: a plant protein discovered by electron microscopy: in *Electron microscopy* 2_1 Ed. by Bresse, S. S. *Proc. Fifth Internat. Congr. Electron Microscopy*, Philadelphia. NY. Acad. Press.

Il'in, S. S. 1939. Effect of mineral fertilizers on development of nodule bacteria and the yield of legumes. *Khimizatsiya sots. zemledeliya*, (7), 47–49.

Ionescu, M. 1958. Activity and virulence of nodule bacteria in different phases of vegetation of vetch and soya. in *Probl. pedol. Bucureşti*, Acad. RPR., 181–187.

Ireland, J. A. 1966. Effect of the level of inoculation and lime pelleting on competition or nodulation of subterranean clover. *Rhizobium Newsletter*, **11** (2), 170–175.

Israilski, V. P. and Starygina, L. P. (Izrail'skii, V. P. and Starygina, L. P.). 1930. Die Dissoziation bei einigen Bakterienarten. *Zentralbl. Bakteriol. Parasitenkunde, Infektionskrankh, und Hygiene*, Part 2, **81**, 1–11.

Itano, A. and Matsuura, A. 1936. Studies on the nodule bacteria. VII. Influence of the extract of nodules on the growth of nodule bacteria. *Ber. Ohara Inst. landwirtsch. Forsch. Kurashiki Japan*, **7**, 379–401 (in English).

Itano, A. and Matsuura, A. 1938. Studies of the nodule bacteria. X. Influence of some stimulating chemicals with special reference to the alkaloids upon the growth and morphology of the nodule bacteria. *Ber. Ohara Inst. landwirtsch. Forsch. Kurashiki Japan*, **8**, 53–68 (in English).

Ivanoff, S. S. 1948. Chlorosis and nodulation of cowpeas as affected by trial applications of sulphur to calcareous soil in the greenhouse. *Plant Physiol.*, **23** (1), 162–164.

Ivanova, L. E. and Sakharnova, M. V. 1965. Microflora of the root system of the broad bean. *Trudy Gor'kovsk. s-kh. in-ta*, **16**, 90–99.

Izrail'skii, V. P. 1928. Races of nodule bacteria. *Vestn. Bakteriol. agron. stantsii.* (25).

Izrail'skii, V. P. 1951. Storage of dried nodule bacteria. *Trudy Vses. n-i. in-ta s-kh. mikrobiologii*, **12**, 171–174.

Izrail'skii, V. P. 1953. Filtrable forms of nodule bacteria. *Mikrobiologiya*, **32**, (6), 645–651.

Izrail'skii, V. P. and Artem'eva, Z. V. 1937. Virulence and activity of nodule bacteria and the immunity of leguminous plants. *Trudy Vses. in-ta udobrenii i agropochvoved*, **2**, (15), 5–22.

Izrail'skii, V. P. and Minyaeva, O. M. 1954. Combined pre-sowing treatment of clover seeds with granosan and nitragin. *Dokl. VASKhNIL*, (3), 39–42.

Izrail'skii, V. P., Runov, E. V. and Bernard, V. V. 1933. *Kluben'kovye bakterii i nitragin* (Nodule bacteria and nitragin), Sel'khozgiz, Moscow-Leningrad.

Izrail'skii, V. P., Ryzhkova, A. S. and Prisyagina, M. G. 1964. Serological studies of nodule bacteria. *Dokl. TSKhA*, (107), 105–113.

Jacks, H. 1956. Effect of seed dressings on damping and nodulation of lucerne. *NZ J. Sci. Technol.*, A.38 (3).

Jackson, C. R. 1965. Peanut-pod mycoflora and kernel infection. *Plant and Soil*, 23 (2), 203–212.

Jacobs, S. E. 1949. The relationship of *Corynebacterium fasciens* (Tilford) Dowson, to the bacteria causing fall and nodule formation. *Proc. Fourth Internat. Congr. Microbiol. Congr. Copenhagen*, 1947, sect. IV, 425.

Jaffe, A., Weyer, F. and Saubert, S. 1961. The role of root temperature in symbiotic nitrogen fixation. *South Afric. J. Sci.*, **51**, 278–280.

Jakubisiak, B. and Gołębiowska, J. 1963. Influence of fungicides on *Rhizobium*. *Acta Microbiol. Polon.*, **12** (3), 196–202 (in English).

Jenkins, H. V., Vincent, J. M. and Water, L. M. 1954. Root nodule bacteria as factors in the establishment of clover in the red basaltic soils of the Lismore districts, New South Wales. 3. Field inoculation trials. *Austral. J. Agric. Res.*, **5**, 77–89.

Jensen, H. L. 1943. *Proc. Linnean Soc. New South Wales*, **68**, 207–220 (quoted by van Schreven, D. 1958).

Jensen, H. L. 1951. The coryneform bacteria. *Annual Rev. Microbiol.*, **6**, 77–90.

Jensen, H. L. 1958. The classification of the bacteria. in Nutrition of the Legumes. Ed. by Hallsworth, E. G. London, 75–86.

Jensen, H. L. 1961. Survival of *Rhizobium meliloti* in soil culture. *Nature*, **192** (4803), 682–683.

Jensen, H. L. 1962. Viability of lucerne nodule bacteria in soil culture. *Tijdsskr. Planteavl.*, A65 (4), 704–715 (in Danish).

Jensen, H. L. 1964. On the relation between hosts and root nodule bacteria in certain leguminous plants. *Tijdsskr. Planteavl.*, A68 (1), 1–22 (in Danish: summary in English).

Jensen, H. L. and Koumaran, K. A. 1965. Transformation microbiologique du biuret. *Ann. Inst. Pasteur*, **109** (3) Suppl., 184–190.

Jensen, H. L. and Petersen, H. I. 1952. Decomposition of hormone herbicides by bacteria. *Acta Agric. Scand.*, **2** (3), (1–3), 215–231 (in English).

Jensen, H. L. and Schroder, M. 1965. Urea and biuret as nitrogen sources for *Rhizobium* Sapp. *J. Appl. Bacteriol.*, **28** (3), 473–478.

Johnson, H. W. 1963. Cooperative soybean breeding research. *Soybean Digest*, 23 (11), 79–83.

Johnson, H. W. and Means, U. M. 1960. Interaction between genotypes of soybeans and genotypes of nodulating bacteria. *Agron. J.*, **52**, 651–654.

Johnson, H. W. and Means, U. M. 1963. Serological groups of *Rhizobium japonicum* recovered from nodules of soybeans (*Glycine max*) in field soils. *Agron. J.*, **55** (3), 269–271.

Johnson, H. W. and Means, U. M. 1964. Selection of competitive strains of soybean nodulating bacteria. *Agron. J.*, **56** (1), 60–62.

Johnson, H. W., Means, U. M. and Clark, F. E. 1958. Factors affecting the expression of bacterial induced chlorosis of soybeans. *Agron. J.*, **50**, 571–575.

Johnson, H. W., Means, U. M. and Clark, F. W. 1959. Responses of seedlings to extracts of soy bean nodules bearing selected strains of *Rhizobium japonicum*. *Nature*, **183**, 308–309.

Johnson, H. W., Means, U. M. and Weber, C. R. 1965. Competition for nodule sites between strains of *Rhizobium japonicum* applied as inoculum and strains in the soil. *Agron. J.*, **57** (2), 179–184.

Johnson, C. M., Stout, P. R., Broyer, T. C. and Carlton, A. B. 1957. Comparative chlorine requirements of different plant species. *Plant and Soil*, **8** (4), 337–353.

Jones, D. G. 1963. Symbiotic variation of *Rhizobium trifolii* with S.100 Nomark white clover (*Trifolium repens* L.). *J. Sci. Food and Agric.*, **14** (10), 740–743.

Jones, D. G. and Evans, A. M. 1966. The response to inoculation of the three chromosome races of *Trifolium ambiguum* sown with and without a companion grass. II. The effect of the method of inoculation on the clover and grass. *J. Agric. Sci.*, **66** (3), 321–326.

Jones, D. G., Munro, J. M. M., Hughes, R. and Davies, W. E. 1964. The contribution of white clover to a mixed upland sward. I. The effect of *Rhizobium* inoculation on the early development of white clover. *Plant and Soil*, **21**, 63–69.

Jordan, D. C. 1952. Studies of legume root nodule bacteria. II. Production and behaviour of colonial mutants produced by X-irradiation. III. Acquisition of growth factors. *Canad. J. Bot.*, **30**, (125), 693–700.

Jordan, D. C. 1955. Observation on the enzymatic degradation and conversion of certain L- and D-amino acids by an effective strain of *Rhizobium meliloti*. *Canad. J. Microbiol.*, **1** (9), 743–748.

Jordan, D. C. 1962. The bacteroides of the genus *Rhizobium*. *Bacteriol. Rev.*, **26** (2), 119–141.

Jordan, D. C. and Coulter, W. H. 1965. On the cytology and synthetic capacities of natural and artificially produced bacteroides of *Rhizobium leguminosarum*. *Canad. J. Microbiol.*, **11** (4), 709–720.

Jordan, D. C. and Garrard, E. H. 1951. Studies of legume root nodule bacteria. I. Detection of effective and ineffective strains. *Canad. J. Bot.*, **29**, 360–372.

Jordan, D. C. and Grinyer, I. 1965. Electron microscopy of the bacteroids and root nodules of *Lupinus luteus*. *Canad. J. Microbiol.*, **11**, 721–725.

Jordan, D. C., Grinyer, I. and Coulter, W. H. 1963. Electron microscopy of infection threads and bacteria in young root nodules of *Medicago sativa*. *J. Bacteriol.*, **86** (1), 125–137.

Jordan, D. C. and San Clemente, C. L. 1955a. Utilization of peptides and L and D-amino acids by effective and ineffective strains of *Rhizobium meliloti*. *Canad. J. Microbiol.*, **1**, 659.

Jordan, D. C. and San Clemente, C. L. 1955b. The utilization of purines, pyrimidines and inorganic nitrogenous compounds by effective and ineffective strains of *Rhizobium meliloti*. *Canad. J. Microbiol.*, **1**, 668.

Juroszek, J. 1963. Nitrogen production (in the symbiotic relationship of *Rhizobium* with leguminous plants). *Acta Microbiol. Polon.*, **12**, 208–210 (in English).

Kalinskaya, T. A. and Il'ina, T. K. 1965. Effect of bound forms of nitrogen and supplementary growth factors on the fixation of nitrogen by Mycobacteria: in *Rol' mikroorganizmov v pitanii rastenii i povyshenii effektivnosi udobrenii* (Role of microorganisms in plant nutrition and in increasing the efficacy of fertilizers). Leningrad, 54–60.

Kalnin'sh, A. D. 1951. Distribution and activity of clover nodule bacteria in the soil of the Latvian SSR: in *Nauchnaya sessiya po voprosam biologii i sel'skogo khozyaistva* (Scientific Meeting on Problems of Biology and Agriculture). *Akad. Nauk LatvSSR*, 54–67.

Kalnin'sh, A. D. 1952. Bacteriophage of nodule bacteria and its effect on the symbiosis of nodule bacteria and leguminous plants. *Trudy In-ta Mikrobiologii AN LatvSSR*, (1), 47–84.

Kalnin'sh, A. D. 1958. Distribution of nodule bacteria of clover in the soils of the Latvian SSR: in *Poluchenie i primenenie bakterial'nykh udobrenii* (Production and application of bacterial fertilizers), Akad. Nauk Ukr.SSR, Kiev, 181–189.

Kalnin'sh, A. D. 1961. Multiplication of random races of nodule bacteria of clover in the soil and rhizosphere of clover and their participation in the formation of nodules: in *Voprosy sel'skokhozyaistvennoi mikrobiologii* (Problems of agricultural microbiology), (5), Riga, 3–21.

Kalnin'sh, A. D. 1964. Serological features of nodule bacteria of clover: in *Mikroorganizmy i rasteniya* (Micro-organisms and plants), 2, Riga, 3–21.

Kalnin'sh, A. D. 1966. Phage of nodule bacteria and its importance in nitrogen fixation in: *Biologicheskii azot i ego rol' v zemledelenii* (Biological nitrogen and its role in farming). Akad. Nauk. SSSR, Moscow.

Kal'ninsh, A. D., Klasens, V. P. and Leimane, I. Ya. 1966. Activity and specificity of nodule bacteria in: *IX Mezhdunarodnyi Mikrobiologicheskii Kongress* (Ninth International Microbiology Congress) summaries of speeches. Moscow, 290–291.

Kamata, E. 1962. Morphological and physiological studies on nodule formation in leguminous crops. 7. Variation in the nodule ability in some strain of *Rhizobium japonicum*. *Proc. Crop. Sci. Soc. Japan*, 37 (1), 78–82 (in English).

Kaszubiak, H. 1964. Spartein-decomposing micro-organisms in the rhizosphere of lupine. *Acta Microbiol. Polon.*, 14 (1), 101–107 (in English).

Kaszubiak, H. 1965. Effect of tingitanine (lathyrine), diaminobutyric acid and homoserine on *Rhizobium*. *Acta Microbiol. Polon.*, 14 (3–4), 309–314 (in English).

Kaszubiak, H. 1966. The effect of herbicides on *Rhizobium*. I. Susceptibility of Rhizobium to herbicides. *Acta Microbiol. Polon.*, 15 (4), 357–365 (in English).

Katznelson, H. 1939. Bacteriophage and the legume bacteria. *Proc. Third Internat. Congr. Soil Sci.*, A, 43–48.

Katznelson, H. and Zagallo, A. C. 1957. Metabolism of rhizobia in relation to effectiveness. *Canad. J. Microbiol.*, 3 (6), 879–884.

Kaufman, D. D. 1964. Microbial degradation of 2,2-dichloropropionic acid in five soils. *Canad. J. Microbiol.*, 10 (6), 843–852.

Kecskés, M. and Manninger, E. 1962. Effect of antibiotics on the growth of rhizobia. *Canad. J. Microbiol.*, 8 (1), 157–159.

Kedrov-Zikhman, O. K. 1951. *Izvestkovanie kislykh dernovo-podzolistykh pochv Latviiskoi SSR* (Liming of acid soddy-podzolic soils of the Latvian SSR, Riga).

Kefford, N. P., Brockwell, J. and Zwar, J. A. 1960. The symbiotic synthesis of auxin by legume and nodule bacteria and its role in nodule development. *Austral. J. Biol. Sci.*, 13 (4), 456–467.

Keilin, D. and Wang, C. L. 1945. Haemoglobin in the root nodules of leguminous plants. *Nature*, 155 (3930), 227–228.

Kellerman, K., Leonard, L. 1913. The prevalence of *Bacillus radicicola* in soil. *Science*, 38, 95–98.

Kern, H. 1965. Untersuchungen zur genetischen Transformation zwischen *Agrobacterium tumefaciens* und *Rhizobium* spec. 1, 2. *Arch. Mikrobiol.*, 51 (2), 140–155; 52 (3), 206–224.

Kerpely, A., Zamory, B. E. and Manninger, E. 1963. Effect of inoculation of legumes with *Rhizobium* strain labelled with phosphorus-32. *Nature*, 198 (4886), 1219.

Khatipova, Kh. M. 1958. Importance of the times of disinfecting seeds with granosan on introduction of bacterial fertilizers. *Byull. Ukr. in-ta oroshaemogo zemledeliya*, (5).

Kirchner, O. 1895. Die Wurzelknöllchen der Soyabohnen. *Beitr. Biol. Pflanzernahr*, 7, 213–224.

Kir'yanova, E. S. 1961. *Nekotorye problemy nematodologii rastenii, pochvy i nasekomykh* (Aspects of the nematology of plants, soils and insects). Samarkand University Press (quoted in Atakhanov, Sh. A., 1964).

Kishinovskii, B. A. 1966. Effectiveness of inoculation of the pea in relation to soil moisture in: *Ispol'zovanie mikroorganizmov dlya povysheniya urozhaya sel'skokhozyaistvennykh kultur* (Use of micro-organisms for increasing the yield of farm crops), Kolos, Moscow, 141–146.

Kleczkowska, J. 1945. A quantitative study of the interaction of bacteriophage in poured plates and the value as a counting method. *J. Bacteriol.*, **50**, 71.

Kleczkowska, J. 1950. A study of phage resistant mutants of *Rhizobium trifolii*. *J. Gen. Microbiol.*, **4**, 298–310.

Kleczkowska, J. 1957. A study of the distribution and effects of bacteriophage of root nodule bacteria in the soil. *Canad. J. Microbiol.*, **3** (2), 171–180.

Kleczkowska, J. 1965. Mutations in symbiotic effectiveness in *Rhizobium trifolii* caused by transforming DNA and other agents. *J. Gen. Microbiol.*, **40** (3), 377–384.

Kleczkowska, J. and Kleczkowski, A. 1952. The effect of specific polysaccharide from the host bacteria and of ribonuclease on the multiplication of *Rhizobium* phages. *J. Gen. Microbiol.*, **7**, 340–350.

Kleczkowski, A. and Kleczkowski, J. 1951. The ability of single phage particles to form plaques and multiply in liquid cultures. *J. Gen. Microbiol.*, **5** (2), 346–356.

Kleczkowski, J. and Kleczkowski, A. 1953. The behaviour of *Rhizobium* bacteriophages during and after exposure to ultraviolet radiation. *J. Gen. Microbiol.*, **8** (1), 135–144.

Kleczkowski, J. and Kleczkowski, A. 1954. The effect of ribonuclease on phage-host interaction. *J. Gen. Microbiol.*, **11** (3), 451–458.

Kleczkowski, J. and Kleczkowski, A. 1956. Effects of clupein and its degradation products on a *Rhizobium* bacteriophage, on its host bacterium and on the interaction between the two. *J. Gen. Microbiol.*, **14** (2), 449–459.

Kleczkowski, J. and Kleczkowski, A. 1959. The effect of infection with bacteriophage on the electrokinetic potential of *Rhizobium leguminosarum*. *J. Gen. Microbiol.*, **21** (2), 308–311.

Kleczkowski, J. and Kleczkowski, A. 1965. Inactivation of a *Rhizobium* bacteriophage by ultraviolet radiation of different wavelengths. *Photochem. Photobiol.*, **4** (2), 201–207.

Kleczkowski, A. and Thornton, H. G. 1944. A serological study of the root nodule bacteria from pea and clover. *J. Bacteriol.*, **48**, 661–672.

Klein, D. T. and Klein, R. M. 1953. Transmittance of tumor-inducing ability to a virulent crown-gall and related bacteria. *J. Bacteriol.*, **66** (2), 220–228.

Kliewer, M. 1961. The effects of various combinations of molybdenum, aluminium, manganese, phosphorus, nitrogen, calcium, hydrogen ion concentration, lime and *Rhizobium* strain on growth, composition and nodulation of several legumes. Ph.D. Thesis, Cornell University, Ithaca, New York.

Kliewer, M. and Evans, H. 1962. B_{12} coenzyme content of the nodules from legumes, alder and of *Rhizobium meliloti*. *Nature*, **194** (4823), 108–109; **195** (4843), 828.

Kliewer, M. and Evans, H. J. 1963a. Cobamide coenzyme contents of soybean nodules and nitrogen fixing bacteria in relation to physiological conditions. *Plant Physiol.*, **38** (1), 99–104.

Kliewer, M. and Evans, H. J. 1963b. Identification of cobamide coenzyme in nodules of symbionts and isolation of the B_{12} coenzyme from *Rhizobium meliloti*. *Plant Physiol.*, **38** (1), 55–59.

Kliewer, M., Lowe, R., Mayeux, P. A. and Evans, H. J. 1964. A biological assay for cobalt using *Rhizobium meliloti*. *Plant and Soil*, **21** (2), 153–162.

Klintsare, A. Y. 1958. Relationships of nodule bacteria and certain groups of soil micro-organisms. *Trudy In-ta mikrobiol. Akad. Nauk LatvSSR*, **7**, 77–83.

Klintsare, A. Ya. 1959a. Change in the effectiveness of symbiosis of *Rhizobium meliloti* in relation to the presence in the soil of micro-organisms influencing their growth. *Trudy In-ta mikrobiol. Akad. Nauk LatvSSR*, (8), 97–104.

Klintsare, A. Ya. 1959b. Effect of surface improvement of carbons on the relationships of nodule bacteria and certain groups of soil micro-organisms. *Trudy. In-ta mikrobiol. Akad. Nauk. LatvSSR*, (8), 105–126.

Klintsare, A. Ya. 1961. Effectiveness of nitragination of legume seeds disinfected with mercuran and TMTD. *Trudy In-ta Mikrobiol.*, (14), Riga.

Klintsare, A. Ya. 1962. Local active races of nitrogen-fixing bacteria and bacterization of disinfected seeds. in *Mikroorganizmy i sreda* (Micro-organisms and the medium), *Izd-vo* Akad. Nauk LatvSSR, Riga.

Klintsare, A. Ya. 1963. Effect of boron and molybdenum on the effectiveness of symbiosis of nodule bacteria and the pea. *Trudy In-ta mikrobiol., Akad. Nauk LatvSSR*, (18), 17–30.

Klintsare, A. Ya. and Klintsare, Ya. 1965. Effect of boron and molybdenum on the symbiosis of nodule bacteria and the pea. *Izv. Akad. Nauk LatvSSR*, (8), 108–114.

Klintsare, A. Ya. and Kreslinya, D. Ya. 1963. Effect of trace elements on the development of nitrogen-fixing bacteria in the rhizosphere of farm plants: in *Pochvennaya i sel'skokhozyaistvennaya mikrobiologiya* (Pedological and agricultural microbiology), 155–162.

Kłosowska, T. 1952. Development of the symbiosis of *Rhizobium leguminosarum* with peas. B.A. Thesis (quoted by Wróbel, T. 1956).

Kluyver, A. J. and van Niel, C. B. 1936. Prospects for a natural system of classification of bacteria. *Zbl. Bakteriol., Parasitenkunde, Infectionskrankh. und Hyg.*, **94**, 369–403.

Knösel, D. 1962. Prüfung von Bakterien auf Fähigkeit zur Sternbildung. *Zbl. Bakteriol., Parasutenkunde, Infektionskrankh. und Hyg.*, Abt. 2, **116** (1), 79–100.

Knyazeva, T. I. 1966. Specificity of certain bacteriophages of clover nodule bacteria: in *Ispol'zovanie mikroorganizmov dlya povysheniya urozhaya sel'skokhozyaistvennykh kultur* (Use of micro-organisms for improving the yield of farm crops). Kolos, Moscow, 136–140.

Kobus, J. 1952. Some morphological and physiological characteristics of the symbiotic bacteria of the Papilionaceae. *Acta Microbiol. Polon.*, **1** (2), 137–150.

Kolosova, N. A. 1965. Nitrogen-fixing capacity of the lupin. *Agrokhimiya*, (6), 61–66.

Kon, S. K. 1955. Other factors related to B_{12}. *Biochem. Soc. Symp.*, **13**, 17–35.

Konishi, K. 1931. Effect of certain soil bacteria on the growth of the root nodule bacteria. *Mem. Coll. Agric. Kyoto Imp. Univ.*, **16**, Ser. Chem., 10.

Konishi, K. and Fukuchi, R. 1935. Effect of certain actinomyces on the growth of the root nodule bacteria. *J. Sci. Soil Manure Japan*, **9**.

Konishi, K., Tsuga, T. and Kawamura, A. 1936. *J. Sci. Soil Tokyo*, **10**, 396–400 (quoted by Van Schreven, D. 1958).

Konokotina, A. G. 1934. Morphological changes in bacteria in the nodules of the chick pea and lupin. *Mikrobiologiya*, (3), 221.

Konokotina, A. G. 1936. Biological factors in the variation of nodule bacteria. *Arkhiv biol. nauk.*, **43**, (2–3), 129–138.

Koontz, F. P. and Faber, J. E. 1961. Somatic antigens of *Rhizobium japonicum*. *Soil Sci.*, **91** (4), 228–232.

Korenyako, A. I. 1942. Effect of root secretions on the development of nodule bacteria. *Mikrobiologiya*, **11**, (3).

Kornilov, A. A. and Verteletskaya, V. 1952. Penetration of sanfoin into dry Steppe regions and the role of nodule bacteria. *Mikrobiologiya*, **20**, (4), 423–428.

Korovin, A. I. and Vorob'ev, V. A. 1965. Effect of low soil temperatures when growth begins on nitrogen fixation by broad beans in relation to doses of nitrogen fertilizers. *Fiziol. rast.*, (6), 1083–1086.

Korsakova, M. P. and Lopatina, G. V. 1934. Relationships between nodule bacteria and leguminous plants. *Mikrobiologiya*, **3**, (2), 204.

Kossowitsch, P. 1892. Durch welche Organe nehmen die Leguminosen den freien Stickstoff auf? *Bot. Z. So.*, 697, 713, 729, 745, 771.

Kováčiková, E. and Ujević, J. 1965. The steeping of lentil seeds against fungal diseases and its influence on the formation of nodules on the plant roots. *Vědecké Práce Ústředniho Výz. Ústavu Rostl. Výroby v Praze-Ruzyni,* 9, Prague, 83–90.

Kowalski, M., Staniewski, K. and Paraniak, M. 1963*a*. The effects of osmotic shock and ultraviolet radiation on *Rhizobium* bacteriophages. *Acta Microbiol. Polon.,* **78** (3), 175–183 (in English).

Kowalski, M., Staniewski, K. and Ziemiecka, J. M. 1963*b*. Recent Polish studies on Rhizobiophages. *Ann. Inst. Pasteur,* **105** (2), 237–241 (in English).

Kozlova, E. I. and Dikareva, T. A. 1963. Effect of herbicides on the microflora of the rhizosphere of certain farm plants. *Agrobiologiya,* (1), 82–87.

Kozlov, I. V. 1962. Effect of bound nitrogen compounds on the nitrogen-fixing activity of nodule bacteria. *Vestn. s-kh. nauki,* (2), 49–54.

Kravtsov, P. V. Effect of weak electric current on the multiplication and activity of nodule bacteria. *Dokl. VASKhNIL,* **11**, 14–16.

Krasil'nikov, N. A. 1941. Variation in nodule bacteria. *Mikrobiologiya,* **10**, (4), 396; *Dokl. Akad. Nauk SSSR,* **31**, (1), 90–92.

Krasil'nikov, N. A. 1945. Grafting of new virulence properties by nodule and certain non-nodule bacteria. *Mikrobiologiya,* **14**, (4), 230–236.

Krasil'nikov, N. A. 1949. *Opreditel' bakterii i aktinomitsetov* (Classification of bacteria and actinomycetes), Akad. Nauk SSSR, Moscow.

Krasil'nikov, N. A. 1958. *Mikroorganizmy pochy i vysshie rastenii* (Soil micro-organisms and higher plants), Akad. Nauk SSSR, Moscow.

Krassil'nikov, N. A. and Asseyva, I. W. (Krasil'nikov, N. A. and Aseeva, I. V.). 1959. The influence of soil bacteria on the free amino acid content of papilionaceous plants. *Folia Microbiol.,* **4**, (1), 45–50 (in English).

Krasil'nikov, N. A., Drobkov, A. A., Shirokov, O. G. and Shevyakova, N. I. 1955. Effect of radioactive elements on the development of nodule bacteria and *Azotobacter.* Trace Elements. *Vsas. sovashchenie po mikroelementam* (All-Union Conference on Trace Elements), summaries of speeches, Riga.

Krasil'nikov, N. A. and Korenyako, A. I. 1940. Methods of quantitative estimation of nodule bacteria in the soil. *Mikrobiologiya,* **9**, (1), 27–31.

Krasil'nikov, N. A. and Korenyako, A. I. 1944. Effect of soil microflora on the virulence and activity of nodule bacteria. *Mikrobiologiya,* **13**, (1), 39–44.

Krasil'nikov, N. A. and Melkumova, T. A. 1963. Variation of nodule bacteria within the nodule of legumes. *Izv. Akad. Nauk SSSR, Ser. Biol.,* (5), 693–706.

Krasheninnikov, F. N. 1916. Assimilation of gaseous nitrogen by the root nodules of legumes: in *Sbornik v chest' 70-letiya K. A. Timiryazeva* (Festschrift for the Seventieth Birthday of K. A. Timiryazev), 307–324.

Kretovich, V. P. and Lyubimov, V. I. 1964. Biochemistry of nitrogen fixation. *Priroda,* (12), 14–21.

Krongauz, E. A., Samsonova, S. P., Verenko, V. D. and Dovgan', T. I. 1966. Importance of the strain of nodule bacteria in the production of dry nitragin: in *Ispol'zovanie mikroorganizmov dlya povysheniya urozhaya sel'skokhozyaistvennykh kul'tur* (Use of micro-organisms for increasing the yield of farm crops), Kolos, Moscow, 141–146.

Krylova, N. B., Krylov, S. V. and Agrafenina, V. I. 1963. Effect of the time of sowing vegetable beans on the formation of nodules. *Dokl. TSKhA,* (93), 89–94.

Kubo, H. 1939. Über das Hamoprotein aus den Wurzelknollchen von Legiminosen. *Acta Phytochim.* (Japan), **11**, 195–200.

Kudzin, Yu., Goncharova, N. and Chukh, G. 1964. On the Steppes of the Ukraine. *Zernobobovye kul'tury,* (5), 17–18.

Kuznetsov, S. V. and Natko, N. V. 1963. Some data on the distribution and activity of nodule bacteria in the soils of the Ashkabad district. *Trudy Turkmensk. s-kh. in-ta,* **12**, 108–110.

Kurbatov, I. M., Rulinskaya, N. S. and Dvoinishnikova, E. I. 1966. Role of peat humus in the nitrogen balance dynamics of the soil: in *IX Mezhdunarodnyi Mikrobiologicheskii kongress* (Ninth International Microbiology Congress), Moscow, Section B-2, summaries of speeches, 276.

Kurkaev, V. T. 1965. Results of study of nitragin in local strains under soya. *Trudy Amursk. s-kh. opytn. st.*, Part 1, 119–122.

Kvasnikov, B. V. and Dolgikh, S. T. 1955. Importance of the assimilation apparatus and the root system in the susceptibility of leguminous plants to nodule bacteria. *Mikrobiologiya*, **24**, (2), 180–187.

Laird, D. G. 1932. Bacteriophage and root nodule bacteria. *Arch. Mikrobiol.*, **3**, 159–193.

Lampovshchikov, P. K. 1951. Effect of mineral fertilizers on the inoculation of clover. *Trudy VASKhNIL*, **12**, 113–122.

Lampovshchikov, P. K. 1956. Effect of disinfectants on nodule bacteria in connection with nitragination of disinfected seeds. *Udobrenie i urozhai*, (8).

Lange, R. T. 1961. Nodule bacteria associated with the indigenous leguminosae of South Western Australia. *J. Gen. Microbiol.*, **26** (2), 351–359.

Lange, R. T. and Alexander, M. 1961. Anomalous infections by *Rhizobium*. *Canad. J. Microbiol.*, **7**, 959–961.

Last, F. T. and Nour, M. A. 1961. Cultivation of *Vicia faba* L. in Northern Sudan. *Emp. J. Exp. Agric.*, **29** (113).

Lasting, V. Microbiological processes in the soil when sweet clover is used as a green fertilizer: in *Sbornik nauchnykh trudov Estonskogo nauchno-issledovatel'nogo instituta zemledeliya i melioratsiya* (Collected scientific works of the Estonian Scientific Research Institute of Agriculture and Soil Improvement), **8**, Tallin (in Estonian).

Lasting, V. and Kuuts, H. 1966. Nitragination. *Sotsialistlik Pollimajandus*, (4), 161–163 (in Estonian).

Lazareva, N. M. 1953. Activity of local strains of clover nodule bacteria. *Trudy Vses. n.-i. in-ta s-kh. mikrobiol.*, **13**, 114–119.

Leonard, L. 1926. A preliminary note on the relation of photosynthetic carbohydrate to formation of soybeans. *J. Amer. Soc. Agron.*, **18**, 1012.

Leizaola, M. de and Dedonder, R. 1955. Etude de quelques polyosides produits par des souche de Rhizobium. *C.R. Acad. Sci. Paris*, **240**, 1825–1827.

Lepapei, A. and Trannov, H. 1934. *Ann. Agron.*, **4** (3) (quoted by B. P. Tsyurupa. 1937).

Levin, A. P., Funk, H. B. and Tendler, M. D. 1954. Vitamin B_{12}, rhizobia and leguminous plants. *Science*, **120**, 784.

Ley, J. de and Rassel, A. 1965. DNA base composition, flagella and taxonomy of the genus *Rhizobium*. *J. Gen. Microbiol.*, **41** (1), 85–92.

Lewis, J. M. 1938. Cell inclusions and the life cycle of rhizobia. *J. Bacteriol.*, **35**, 573–587.

Lie, T. A. 1964. Nodulation of leguminous plants as affected by root secretions and red light. Thesis. Wageningen. H. Veenman and N. V. Zonen, *Netherl. J. Agric.*, 1965, **13** (1).

Lilly, V. G. and Leonian, L. H. 1945. The interrelationship of iron and certain accessory factors in the growth of *Rhizobium trifolii* strain 205. *J. Bacteriol.*, **50** (3), 383–395.

Lim, G. 1963. Studies of the physiology of nodule formation. 8. The influence of the size of the rhizosphere population of nodule bacteria on root hair infection in clover. *Ann. Bot.*, **27** (105), 55–67.

Link, G. K. K. and Klein, R. M. 1951. Inhibitory and stimulatory effects of indole-acetic acid on the development of the bean hypocotyl. *Bot. Gaz.*, **112** (4), 400–417.

Lipman, J. G. 1912. The associative growth of legumes and non-legumes. *New Jersey Agric. Exptl. Stat. Bull.*, **253**, 48.

Lipmann, Ch. B. and Fouler, L. 1915. Isolation of *Bacterium radicicola* from soil. *Science*, **41**, 256.

Ljunggren, H. 1961. Transfer of virulence in *Rhizobium trifolii*. *Nature*, **191** (4788), 623.

Ljunggren, H. and Fahraeus, G. 1959. Effect of *Rhizobium* polysaccharide on the formation of polygalacturonase in lucerne and clover. *Nature*, **184** (4698), 1578–1579.

Ljunggren, H. and Fahraeus, G. 1961. The role of polygalacturonase in root hair invasion by nodule bacteria. *J. Gen. Microbiol.*, **26** (3), 521–528.

Loginov, Yu. M. Investigation of nitrogen atmospheric fixation by the lupin using molecular nitrogen-15. *Agrokhimiya*, (11), 21–28.

Löhnis, F. and Hansen, R. 1921. Nodule bacteria of leguminous plants. *J. Agric. Res.*, **20**, 543.

Löhnis, M. P. 1930. Can *Bacterium radicicola* assimilate nitrogen in the absence of the host plant? *Soil Sci.*, **20**, 37–57.

Lo Min-dyan' (Lo Ming-tan). 1961. *Obrazovanie kluben'kov u masha (Phaseolus aureus)* (Formation of nodules in the Oregon pea (*Phaseolus aureus*)). (Author's abstract of M.Sc. thesis). Leningrad.

Loneragan, J. F. and Douling, E. J. 1958. The interaction of calcium and hydrogen ions in the nodulation of subterranean clover. *Austral. J. Agric. Res.*, **9**, 464–472.

Longley, B. J., Berge, T. O., Lanen, J. M. and van Baldwin, I. L. 1957. Changes in the infective ability of rhizobia and *Phytomonas tumefaciens* induced by culturing on media containing glycine. *J. Bacteriol.*, **33**, 29–30.

Loos, M. A. and Louw, H. A. 1964. A study of the clover root nodule bacteria in soils of the George District. *S. Afric. J. Agric. Sci.*, **7**, 135–146.

Loos, M. A. and Louw, H. A. 1965. The influence of calcium carbonate amendments on the nodulation of white clover in the acid soils of the George district. *S. Afric. J. Agric. Sci.*, **8**, 729–736.

Lopatina, G. V. 1960. Selection of active cultures of saintoin nodule bacteria. *Byull. nauchno-tekhn inf. po s-kh. mikrobiol.*, No. 7, (1), 7–12.

Lopatina, G. V. and Lazareva, N. M. 1957. Cultivation and storage of nodule bacteria at low temperatures. *Byull. nauchno-tekhn. inf. po s-kh. mikrobiol.*, (3), 3–5.

Lorkiewicz, L. and Dusiński, M. 1963. The antigenic structure of *Rhizobium trifolii* mutants. *Acta Microbiol. Polon.*, **12** (3), 165 (in English).

Lorkiewicz, L., Żelazna, I. and Przybojewska, B. 1965. Alkaline phosphatase activity of *Rhizobium trifolii* mutants. *Acta Microbiol. Polon.*, **14** (3–4), 225–230 (in English).

Loustalot, A. J. and Tefford, E. A. 1948. Physiological experiments with tropical kudzu. *J. Amer. Soc. Argon.*, **40**, 503–511.

Lowe, R. H., Evans, H. J. and Ahmed, Sh. 1960. The effect of cobalt on the growth of *Rhizobium japonicum*. *Biochem. and Biophys. Res. Commun.*, **3** (6), 675–678.

Lowe, R. H. and Evans, H. J. 1962a. Cobalt requirement for the growth of rhizobia. *J. Bacteriol.*, **83**, 210–211.

Lowe, R. H. and Evans, H. J. 1962b. Carbon dioxide requirement for growth of legume nodule bacteria. *Soil Sci.*, **94** (6), 351–356.

Lynch, D. L. and Sears, O. H. 1950. Differential response of strains of *Lotus* nodule bacteria to soil treatment practices. *Soil Sci. Soc. America Proc.*, **15**, 176–180.

Lynch, D. L. and Sears, O. H. 1952. *Amer. Soc. Agron. Abstrs.*, 68 (quoted by Vincent, J. 1966).

Lyon, T. L. and Bizell, J. A. 1911. A previously unknown benefit from the growth of legumes. New York (Cornell) *Agric. Exptl. Sta. Bull.*, 294.

Machavariani, M. Z. 1951. Distribution of nodule bacteria of *Phaseolus* in the soils of Georgia, *Mikrobiologiya*, **20**, (6), 500–505.

Maeda, K. 1960. Study of the influence of the temperature of the rhizosphere on the symbiosis of legumes and nodule bacteria. *Proc. Crop. Sci Soc.*, Japan, **29**, (1), 158–160 (in Japanese; summary in English).

Madhok, M. 1940. Association of legumes and non-legumes. *Soil Sci.*, **49** (6).

Maisuryan, N. A. 1962. Problem of the production of plant proteins. *Izv. TSKhA*, (2), 7–18.

Makrinov, I. A. 1915. How and in what conditions to use bacterial earth-fertilizer preparations. *Sel'skii khozayin Petrograd*, 1–24.

Malińska, E. and Pęęziwilk, Z. 1958. Investigations on the production and storage of cobalamine by *Rhizobium* bacteria. *Acta Microbiol. Polon.*, **7** (2), 125–130 (in Polish; summary in English).

Manil, P. 1958. The legume—rhizobia symbiosis. in *Nutrition of the Legumes*. Ed. by Hallsworth, E. G. London, 124–133.

Manil, P. 1960. Essais de 'transformations' d'une souche de *Rhizobium* par l'DNA extrait de Agrobacterium tumefaciens. *Bull. Inst. agron. et Stat. rech. Gembloux*, **28** (3), 272–275.

Manil, P. 1963. Le *Rhizobium* et la fixation symbiotique de l'azote. A propos de la taxonomie et de la classification des Rhizobium. Quelques données recentes sur le determinisme biochimique de la fixation. *Ann. Inst. Pasteur.*, **105** (1), 19–45.

Manil, P. and Bonnier, Ch. 1951. *Bull. Inst. agron. et Stat. rech. Gembloux*, **19,** 15–32 (quoted by Hamatová-Hlaváčková, E. 1965).

Manil, P. and Bonnier, Ch. 1955. Note concernant l'inoculation du semences de *Trifolium pratense* en Belgique. *Bull. Inst. agron. et Stat. rech. Gembloux*, **23** (4).

Manil, P. and Brakel, J. 1961. A propos de la repartition geographique de *Rhizobium trifolii* et de *Rhizobium meliloti*, en Belgique. *Bull. Inst. Agron. et Stat. rech. Gembloux*, **29** (3–4), 328–334.

Manil, P. and Wernimont, H. 1961. Action differentielle de diverses souches de *Rhizobium* sur des indicateurs de rH. *Antonie van Leeuwenhoek. J. Microbiol. and Serol.*, **27** (1), 113–120.

Manninger, E. 1962. Biochemical examination of *Rhizobium* strains. *Acta. microbiol. Acad. Sci. Hung.*, **9** (3), 219–225 (in English).

Manninger, E. 1963–1964. Biochemische Untersuchungen von *Rhizobium*-arten (Gruppen). *Acta microbiol. Acad. Sci. Hung.*, **10** (2).

Manninger, E. 1964. Preservation of *Rhizobium* strains in a lyophilized condition. *Agrokém. és talaj*, **13** (3–4), 287–290 (in Hungarian; summaries in Russian and English).

Marnauza, A. A. 1966. *Vliyanie azotnogo i kaliinogo pitaniya kormovykh bobov na effektivnost' ikh simbioza s kluben'kovými bakteriyami* (Effect of nitrogen and potassium nutrition of broad beans on the efficiency of their symbiosis with nodule bacteria). (Author's abstract of M.Sc. thesis). Elgava.

Marshall, K. C. 1956. A lysogenic strain of *Rhizobium trifolii. Nature*, **177,** 92.

Marshall, K. C. 1964. Survival of root nodule bacteria in dry soils exposed to high temperatures. *Austral. J. Agric. Res.*, **15** (2), 273–281.

Marshall, K. C. and Roberts, F. J. 1963. Influence of fine particle materials on survival of *Rhizobium trifolii* in sandy soil. *Nature*, **198** (4878), 410–411.

Marshall, K. C. and Vincent, J. M. 1954. Relationship between the somatic antigens of *Rhizobium trifolii* and susceptibility to bacteriophage. *Austral. J. Sci.*, **17,** 68–69.

Marszewska-Ziemiecka, J., Nowotnówa, A. and Klukowska, W. 1938. The inoculation of leguminous plants. *Pamiętnik Panstwowego Inst. Naukow. Gospodarstwa Wiejskiego w Pulawach*, **17** (1), 235–286 (in English).

Marszewska-Ziemiecka, J. and Gołębiowska, J. 1948. The influence of inoculation on the alkaloid and protein content of sweet yellow lupine. *Ann. Univ. M. Curie-Sklodowska, Lublin*, **3** (8), 173–195 (in English).

Masefield, G. B. 1958. Some factors affecting nodulation in the tropics. in *Nutrition of the Legumes*. Ed. by Hallsworth, E. G. London, 202–212.

Masefield, G. B. 1961. The effect of irrigation on nodulation of some leguminous crops. *Empire J. Exptl. Agric.*, **29** (113), 51–59.

Masefield, G. B. 1965. The effect of organic matter in soil on legume nodulation. *Exptl. Agriculture*, **1** (2), 113–119.

Masterson, C. L. 1965. Studies on the toxicity of legume seeds towards Rhizobium. *Ann. Inst. Pasteur*, **109** (3) Suppl., 216–217 (in English).

Materassi, R. 1956. Ricerche sui metodi pratici di selezione ed inoculazione dei batteri simbionti della leguminose. *Agric. Ital.*, **56** (7), 244–266.

Maze, M. 1898. Les microbes des nodosites des Legumineuses. 3. Morphologie du microbe des nodosites. *Ann. Inst. Pasteur*, **12** (1–25), 128–155.

Mavritskii, N. V. 1947. Precursors and the development of nodules of legumines. *Nauchn. trudy Ukr. in-ta. ovoshchevodstva*, **1**, 120–126.

McConnel, J. T. and Bond, G. 1957a. A comparison of the effect of combined nitrogen on nodulation in non-legumes and legumes. *Plant and Soil*, **8** (4), 378–388.

McConnel, J. T. and Bond, G. 1957b. Nitrogen fixation in wild legumes. *Ann. Bot.*, New Series, **21** (81), 185–192.

McCoy, E. 1932. Infection by *Bacterium radicicola* in relation to the micro-chemistry of the hosts cell walls. *Proc. Roy. Soc.*, B, **110**, 514–533.

McGonagle, M. P. 1944. Cultures of exised leguminous roots. *Nature*, **153**, 538–529.

McGonagle, M. P. 1949. *Proc. Roy. Soc. Edinburgh*, B, **63**, 219–229 (quoted by van Schreven, D. 1958).

McKell, C. M. and Whalley, D. B. 1964. Compatibility of 2-chloro-6-(trichloro-methyl)-pyridine with *Medicago sativa* L. inoculated with *Rhizobium meliloti*. *Agron. J.*, **56** (1), 26–28.

McLeod, L. B. and Jackson, L. P. 1965. Effect of concentration of the aluminium ion on root development and establishment of legume seedlings. *Canad. J. Soil Sci.*, **45** (2), 221–245.

Meisel', M. N., Biryuzova, V. I., Volkova, T. M., Malatyan, M. N. and Medvedeva, G. A. 1964. Functional morphology and cytochemistry of the mitochondrial apparatus of micro-organisms: in *Elektronnaya i fluorestsentnaya mikroskopiya kletki* (Electron and fluorescent microscopy of the cell). Nauka. Moscow, 1–15.

Melkumova, T. A. 1957. *Biologicheskie osobennosti kluben'kovykh bakterii, vydelennykh iz raznykh sortov lyutserny, vozdelyvaemykh v usloviyakh Azerbaidzhanskoi SSR* (Biological features of nodule bacteria isolated from different varieties of lucerne grown in the Azerbaijan USSR). (Author's abstract of M.Sc. thesis), Moscow.

Melkumova, T. A. 1961. Distribution of nodule bacteria in cotton-lucerne crop rotation in the irrigated soils of the Kura-Araksin lowlands. *Izv. Akad. Nauk AzSSR, Ser. Biol.*, (11), 91–95.

Melkumova, T. A. and Gazanchyan, Zh. M. 1964. Effect of trace elements on the activity and virulence of lucerne nodule bacteria. *Dokl. Akad. Nauk AzSSR*, **20**, (2), 53–57.

Mes, M. G. 1959. Influence of temperature on the symbiotic nitrogen fixation of legumes. *Nature*, **184** (4704), 2032–2033.

Meshkov, N. V. 1963. Effect of bacterial fertilizers on the balance of soddy-podzolic soils and the crop yield in: *Vsesoyuznaya konferentsiya po sel'sko-khozyajstvennoi mikrobiologii, 26–30 marta, 1963g* (All Union Conference on Agricultural Biology, 26–30 March, 1963), summaries of speeches. Leningrad, 91–94.

Meshkov, N. V. 1966. Productivity of fixation of atmospheric nitrogen by leguminous plants in soddy-podzolic soils in a vegetative experiment: in *Balans azota v dernovo-podzolistykh pochvakh* (Nitrogen balance of soddy-podzolic soils), Nauka, Moscow, 183–210.

Meshkov, N. V. and Khodakova, R. N. 1966a. Effect of biological nitrogen accumulated in the roots of legume precursors on the nitrogen nutrition of graminaceous plants. in *Balans azota v dernovo-podzotshykh pochvakh* (Nitrogen balance of soddy-podzolic soils), Nauka, Moscow, 211–222.

Meshkov, N. V. and Khodakova, R. N. 1966*b*. Biological fixation of atmospheric nitrogen in a vegetation experiment in soddy-podzolic soils: in *IX Mezhdynarodnyi mikrobiologicheskii kongress* (Ninth All-Union Microbiology Congress), summaries of speeches. Moscow, 297.

Meyer, D. R. and Anderson, A. J. 1959. Temperature and symbiotic nitrogen fixation. *Nature*, **183** (4653), 61.

Mezharaupe, V. A. 1964. Effect of cultivated plants on the activity of nodule bacteria. in *Mikroorganizmy i rastenii* (Micro-organisms and plants), **2**, 41–52, Riga.

Michael, G. 1941. *Z. Bodenkunde Pflauzernahr.*, **25**, 65–120 (quoted by van Schreven, D. 1958).

Miettinen, J. K. 1955. Free amino acids in the pea plant (*Pisum sativum*). *Ann. Acad. Sci. Fennicae*, Series A11, (60), 520–535 (in English).

Migahid, A. M., El Nady, A. F. and Abd El Tahman, A. A. 1959. The effect of gamma radiation on bacterial nodule formation. *Plant and Soil*, **11** (2), 139–144.

Mikhaleva, V. V. 1952. *Antagonisticheskoe deistvie pochvennykh aktinomitsetov na kluben'kovye bakterii* (Antagonistic effect of soil actinomycetes on nodule bacteria). (Author's abstract of M.Sc. thesis). Moscow.

Mikhaleva, V. V., Borodulina, Yu. S., Kochunova, T. A. and Shirokova, T. L. 1965. Selection of antibiotics to suppress the development of *Bac. megaterium* contaminating cultures of nodule bacteria: in *Rol' mikroorganizmov v pitanii rastenii i povyshenii effektivnosti udobrenii* (Role of micro-organisms in plant nutrition and in increasing the effectiveness of fertilizers), Kolos, Moscow, 126–130.

Miller, R. H. and Schmidt, E. L. 1965. A technique for maintaining a sterile soil plant root environment and its application to the study of amino-acids in the rhizosphere. *Soil Sci.*, **100** (4), 267–273.

Millington, A. J. 1955. Deep placement of rhizobial cultures as an aid to legume inoculation. *J. Austral. Inst. Agric. Sci.*, **21** (2), 102–103.

Milovidov, P. F. 1926. Über einige neue Beobachtungen an den Lupinenknöllchen. *Zbl. Bakteriol., Parasitenkunde, Infektionskrankh. und Hyg.*, Abt. 2, **68**, 333–345.

Milovidov, P. F. 1928. Recherches sur les tubercules du lupin. *Rev. gen. bot.*, **40**, 193–205.

Milovidov, P. F. 1935. Ergebnisse der Nuclealfarbung bei den Myxobacterien und einigen anderen Bakterien. *Arch. Mikrobiol.*, **6**, 475–509.

Mishustin, E. N. and Bernard, V. V. 1938. Nitragin and its uses. *Khimizatsiya sots. zemled.*, (11), 28–49.

Mishustin, E. N. and Erofeev, N. S. 1965. Elimination of a nitrogen deficit in the soil using straw as organic fertilizer. *Mikrobiologiya*, **34**, (6), 1056–1062.

Mishustin, E. N. and Erofeev, N. S. 1966. Nature of toxic compounds of straw decomposing in the soil. *Mikrobiologiya*, **35**, (1), 150–154.

Mishustin, E. N. and Naumova, A. N. 1955. Isolation of the toxic substances of lucerne and their effect on cotton and the soil microflora. *Izv. Akad. Nauk SSSR, Ser. Biol.*, (6), 3–9.

Mishustin, E. N. and Shil'nikova, V. K. 1966. Nitrogen fixing activity of nodule bacteria and its indices: in *IX Mezhdunarodnyi Mikrobiologicheskii Kongress* (Ninth International Microbiology Congress), summaries of speeches. Moscow, 292.

Mityushova, N. M. 1955. Nitrogen content in the bleeding sap of *Phaseolus* with and without nodules. *Uch. zapiski LGU*, No. **186**, (39), 233–240.

Mocrić, A. and Strunjak, R. 1963. Further information on the metabolism of nodule bacteria with soya, lupin, lucerne and clover, *Arh. poljoprivredne nauke*, **16** (52), 84–93 (in Serbo-Croat, summary in German).

Moore, C. T. 1905. Soil inoculation for legumes with reports upon the successful use of artificial cultures by practical farmers. *Bull. US Bur. Plant Dept. Agric. Industr.*, **71**, 1–72.

Moore, A. W. 1960. Symbiotic nitrogen fixation in a grazed tropical grass-legume pasture. *Nature*, **185** (4713), 638.

Mosse, B. 1964. Electron-microscopic studies of nodule development in some clover species, *J. gen. Microbiol.*, **36** (1), 49–63.

Moustafa, E. 1964. Note on species differences and developmental changes in root nodule esterase and phosphatase isoenzymes. *NZ J. Sci.*, **7** (4), 608–610.

Mulder, E. Dzh., Lai, T. A., Dilz, K. and Khauers, A. (Mulder, E. G., Lie, T. A., Dilz, K. and Howers, A.). 1966. Effect of *p*H on symbiotic nitrogen fixation by certain leguminous plants: in *IX Mezhdunarodnyi Mikrobiologicheskii Kongress* (Ninth international Microbiology Congress). Symposium V.-1, summaries of speeches. Moscow, 98–113.

Mulder, E. G. 1948*a*. Importance of molybdenum in the nitrogen metabolism of micro-organisms and higher plants, *Plant and Soil*, **1** (1), 94–120.

Mulder, E. G. 1948*b*. Investigations on the nitrogen nutrition of pea plants, *Plant and Soil*, **1** (2), 179–212.

Mulder, E. G. 1954. Molybdenum in relation to growth of higher plants and micro-organisms. *Plant and Soil*, **5** (4), 368–415.

Mulder, E. G., Bakema, K. and van Veen, W. L. 1959. Molybdenum in symbiotic nitrogen fixation and in nitrate assimilation, *Plant and Soil*, **10** (4), 319–334.

Mulder, E. G. and van Veen, W. L. 1960*a*. The influence of carbon dioxide on symbiotic nitrogen fixation, *Plant and Soil*, **13** (3), 265–278.

Mulder, E. G. and van Veen, W. L. 1960*b*. Effect of *p*H and organic compounds on nitrogen fixation by red clover, *Plant and Soil*, **13** (2), 91–113.

Müller, A. and Stapp, C. 1925. Beiträge zur Biologie der Leguminoseknöllchenbakterien mit besonderer Berucksichtigung ihrer Artverschiedenheit, *Arb. Biol. Reichsanst. Landwirtsch. und Forstwesen*, **14**, 455–554.

Murphy, S. G. and Elkan, G. H. 1963. Growth inhibition by biotin in a strain of *Rhizobium japonicum*, *J. Bacteriol.*, **86** (4), 884–885.

Murphy, S. G. and Elkan, G. H. 1965. Nitrogen metabolism of some strains of *Rhizobium japonicum* having different nodulating capacities, *Canad. J. Microbiol.*, **11** (6), 1039–1041.

Myskow, W. and Ziemiecka, J. (Myshov, V. and Ziemitska, I.). 1966. Conversion in the soil of nitrogenous compounds: in *IX Mezhdunarodnyi Mikrobiologicheskii Kongress* (Ninth International Microbiology Congress), summaries of speeches. Moscow, 273.

Myskow, W. and Morrison, R. 1963. Decomposition of leguminous plant roots in sand. I. Transformation of nitrogen compounds, *J. Sci. Food and Agric.*, **11**, 813–821.

Natman (Nutman, P. S.). 1966. Genetic and physiological factors influencing the formation of clover nodules and nitrogen fixation: in *IX Mezhdunarodnyi Mikrobiologicheskii Kongress* (Ninth International Microbiology Congress). Symposium V.-1, summaries of speeches. Moscow, 78–89.

Naumova, A. N. 1966. Biologically active substances in a culture of *Rhizobium leguminosarum*: in *IX Mezhdunarodnyi Mikrobiologicheskii Kongress* (Ninth International Microbiology Congress), summaries of speeches. Moscow, 299.

Naundorf, G., Nilsson, R. 1942. Über formbildende Wirkstoffe bei *Azotobacter chroococcum* und der Einflussdieser formativen Wirkstoff auf die Bakteroiden-Bildung von *Bacterium radicicola*, *Naturwissenschaften*, **30**, 753.

Naundorf, G. and Nilsson, R. 1943. Über formbildende Wirkstoffe bei *Azotobacter chroococcum* und der Einflus dieser formativen Wirkstoffe auf die Bildung von Gigasformen bein *Bacterium radicicola*, *Naturwissenschaften*, **31**, 346.

Negryanu, V., Shtefanik, G., Dumitru, I. and Gontsa, E. (Negranu, V., Stefanic, G.,

Dumitru, I. and Gontsa, E.). 1961. Effectiveness of 'nitragin-soya' fertilizer prepared in a culture medium containing various energy substances. *Lucrări ştiintifice*, **3**, 111.

Němec, B. 1915. Über die Bakterienknöllchen von *Ornithopus sativus*, *Bull. Internat. Acad. Sci. Bohěme*, 1–13.

Nichiporovich, A. A. and Perevozchikova, M. F. 1930. Application of nitragin under soya in the soils of the North Caucasus. *Izv. po opytn. delu sev Kavkaza*, (5), 22.

Nicholas, D. J. 1957. The function of trace metals in the nitrogen metabolism of plants, *Ann. Bot.*, **21** (84), 587–598.

Nicholas, D. J. 1958a. in *Metals and Enzyme Activity*, *Biochem. Soc. Sympos.*, No. 15, Cambridge University Press.

Nicholas, D. J. 1958b. Discussion: in *Nutrition of the Legumes*. Ed. by Hallsworth, E. G. London, 182.

Nicholas, D. J. 1962. N^{13}, N^{15} support theory. Mechanism studies show that ammonia is preferred nitrogen form; it can replace nitrogen and be used immediately, *Chem. and Engng. News*, **40** (19), 43.

Nicholas, D. J. 1963a. How do microbes 'fix' nitrogen from the air? *New Scientist*, **20** (369), 680–683.

Nicholas, D. J. 1963b. The biochemistry of nitrogen fixation, *Symbiot. Assoc.* Cambridge University Press, 92–124.

Nicol, H. 1934. The derivation of the nitrogen of crop plants with special reference to associated growth, *Biol. Rev.*, **9** (4), 383.

Nicol, H. and Thornton, H. C. 1941. Competition between related strains of nodule bacteria and its influence on infection of the legume host. *Proc. Roy. Soc.*, B, **130** (858), 32–59.

Nielsen, N. 1940. Untersuchungen über biologische stickstoffbindung. I. Der Wert verschiedener Aminosauren als Stickstoffquelle fur *Bacterium radicicola*, *CR trav. Lab. Carlsberg. ser. physiol.*, **23**, 115–134.

Nilsson, R., Bjälfve, G. and Burstrom, D. 1938. Vitamin B_1 als Zuwachsfaktor fur Bacterium radicicola. I, II. *Naturwisshenschaften*, **26**, 284, 661.

Nilsson, P. E. 1957. The influence of antibiotics and antagonists on symbiotic nitrogen fixation in legume cultures. *Ann. Roy. Agric. Coll. Sweden*, **23**, 219–253 (in English).

Nilsson, P. E. and Rydin, C. 1954. Studies on symbiotic nitrogen fixation by a new strain of tetraploid red clover. *Arch. Mikrobiol.*, **20**, 398–403.

Nitá, L. 1963. Physiological properties of pea and vetch nodule races of *Rhizobium* with large nitrogen-fixing capacity. *Agrokém. és talaj*, **12** (4), 647–660 (in Hungarian; summaries in Russian and French).

Nitse, L. (Nitá, L.). 1958. *Azotfiksiruyushchaya aktivnost' kluben'kovykh bakterii gorokha i viki i dinamika nakopleniya azota v bobovom rastenii* (The nitrogen-fixing activity of pea and vetch nodule bacteria and accumulation of nitrogen in the legume. (Author's abstract of M.Sc. thesis), Moscow.

Nobbe, F. and Hiltner, L. 1893. Wodurch werden die knollchenbesitzenden Leguminosen befahigt, den freien atmospharischen Stickstoff fur sich zu werwerten? *Landwirtsch. Vers. Stat.*, **42**, 459–478.

Nobbe, F. and Hiltner, L. 1896. Bodenimpfung fur Anbau von Leguminosen. *Sachs. Landwirtsch. Z.*, **44**, 90–92.

Norman, A. G. 1950. The fate of complex organic compounds in soil. *Trans. Fourth Internat. Congr. Soil Sci.* Amsterdam, **3**, 100–102.

Norris, D. O. 1956. Legumes and the *Rhizobium* symbiosis. *Empire J. Exptl. Agric.*, **24**, 247–270.

Norris, D. O. 1958a. A red strain of *Rhizobium* from *Lotononius bainesii* Baker. *Austral. J. Agric. Res.*, **9**, 629–632.

Norris, D. O. 1958b. Lime in relation to the nodulation of tropical legumes: in *Nutrition of the Legumes*. Ed. by Hallsworth, E. G. London, 164–182.

Norris, D. O. 1959a. The role of calcium and magnesium in the nutrition of *Rhizobium*. *Austral. J. Agric. Res.*, **10** (5), 651–698.

Norris, D. O. 1959b. *Rhizobium* affinities of African species of *Trifolium*. *Empire J. Exptl. Agric.*, **27** (106), 87–97.

Norris, D. O. 1959c. Nodulation problems. *Rural Res. CSIRO*, (29), 15–17.

Norris, D. O. 1963. A porcelain-bead method for storing *Rhizobium*. *Empire J. Exptl. Agric.*, **31** (123), 255–258.

Norris, D. O. 1965. Acid production by *Rhizobium*. A unifying concept. *Plant and Soil*, **22** (2), 143–166.

Norris, D. O. and t'Mannetje, L. 1964. The symbiotic specialization of African *Trifolium* spp. in relation to their taxonomy and their agronomic use. *East Afric. Agric. J.*, **29**, 214–235; *Rhizobium Newsletter*, 1965, **10** (1), 89–90.

Novikova, A. T. and Irtuganova, L. D. 1966. Formation of heteroauxin by soil micro-organisms. *Mikrobiologiya*, **35**, (4), 707–711.

Nowacki, E. and Przybylska, J. 1961. Tingitanine, a new free amino acid from seeds of Tanger pea (*Lathyrus tingitanus*). *Bull. Acad. Polon.*, *Sci. Ser. Sci. Biol.*, **9**, 279 (in English).

Nowak, W. 1966. Über ein Sternchenbilden des Bodenbakterium aus dem Formenkreis der Chromobacteriaceae. *Naturwissenschaften*, **53** (13), 338–339.

Nowak, W. and Netzsch-Lehner, A. 1965. Reisolierungsversuche von Knollchenbakterien (*Rhizobium* sp.) aus Boden. *Zbl. Bakteriol.*, *Parasitenkunde, Infektionskrankh. und Hyg.*, Abt. 2, **119** (6), 570–578.

Nowotny-Mieczyńska, A. 1952. Effect of certain factors on the pigmentation of the Papilionaceae. *Acta Microbiol. Polon.*, **1** (1), 42–51.

Nowotny-Mieczyńska, A. and Zinkiewiez, J. 1959. Effect of plant nutrition on the activity of *Rhizobium trifolii* in symbiosis with clover. *Acta Microbiol. Polon.*, **8**, (3–4), 309–313, Dyskus. 313.

Nutman, P. S. 1946. Variation within strains of clover nodule bacteria in the size of nodule produced and in the effectivity of the symbiosis. *J. Bacteriol.*, **51** (4), 411–432 (in English).

Nutman, P. S. 1948. Physiological studies on nodule formation. I. The relation between nodulation and lateral root formation in red clover. *Ann. Bot.*, New Series, **12** (46), 81–96.

Nutman, P. S. 1949. Physiological studies on nodule formation. 2. The influence of delaying inoculation on the rate of nodulation in red clover. *Ann. Bot.*, New Series, **13**, 261–283.

Nutman, P. S. 1952. Studies on the physiology of nodule formation. 3. Experiments on the excision of root tips and nodules. *Ann. Bot.*, New Series, **16**, 80–101.

Nutman, P. S. 1953. Studies on the physiology of nodule formation. 4. The mutual inhibitory effects on nodule production of plants grown in association. *Ann. Bot.*, New Series, **17**, 95–126.

Nutman, P. S. 1956. The influence of the legume in root nodule symbiosis. A comparative study of host determinants and functions. *Biol. Rev.*, **31** (2), 109–152.

Nutman, P. S. 1957. Studies on the physiology of nodule formation. 5. Further experiments on the stimulating and inhibitory effects of root secretions. *Ann. Bot.*, New Series, **21** (83), 321–337.

Nutman, P. S. 1959a. *Soc. Exptl. Biol. Sympos.*, **13**, 42 (quoted by Vincent, J. 1966a).

Nutman, P. S. 1959b. Some observations on root-hair infection by nodule bacteria. *J. Exptl. Bot.*, **10** (29), 250–263.

Nutman, P. S. 1962. The relation between root hair infection by *Rhizobium* and nodulation in *Trifolium* and *Vicia*. *Proc. Roy. Soc.*, B, **156** (962), 122–137.

Nutman, P. S. 1963. Factor influencing the balance of mutual advantage in legume symbiosis. in *Thirteenth Sympos. Soc. Gen. Microbiol.*, *London, April* 1963, 51–71.

Nutman, P. S., Thornton, H. G. and Quastel, J. H. 1945. Inhibition of plant growth by 2,4-dichlorophenoxyacetic acid and other plant growth substances. *Nature*, **155** (3939), 498–500.

Obaton, M. and Blachere, H. 1963. Observations sur l'inoculation des legumineuses en France. *Ann. Inst. Pasteur*, **105** (2), 282–290.

Obaton, M. and Blachere, H. 1965. L'inoculation de la lucerne. 2. Resultats de l'experimentation. *Ann. Agron.*, **16** (1), 25–52.

Obraztsova, A. A., Minenkov, A. R., Revyakina, E. V., Golland, D. M. and Krasil'-nikova, A. I. 1937. Trace elements as a factor enhacing the efficacy of nitragin. *Mikrobiologiya*, **6**, (7).

Okuda, A. and Yamaguchi, M. 1960. Nitrogen-fixing micro-organisms in paddy soils. 6. *Soil and Plant Food*, **6** (2), 76–85.

Oplištilová, K. and Vančura, V. 1963. Growth promoting elements in a culture of *Rhizobium. Sbor. Českosl. akad. zeměd. věd. Rostl. výroba.*, **9** (7–8), 734–736.

Orcutt, F. S. 1937. Nitrogen metabolism of soybeans in relation to the symbiotic nitrogen fixation process. *Soil Sci.*, **44**, 203–215.

Orcutt, F. S. and Fred, E. B. 1935. Light intensity as an inhibiting factor in the fixation of atmospheric nitrogen by Manchu soybeans. *J. Amer. Soc. Agron.*, **27**, 550–558.

Ouellette, G. J. and Desseureaux, L. 1958. Chemical composition of alfalfa as related to degree of tolerance to manganese and aluminium. *Canad. J. Plant. Sci.*, **38**, 206–218.

Owens, L. and Wright, D. 1965. Rhizobial-induced chlorosis in soy beans–isolation production in nodules, and varietal specificity of the toxin. *Plant Physiol.*, **40** (5), 927–933.

Owens, L. D. and Thompson, J. F. 1966. Rhizobial-induced chlorosis in soy beans. *Rhizobium Newsletter*, **11** (2), 168–169.

Parker, C. A. and Oakley, A. E. 1963. Nodule bacteria for two species of seradella *Ornithopus sativus* and *Ornithopus compressus. Austral. J. Exptl. Agric. and Animal Husbandry*, **3** (8), 9–10.

Parker, C. and Oakley, A. 1965. Reduced nodulation of lupins and seradella due to lime pelleting. *Austral. J. Exptl. Agric. and Animal Husbandry*, **5** (17), 144–146.

Parle, J. 1958. Field observations of copper deficiency in legumes: in *Nutrition of the Legumes*. Ed. by Hallsworth, E. G. London, 280–283.

Pate, J. 1958. Studies of the growth substances of legume nodules using paper chromatography. *Austral. J. Biol. Sci.*, **11**, 516–528.

Pate, J. 1961. Temperature characteristics of bacterial variation in legume symbiosis. *Nature*, **192** (4803), 637–639.

Payne, M. G. and Fults, J. L. 1947. Some effects of 2,4D, DDT and colorado 9 on root nodulation in the common bean. *J. Amer. Soc. Agron.*, **39** (1), 52–55.

Peive, Ya. V. 1965. Biochemistry of trace elements and problems of the nitrogen nutrition of plants. *Vestn. Akad. Nauk SSSR*, (1), 42–50.

Peive, Ya. V. 1967. Role of trace metals in biochemical processes catalysed by enzymes. *Izv. Akad. Nauk SSSR, Ser. Biol.*, (1), 11–19.

Peive, Ya. V. and Zhiznevskaya, G. Ya. 1966. Haemoglobin in the nodules of leguminous crops, trace elements and the fixation of molecular nitrogen. *Izv. Akad. Nauk SSSR, Ser. Biol.*, (5), 644–652.

Peive, Ya. V., Zhiznevskaya, G. Ya. and Tenisone, I. V. 1965. Comparative study of the activity of nitrate reductase and dehydrogenase in nodules of the lupin and broad bean: in *Mikroelementy i produktivnost' rastenii* (Trace elements and plant productivity), *Riga*, 27–49.

Peive, Ya. V., Safonov, V. I., Zhiznevskaya, G. Ya., Dmitrieva, M. I. and Nemchenko, G. I. 1966. *Razdelenie gemoglobina iz kluben'kov lyupina metodom diskovogo elektroforeza na poliakrilamidnom gele. Tezisy soveshchaniya po khimii rastitel'nykh belkov*

(Extraction of haemoglobin from lupin by disc electrophoresis on polyacrylamide gel. Papers presented at a conference on the chemistry of plant proteins). Kishenev.

Peterburgskii, A. V. 1964. Effect of lime, molybdenum and vanadium on legumes in acid soils. *Izv. TSKhA*, (2), 49–64.

Peterburgskii, A. V. 1965. Vanadium—a necessary trace element for leguminous crops: in *IX Mendeleevskii s"ezd* (Ninth Mendeleev Congress). Agrochemistry section, 141–143.

Peterburgskii, A. V. and Sabo, B. 1963. Effect of lime, molybdenum and vanadium on the yield and chemical composition of the pea. *Dokl. TSKhA*, (94), 73–82.

Petersen, H. I. 1950. Experimental use of chemicals against weeds in spring-sown cereals and stubble fields. *Tidsskr. Planteavl.*, A, **53**, 678–708 (in Danish).

Petersen, H. I. 1952. Remarks on the chemical control of weeds. *Ugeskr. Landm.*, A, **97** (18), 266–267 (in Danish).

Peterson, N. V. 1954. Interaction between nodule bacteria and certain soil micro-organisms. *Mikrobiol. zh.*, **16** (3), 14–17 (in Ukrainian).

Petrenko, G. Ya. 1953. Specific and local races of nodule bacteria and *Azotobacter*. in *Rol' mikroorganizmov v pitanii rastenii* (Role of micro-organisms in plant nutrition). Sel'khozgiz, Moscow, 84–90.

Petrosyan, A. P. 1953. Effect of ecotypes of nodules bacteria on the yield of leguminous plants: in *Voprosy sel'skokhozyaistvennoi i promyshlennoi mikrobiologii* (Problems of agricultural and industrial microbiology) part 1 (VII). Izd-vo Akad. Nauk SSSR, Moscow, 3–13.

Petrosyan, A. P. 1959. *Ekologicheskie osobennosti kluben'kovykh bakterii v Armyanskoi SSR* (Ecological features of nodule bacteria in the Armenian SSR). Izd-vo Ministerstva sel'skogo khozyaistva ArmSSR.

Petrosyan, A. P. and Avvakumova, E. N. 1963. Cytochemical features of nodule bacteria: in *Pochvennaya i sel'skokhozyaistvennaya mikrobiologiya* (Pedological and agricultural microbiology). Izd-vo Akad. Nauk. UzbSSR, Tashkent, 205–209.

Petrosyan, A. P. and Avvakumova, E. N. 1964. Cytological and cytochemical changes in nodule bacteria within the nodules. *Dokl. Akad. ArmSSR*, **39**, (1), 49–52.

Petrović, V. 1963. Testing the capacity for competition between different strains of *Rhizobium meliloti*. *Zemljište i biljka*, **12** (1–3), 305–309.

Petrushenko, O. P. and Salakhov, Kh. 1965. Nodule bacteria in the soil, their activity and virulence. *Uzb. biol. zh.*, (6), 20–24.

Pieters, A. I. 1927. *Green manuring principles and practice*. New York, John Wiley, 356.

Pietz, J. 1938. Beitrag zur Physiologie des Wurzelknollchenbakteriums. *Zbl. Bakteriol., Parasitenkunde, Infektionskrankh. und Hyg.* Abt. II, **99**, 1–32.

Pillai, R. W. and Sen Abhiswar. 1966. Salt tolerance of *Rhizobium trifolii*. *Indian J. Agric. Sci.*, **36** (2), 80–85.

Pinck, L. A., Allison, F. E. and Gaddy, V. L. 1946. The effect of straw and nitrogen on the yield and quantity of nitrogen fixed by soybeans. *J. Amer. Soc. Agron.*, **38**, 421–431.

Pirson, A. 1937. *Z. Angew. Bot.*, **31**, 193–267 (quoted by van Schreven, D. 1958).

Pochon, J. and Barjac, H. de. 1958. Traite de microbiologie des soils. Dunod, Paris.

Poglazova, M. N. 1964. Luminescent microscopic study of the structure of Actinomycetes. *Mikrobiologiya*, **33**, (3), 459–462.

Pohlman, G. G. 1931a. Nitrogen fixation by *Rhizobium meliloti* and *Rhizobium japonicum*. *J. Amer. Soc. Agron.*, **23**, 22–27.

Pohlman, G. G. 1931b. Changes produced in nitrogenous compounds by *Rhizobium meliloti* and *Rhizobium japonicum*. *Soil Sci.*, **31**, 385–406.

Porter, F. E., McAlpine, V. M. and Kaerwer, H. E., Jr. 1962. Seed inoculation (Battelle Memorial Institute). *USA Patent*, Cl 47-1, No. 3054219, 20.02.59, 18.09.62.

Pochon, J. and de Barjac, H. (Poshon, Zh. and de Barzhak, G.). 1960. *Pochvennaya mikrobiologiya* (Soil microbiology). Inostrannaya literatura. Moscow.

Possingham, J. V., Moye, D. V. and Anderson, A. J. 1964. Influence of elevated shoot and root temperature on nitrogen fixation. *Plant Physiol.*, **39**, 561–563.

Powrie, J. K. 1964. The effect of cobalt on the growth of young lucerne on a siliceous sand. *Plant and Soil*, **21**, 81–93.

Pražmowski, A. 1888. Über die Wurzelknöllchen der Leguminosen. *Bot. Zbl.*, **36**, 241–257.

Pražmowski, A. 1889. Das Wesen und die biologische Bedeutung der Wurzelknöllchen der Erbse. *Bot. Zbl.*, **39**, 356.

Pražmowski, A. 1890. Über die Wurzelknöllchen der Leguminosen. *Bot. Zbl.*, **39**, 161–179.

Proctor, H. 1963. Metabolism of low molecular weight compounds by rhizobia. A survey. *NZ J. Agric. Res.*, **6** (1), 17–26.

Pronin, V. A. 1966. Effect of fertilizers on the nitrogen-fixing activity of nodule bacteria and the yield of the leguminous plant: in *Govoryat molodye uchenye* (Young scientists speak), **1**, Moscow, 294–299.

Pronchenko, T. S. and Andreeva, E. I. 1964. Toxicity of certain chemical preparations for nodule bacteria. *Khimiya v sel'skom khozyaistve*, (8), 17–20.

Prucha, M. I. 1915. Physiological studies of *Bac. radicicola* of Canada field pea. *NJ (Cornell) Agric. Coll.*, (5), 1–83.

Pumpyanskaya, L. V. 1959. Preparation of pure dry cultures of nodule bacteria. in *Mikrobiologiya na sluzhbe sel'skogo khozyaistva* (Microbiology in the service of agriculture). Sel'khozgiz, Moscow, 103–104.

Purchase, H. F. 1953. Nodule bacteria in the rhizosphere. *Reprint Rothamsted Experimental Station*, 66–67.

Purchase, H. F. 1958. Restriction of infection threads in nodulation of clover and lucerne. *Austral. J. Biol. Sci.*, **11**, 155–161.

Purchase, H. F. and Nutman, P. S. 1957. Studies on the physiology of nodule formation. 4. The influence of bacterial numbers in the rhizosphere on nodule initiation. *Ann. Bot.*, New Series, **21**, 439–454.

Purchase, H. F., Vincent, J. M. and Ward, L. M. 1951a. The field distribution of strains of nodule bacteria from five species of *Medicago*. *Austral. J. Agric. Res.*, (2), 261–272.

Purchase, H. F., Vincent, J. M. and Ward, L. M. 1951b. Serological studies of the root nodule bacteria. *Proc. Linnean Soc. NS Wales*, **76**, 1–6.

Pyarsim, E. 1966. Distribution of nodules and their morphology in leguminous plants in the Estonian SSR. *Izv. Akad. Nauk EstSSR, Ser. Biol.*, 15–28.

Rabotnova, I. L. 1936. Redox regime of the nitrogen assimilating group *Rhizobium*. *Mikrobiologiya*, **5**, (2).

Rabotnova, I. L. 1938. Can rhizobia fix nitrogen in pure culture? *Mikrobiologiya*, **7**, (6), 673–682.

Rabotnova, I. L. 1957 *Rol' fiziko-khimicheskikh usloviyakh* (pH i rH_2) *v zhiznedeyatel'nosti mikroorganizmor* (Role of physicochemical conditions (pH and rH_2) in the vital activity of micro-organisms). *Akad. Nauk SSSR*, Moscow.

Radcliffe, J. C. 1964. The survival of rhizobia on seeds of *Trifolium subterraneum* L. *Diss. Abstr.*, **25**, 3181.

Radulović, V. 1959–1960. Note on the investigation of nodule bacteria (*Rhizobium* spp.) on Leguminosae in Bosnia and Hercegovina. *Radovi Poljopr. fak. Univ. Sarajevu*, **8–9** (10–11), 269–288.

Radulović, V. 1963. Effect of inoculation of seeds and calcification of soil acids on the numbers of *Rhizobium radicicola* var. *leguminosarum* in the rhizosphere of peas. *Vicia sativa* Roth. *Zemljište i biljka*, **12** (1–3), 301–304.

Radulovich, V. (Radulović, V.). 1966. Effect of growth rate of root systems of plants on the nitrogen-fixing capacity of nodule bacteria: in *IX Mezhdunarodnyi mikrobiologicheskii Kongress* (Ninth International Microbiology Congress), summaries of speeches. Moscow, 292.

Raggio, M. and Raggio, N. 1956. Relacion entre cotiledones y nodulacion y factores que la afectan. *Phyton* (Buenos Aires), **7**, 103–119.

Raggio, M., Raggio, N. and Burris, R. H. 1959. Enhancement by inositol of the nodulation of isolated bean roots. *Science*, **129**, 211–212.

Raggio, M., Raggio, N. and Torrey, J. G. 1957. The nodulation of isolated leguminous roots. *Amer. J. Bot.*, **44**, 325–334.

Raggio, M., Raggio, N. and Torrey, J. G. 1965. The interaction of nitrate and carbohydrates in rhizobial root-nodule formation. *Plant Physiol.*, **40**, 601–606.

Raicheva, L. B. 1957a. Nodule bacteria of the genus *Rhizobium meliloti* in the rhizosphere of wheat, oats, maize and alfalfa. *Nauchni trudove In-ta pochveni izsledovaneto 'N. Pushkarov'*, 3, Zemizdat, Sofia, 493–502.

Raicheva, L. B. 1957b. Investigation of nodule bacteria in Bulgarian conditions. *Selekostop. misăl*, **2** (1), 17–22 (in Bulgarian).

Raicheva, L. B. 1959. Experiments with bacterial fertilization of alfalfa in leached black earth-smolnits. *Nauchni trudove In-ta pochveni izsledovaneto 'N. Pushkarov'*, 5, Zemizdat, Sofia, 113–127.

Raicheva, L. B. 1962. Propagation and virulence of vetch nodule bacteria in Bulgaria. *Izvestiya Tsentr. n.-i. in-ta pochvoznanie i agrotekhn*, **4**, 167–185 (in Bulgarian).

Raicheva, L. B. 1966. Effectiveness of nitragin in Bulgaria: in *IX Mezhdunarodnyi Mikrobiologicheskii kongress* (Ninth International Microbiology Congress), summaries of speeches. Moscow, 299.

Ramaswami, P. P. and Nair, K. S. 1965. Symbiotic variation of *Rhizobium* from nodules of redgram (*Cajanus cajan*). *Madras Agric. J.*, 52 (5), 239–240.

Rangaswami, G. and Oblisami, G. 1962a. Studies on some legume root nodule bacteria. *J. Indian Soc. Soil. Sci.*, **10** (3), 175–185.

Rangaswami, G. and Oblisami, G. 1962b. Quantitative studies on nitrogen fixation by five legume nodule bacteria. *J. Annamalia Univ.*, **23**, 119–126.

Rankov, V., Elenkov, E., Surlenkov, P. and Velev, B. 1966. Effect of certain herbicides on the development of nitrogen-fixing bacteria. *Agrokhimiya*, (4), 115–120.

Ratner, E. I. 1965. *Pitanie rastenii i primenenie udobrenii* (Plant nutrition and the use of fertilizers). Nauka, Moscow.

Ratner, E. I. and Akimochkina, T. A. 1964. Combined effect of molybdenum and certain vitamins of the B group on the productivity of symbiotic nitrogen fixation in soya. *Agrokhimiya*, (4), 59–68.

Rautanen, N. and Saubert, S. 1955. Root nodules of leguminous plants. A chemical study. *Suomen Kemistilehti*, **28B**, 66–70 (in English).

Rautenshtein, Ya. I. 1963. Importance of the problem of bacteriophage in agricultural microbiology: in *Rol' mikroorganizmov v pitanii rastenii i povyshenii udobrenii* (Role of micro-organisms in plant nutrition and in improving the efficacy of fertilizers). Leningrad, 22–28.

Razumovskaya, Z. G. 1932. Problem of bacteriophage. *Arkhiv biol. nauk.*, **32**, (4), 304.

Razumovskaya, Z. G. 1963. Nodule bacteria and cross-infected groups of legumes: in *Tezisy dokladov Vsesoyuznoi konferentsii po sel'sko-khozyaistvennoi mikrobiologii* (Summaries of speeches of the All-Union Conference on Agricultural Microbiology). Leningrad, 91.

Razumovskaya, Z. G. 1965. Cross-infected groups of legumes and species of nodule bacteria. *Trudy Petergofskogo biologicheskogo instituta*: in *Osnovy mikrobiologii* (Principles of microbiology). Leningrad University Press, part 19, 3–11.

Razumovskaya, Z. G. and Vasil'eva, O. A. 1956a. Effect of nodule bacteria on the chemical composition of legumes. *Uch. zapiski LGU*, No. 216, (41), 196–201.

Razumovskaya, Z. G. and Vasil'eva, O. A. 1956b. Some data on the structure of nodules of the lupin infected with active and inactive strains of nodule bacteria. *Uch. zapiski LGU*, No. 216, (41), 202–210.

Razumovskaya, Z. G., Isakova, N. P. and Petrova, G. N. 1952. Observations on the development of clover nodules. *Vestn. LGU*, (4), 122–127.

Razumovskaya, Z. G. and Fan' Yun'-lyu (Fan Yun-liu). 1961. Active and inactive nodules of legumes. *Trudy in-ta mikrobiologii Akad, Nauk SSSR*. (11), 169–176.

Read, M. 1953. The establishment of serologically identifiable strains of *Rhizobium trifolii* in field soils in competition with the native micro-flora. *J. Gen. Microbiol.*, 9 (1), 1–14.

Reisenauer, H. M. 1960. Cobalt in nitrogen fixation by a legume. *Nature*, 186, 375–376.

Richardson, D. A., Jordan, D. C. and Garrard, E. H. 1957. The influence of combined nitrogen on nodulation and nitrogen fixation by *Rhizobium meliloti* Dangeard. *Canad. J. Plant Sci.*, 37 (3), 205–214.

Rigaud, J. 1965. Contribution a l'etude d'un milieu synthetique pour la croissance de Rhizobium. *Ann. Inst. Pasteur*, 709 (3) Suppl., 272–279.

Rigaud, J. and Bulard, C. 1965. Sur la presence d'indolyl-3-aldehyde et d'acide indolyl-3-carboxylique dans les milieux de culture de Rhizobium. *CR Acad. Sci.*, 261 (3), 784–786.

Roberts, J. L. and Olson, F. R. 1942. The relative efficiency of strains of *Rhizobium trifolii* as influenced by soil fertility. *Science*, 95 (2468), 413–414.

Robinson, R. 1945. The antagonistic action of the by-products of several soil micro-organisms on the activities of the legume bacteria. *Soil Sci. Soc. America Proc.*, 10,.

Robinson, R. 1957. Legume inoculation a method of testing *Rhizobium* cultures. *East Afric. Agric. J.*, 22 (3), 130–132.

Roizin, M. B. 1959. Nodules on leguminous plants of the Kolsk peninsula. *Bot. zh.*, 44, (4), 467–474.

Roizin, M. B. 1966. Effect of nitragin on the yield of leguminous plants in polar farming. *Agrokhimiya*, (5), 34–39.

Roizin, M. B. and Buleishvili, Ya. M. 1966. Economic advantages of using nitragin in polar farming with sowings of annual leguminous plants. *Agrokhimiya*, (7), 70–72.

Rolitski, E. B. 1966. Bacteriocins of nodule bacteria: in *IX Mezhdunarodnyi Mikro-biologicheskii Kongress* (Ninth International Microbiological Congress), summaries of speeches. Moscow, 236.

Romashev, P. I. 1936. Utilization of the nitrogen of legumes by graminaceous grasses in mixed sowings. *Khimizatsiya sots. zemledeliya*, (11), 28–40.

Rorison, J. H., Sutton, C. D. and Hallsworth, E. G. 1958. The effect of climatic conditions on aluminium and manganese toxicities: in *Nutrition of the Legumes*. Ed. by Hallsworth, E. G. London, 62–68.

Rothschild, D. J. de. 1963. Anatomia del nodulo de algunas leguminosas cultivadas. *Rev. Inst. Municip. Bot.*, 3 (1), 3–32.

Rougieux, R. 1966. Sur la presence du *Rhizobium* (*Rhizobium meliloti*) dans les nodules de luzernes cultivées en basis Sahariennes (El Afiane, Sud-Est Algerien). *Ann. Inst. Pasteur*, 111, 100–101.

Rovira, A. D. 1961. *Rhizobium* numbers in the rhizospheres of red clover and paspalum in relation to soil treatment and the numbers of bacteria and fungi. *Austral. J. Agric. Res.*, 12 (1), 77–83.

Rovira, A. D. and Greacen, E. L. 1957. The effect of aggregate disruption on the activity of micro-organisms in the soil. *Austral. J. Agric. Res.*, 8 (6), 659–673.

Rubenchik, L. I., Bershova, O. I., Smalii, V. T., Zinov'eva, Kh. G., Andreyuk, E. I., Mal'tseva, N. N., Knizhnik, Zh. P., Kozlova, I. A., Smirnova, E. N. and Kogan,

S. B. 1966. Nitrogen-fixing micro-organisms of the soils of the Ukrainian SSR: in *IX Mezhdunarodnyi Mikrobiologicheskii kongress* (Ninth International Microbiology Congress), Moscow, Section V-2, summaries of speeches, 299–300.

Rudakov, K. I. and Birkel', M. R. 1954. Formation of nodules and protopectinase bacteria. *Trudy In-ta mikrobiol. Akad. Nauk SSSR*, (3), 125–143.

Ryumin, N. Nitragin, 1938 *Len i konoplya*, (1).

Sabinin, D. A. and Vyalovskaya, A. E. 1931. Guza-pai as a fertilizer in cotton growing. *Za khlopkovuyu nezavisimost'* (2), 55–62.

Sabo, B. 1964. *Sravnitel'noe deistvie izveski, molibdena i vanadiya na urozhai i kachestvo bobovykh kultur* (Comparative effect of lime, molybdenum and vanadium on the yield and quality of leguminous crops). (Author's abstract of M.Sc. thesis). Moscow.

Sahlman, K. and Fähraeus, G. 1962. Microscopic observations on the effect of indole-3-acetic acid upon root hairs of *Trifolium repens*. *Kgl. Lantbrukshögskol. Ann.*, 28, 261–266 (in English).

Sahlman, K. and Fähraeus, G. 1963. An electron microscope study of root hair infection by *Rhizobium*. *J. Gen. Microbiol.*, 33 (3), 425–427.

Sakhautdinov, M. 1939. Effect of inoculation on the course of accumulation of nitrogen, ash and other substances in the green mass of soya. *Trudy Bashkirsk s-kh. in-ta*, (2), Ufa.

Samsonova, S. P. 1959. *Vliyanie vysushivaniya liofil'nym metodom na kluben'kovye bakterii* (Effect of lyophilic drying on nodule bacteria). (Author's abstract of M.Sc. thesis). Moscow.

Samsonova, S. P. and Finkel'shtein, M. Ya. 1965. Nitragin-lucerne. *Zernobobovye kul'tury*, (11), 10–11.

Samtsevich, A. S. 1939. Liming of soils and adsorption of bacteria. *Khimizatsiya sots. zemledeliya*, (12).

Samtsevych, A. S., Zynov'eva, Kh. H., Heller, I. A. and Zaremba, V. P. (Samtsevich, A. S., Zinov'eva, Kh.G., Geller, I. A., Zaremba, V. P.). 1959. *Effektyvnist' bakteryal'nykh dobryv na Ukrayini* (Effectiveness of bacterial fertilizers in Ukraine), ULSHY, Kiev (in Ukrainian).

Sankaram, A. 1960a. The future rhizobiology of tropical legumes. *Sci. and Culture*, 25 (8), 464–467.

Sankaram, A. 1960b. Survival of rhizobia in peat-based legume inoculants. *Current Sci.*, 29 (3), 97.

Sankaram, A., Raju, P. V. L. N. and Kasaiah, S. 1963. Effect of phosphate on nitrogen fixing power of root nodule bacteria. *Sci. and Culture*, 29 (8), 406–407.

Sarić, Z. 1953. The effect of nitragination on the yield of soya. *Zemljiste i biljka*, (1), 157–168.

Sarić, Z. 1959. Nodule bacteria on certain strains of peas. *Letopis Naučnih Radov (Poljóprivredni Fak. Novy Sad)*, 3, 100–106.

Sarić, Z. 1963a. Nature of the virulence of various strains of nodule bacteria against peanuts. *Arh. Poljopr. Nauke*, 16 (53), 90–105.

Sarić, Z. 1963b. Specificity of strains of the nodule bacteria *Rhizobium* sp. isolated from peanuts in some parts of Yugoslavia. *J. Sci. Agric. Res.*, 16 (51), 60–76.

Saubert von Hausen, S. and Gylswyk, N. O. van. 1957. *South Afric. J. Sci.*, 53, 233–236 (quoted by Schreven, D. A. van. 1958).

Schaede, R. 1941. Untersuchungen in den Wurzelknöllchen von *Vicia faba* und *Pisum sativum*. *Beitr. Biol. Pflanz.*, 27 (11), 165–188.

Schiel, E., Olivero, E. L. G. de, Diéguez, R. N., Pacheco, J. C. and Enokida, E. 1963. Effect de la colchicine sur la virulence et l'efficacite de *Rhizobium meliloti* sur la luzerne. *Ann. Inst. Pasteur*, 105 (2), 332–340.

Schlegel, H. G. 1962. Die Isolierung von Poly-β-hydroxybuttersaure aus Wurzelknöllchen von Leguminosen. *Flora*, 152 (2), 236–240.

Schluchterer, E. and Stacey, M. 1945. The capsular polysaccharide of *Rhizobium radicicolum. J. Chem. Soc.*, **776**.

Schmehl, W. R., Peech, M. and Bradfield, R. 1950. Causes of poor growth of plants on acid soils and beneficial effects of liming. I. Evaluation of factors responsible for acid soil injury. *Soil. Sci.*, **70**, 393–410.

Schreven, D. A. van. 1958. Some factors affecting the uptake of nitrogen by legumes. in *Nutrition of the Legumes*. Ed. by Hallsworth, E. G. London, 137–163, 212.

Schreven, D. A. van. 1959. Effects of added sugars and nitrogen on nodulation of legumes. *Plant and Soil*, **11** (2), 93–112.

Schreven, D. A. van. 1964. The effect of some actinomycetes on rhizobia and *Agrobacterium* radiobacter. *Plant and Soil*, **21**, 283–302.

Schreven, D. A. van, Harmsen, G. W., Lindenbergh, D. J. and Otzen, D. 1953. Experiments on the cultivation of *Rhizobium* in liquid media for use on the Zuidersee polders. *Antonie van Leeuwenhoek, J. Microbiol. and Serol.*, **19** (4), 300–308.

Schreven, D. A. van, Otzen, D. and Lindenbergh, D. J. 1954. On the production of legume inoculands in a mixture of peat and soil. *Antonie van Leeuwenhoek. J. Microbiol. and Serol.*, **20** (1), 33–57.

Schwendimann, F. 1955. Versuche zur Herstellung eines *Rhizobium*—Impimaterials fur Luzerne unter besonderer Berucksichtigung schweizerischer Verhaltnisse. Promotionsarbeit, Eidg. Techn. Hochsch. Zurich.

Schwinghamer, E. A. 1964. Association between antibiotic resistance and ineffectiveness in mutant strains of *Rhizobium* spp. *Canad. J. Microbiol.*, **10** (2), 221–233.

Schwinghamer, E. A. 1965. Host controlled modification of *Rhizobium* bacteriophage. *Austral. J. Biol. Sci.*, **18** (2), 333–343.

Schwinghamer, E. A. 1956. Factors affecting phage-restricting ability in *Rhizobium leguminosarum* strain L.4. *Canad. J. Microbiol.*, **12** (2), 395–408.

Schwinghamer, E. A., Reinhard, D. J. 1963. Lysogeny in *Rhizobium leguminosarum* and *Rhizobium trifolii. Austral. J. Biol. Sci.*, **16** (3), 597–605.

Segi, I. and Manninger, E. (Szegi, I. and Manninger, E.). 1964. Use of bacterial fertilizers in Hungary: in *Bakterial'nye udobreniya* (Bacterial fertilizers). Kolos. Moscow, 88–94.

Sembrat, Z. 1934. Influence of caffeine on morphological changes of the nodule bacteria. *Acta Soc. Bot. Polon.*, **11**, 333–346 (in English).

Sen, A. N. 1965. Relationship between the efficiency of different strains of soybean nodule organisms (*Rhizobium japonicum*) with their ability to consume glucose in pure culture. *Sci. and Culture*, **31** (8), 429–430.

Sen Arindra Nath. 1966. Inoculation of legumes influenced by soil and climatic conditions. *Indian J. Agric. Sci.*, **36** (1), 1–8.

Senn, H. A. 1938. Chromosome number relationships in the Leguminosae. *Bibliogr. Genet.*, **12**, 175–236.

Shevchuk, V. E. 1962. Effectiveness of applying nitragin under leguminous crops. *Izv. Irkutsk. s-kh. in-ta.*, (24).

Shevchuk, V. E. *Bakterial'nye udobreniya v Vostochnoi Sibiri* (Bacterial fertilizers in Eastern Siberia). Vost.-Sib. kn. izd-vo.

Shevchuk, V. E. 1965. Effectiveness of inoculation and the level of biological fixation of atmospheric nitrogen. *Izv. Irkutsk. s-kh. in-ta.*, **3** (25), 119–138.

Shevchuk, V. E. 1966. Effectiveness of inoculation of leguminous plants and the level of biological fixation of molecular nitrogen: in *IX Mezhdunarodnyi Mikrobiologicheskii Kongress* (Ninth International Microbiology Congress), summaries of speeches. Moscow, 287.

Shemakhanova, N. M. and Bun'ko, I. P. 1966a. Accumulation of group B vitamins in the nodules of *Phaseolus*: in *IX Mezhdunarodnyi Mikrobiologicheskii Kongress*

(Ninth International Microbiology Congress), summaries of speeches. Moscow, 300–301.

Shemakhanova, N. M. and Bun'ko, I. P. 1966b. Group B vitamins in the nodules of certain leguminous plants: in *Biologicheskii azot i ego rol' v zemledelii* (Biological nitrogen and its role in agriculture). Akad. Nauk. SSSR, Moscow.

Shil'nikova, V. K. 1962. The nitrogen composition of the bleeding sap of pea plants infected with races of nodule bacteria differing in efficacy. *Izv. Akad. Nauk SSSR, Ser. Biol.*, (6), 840–844.

Shil'nikova, V. K. 1963. *Issledovanie azotnogo sostava bakterizovannykh rastenii i izoelektricheskoi tochki kluben'kovoi tkani kak vozmozhnykh kriteriev effektivnosti azotfiksiruyushchikh bakterii* (Study of the nitrogen composition of bacterized plants and the isoelectric point of nodular tissue as possible criteria of the effectiveness of nitrogen-fixing bacteria). (Author's abstract of M.Sc. thesis). Moscow.

Shil'nikova, V. K. and Agadzhanyan, K. G. 1965. Respiratory activity of various strains of nodule bacteria. *Dokl. TSKhA*, 109, 329–337.

Shil'nikova, V. K. and Agadzhanyan, K. G. 1965. Study of the dehydrogenase activity of strains of nodule bacteria differing in their effectiveness. *Izv. TSKhA*, (3), 126–130.

Shil'nikova, V. K., Poglazova, M. N. and Agadzhanyan, K. G. 1966. Dehydrogenase activity of symbiotic nodule bacteria. *Izv. TSKhA*, (5), 41–44.

Shkol'nik, M. Ya. and Bozhenko, V. P. 1956. Effect of various methods of trace element supply on the development, yield and food qualities of red clover and its trace element content. *Izv. Akad. Nauk SSSR, ser. biol.*, (4), 39–57.

Shmidt, E. F. 1964. Differentiation of strains of nodule bacteria on the basis of determination of the activity of the 'substrate-free' dehydrases. *Mikrobiologiya*, 33 (2), 284–291.

Shtern, E. A. 1953. Serological characterization of nodule bacteria. *Mikrobiologiya*, 22 (4), 423–430.

Shtina, E. A., Pankratova, E. M., Perminova, G. N., Tret'yakova, A. N. and Yung, L. A. 1966. Participation of blue-green algae in the accumulation of nitrogen in the soil: in *IX Mezhdunarodnyi Mikrobiologicheskii Kongress* (Ninth International Microbiology Congress), summaries of speeches. Moscow, 277–278.

Shulyndin, A. F. 1953. Improving the effectiveness of soya nitragin by selecting soya varieties with high susceptibility to nodule bacteria. *Mikrobiologiya*, 22 (3), 288–294.

Shumilin, L. G. and Muravin, E. A. 1965. Effectiveness of applying phosphorus fertilizers, molybdenum and nitragin under pulse legumes in leached black earth. *Dokl. TSKhA*, (115), 19–22.

Shchepkina, O. I. 1959. Study of techniques of improving the effectiveness of nodule bacteria in a grass mixture: in *Mikrobiologiya na sluzhbe sel'skogo khozyaistva* (Microbiology in the service of agriculture). Sel'khozgiz. Moscow, 104–109.

Sidorenko, A. I. and Fedorova, L. N. 1965. Effect of certain pea seed disinfectants on the development of the bacterium *Rhizobium leguminosarum*: in *Selektsiya i semenovodstvo* (Selection and seed growing). (4), Urozhai, Kiev, 145–149.

Sironval, C. 1958. Relation between chlorophyll metabolism and nodule formation in soya bean. *Nature*, 181 (4618), 1272–1273.

Škrdleta, V. 1965a. Somatic sero-groups of *Rhizobium japonicum*. *Plant and Soil*, 23 (1), 43–48.

Škrdleta, V. 1965b. Lyophilization of rhizobia. *Acta Microbiol. Polon.*, 14 (1), 109–113 (in English).

Škrdleta, V., Vintikova, H. and Stogl, M. 1964. The sensitiveness of some rhizobia to some herbicides. *Czech. Min. Zemeded Les. Vond. Hospodarstwa.*, 37 (8), 827–834 (in English).

Smith, F. B., Blaser, R. E. and Thornton, G. T. 1945. Legume inoculation. *Univ. Florida Bull.*, 417, 1–32.

Smith, J. D. 1949. Haemoglobin and the oxygen uptake of leguminous-root nodules. *Biochem. J.*, **44** (5), 591–599.

Smith, J. and Jordan, D. 1949. The concentration and distribution of haemoglobin in the root nodules of leguminous plants. *Biochem. J.*, **44** (5), 585–591.

Snieszko, S. 1928. L'influence exercée par la concentration des ions d'hydrogene du milieu nutritif sur le developpement des bacteries des nodosites du haricot. *Bull. Acad. Polonica Sci.*, *Ser. B Science (Bot.)*.

Sobachkin, A. A. and Muravin, E. A. 1963. Effect of molybdenum on the yield of broad beans. *Dokl. TSKhA*, (89), 67–70.

Sokolov, A. V. 1957. Utilization of nitrogen by leguminous grasses in agriculture. *Trudy. Pochv. in-ta Akad. Nauk SSSR*, 50.

Sokurova, E. N. 1956. Certain patterns of the action of penetrating radiations on micro-organisms. *Izv. Akad. Nauk SSSR, Ser. Biol.*, (6), 35–53.

Spencer, J. F. T. 1954. Oxygen uptake by rhizobia in soil. *Canad. J. Bot.*, **32** (2), 380–385.

Spicher, G. 1954. Lebensdauer und Stickstoffbindung der Knöllchenbakterien von Lupine, Seradella und Klee in Abhängigkeit von ihrer Gestalt. *Zbl. Bakteriol.*, *Parasitenkunde, Infektionskrankh. und Hyg.*, Abt. 2, **107**, 383–418.

Stalder, L. 1952. Über Dispositionsverschiedungen bei der Bildung von Wurzel-knöllchen. *Phytopathol., Z.*, **18** (4), 376–403.

Staniewski, P. and Kowalski, M. 1965. Effect of lysogenization on variability of phage type in *Rhizobium meliloti*. *Acta Microbiol. Polon.*, **14** (3–4), 231–236 (in English).

Staniewski, R., Kowalski, M., Gogacz, E. and Sokolowska, F. 1962. Susceptibility of *Rhizobium* strains to phages. *Acta Microbiol. Polon.*, **11** (3), 245–254.

Staniewski, R., Kowalski, M. and Gorzkowska, K. 1963. The rate of phage adsorption on *Rhizobium* cells. *Acta Microbiol. Polon.*, **12** (3), 184–187 (in English).

Stenz, E. 1962. Über den Einfluss von Bakterienfiltraten und Wuchsstoffen auf Wurzelhaare. *Wiss. Z. Karl-Marx-Univ. Leipzig. Math. Naturwiss. Reige*, **11** (4), 641–646.

Stevens, I. W. 1923. Can all strains of a specific organism be recognized by agglutination? *J. Infect. Diseases*, **33**, 557.

Stevens, I. W. 1925. A study of various strains of *Bac. radicicola* from nodules of alfalfa and sweet clover. *Soil Sci.*, **20**, 45–66.

Stoklasa, J. 1895. Studien über die Assimilation elementaren Stickstoffs durch die Pflanzen. *Landwirtsch. Jahresh.*, **24**, 827–863.

Stolp, H. 1952. Beitrage zur Frage der Beziejungen zwischen Mikroorganismen und höheren Pflanzen. *Arch. Mikrobiol.*, **17**, 1–14.

Strijfom, B. W. and Allen, O. N. 1966. Medium supplementation with l- and d-amino acids relative to growth and efficiency of *Rhizobium meliloti*. *Canad. J. Microbiol.*, **12** (2), 275–284.

Stul'neva, A. M. 1965. Effect of nitraginization and trace elements on the yield of the green mass and blue-hybrid lucerne seeds in the second year. *Izv. Irkutsk. s-kh. in-ta*, **3**, (25), 377–390.

Szende, K. and Ördögh, F. 1960. Die Lysogenie von *Rhizobium meliloti*. *Naturwissenschaften, Jahresb.*, **17**, 404–405.

Szende, K., Sik, T., Ördögh, G. and Gyorffy, B. 1961. Transfer of immunity by nucleic acids of a lysogenic *Rhizobium* strain. *Biochim. Biophys. Acta*, **47**, 215–217.

Subba-Rao, N. S. and Vasantha, P. 1965a. Nodulation of *Trifolium alexandrinum in vitro* and nitrate effect on the amino acid composition of the plant and its root exudate. *Canad. J. Bot.*, **43** (10), 1189–1194.

Subba-Rao, N. S. and Vasantha, P. 1965b. Fungi on the nodular surface of some legumes. *Naturwissenschaften*, **52**, 44–45.

Suchting, H. 1904. Kritische Studien uber die Knollchen bakterien. *Zbl. Bakteriol., Parasitenkunde, Infektionskrankh. und Hyg.*, Abt. 2, **11**, 377–388.

Tagiev, V. D. 1965. Effect of heteroauxin on the activity and virulence of lucerne nodule bacteria. *Izv. Akad. Nauk SSSR, Ser. Biol.*, (2), 291–292.

Tagiev, V. D. 1966. Effect of heteroauxin on the activity of pea nodule bacteria in different phases of development. *Izv. Akad. Nauk SSSR, Ser. Biol.*, (3), 448–451.

Takahashi, J. and Quadling, C. 1961. Lysogeny in *Rhizobium trifolii. Canad. J. Microbiol.*, **7**, 455–465.

Tanner, J. W. and Anderson, J. C. 1963. An external effect of inorganic nitrogen in root nodulation. *Nature*, **198** (4877), 303–304.

Tanner, J. W. and Anderson, J. C. 1964. External effect of combined nitrogen on nodulation. *Plant Physiol.*, **39** (6), 1039–1043.

Taranovskaya, V. G. 1931. Lime and legumes in improving the nitrogen balance of solodized soils. *Udobrenie i urozhai*, (6), 521–531.

Taylor, G. S. and Parkinson, D. Studies on fungi in the root region. IV. Fungi associated with the roots of *Phaseolus vulgaris* L. *Plant and Soil*, **22**, 1–20.

Tešić, Z. and Todorović, M. 1963. Note on the classification of nodule bacteria. *Zemljište i biljka*, **12** (1–3), 279–285 (in Serbo-Croat; abstract in English).

Tewari, G. P. 1963. Note on effects of soil sterilization and some mineral nutrients on commercial strains of cowpea *Rhizobium* in Western Nigeria. *Empire J. Exptl Agric.*, **31** (121), 50–52.

Tewari, G. P. 1966. Effect of planting-date on nodulation and dry matter yield of cowpea in Nigeria. *Exptl Agric.*, **2**, 45–47.

Thimann, K. V. 1936. On the physiology of the formation of nodules on legume roots. *Proc. Nat. Acad. Sci. USA*, **22**, 511–514.

Thompson, J. A. 1960. Inhibition of nodule bacteria by an antibiotic from legume seed coats. *Nature*, **187**, 619–620.

Thorne, D. W. and Brown, P. E. 1937. The growth and respiration of some soil bacteria in juices of leguminous and nonleguminous plants. *J. Bacteriol.*, **34**, 578–580.

Thornton, H. G. 1929a. The effect of fresh straw on the growth of certain legumes. *J. Agric. Sci.*, **19**, 563–570.

Thornton, H. G. 1929b. The role of the young lucerne plant in determining the infection of the root by the nodule forming bacteria. *Proc. Roy. Soc.*, B, **104**, 481–492.

Thornton, H. G. 1929c. The influence of the number of nodule bacteria applied to the seed on nodule formation in legumes. *J. Agric. Sci.*, **19**, 373–381.

Thornton, H. G. 1930. The early development of the root nodule of lucerne (*Medicago sativa* L.). *Ann. Bot.*, New Series, **44**, 385–392.

Thornton, H. G. 1936. The action of sodium nitrate upon the infection of lucerne root hairs by nodule bacteria. *Proc. Roy. Soc.*, B, **119**, 474–492.

Thornton, H. G. 1947. Report of the Department of Soil Microbiology (Rothamsted Experimental Station) for the years 1939–1945. Gibbs and Bamforth Ltd., St. Albans.

Thornton, H. G. 1947. The biological interactions of *Rhizobium* with its host legume. *Antonie van Leeuwenhoek. J. Microbiol. and Serol.*, **12**, 85–96.

Thornton, H. G. 1952. The symbiosis between rhizobia and leguminous plants and the influence of this on the bacterial strain. *Proc. Roy. Soc.*, B, **139** (895), 170–176.

Thornton, H. G. and Gangulee, N. 1926. The life cycle of the nodule organism *Bacterium radicicola* (Beij) in soil and its relation to the infection of the host plant *Proc. Roy. Soc.*, B, **99**, 427–451.

Thornton, H. G. and Kleczkowski, J. 1950. Use of antisera to identify nodules produced by the inoculation of legumes in the field. *Nature*, **166**, 1117–1118.

Thorogood, E. 1957. Oxygenated ferroheme proteins from soybean nodules. *Science*, **126**, 1011–1012.

Til'ba, V. A. and Golodyaev, G. M. 1966. Nitrogen-fixing bacteria of the arable soils of the *Primor'e*. in *Problemy biologii na Dal'nym Vostoke* (Problems of biology in the Soviet Far East), Vladivostok, 152–154.

Timiryazev, K. A. 1893. *Istochniki azota dlya rastenii. Izbrannye sochineniya* (Sources of nitrogen for plants. Selected Works), **2**, 1948.

Tittsler, R. and Lisse, M. 1936. The electrophoretic potential of *Rhizobium meliloti*. *J. Bacteriol.*, **6**.

Tokhver, V. and Lokk, E. 1966. Problem of prototrophicity and the radio-resistance of certain species of nodule bacteria. *Izv. Akad. Nauk EstSSR, Ser. Biol.*, **15**, (3), 347–356.

Tonhazy, N. E. and Pelczar, M. J., Jr. 1954. Oxidation of indoleacetic acid by an extracellular enzyme from *Polyporus versicolor* and a similar oxidation catalyzed by nitrite. *Science*, **120**, 141–142.

Tove, S. R. and Wilson, P. W. 1948. Isotopic studies of fixation by rhizobia in presence of hemoprotein. *Proc. Soc. Exptl Biol.*, **69**, 184–186.

Trepachev, E. P. and Artashkova, N. A. 1965. Effect of magnesium and calcium on the lupin. *Agrokhimiya*, (9), 65–71.

Trepachev, E. P. and Khabarova, A. I. 1966. Determination of true nitrogen fixation by legumes. *Vestn. s-kh. nauki*, (12), 105–108.

Trinick, M. 1965. *Medicago sativa* nodulation with *Leucana leucacephala* root nodule bacteria. *Austral. J. Sci.*, **27** (9), 263–264; *Rhizobium Newsletter*, 1965, **10** (2), 164–167, 209.

Tsyurupa, B. N. 1937. Effect of inoculation on the yield of various varieties of soya and *Phaseolus*. in *Sbornik nauchno-issledovatel'nykh rabot Azovo-Chernomorskogo sel'sko-khozyaistvennogo instituta* (Collected scientific works of the Azov-Black Sea Agricultural Institute) (5), 14–30.

Tulaganov, A. T. 1954. Harmful nematodes of cultivated plants of Uzbekistan. *Trudy In-ta zoologii i parazitologii Akad. Nauk UzbSSR.*, Tashkent.

Turchin, F. B. 1956. Study of biological fixation of atmospheric nitrogen by nodule bacteria using [15]N. *Tezisy dokladov soveshchaniya po voprosam izucheniya s pomoshch'yu metodov mechenykh atomov pitaniya rastenii i primeneniya udobrenii* (Summaries of speeches at the conference on the study of plant nutrition and the use of fertilizers using isotopes). Moscow, 27–28.

Turchin, F. V. New data on the mechanism of fixation of atmospheric nitrogen in the nodules of leguminous plants. *Pochvovedenie*, (10), 14–24.

Turner, E. R. 1955. The effect of certain adsorbents on the nodulation of clover plants. *Ann. Bot.*, New Series, **19** (73), (65), 148–160.

Tuzimura, K. 1950. Studies on the respiration of the root-nodules of leguminous plants. *J. Sci. Soil Manure*, **21**, 111–115, 283–287; *J. Agric. Chem. Soc. Japan*, **24**, 97–100.

Tuzimura, K. and Watanabe, J. 1960. Saprophytic life of *Rhizobium* in soils free from the host plants. Ecological studies of *Rhizobium* in soils. *Soil and Plant Food*, **6** (1), 44.

Tuzimura, K. and Watanabe, J. 1961a. Multiplication of *Rhizobium* in the rhizosphere of host plants. Ecological studies of *Rhizobium* in soil. 3. *Soil and Plant Food*, **6** (3), 138.

Tuzimura, K. and Watanabe, J. 1961b. Estimation of number of root-nodule bacteria by a nodulation-dilution frequency method. Ecological studies of *Rhizobium* in soils. 1. *Soil Sci. and Plant Nutr.*, **7** (2), 61–65.

Tuzimura, K. and Watanabe, J. 1961c. Root nodule bacteria as members of rhizospheral microflora. Ecology of root nodule bacteria in soils. 4. *Soil Sci. and Plant Nutr.*, **7** (1), 34.

Tuzimura, K. and Watanabe, J. 1962a. Ecological studies of root nodule bacteria.

2. The growth of *Rhizobium* in the rhizosphere of the host plant. *Soil Sci. and Plant Nutr.*, **8** (2), 19–24.

Tuzimura, K. and Watanabe, J. 1962*b*. The effect of the rhizosphere of various plants on the growth of *Rhizobium*. Ecological studies of root nodule bacteria. 3. *Soil Sci. and Plant Nutr.*, **8** (4), 153–157.

Tuzimuts, K. and Watanabe, J. 1964. Difference in the rhizosphere effect on *Rhizobium trifolii* and *R. meliloti* between soils. 6. Ecology of root nodule bacteria in soil. *Soil Sci. and Plant Nutr.*, **10** (3), 134.

Uher, M. 1937. Ein Beitrag zum Problem des Bakterienkernes. *Ann. Acad. Tchécosl. Agric.*, **12**, 474–478.

Ukrainskii, V. T. 1954. Nodule bacteria of the roots of leguminous plants. *Mikrobiologiya*, **23**, (3), 291–296.

Umbreit, W. W. 1944. Three more reasons for soybean inoculation. *Soybean Digest*, **4** (6), 9–10.

Uspenskaya, T. A. 1953. *Azotfikziruyushchaya aktivnost' kluben'konykh bakterii v kluben'kakh gorokha i soi pri razlichnoi prodolzhnitel'nosti i intensivnosti osveshcheniya rastenii i v chistykh kul'turakh* (Nitrogen-fixing activity of nodule bacteria in pea and soya nodules for different times and intensities of illumination of plants and in pure cultures). (Author's abstract of M.Sc. thesis). Moscow.

Uziak, Z. 1964. The utilization of inorganic or organic nitrogen by leguminous plants at various C:N ratios in the plants. I, II. *Ann. Univ. M. Curie-Sklodowska, Lublin*, Sect. E, **19**, 163–187; 499–535 (in English).

Uziakowa, L. 1959. The influence of different forms of nitrogen-nutrition upon the growth and symbiosis of soybean. *Acta Microbiol. Polon.*, **8** (3–4), 315–318 (in English).

Valera, L. and Alexander, M. 1965*a*. Nodulation factor for *Rhizobium* legume symbiosis. *J. Bacteriol.*, **89** (4), 1134–1139.

Valera Concepsion, L. and Alexander, M. 1965*b*. Reversal of nitrate inhibition of nodulation by indolyl-3-acetic acid. *Nature*, **206** (4981), 326.

Van Chzhi-Tun (Wang Chih T'ung). 1963. Effect of the nodules on the free amino acid content and the intensity of photosynthesis in the case of vetch. *Izvestiya in-ta pochvoznanie i agrotekhn. 'N. Pushkarov'*, **7**, Sofia.

Vandecaveye, S. 1927. *Soil Sci.*, **23**, 355–362 (quoted by Schreven, D. A. van. 1958).

Vantsis, J. T. and Bond, G. 1950. The effect of charcoal on the growth of leguminous plants in sand culture. *Ann. Appl. Biol.*, **37**, 159–168.

Varaksina, A. S. and Pozharinskaya, N. A. 1965. Effects of nitragination of seeds on the microflora and yield of broad beans. *Agrobiologiya*, (1), 25–31.

Vartiovaara, U. 1933. Untersuchungen über die Leguminosen-Bakterien und Pfanzen. 12. Über die Stickstoffhaushalt des Hafers bei feldmässigen Mischkulturen zusammen mit der Erbse. *Z. Pflanzenernähr., Düng und Bodenkunde*, A. **37**, 253–259.

Vartiovaara, M. 1937. Investigations on the root nodule bacteria of leguminous plants. 21. The growth of the root nodule organisms and inoculated peas at low temperature. *J. Agric. Sci.*, **27**.

Vass, A. F. 1919. The influence of low temperature on soil bacteria. *NJ (Cornell) Agric. Exptl Stat. Mem.,*, **27** 1039–1074.

Verbolovich, P. A. 1949. Haemoglobin and the assimilation of atmospheric nitrogen by the root nodules of leguminous plants. *Izv. Akad. Nauk KazSSR, Ser. Mikrobiol.*, (1), 77–85.

Verona, O., Gambogi, P. and Mazzanti, P. 1966. Ricerche sulla rizosfera di *Vicia faba* L. *Agric. Ital.*, **66** (1), 28–44.

Vidal, G. and Visona, L. 1965. Étude de la mycoflore de la rhizosphère de trois légumineuses. *Ann. Inst. Pasteur*, **108** (4), 535–540.

Viermann, H. 1929. Die Wurzelknöllchen der Lupine. *Bot. Arch.*, **25**, 45–86.

Vikulina, L. A. and Krylova, L. N. 1966. Natural inoculation of the pea by nodule bacteria and its yield in the Udmur ASSR. in *Ispol'zovanie mikroorganizmov dlya povysheniya urozhaya sel'skokhozyaistvennykh kultur* (Use of micro-organisms for increasing the yield of farm crops), Kolos, Moscow, 125–130.

Vincent, J. M. 1941. Serological studies of the root nodule bacteria. 1. *Proc. Linnean Soc. NS Wales*, **66**, 145–154.

Vincent, J. M. 1942. Serological studies of the root nodule bacteria. 2. *Proc. Linnean Soc. NS Wales*, **67**, 82–86.

Vincent, J. M. 1945. Host specificity amongst root-nodule bacteria, isolated from several clover species. *J. Austral. Inst. Agric. Sci.*, **11**, 121–127.

Vincent, J. M. 1953. Vi-like antigen in clover nodule bacteria. *Austral. J. Sci.*, **15** (4), 133–134.

Vincent, L. M. 1958. Survival of the root nodule bacteria. in *Nutrition of the Legumes*. Ed. by Hallsworth, E. G. London, 108–123.

Vincent, J. M. 1962. Influence of calcium and magnesium on the growth of *Rhizobium*. *J. Gen. Microbiol.*, **28** (4), 653–663.

Vincent, J. M. 1966a. Legume inoculation as an exercise in applied microbiology. *Report Rothamsted Experimental Station*, 1–27.

Vincent, J. M. 1966b. Manual on the Root-Nodule Bacteria. Rothamsted Experimental Station.

Vincent, J. M. (Vintsent, Dzh. M.). 1966. Inoculation of seeds of leguminous plants; a task in applied microbiology: in *IX Mezhdunarodnyi Mikrobiologicheskii kongress* (*Ninth International Microbiology Congress*), Symposium B-1, Moscow, 56–57.

Vincent, J. M. and Colburn, J. R. 1961. Cytological abnormalities in *Rhizobium trifolii* due to a deficiency of calcium and magnesium. *Austral. J. Sci.*, **23**, 269.

Vincent, J. M., Humphreys, B. and North, R. J. 1962. Some features of the fine structure and chemical composition of *Rhizobium trifolii*. *J. Gen. Microbiol.*, **29**, 551–555.

Vincent, J. M. and Jancey, C. H. 1962. A calcium-sensitive strain of *Rhizobium trifolii*. *Nature*, **195** (4836), 99–100.

Vincent, J. M., Thompson, J. A. and Donovan, K. O. 1962. Death of root-nodule bacteria on drying. *Austral. J. Agric. Res.*, **13** (2), 258–270.

Vincent, J. M. and Waters, L. M. 1954. The root-nodule bacteria as factors in clover establishment in the Lismore district, New South Wales. *Austral. J. Agric. Res.*, **5** (1), 61–76.

Vinogradova, Kh. G. 1952. Molybdenum and its biological role. *Mikroelementy v zhizni rastenii i zhivotnykh* (Trace elements in the life of plants and animals), Moscow, 515–538.

Vinogradsky (Vinogradskii, S. N.). 1936. Etudes sur la microbiologie du sol. Recherches sur les bacteries radicicoles des legumineuses. *Ann. Inst. Pasteur*, 56–69.

Vintiková, J., Šrogl, M. and Skrdleta, V. 1961. A contribution to the serological identification of the *Rhizobia*. *Folia Microbiol.*, **6** (4), 243–249.

Virtanen, A. I. 1928. Über die Einwirkung und die Zusammensetzung der Leguminosepflanzen. *Biochem. J.*, **193**, 300–312.

Virtanen, A. I. 1945. Symbiotic nitrogen fixation. *Nature*, **155** (3947), 747–748.

Virtanen, A. I. 1953. Microbiology and chemistry of symbiotic nitrogen fixation. *Internat. Bot. Congr. Proc.*, *Stockholm*, **7**, 156–159 (in English).

Virtanen, A. I. and Hausen, S. von. 1931. Untersuchungen über die Leguminosen-Bakterien und Pflanzen. Die Ausnutzung verschiedener Stickstoffsverbindung durch Leguminosepflanzen. *Biochem. Z.*, **232**, 1–14.

Virtanen, A. I. and Hausen, S. von. 1936. Investigations on the root nodule bacteria of leguminous plants. 17. Continued investigations on the effect of air content of the medium on the development and function of the nodule. *J. Agric. Sci.*, **26**, Pt. II, 281–287.

Virtanen, A. I., Jorma, J., Linkola, H. and Linnasalmi, A. 1947. On the relation between nitrogen fixation and leghaemoglobin content of leguminous root nodules. *Acta Chim. Scand.*, **1**, 90–111 (in English).

Virtanen, A. I. and Laine, T. 1936. Investigations on the root nodule bacteria of leguminous plants. 18. Breakdown of proteins by the root nodule bacteria. *Biochem. J.*, **30**, 377–381.

Virtanen, A. I. and Laine, T. 1937a. N-fixation by excised root nodules. *Suomen Kemistilehti*, **10**, 24 (in English).

Virtanen, A. I. and Laine, T. 1937b. The decarboxylation of d-lysine and l-aspartic acid. *Enzymologia*, **3**, 266–270.

Virtanen, A. I. and Laine, T. 1946. Red, brown and green pigments in leguminous root nodules. *Nature*, **157**, 25–26.

Virtanen, A. I. and Linkola, H. 1947. Competition of *Rhizobium* strains in nodule-formation. *Antonie van Leeuwenhoek. J. Microbiol. and Serol.*, **12**, 65–77.

Virtanen, A. I. and Linkola, H. 1948. On the antibacterial effect of spore-forming soil bacteria on the legume bacteria. *Suomen Kemistilehte*, **21**, 12–13 (in English).

Virtanen, A. I. and Miettinen, J. K. 1963. Biological nitrogen fixation: in *Plant Physiol.*, Ed. by Steward, F. C., III, 539–669.

Virtanen, A. I., Nordlund, M. and Hollo, E. 1934. Fermentation of sugar by the root nodule bacteria. *Biochem. J.*, **28**, 796–802.

Virtanen, A. I. and Santaoja, I. M. 1959. Annual Report of the Foundation for Chemical Research, 1959 (in Finnish).

Virtanen, A. I., Synnove, E. and Karstrom, H. 1933. Untersuchungen über die leguminose Bakterien und Pflanzen. *Biochem. Z.*, **258**, 106–112.

Virtanen, A. I. and Torniainen, M. A. 1940. A factor influencing nitrogen excretion from leguminous root nodules. *Nature*, **145** (3662), 25–26.

Visona, L. and Tardieux, P. 1964. Antagonists des *Rhizobium* dans la rhizosphere du trèfle et de la lucerne. *Ann. Inst. Pasteur, Suppl.*, (3), 297–302.

Voets, J. 1949. Investigation of the nuclear structure of *Rhizobium*, Landbouwhogeschool en Opzoekingssta Staat Gent, **14**, 235–250 (in Flemish).

Voronin, M. S. 1867. Observation sur certains exercissances que présentent les racines de l'aune et du lupin des jardins. *Ann. Sci. natur Bot. ser.* 5, **7**, 73–86.

Voronin, M. S. 1886. Mycorrhiza. *Zapis. Peterb. Akad. nauk.*

Voronkevich, I. V. 1966. Survival of phytopathogenic bacteria outside plant tissues in relation to their origin and the evolution of their parasitic properties. *Zh. obshch. biol.*, **27**, (6), 633–646.

Voroshilova, E. A., Lopatin, N. G. and Pashkina, T. S. 1964. Effect of local strains of nodule bacteria on the yield of soybean. *Agrokhimiya*, (8), 131–137.

Vyas, N. D. and Desay, J. R. 1953. Effect of different doses of superphosphate on the fixation of atmospheric nitrogen through pea. *J. Indian Soc. Soil Sci.*, **1**, 32–46.

Vyas, S. R. and Prasad, N. 1959. Studies on *Rhizobium* from *Lornia diphylla*. *Proc. Indian Acad. Sci.*, B, **49** (3), 156–160.

Vyval'ko, I. G. and Sokorenko, N. V. 1966. Effect of inoculation with nodule bacteria on the amino acid composition of the vegetative organs of soya. *Agrokhimiya*, (3), 61–69.

Wagenbreth, D. 1961a. Ein Beitrag zur systematischen Einordnung der Knöllchenbakterien durch Bestimmung des relativen Basengehaltes ihrér Desocyribonucleinsäuren. *Flora*, **151** (2), 219–230.

Wagenbreth, D. 1961b. Über das Wachstum und die Knöllchenbildungsfähigkeit isolierter Erbsenwurzein in geteiltem Medium. *Flora*, **151** (4), 607–620.

Wagenbreth, D. 1965. Zur Transformierbarkeit den Knöllchenbakterien. *Arch. Mikrobiol.*, **52** (2), 147–168.

Ward, H. M. 1887. On the tubercular swellings on the roots of *Vicia faba. Phil. Trans. Roy. Soc.*, B, **178**, 538–562.

Ward, H. M. 1889. On the tubercles on the roots of leguminous plants with special reference to the pea and bean. *Proc. Roy. Soc.*, **46**, 431–443.

Watanabe, J. and Tuzimura, K. 1962. Root nodule bacteria isolated from *Trifolium lupinaster. Soil Sci. and Plant Nutr.*, **8** (1), 43–44.

Weber, C. R. 1966. Nodulating and non-nodulating soy bean isolines. I. Agronomic and chemical attributes. II. Response to applied nitrogen and modified soil conditions. *Agron. J.*, **58** (1), 43–48.

Weber, E. 1930. Salpeterdüngung als Beeinträchtigung der Stickstoffsammlung durch Leguminosen. *Zbl. Bakteriol., Parasitenkunde, Infektionskrankh. und Hyg.*, Abt. 2, **82**, 353–359.

Weichsel, G. 1961. Untersuchungen über die Ausscheidung von Stickstoffverbindungen aus den Würzelknöllchen von Leguminosen. *Flora*, **171** (4), 535–571.

Weir, J. B. 1964. The effect of inositol on the growth and nodulation of diploid and tetraploid white clover grown in pot culture. *Plant and Soil*, **20**, 175–183.

Whyte, R. O., Nilsson-Leissner, G. and Trumble, H. C. 1953. Legumes in agriculture. *Agric. Studies*, **21**, FAO, Rome (in English).

Wieringa, K. T. 1963. Organismes isolés du sol des Apennins producteurs d'antibiotiques envers diverses souches de *Rhizobium. Ann. Inst. Pasteur*, **105** (2), 417–425.

Wieringa, K. T. and Bakhuis, J. A. 1957. Chromatography as a means of selecting effective strains of rhizobia. *Plant and Soil*, **8** (3), 254–262.

Wiken, T. 1956. Möglichkeiten der Produktionssteigerung in der schweizerischen Landwirtschaft durch neuzeitliche bakterielle impfverfahren. *Schweiz. landswirtsch. Monatsh.*, **34** (11), 12, 3–31.

Williams, L. F. and Lynch, D. L. 1954. Inheritance of a non-nodulation character in the soybean. *Agron. J.*, **46** (1), 28–29.

Wilson, J. K. 1917. Physiological studies of the *Bacillus radicicola* of the soy bean (soja Max piper) and the factors influencing nodule production. *NJ (Cornell) Agric. Exptl Stat. Bull.*, **386**, 369.

Wilson, J. K. 1926. Legume bacteria population of the soil. *J. Amer. Soc. Agron.*, **18**, 911–919.

Wilson, J. K. 1930a. Seasonal variation in the numbers of two species of *Rhizobium* in soil. *Soil Sci.*, **30**, 289–296.

Wilson, J. K. 1930b. *Abstr. Proc. Second Internat. Congr. Soil Sci.*, 3 Commiss., 89 (quoted by Dorosinskii, L. M., 1965).

Wilson, J. K. 1939a. Symbiotic promiscuity in the Leguminosae. *Third Internat. Soil Sci. Soc. Trans.*, A, 49–57; *Amer. Soc. Agron.*, **31**, 159–170.

Wilson, J. K. 1939b. Leguminous plants and their associated organisms. *Cornell Univ. Agric. Exptl Stat. Mem.*, (221), 1–48.

Wilson, J. K. 1939c. Relationship between pollination and nodulation of the Leguminosae. *J. Amer. Soc. Agron.*, **31** (2), 159–171.

Wilson, J. K. 1944. Over five hundred reasons for abandoning the cross inoculation groups of legumes. *Soil Sci.*, **58**, 61–69.

Wilson, J. and Wilson, P. 1942. Biotin as a growth factor for rhizobia. *J. Bacteriol.*, **43**, 329.

Wilson, P. W. 1939. Mechanism of symbiotic nitrogen fixation. *Ergebn. Enzymol.*, **8**, 13–27.

Wilson, P. W. 1940. The biochemistry of symbiotic nitrogen fixation. Madison, University of Wisconsin Press.

Wilson, P. W. 1958. Asymbiotic nitrogen fixation: in *Handbuch der Pflanzenphysiologie*. Ed. by W. Ruhland, **8**, 9–47. Berlin, Springer.

Wilson, P. W. and Burton, J. C. 1938. Excretion of nitrogen by leguminous plants. *J. Agric. Sci.*, **28**, 307–323.

Wilson, P. W., Fred, F. B. and Salmon, M. R. 1933. *Soil Sci.*, **35**, 145–165 (quoted by Schreven, D. A. van. 1958).

Wilson, P. W. and Wyss, O. 1937. Mixed cropping and the excretion of nitrogen by leguminous plants. *Contrib. Dept. Agron. Bacteriol. and Agron. Chem. Univ. Wisconsin*, (147).

Wilson, S. B. and Hallsworth, E. G. 1965. Studies on the nutrition of the forage legumes. 4. The effect of cobalt on the growth of nodulate and non-nodulated *Trifolium subterraneum* L. *Plant and Soil*, **22** (2), 260–279; **23**, 60–78.

Wipf, L. 1939. Chromosome numbers in root nodules and root tips of certain Leguminosae. *Bot. Gaz.*, **101**, 51–67.

Wipf, L. and Cooper, D. C. 1938. Chromosome numbers in nodules and roots of red clover, common vetch and pea. *Proc. Nat. Acad. Sci. USA*, **24**, 87–91.

Wipf, L. and Cooper, D. C. 1940. Somatic doubling of chromosomes and nodular infection in certain Leguminosae. *Amer. J. Bot.*, **27** (9), 821–824.

Wolf, M. and Baldwin, J. L. 1940. The effect of glycine on the rhizobia. *J. Bacteriol.*, **34** (3), 344.

Worth, W. A. and McCabe, A. 1948. Differential effects of 2,4-D on aerobic, anaerobic and facultative anaerobic micro-organisms. *Science*, **108**, 16–18.

Wright, W. H. 1925. The nodule bacteria of soybeans. *Soil Sci.*, **20**, 95–141.

Wróbel, T. 1956. A rapid method for determining the activity of strains of the root-nodule bacteria. *Acta Microbiol. Polon.*, **5** (1–2), 117–119 (in English).

Wróbel, T. 1959. Production of cultures of *Rhizobium* for leguminous plants and the results of their application in field experiments. *Acta Microbiol. Polon.*, **8** (3–4), 321–332 (in English).

Wróbel, T. and Allen, O. N. 1961. *Pamiętnik Pulawski-Prace JUNG*, 4, 15–24 (quoted by Hamatová-Hlaváčková, E. 1965).

Wróbel, T. and Gołębiowska, J. (Vrubel' and Golembevska) 1956. Inoculation of Papilionaceae by papillary bacteria in different types of soil. *Acta Microbiol. Polon.*, **5** (1–2), 121–123 (in Polish; abstract in Russian).

Wróbel, T. and Gołębiowska, J. (Vrubel', T. and Golembevska, Yu). 1956. Inoculation of leguminous plants by nodule bacteria from various soils. *Acta Microbiol. Polonica*, **5**, (1–2), 121–124 (in Russian).

Yagodin, B. A. 1964. *Mikroelementy v ovoshchevodstve*. (Trace elements in vegetable growing). Kolos. Moscow.

Yakovleva, Z. M. 1959. Isoelectric point of nodule bacteria. *Izv. Akad. Nauk SSSR, Ser. Biol.*, (4), 595–598.

Yakoleva, Z. M. 1961. Properties of symbiotic nodule bacteria. *Trudy In-ta mikrobiologii Akad. Nauk SSSR*, (11), 198–201.

Yakovleva, Z. M. 1963. Some data on the properties of symbiotic nodule bacteria. *Trudy In-ta ? ochoved. Akad. Nauk KazSSR*, (14), 56–61.

Yakovleva, Z. M. 1964. Bacteroids of nodule bacteria (reproduction). in *Pochvennye issledovanie v Kazakhstane* (Soil Studies in Kazakhstan). Alma-Ata, 173–178.

Yakovlev, V. A. and Levchenko, L. A. 1964. Localization of dehydrogenases associated with nitrogen fixation in *Azotobacter vinelandii*. *Dokl. Akad. Nauk SSSR, ser. biol.*, **159** (5), 1173–1174.

Yamane, G. and Higashi, S. 1963. Interspecific transformation of host-specificity in root nodule bacteria. *Bot. Mag. Tokyo*, **76** (898), 149–154 (in English).

Yates, M. G. and Hallsworth, E. G. 1963. Some effects of copper in the metabolism of nodulated subterranean clover. *Plant and Soil*, **19** (2), 265–284.

Yukhimchuk, F. F. 1957. *Azotnyi obmen i vozrastnye izmeneniya bobovykh rastenii* (Nitrogen exchange and age changes in legumes). Izd-vo s.-kh. lit-ry, Kiev.

Zagor'e, I. V. 1966. Selection of active variants of nodule bacteria of pea obtained on exposure to ultraviolet radiation: in *Ispol'zovanie mikroorganizmov dlya povysheniya urozhaya sel'skokhozyaistvennykh kultur* (Use of micro-organisms for increasing the yield of farm crops), Kolos. Moscow, 113–116.

Zaremba, V. P. 1953. Natural inoculation of leguminous crops by nodule bacteria and nitragin: in *Voprosy primeneniya bakterial'nykh udobrenii* (Problems in the use of bacterial fertilizers), Kiev, 51–58.

Zaremba, V. P. 1959. Effect of disinfectants on development of nitrogen-fixing bacteria. in *Mikrobiologiya na sluzhbe sel'skomu khozyaistvu* (Microbiology in the service of agriculture), 110–114.

Zaremba, V. P. and Malins'ka, S. M. 1966. Productivity of leguminous crops in connection with their natural and supplementary inoculation by nodule bacteria: in *Mykrobiolohiya dlya narodnoho hospodarstva i medytsyny* (Microbiology for the national economy and medicine). Naukova Dumka, Kiev, 83–89 (in Ukrainian).

Zaremba, V. P., Tomashevskaya, E. G. and Malinskaya, S. M. 1966. Ability of nodule bacteria to assimilate poorly-soluble calcium phosphates. *S-kh biologiya*, **1**, (1), 842–850.

Zavertailo. Relationship of host plant variety and geographical origin of strains of nodule bacteria to the effectiveness of bacterization of lucerne. *Byull. nauchno-tekhn. inf. po s-kh. mikrobiol.*, (1), (13), Leningrad, 27–30.

Żelazna, I. 1962. Phosphatase activity of *Rhizobium. Acta Microbiol. Polon.*, **9** (4), 329–334 (in English).

Żelazna, I. 1963. Transformation of *Rhizobium. Acta Microbiol. Polon.*, **12** (3), 166–174 (in English).

Żelazna, I. 1964*a*. Transformation in *Rhizobium trifolii*. I. The influence of some factors on the transformation. *Acta Microbiol. Polon.*, **13** (4), 283–291 (in English).

Żelazna, I. 1964*b*. Transformation in *Rhizobium trifolii*. 2. Development of competence. *Acta Microbiol. Polon.*, **13** (4), 291–298.

Zhvachkina, A. A. 1960. Effect of salination on microbiological processes in the soil: in *Materialy po izucheniyu pochv Urala i Povolzh'ya* (Material for study of soils of the Urals and Volga region), Ufa, 239–245.

Ziemiecka, J. 1963. Recherches récentes sur les *Rhizobium. Ann. Inst. Nat. Agron.*, **1**, 65–72.

Zycha, H. 1932. Sauerstoffoptimum und Nährböden 'aerober' Bakterien. *Arch. Mikrobiol.*, **3** (2), 194–204.

3 Symbiotic Assimilation of Nitrogen by Non-Leguminous Plants

Nodulation of Plants other than Leguminosae

Nodules or nodule-like formations on the roots of plants other than the legumes are quite a widespread phenomenon in both gymnosperms and angiosperms. Among the angiosperms nodules are detected mostly on dicotyledons, but they have been reported on monocotyledons. In most cases the nodules form on the roots of trees and semi-shrubs, but they are sometimes found on herbaceous plants (families Cruciferae, Scrophulariacae and Gramineae. Among the gymnosperms, nodules have been found in the root systems of many of the Coniferales and Cycadales. In dicotyledons the nodules are relatively frequent in the roots of members of the families Coriariaceae, Myricaceae, Betulaceae, Casuarinaccac, Elaeagnaceae and Rhamnaceae (Bond, 1963) and considerably less frequent in the Zygophyllaceae. In individual cases they have been found on the roots of members of the Cruciferae, Rubiaceae, Ericaceae, Rosaceae and Scrophulariaceae. Among monocotyledons, nodules have been described only in the Gramineae.

Nodules are found on leaves as well as roots. In particular, leaf nodules are known in tropical members of the Dioscoraceae, Eubiaceae and Myrsinaceae.

Information about the nodules of gymnosperms has been reviewed by E. Allen and O. Allen (1964) and, in part, in the monograph by Kelly (1952). The nodules of shrubs and trees belonging to the Coriariaceae, Myricaceae, Betulaceae, Casuarinaceae, Elaeagnaceae and Rhamnaceae have been reviewed by Bond (1963). Nodule formations on the roots of Zygophyllaceae are considered by E. Allen and O. Allen (1958). Schwartz (1952) summarized information about nodules on leaves. We shall analyse the results of these studies and of several new investigations.

Nodules of Gymnosperms

Root nodules of gymnosperms are found in respentatives of the orders Cycadales, Ginkgoales and Coniferales. This group is set out in Table 3.1 (from the material of E. Allen and O. Allen (1964)).

155

Table 3.1. Gymnosperms Bearing Nodules

Order	Cycadales	Genus	Agathis
Family	Cycadaceae		Araucaria
Genus	Bowenia	Family	Cupressaceae
	Ceratozamia	Genus	Libocedrus
	Cycas	Family	Podocarpaceae
	Dioon	Genus	Acmophyle
	Encephalartos		Dacridium
	Macrozamia		Microcachrys
	Stangeria		Phyllocladus
	Zamia		Pherosphaera
Order	Ginkgoales		Podocarpas
Family	Ginkgoaceae		Saxegothaea
Genus	Ginkgo (biloba)	Family	Taxodiaceae
Order	Coniferales	Genus	Sciadopitys
Family	Araucariaceae		

The order Cycadales has one family Cycadaceae, which includes nine genera of tropical trees. On the roots of between eighty-five and ninety species belonging to eight genera (*Mycrocycas* has not been investigated) root swellings are found—dichotomously branching coralloid nodules (E. Allen and O. Allen, 1964). They are considered to be thickened, modified roots (Schacht, 1893).

Allen and Allen (1964) believe that the organisms causing the nodules to form in this family are ubiquitous in the wild. However, the nature of the agent has not yet been established conclusively.

Drawing on several sets of results (Reinke, 1879; Hariet, 1892; Winter, 1935; Takesige, 1937; Douin, 1953), the Allens consider that these organisms are algae, in particular blue-green algae—various species of *Nostoc* and *Anabaena*. Ferwandez and Bhat (1945) also thought that the agents are algae, but green algae, namely *Chlorococcum humicolum*. Algae were also found by Schact (1853) in the tissues of coralloid formations in *Zamia pumila*, *Cycas mexicana*, *C. cireinalis*, *Ceratozamia mexicana* and *Cer. muricata*.

There are very few who have favoured a bacterial origin for these nodules. Among them were Life (1901) and McLuckie (1922) who believed that the formation of nodules in *Cycas* is due to *Rhizobium*, and Kellerman (1910) who isolated nitrogen-fixing bacteria from the nodules. On the basis of McLuckie's findings (1922) Kelly (1952) considered the nodules of *Macrozamia* to be bacterial formations. He regarded the nodules of *Cycas* and *Encephalartos* as mycorrhiza. Spratt (1915) thought that in some species of *Cycas* three organisms—*Azotobacter*, *Rhizobium* and algae—are in symbiosis with the host tissues.

The function of the nodules in the Cycadaceae has not been established. It can be assumed that they fix atmospheric nitrogen (Life, 1901; Bottomley, 1907; Douin, 1953). There is no direct evidence, however, that these symbiotic systems fix nitrogen in sterile conditions. We shall mention only three sets of experiments.

First, Douin (1954) established that there is a higher content of nitrogen in the nodules of *Cycas* than in the roots. Second, Bond (1959) found vigorous nitrogen fixation in the excised nodules of *Ceratozamia* and *Encephalartos* kept in an atmosphere of nitrogen-15, but not in roots without nodules. Third, Bergersen *et al.* (1965) found that nitrogen was fixed (in an atmosphere of nitrogen-15) by the coralloid formations of *Macrozamia communis* in symbiosis with endophytic blue-green algae.

It is also probable that the endophyte stimulates the negative geotropic reaction of the small lateral roots (E. Allen and O. Allen, 1964). Aeration is also a possible function. Roots of species of *Cycas* situated 20 cm below the surface usually have no endophyte cells (E. Allen and O. Allen, 1964).

There is very little information about nodulation in the Ginkgoaceae and, moreover, in most cases it is conflicting. *Ginko biloba* which alone has survived to the present was regarded by Kelly (1935), referring to the work of Reinke (1879) and Klečka and Vukolov (1935), as a mycorrhizal plant. In the same species Hiltner (1903) observed nodules similar to those found in *Podocarpus*. But the Allens (1964) found no root nodules on *Ginko biloba*.

Of the Coniferales we shall consider the most studied species—*Podocarpus* in the family Podocarpaceae (nodules are found in twenty-six species of this genus). These nodules measure 1–2 mm across and are for the most part spherical or semi-spherical. They usually develop in opposite rows on the surface of the root and often in large numbers. The nodulated roots are usually arranged in the form of cushions of different dimensions, which are different sizes in transverse and longitudinal sections, in the surface layer of the litter (Yeates, 1924). The nodules, similar to those of *Alnus*, discussed later, are lateral roots which have been arrested in growth, but they differ from those of *Alnus* in not being branched.

The agent of nodule formation in podocarps is unknown. Most investigators (Janse, 1897, Nobbe and Hiltner, 1899; Shibata, 1902; Yeates, 1924; Shaede, 1943; Bergersen and Costin, 1964) observed fungal hyphae in the nodules of *Podocarpus*. Nobbe and Hiltner (1899), Shaede (1943) and Becking (1965*a*) identified them as Phycomycetes. In some species of *Podocarpus* vesicles, spore cases and arbuscles have been found in the nodules (Janse, 1897; Shaede, 1944; Ferreira dos Santos, 1947). It is possible that the nodules are mycorrhiza. Kelly (1952) writes that '... the general view is that genuine mycorrhiza are encountered in *Podocarpus*.' Saxton (1930) apparently supported this view. In one paper on the nodules of *Podocarpus* Baylis *et al.* (1963) also regarded the nodules as mycorrhizal. But Spratt (1912*b*), McLuckie (1923) and Phillips (1932) considered that the formation of these nodules was due to *Rhizobium*, while Uemura (1964) regularly isolated Actinomycetes from them.

On the surface of young nodules of most podocarp species are many hairs which are anatomically similar to typical root hairs. Their function has not been investigated. The endophyte penetrating the plant cells causes the nuclei to swell. Later, when the cells of the endophyte are dissolved by the proteolytic enzymes formed by the nodular tissue, the nuclei resume their original size. Shibata (1902) observed the formation of proteolytic enzymes in large amounts in the tissues of the nodule but not the roots (even in the adjacent tissues), apparently connected with the reaction of the plant tissues to the penetration of the endophyte. Infected cells of the cortex of the host plant are multi-nucleate, and are often more than twice the size of uninfected cells.

In the pericycle of the stele* each year a new nodule is laid down which develops at first within the cortex of the old nodule, and then as it grows it moves outward. When such a process is repeated several times, nodules form with a rosary appearance (Yeates, 1924).

Yeates (quoted by E. and O. Allen, 1964) observed four concentric zones of cortex within the secondary nodules of *Podocarpus ferrugineus*, corresponding, in his view, to the four periods of development of the nodule. It is interesting that only the periphery of the cortex of the secondary nodule contained live endophyte cells. It is not known how the endophyte penetrates the tissue of the secondary nodule; it could be a result of infection from the soil.

So far insufficient work has been done to show the importance of nodules to podocarps. Investigators first tried to show that the nodules fix molecular nitrogen (Hooker, 1854; Nobbe and Hiltner, 1899, Spratt, 1912; McLuckie, 1923; Kondo, 1931; Bond, 1959; Bergersen and Costin, 1964). Thus, Nobbe and Hiltner (1899) decided, on the basis of five years development of nodulated podocarps in a medium lacking nitrogen, that the nodules play a significant role in the fixation of atmospheric nitrogen. Kondo (1931) considered the nodules important because they contained more nitrogen than the other parts of the roots. Spratt (1912*a*) and McLuckie (1923), after observing the fixation of molecular nitrogen by cultures of bacteria isolated from nodules, noted the great importance of the nodules to the life of the plants.

Bergersen and Costin (1964) recorded the fixation of nitrogen in the nodules of *Podocarpus lawrencei* and Becking (1965) did so for the nodules of *P. rospigliosii* when exposed to nitrogen-15. The increase in nitrogen content was, however, very small. According to Bergersen and Costin it was 0·5 γ per 1·9 g of nodules in 4 hours and according to Becking (1965) 2·2 γ per 0·5 g of the fresh weight of the nodules in 24 hours.

In these conditions the transfer of fixed nitrogen to the plant is limited; nevertheless, Becking concludes that some molecular nitrogen is fixed in the symbiotic system. He justifies his views not from the results obtained with the isotope, but from evidence that plants with nodules, cultivated in a nitrogen-free medium, survive for 3–4 years in spite of signs of a serious nitrogen deficiency. However, Bond (1959) exposed the roots of species of *Podocarpus* with nodules to nitrogen-15 and found no fixation of atmospheric nitrogen.

And there is also the hypothesis of McLuckie (1923) and Yeates (1924), that the nodules of podocarps serve as water reservoirs, because, after dissolution

* The stele is the central cylinder of stem or root and the pericycle is the outer layer of the stele.

of the endophyte, a thickened layer of tracheidal cells can be seen in the cortex of the nodules.

Coriariaceae, Myricaceae, Betulaceae, Casuarinaceae, Elaeagnaceae and Rhamnaceae

More than a hundred years ago root nodules were found in several representatives of these families—trees and semi-shrubs—and there is extensive literature on the subject (for example Miehe, 1918; Snyder, 1925; Jepson, 1936; Fletcher, 1955; Bond, 1956, 1959, 1964; Gardner, 1965; Taubert, 1956; Morrison and Harris, 1958; E. Allen and O. Allen, 1958, Becking 1961; Rossi, 1964; Becking *et al.*, 1964; Silver, 1964; Moore, 1964). Bond *et al.* particularly have carried out a large number of investigations (for example 1954, 1955, 1961, 1962).

The root nodules in these plants are found only in species of *Coriaria* (order Coriariales, family Coriariaceae), *Myrica gale* and *Comptonia* (order Myricales, family Myricaceae), *Alnus* (order Fagales, family Betulaceae), *Casuarina* (order Casuarinales, family Casuarinaceae), *Elaeagnus*, *Hippophaë*, *Shepherdia* (order Rhamnales, family Elaeagnaceae) and, *Ceanothus* and *Discaria* (order Rhamnales, family Rhamnaceae).

In nature there are far more genera belonging to these families and orders, but no nodules have been found on them. According to Bond (1963), for example, the family Rhamnaceae includes forty genera, but only in two—*Ceanothus* and *Discaria*—have nodules been found. Not all the species of a genus are nodulated.

The nodules of most representatives of this group are pinkish red, but with age they acquire a brownish tint. In species of *Elaeagnus*, they are white. Originally the colour was thought to be due to the presence of an anthocyanin pigment (Bond, 1951). But recent information (Davenport, 1960) has indicated the presence of haemoglobin in the nodules of *Casuarina*, *Alnus* and *Myrica*.

Nodule-forming plants are encountered in different climatic zones, but some are associated with a definite territory. Thus, for example *Shepherdia* and *Ceanothus* are found only in North America. Plants of the genus *Casuarina* prefer tropical regions (Australia, the Pacific Islands). Species of *Elaeagnus* and *Hippophaë*—the sea buckthorn—are much more widespread; some are found as far north as the Arctic. Plants that form nodules usually grow in soils deficient in nutrient elements—sand dunes, craggy rocks, marshes and so on.

The best studies carried out on plants of this group have been with the genus *Alnus* and in particular, *Alnus glutinosa*, on which Voronin found nodules as long ago as the 1870's. It is thought (Karavaev, 1959) that nodules are peculiar not only to existing but to extinct species of alder, for they have been found in the roots of the fossil alder, *Alnus* sp. in tertiary deposits in the valley of the River Aldan in Yakutia. Considerably more work has been done on *Myrica gale*.

The least studied genius is *Discaria*, on which Morrison and Harris found nodules in 1958.

In these plants nodules form, sometimes reaching a considerable length, as a result of repeated dichotomous or trichotomous branchings of the apices, in a corraloid cluster. The nodules of *Casuarina equisetifolia* are often as long as 15 cm and function for several years.

The nodulated roots of *Myrica* and *Casuarina* are negatively geotropic, while the roots of the corresponding uninfected plants are positively geotropic. Positive geotropism is also characteristic of the nodulated roots of the alder.

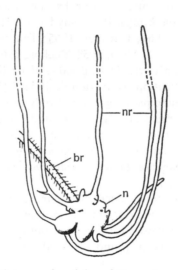

Fig. 20. Diagram of nodules of *Casuarina equisetifolia*

n, Nodules; br, root hairs on root bearing nodules; nr, rootlets radiating from the nodule (× 0·75).

Silver *et al.* (1966) felt that this was connected with the content of auxin (β-indolylacetic acid) in the roots. Thus, the presence of endogenous auxin in alder nodules (20 mg/kg) and the roots of uninoculated *Myrica* and *Casuarina* determines the positive geotropic reaction. Destruction of the auxin by the oxidase system in the nodules of *Myrica* and *Casuarina* (this system is much more active in the nodules than in the rest of these plants) accounts for the negative geotropism of the roots.

In *Myrica gale* and *Casuarina equisetifolia* the nodules are covered with rootlets reaching a length of 10 cm (Fig. 20); their tips have root caps. In some cases, root hairs develop on the rootlets, which are never infected with endophyte. Both these and the new lobes of the nodules develop either endogenously from the pericycle of the stele as in *Myrica gale* and *Ceanothus* or, more often, from the apical meristem, as in *Alnus, Elaeagnus* and *Casuarina*. Conical outgrowths form on the nodules of *Alnus* when kept in an aqueous medium. These and the rootlets in *Myrica gale* apparently serve to supply the roots with air in anaerobic conditions (Bond, 1952; Fletcher, 1955).

Unlike the roots, the nodules have no root caps or root hairs, but a suberous surface layer of cells. The time taken from the moment of infection to the formation of nodules varies in different species. For example, in individual species of sea buckthorn it is 11 days, in species of *Ceanothus* 26 days (Bond, 1963). In some species of *Casuarina* the interval is as long as 70 days (E. Allen and O. Allen). In *Elaeagnus* and *Shepherdia* it is 40–60 days (Nadson, 1955; Verner, 1965), and in alder species 3–15 months (Roberg, 1934; von Plotho, 1941; Ferguson and Bond, 1953).

There is no unanimity of opinion on the site of formation of the nodules. Pommer (1956) considers that they originate in the cortex of the root whereas Taubert (1956) believes that they originate from a modified lateral root, particularly in alders.

The initial sign of infection, as in leguminous plants, is the bending, curling or branching of the root hairs. This was noted in *Myrica gale* by Fletcher (1955), in *Casuarina* by Bond (1963), in *Alnus* by Hiltner (1903), Krebber (932) and Pommer (1956). Deformation of the root hairs begins 4–5 days after infection and can be seen with the naked eye (Fletcher, 1955). Uninfected root hairs remain straight. The reasons for the deformation of the infected root hairs are obscure. They are assumed to be similar to the causes in leguminous plants (Bond, 1963).

The initial stages of penetration of the endophyte into the root hairs and the cells of the roots have not been studied adequately. After the endophyte has penetrated the cortical tissue cell division is restricted in the meristem. A lateral root developing from the pericycle near the focus of infection, through the infected region of the cortex, is modified and infected. A nodule develops and when it has aged it is covered with periderm with a well developed suberous layer (Bond, 1963). The nodules in question constitute, in effect, modified roots, or roots with arrested development. Young nodules on plants of this group, in particular, *Alnus glutinosa*, have the following structure in cross sections from the centre to the periphery. The stele is surrounded by endoderm, then there is a layer of hypertrophied and hyperplastic cortical parenchyma. The endophyte is found in most of the cells of this zone. The uninfected cells or the cortical parenchyma often contain starch or tannins. On the outside there is a layer of apical meristem (E. Allen and O. Allen, 1958; Hawker and Fraymouth, 1951; Bond, 1963; Becking *et al.*, 1964; Gardner, 1965).

The endophyte is never found in the cells of the apical meristem or the stele (for example, Shibata, 1902; Arzberger, 1910; Bottomley, 1912; Silver, 1964).

Nodules of plants belonging to the genera *Hippophaë*, *Elaeagnus*, *Ceanothus*, *Myrica*, *Coriaria* and *Casuarina* have a similar structure although there are some differences. In *Ceanothus* and *Myrica cerifera* the infected cells tend either to be at the centre or the periphery of the cortex of the nodule while in *Myrica gale* they are scattered over the whole cortical parenchyma, and in *Coriaria* the infected hypertrophied cells lie on one side of the stele so that the stele is not central in the root (Bond, 1963).

The nature of the endophyte infecting the various species in question has not yet been established conclusively.

Electron microscopy with ultrathin sections has increased considerably knowledge of the development of the endophyte and its taxonomy (Silver, 1964; Becking et al., 1965; Gardner, 1965).

In the nodules of plants of this group the endophyte is known to occur as hyphae, vesicles and bacteria-like cells or bacteroids as Gardner (1965) and some others called them. According to Becking (1966) the bacteria-like cells should not be called bacteroids because this term has a special meaning in relation to nodule bacteria. Von Plotho (1941) and Shaede (1961), having found all these forms, suggested that the endophyte of Alnus glutinosa was polymorphous, developing in three stages; septate hyphae, vesicles and bacteria-like cells.

Becking et al. (1964) found fine septate hyphae in the cells of the host, especially in the initial period of infection, and so did Silver (1964), Gardner (1965) and others. According to Gardner, the hyphae are branched and surrounded by a 'capsular' substance. At the ends of the hyphae round, bottle or club-shaped inflations usually form vesicles. As the vesicles mature the hyphae are destroyed. As a result, the link between the vesicles is lost. The ends of the hyphae in Myrica cerifera however, are thickened like a club during the whole of the development of the nodule (Silver, 1964).

Allen et al. (1966) noticed a mycelial endophyte with hyphae less than 1 μm in diameter in all species of Coriaria. Silver made the same observation in Myrica cerifera (1964). The diameter of the club-shaped ending in M. cerifera is 1·5–2 μm. However, Bond (1963) was not able to confirm the several reports of mucilaginous threads or hyphae in the cell of the host plant. Hawker and Fraymouth (1951) also noted the absence of hyphae in the nodules of this group of plants.

The vesicles of Alnus, Hippophaë, Elaeagnus and Ceanothus are round and those of Coriaria club-shaped. They measure between 2 and 5 μm across. They have a tight membrane envelope and their internal cavity is divided by arbitrarily distributed septa—membranes of plasmatic origin. Often the septa do not stretch right across the vesicle (Becking, 1966). Gardner (1965) called the cavities formed by the septa subunits of the vesicles. At the centre of a subunit is a nucleus-like body surrounded by granular cytoplasm 8 nm thick, which is closely contiguous with the thin plasma membrane. According to Gardner (1965) and Becking (1966) the formation of the septa in the vesicles is accompanied by the formation of special organelles—plasmalemmasomes or 'peripheral bodies'. They are either of irregular shape or onion shaped. Similar organelles are observed in Actinomycetes.

It is interesting that, as well as the round vesicles, Gardner (1965) described vesicles of irregular shape with very dense contents. They characteristically had narrow projections like germ tubes. In Gardner's view, such vesicles are in the active state. Each vesicle is surrounded by a capsular substance which approaches the membrane only at certain points (Gardner, 1965).

Hawker and Fraymouth (1951) as well as Gardner (1965) consider that the vesicles are analogous to sporangia. Hawker and Fraymouth consider that the germination of sporangia is connected with the release of motile spores. But Gardner takes a different view, he sees the vesicles as sporangia because they contain subunits and growth structures and because they predominate in the winter. On this basis he considers the vesicles to be resting sporangia. In senescent nodules the vesicles are deformed and dissolved by the host cell.

Vesicles and bacteria-like cells are not usually encountered in the same host cell. The host cells containing vesicles are always alive and have a normal cytoplasmic structure including nuclei and mitochondria. Plant cells containing the bacteria-like cells are usually dead and have disorganized cytoplasmic structures. On the other hand, most bacteroid cells retain their structure when the nodules age. Some of them, however, are dissolved (Becking et al., 1964). The bacteroids may be rod-shaped or bent. Their dimensions vary within the limits 0.8×8 to 10μm.

The internal structure of the bacteroids is close to that of the subunits of the vesicles (Gardner, 1965). There is a well marked central nuclear region surrounded by granulated cytoplasm, plasma membranes, external membrane and capsular substance (Gardner, 1965). According to Becking (1966) the mucilage of the capsule is of a polysaccharide nature. Here, as in the vesicles, the formation of plasmalemmasomes has been observed (Becking et al., 1964; Gardner, 1965).

It has been suggested that the bacteria-like bodies occur as a result of either digestion of hyphae and vesicles by the plant cell or by division of the hyphae. They are possibly spore-like formations involved in the multiplication of the endophyte in the soil (Becking, 1966). At any rate, once in the soil after the nodule has been discharged they can infect new plants (Becking, 1966).

Bacteroids are most numerous in nodules in spring and summer. They are possibly then most active metabolically. The taxonomic position of the endophytes of this group of plants remains to be decided. The possibility of identifying the endophytes depends on the purity of the culture isolated and its ability to form nodules when manufactured in sterile conditions. Numerous attempts made along these lines, however, have mostly been unsuccessful (Lieske, 1921; Krebber, 1932; Bouwens, 1943; Quispel, 1954; Fletcher, 1955; Silver, 1964; Uemura, 1952, 1961). Uemura isolated a large collection of Actinomycetes but none of the cultures induced nodules in test plants. Equally unsuccessful was Rossi's attempt (1964) to inoculate Alnus glutinosa with Actinomycetes isolated from the nodules of this plant.

Many investigators believe that cultures of micro-organisms are obligate symbionts and cannot be isolated even with the use of a large range of nutrient media. This, of course, makes it much more difficult to establish the systematic position of the organisms responsible for the formation of nodules in non-leguminous plants.

From this point of view it may be worth cultivating the endophyte in tissue cultures of the nodules. In this way Becking (1965, 1966) isolated several Actinomycetes.

It is interesting to note that when he cultivated Actinomycetes in a medium containing organic nitrogen-containing compounds, Becking (1966) was able to induce vesicle formation. In artificial media, however, there were no uninfected septa characteristic of the vesicles of the nodules. In Becking's view vesicle formation, with the membranes and septa peculiar to them, in nodules is in some way connected with nitrogen fixation.

Explaining unsuccessful attempts to isolate the endophyte from nodules, Silver (1964) argued that it may require obligatory passage through the soil. Some investigators, however, present positive results of experiments with pure cultures of Actinomycetes. Thus, for example, alder nodules were obtained when plants were infected with pure cultures of Actinomycetes by Peklo (1910), von Plotho (1941) and also by Pommer (1959), who had previously been unable to do this (1956). Similar results were obtained by Jounken (1919) with *Myrica cerifera* and by Niewiarowska (1959, 1961) with *Hippophaë*. Bond (1963), however, considers this work not to be very convincing because of the insufficient number of controls and because negative results were obtained when experiments were repeated by other investigators. In 1965, Danilewicz published results confirming the possibility of the formation of nodules on alder roots by a culture of Actinomycetes.

Most investigators conclude that the endophytes of the group of plants we are considering are Actinomycetes (for example Peklo, 1910; Miehe, 1918; Krebber, 1932; Schaede, 1933, 1962; Fletcher, 1955; Kämppel and Wartenberg, 1958; Schwarz, 1959; Furman, 1959; Bond, 1963; Silver, 1964; Becking et al., 1964, 1966; Gardner, 1965). Some early investigators also believed that the formation of these nodules involved fungi (for example Woronin, 1867; Frank, 1892; Shibata, 1902; Hiltner, 1898). Several others (for example Jendo and Takase, 1932; Hawker and Fraymouth, 1951; Taubert, 1956) consider that the endophytes of certain plants are mucilaginous fungi—Myxomycetes (order Plasmodiophorales).

The suggestion has been made that the nodules of these plants are formed by bacteria (for example Maire and Tison, 1909; Bottomley, 1906; Spratt, 1912; Bose, 1947; Virtanen and Saastomoinen, 1936). This idea has not been confirmed for the entire group. Hoak (1964) attempted to infect dicotyledons of this group with a culture of *Rhizobium*. In no case did he obtain the characteristic bending of the root hair, which indicates that the root system was not responding to inoculation with bacteria. In nodules of the sea buckthorn *Hippophaë rhamnoides* however, Koslova et al. (1966) found three cultures of bacteria resembling *Rhizobium*. Similar cultures were described by Panosyan (1943) in nodules of *Elaeagnus angustifolia*. When sea buckthorn plants were infected with the bacteria that had been isolated, nodules formed in the roots. In synthetic media the bacteria developed weakly and did not fix nitrogen, whereas, when they were in symbiosis a considerable amount of nitrogen was fixed (Koslova et al., 1966).

Finally, there is the view that the roots have a mixed infection involving at least two endophytes (Arcularius, 1928; Lieske, 1921; Uemura, 1952).

Although the nature of the endophytes in the plants in question is not clear, it is quite obvious that they are specific for certain species or related systematic groups and possess differing degrees of virulence. Their specificity can be judged from the fact that attempts at cross infection between different genera using the crushed contents of the nodules have usually proved unsuccessful. Nodules formed only in the species from which material was taken for inoculation, or sometimes in related species (Roberg, 1938; Bond, 1963). But representatives of certain genera, in particular Elaeagnus, Hippophaë and Shepherdia, can be cross infected (Gardner and Bond, 1957; Moore, 1964). This is apparently an example of the same cross infectivity encountered in leguminous plants. The degree of virulence of the endophyte differs appreciably in relation to representatives of the cross infected group (Roberg, 1934; Bond, 1963). We cannot fail, however, to mention that Kravtsov's experiments (1950) did not reveal cross infection of Hippophaë and Elaeagnus.

Nodules usually form when the medium is neutral or weakly acid. The presence, concentration and form of the salts of the nitrogen-containing compound in the medium is also very important (Stewart, 1963). Species differ in their sensitivity to nitrogen; species of Shepherdia and Hippophaë are more sensitive than those of Myrica and Alnus to the same concentrations of nitrogen. The introduction of KNO_3 and $(NH_4)_2SO_4$ into a nitrogen-free culture medium containing infected alder plants reduces the size and number of nodules. The introduction of potassium superphosphate, on the other hand, leads to an increase in the number of nodules (quoted from E. Allen and O. Allen, 1958). A decrease in light intensity and reduction in day length also reduce considerably the weight and number of nodules formed.

There is now a good deal of detailed evidence indicating that the nodules on plants of the Coriariaceae, Myricaceae, Casuarinaceae, Elaeagnaceae and Rhammaceae can fix molecular nitrogen. There is also some indication that the micro-organisms isolated from the nodules of these plants assimilate atmospheric nitrogen in vitro. For example, according to Fiuczek (1959), the Actinomycetes (Streptomyces alni) he isolated from Alnus glutinosa in a mineral medium with starch, fixed about 2·4 mg nitrogen per 100 mg of medium. But, as E. and O. Allen rightly pointed out (1958), it has still not been demonstrated that these micro-organisms are in fact plant symbionts.

Experiments to establish nitrogen fixation have been carried out with many plants of the genera Hippophaë, Ceanothus, Alnus, Casuarina, Coriaria and Myrica, invariably with positive results. The activity of nodules formed can practically cover the plant's requirements for nitrogen (Bond, 1959, 1963).

We shall give a few examples. In Aldrich-Blake's experiments (1932) the control *Casuarina* plants were 42·5 cm tall at the end of the season, while inoculated plants were 137·5 cm tall. Bond (1963) established the quantity of nitrogen accumulation by various species of nodulated plants (Table 3.2).

When the nodules are removed from the roots of the test plants, symptoms of nitrogen starvation appear; they go when new nodules are formed.

Several investigators (Bond, 1957, 1963; Morrison and Harris, 1958; Morrison, 1961; Vlamis *et al.*, 1958; Sloger and Silver, 1965, 1966) have used nitrogen to verify the ability of a considerable number of symbiotic plants of

Table 3.2. Accumulation of Nitrogen by Various Species of Nodulated Plants in a Medium without Nitrogen

	Amount of nitrogen fixed in first growing season (mg)	Amount of nitrogen fixed in subsequent growing season	
		(mg)	Recording time
			End of season:
Alnus glutinosa	300	2,500	Second
Myrica gale	146	146	Third
Hippophaë rhamnoides	26	200	Second
Casuarina cunninghamiana	70	1,400	After 2·5
Elaeagnus angustifolia	36	186	After 1·5

the genera *Alnus*, *Myrica*, *Hippophaë*, *Casuarina*, *Elaeagnus*, *Discaria*, *Shepherdia* and *Ceanothus* to fix molecular nitrogen. The investigations were carried out with intact plants and with the nodules separated from them. They showed that nitrogen-15 is bound only in the presence of nodules. Analysis of the nodules and roots showed maximum nitrogen to be present in the nodules (Bond, 1964; Delwiche *et al.*, 1965).

According to Becking (1966), nodulated alder fixes as much as 630 mg of nitrogen per plant in 48 weeks. Excised nodules incubated in an atmosphere of nitrogen-15 fix considerable quantities of nitrogen in the first twelve hours. Very young and very old nodules fix practically no nitrogen.

A considerable part of the assimilated nitrogen is transmitted to the aerial part of the plant. Transport of bound nitrogen from the nodules begins a certain time (approximately 6 hours) after the start of fixation (Bond, 1963). During the growing season about 90 per cent of the nitrogen fixed is constantly transported to the plant (Stewart, 1962).

Leaf *et al.* (1959) established by chromatography that fixed nitrogen accumulates in the nodules of *Alnus* in the form of glutamic acid and citrulin, and in *Myrica* in the form of amides. In their view binding of molecular nitrogen proceeds through ammonia.

Aseeva *et al.* (1966) also noted a high content of glutamic acid and citrulin in the nodules of the alder, as well as glutamine, alanine and γ-aminobutyric acid. Since large quantities of glutamine synthetase are also found in the nodules, they assumed that the glutamine formed is rapidly utilized in the synthesis of other nitrogen-containing compounds.

Further study of the mechanism of nitrogen fixation may well involve sterile culture of the nodular tissue containing the endophyte. In this way the substances normally produced by the aerial parts and root system of the host can be excluded while those to be investigated can be added to the medium in definite quantities (Becking, 1965c).

Nitrogen fixation in plants of the group in question is stimulated sharply in the presence of molybdenum and cobalt. In Becking's experiment (1961) the content of nitrogen in the aerial parts of *Alnus glutinosa* was 370 per cent greater in plants treated with molybdenum compounds than in untreated control plants. Molybdenum compounds accumulate mostly in the nodules. Bond observed (1963) a favourable effect of cobalt compounds on nitrogen fixation, and so did Hewitt and Bond (1966). Factors such as hydrogen and carbon dioxide depress the binding of molecular nitrogen, as in leguminous plants.

When there is a low content of atmospheric oxygen, nitrogen fixation is suppressed. It proceeds with maximum intensity in the presence of 12–15 per cent atmospheric oxygen and stops if the content increases to 30–50 per cent (Bond, 1963). The ratio of oxygen to nitrogen in the atmosphere has an important effect (Bond, 1964).

E. and O. Allen (1964) have collected exhaustive information about the possible practical significance of this nitrogen fixing group of plants. Most representatives of the group grow in regions where cultivation of farm plants is not an economic proposition. Naturally in these regions the plants in question are important as pioneer perennials (Kohnke, 1941; Stone, 1955).

The ability to fix molecular nitrogen constitutes the fundamental importance of this group of plants. According to Dommergues (1963) the annual nitrogen increment in the soil of dunes colonized by *Casuarina equisetifolia* in the Cape Verde Islands is 143 kg/hectare. According to Delwiche *et al.* (1965) *Ceanothus*, even if only constituting a tenth of the flora, gives a yearly nitrogen increment of 60 kg/hectare. The nitrogen content of the soil under alder is usually 30–50 per cent higher than birch, pine or willow (Ovegton, 1965) and 65 per cent higher than under Douglas fir (Tarrant, 1961). According to Virtanen's calculations (1962) an alder grove (on average five plants per square metre) gives in seven years a nitrogen increment of 700 kg/hectare. The annual nitrogen increment due just to shed leaves is, according to Lawrence (1958), about 155 kg/hectare (for five year old alder).

Tarrant *et al.* (1951) and Virtanen (1962) mentioned the value of foliage and leaf litter of alder, which is due to its large content of nitrogen. According to Mikola (1958) the leaf litter of *Alnus incana* and *Alnus glutinosa* stimulates the growth of *Pinus silvestris* more than that of other species of woody plants.

These nitrogen-fixing plants, notably alder, are particularly useful in mixed forest plantations. Thus the weight of three year old seedlings of *Populus trichocarpa* in association with alder (Alaska) was twenty-two times greater than that of plants of the same age growing separately (Lawrence *et al.*, 1950; Lawrence, 1958). The same phenomenon was observed in mixed forest plantations of Douglas fir and alder (Lawrence, 1958); needles of pine and fir growing in association with alder contained considerably more nitrogen than in plantations of pine and birch (Holmsgaard, 1960) or in a pure plantation of Douglas fir (Tarrant, 1961).

Different species of *Ceanothus* have been reported to have a favourable effect and after-effect on the development of certain plants of other orders (for example Wahlenberg, 1930; Quick, 1944; Hellmers and Kellcher, 1959). Golman (1961) thought that the effect of dense alder plantings on the banks of Californian lakes was reflected not only in an increase in the content of nitrogen in the lake deposits but even in an increase in the planktonic biomass.

Zygophyllaceae, Cruciferae, Scrophulariaceae, Rubiaceae, Rosaceae and Ericaceae

Nodules were first found on the roots of the Zygophyllaceae (order Cruinales) by Isachenko (1913) looking at *Tribulus terrestris*. Nodules were found later on other species of *Tribulus* (Sabet, 1946; E. and O. Allen, 1961).

Most of the Zygophyllaceae are xerophytic bushes of perennial grasses, distributed in tropical and sub-tropical deserts, but also growing in sandy dunes, waste lands and marshes in the temperate belt.

It is interesting to note that such tropical plants as *Zygophyllum coccineum* form nodules only in conditions of high temperature and low soil moisture (Montasir and Sidrak, 1952). Soil moisture up to 80 per cent of the total water capacity impedes the formation of nodules. As is well known, in temperate leguminous plants the reverse is true; with insufficient moisture, nodules do not form. Nodules differ in size and position on the root system within the Zygophyllacea (on *Fagonia arabica*, *Tribulus terrestris*, *Tribulus alatus*). Large nodules usually develop on the main root close to the surface of the soil, and the smaller nodules on the lateral roots at a greater depth. Sometimes nodules form on the stems if they lie on the surface of the soil (Sabet, 1946).

According to Isachenko (1913) nodules in puncture vines (*Tribulus terrestris*) growing on the sands of the bank of the Southern Bug, in the south of Russia, are shaped like small, white, slightly pointed or round warts. They are usually covered with a plexus of fungal hyphae penetrating within the root cortex. Isachenko classified the endophyte as a fungus.

Sabet (1946) studied the nodules in *Zygophyllum coccineum* in much greater detail and decided that they were terminal thickenings of the roots. He repeatedly found bacteroids in the nodules and was able to isolate a culture of bacteria closely resembling *Rhizobium*. In Sabet's experiments inoculation of aseptically cultivated plants led to normal nodule formation and appreciable improvement in plant growth. Uninfected control plants usually died after a few months. There is reason to believe that the nodules provided the plants with nitrogen. The bacteria isolated did not form nodules on plants of the genus *Tribulus*.

According to E. and O. Allen (1949) the nodules of the tropical plants *Tribulus cistoides* (grown from seeds obtained from Haiti) are firm and round with a diameter of about 1 mm, and joined to the root by a wide base. They are usually arranged alternately or bilaterally (Plate 5), are not branched and on old roots they are frequently verticillate. There is characteristically no meristematic zone in the nodules. The situation is similar with regard to the formation of nodules in conifers (E. and O. Allen, 1964), and means that nodules originate from division of the cells of the pericycle of the stele.

Nodules of *Tribulus cistoides* at various stages of development have been found to contain no micro-organisms. On this basis and also because of the accumulation of large amounts of starch in the nodules E. and O. Allen (1949) considered that they provided the plants with nutrient stores.

Clearly nodules can differ greatly in different members of the Zygophyllaceae. Many of these formations have only been investigated superficially. It is to be hoped that the gaps in our knowledge will be filled in the not too distant future.

Nodules have been recorded on the roots of plants of other families. They have been found on *Brassica* and *Raphanus* (of the family Cruciferae (quoted by Schwartz, 1959)). It is assumed that they are formed by bacteria which are able to fix molecular nitrogen.

Beijerinck (1888) found nodules in *Melampyrum pratense* and *Rhinanthus major* (of the family Scrophulariaceae). In the nodules of *M. pratense* he observed only bacteria, while in those of *Rh. major*, he also observed infection threads and bacteroids. According to his findings the structure of the nodules is identical to that in leguminous plants.

Among the Rubiaceae nodules are encountered in *Coffea rubusta* and *Coffea klainii* (Steyaert, 1932). They are dichotomously branched, sometimes flattened and fan shaped. Bacteria and bacteroids are encounted in the nodular tissues. In Steyaert's view the bacteria belong to *Rhizobium* but he called them *Bacillus coffeicola*.

In the Rosaceae nodules have been found on dryad (or *Dryas octopetala*) and *Dryas drummondii* (Lawrence, 1953). Bond (1963) considered these data unreliable and did not refer to them in his review. However, E. and O. Allen (1964), listing plants with nodules, also mentioned *D. drummondii*, referring back to Crocker and Major (1955).

Dryas dummondii has practical value in the reclamation of northern areas where glaciers are receding and also because it grows well on glacial detritus (E. and O. Allen, 1964). Lawrence and Hulbert (1950) and Schoenke (1957) also referred to the good growth and increase in the content of nitrogen in the leaves of *Populus trichocarpa* when in association with *Dryas*.

Typical coralloid nodules have been described in two other representatives of the Rosaceae-*Purshia tridentata* and *Cercocarpus betuloides* (Wagle and Vlamis, 1961; Vlamis *et al.*, 1964; E. and O. Allen, 1964). With nodules, these two species grow well in soils devoid of nitrogen. There is no published information, however, about the structure of these nodules or their cause.

In the family Ericaceae we can mention only *Arctostaphylos uva-ursi* (bear's ear or ptarmiganberry) which has a nodulated root system (E. and O. Allen, 1964). Christoph (1921) and Cooper (1922) considered that these were coralloid ectotrophic mycorrhizae (quoted from Kelly, 1952).

Nodules in Gramineae

Nodules were first discovered on the roots of graminaceous plants by Nogtev (1938) investigating the meadow foxtail—*Alopecurus pratensis* (Fig. 21). Later, many Soviet investigators worked with this plant (Mudrova, 1937; Rabotnova and Ponomarenko, 1949; Silina *et al.*, 1958; Krasilinikova-Krainova, 1962). Nodules have also been found on Kentucky blue grass, *Poa pratensis*, Siberian lymegrass *Clinelymus sibiricus* and *Clinelymus ventricosus* (Savel'ev *et al.*, 1958).

The nodules of graminaceous plants vary in shape and size. The largest are found in foxtail (Savel'ev *et al.*, 1958), forming at the root tips (Fig. 21). In the first stages of development they have well developed hairs and a root cap. Later, as a result of abnormal proliferation of the parenchymatous cells, they assume an oblong, spindle-like or round shape. The parenchymatous tissues always contain many starch grains (for example Savel'ev *et al.*, 1958; Krasil'-nikova-Krainova, 1962). The vascular bundle is located at the centre of the nodule. The structure of the roots and nodules as a whole is identical.

At an early age the nodules of meadow foxtail are bright and transparent or semi-transparent, but with age they become brown or black. The formation of nodules does not depend on the type of soil, but is intensified by the introduction of manure and mineral fertilizers.

Some claim (Nogtev, 1958; Krasil'nikova-Krainova, 1962) that bacteria can be found in the cells of the parenchyma, but others say there are no micro-organisms in the cells of the nodule (Rabotnova and Ponomarenko, 1949; Silina, 1955).

Silina (1955) who cultivated meadow foxtail in sterile conditions and found that nodules formed, considered them to be root thickenings acting as deposits of reserve substances, including starch. But when the nutrient media were inoculated with an aqueous suspension of crushed nodules of foxtail, previously surface sterilized, all investigators obtained a growth of bacteria (Nogtev,

1938; Rabotnova and Ponomarenko, 1949; Silina, 1955; Krasil'nikova-Krainova, 1962). All the bacteria isolated were oligonitrophyllic and able to grow on Ashby's nitrogen-free medium. Nitrogen was not observed to accumulate (Rabotnova and Ponomarenko, 1949). But in Krasil'nikov-Krainova's experiments (1962) pure cultures of bacteria isolated from the nodules of foxtail fixed about 1 mg of nitrogen per 100 ml of medium.

Fig. 21. Nodules of meadow foxtail (after Nogtev)

Inoculating foxtail and Siberian lymegrass with aqueous suspensions of crushed nodules of the corresponding species in the absence of nitrogen, Savel'ev (1958) found a considerable gain in yield compared with controls. In these experiments the nitrogen balance was not determined and so it is difficult to identify the effect of the bacteria on the plant. Their effect may to some extent be due to the biologically active substances they produce. At any rate, such a conclusion is indicated by the experiments of Rabotnova and Ponomarenko (1949), who denied that the bacteria they isolated from foxtail nodules fixed nitrogen, but they obtained a 1·5-fold increase in the yield when they used them for bacterization.

Leaf Nodules

Nodules have been found on the leaves of some tropical dicotyledons. The first reference to a symbiotic formation of this type, on the leaves of *Pavetta* (of the family Rubiaceae), is to be found in the work of Zimmerman (1902). Later, leaf nodules were found in representatives of *Grumilea*, *Psychotria*, *Heterophylla* and *Lecanosperma* (of the family Rubiaceae), *Ardisia* (of the family Myrsinaceae) and *Dioscorea* (of the family Dioscoraceae) (for example Boas, 1911; Miehe, 1912; Faber, 1912; Orr, 1923; Milovidov, 1928).

Fig. 22. Leaf nodules of *Pavetta*. Bacterial preparations are shown in boxes.

In 1933, Bremekamp counted more than forty-two species of *Psychotria* with leaf nodules. Now some 400 different species are known (Schwartz, 1959). Leaf nodules have been studied in greatest detail for *Pavetta* (Fig. 22) and *Psychotria*. Anatomically and morphologically they are very similar.

According to Adjanohoun (1957) in *Psychotria* the nodules are located in the leaf parenchyma (mesophyll), which has intercellular spaces on the lower surface of the leaf, arranged in lines over the main vein or scattered between the side veins. The nodules have an intense green colour, with chloroplasts and tannins concentrated in them. Their shapes may differ, and when they age longitudinal cracks often appear.

When fully developed a nodule consists of many cells formed as a result of hyperplasia. The intercellular spaces are filled with bacterial cells present in the mucilaginous mass. The nodules are shed with the leaf. The bacteria usually die when the leaf decays (Pillai and Sen, 1966). The primary infection of the plants with the bacteria evidently originates in the seeds, for plants grown from sterile seeds do not have nodules and, as a rule, are chlorotic (Pillai and Sen, 1966). Apparently the seeds are infected with bacteria present on the plant during ripening. It is still not clear, however, how the endophyte spreads through the plant. Bacteria can be recovered readily from the nodules of *Pavetta* and *Psychotria*.

According to Pillai and Sen (1966) they are small, usually bent, immobile, Gram-negative rods, forming milky white or semi-transparent mucilaginous colonies on nutrient agar. As they age the colonies become adhesive.

Silver *et al.* (1961, 1963) isolated bacteria from the leaf nodules of *Psychotria bacteriophylla* and described them as Gram-negative, non-sporulating rods unresistant to acid. The young cells are motile. In many properties the bacteria isolated resemble *Mycobacterium rubiacearum*, but they differ in certain biochemical properties, in particular the lack of acid resistance. The bacteria are assigned to the group *Klebsiella-Aerobacter* and are named *Klebsiella rubiacearum*. When seeds of *Psychotria* were inoculated with these bacteria, nodules formed on the leaves. Artificial infection of the leaves of *Psychotria* (quoted by E. and O. Allen, 1958) with bacteria from the nodules also causes nodules to form on newly-formed leaves.

The bacteria isolated from *Pavetta* and *Psychotria* fix molecular nitrogen in pure culture. This ability was also possessed by a cell-free mass obtained from bacteria of this genus (Centifanto, 1965). Rao (1923) also noted definite nitrogen fixation in cultures of the bacteria isolated from the nodules of *Pavetta indica* and *Chomelia asiatica*. Pillai and Sen (1966) gave figures of the order of 25 mg of nitrogen per gram of sugar utilized as characterizing the nitrogen-fixing activity of bacteria from the nodules of *Dioscorea*.

Apparently the bacteria fix nitrogen particularly successfully on the leaves of *Psychotria* in conditions of symbiosis (Faber, 1912, 1914; Silver *et al.*, 1961; Pillai and Sen, 1966). It is interesting, however, that plants without nodules develop less well when treated with mineral nitrogen than those with nodules (Silver *et al.*, 1961). This suggests that the bacteria supply the plants not only with nitrogen but also with certain biologically active substances.

In 1953, Stevenson found nodules on the stipules of several species of *Coprosoma*. He was able to isolate bacteria from the nodules in pure culture and demonstrate their ability to fix nitrogen. In Stevenson's view these bacteria play quite an important role in the nitrogen nutrition of plants in poorly fertile soils.

Closely connected with the leaf nodules are the microflora of the phyllosphere—epiphytic microflora of the leaf surface (Ruinen, 1956, 1961, 1965; di Menna, 1958; Greenland, 1958; Kerling, 1958; Pillai and Sen, 1966). The phyllosphere can be seen with the naked eye—brilliant films or differently coloured spots on the surface of the leaf (Pillai and Sen, 1966). The bacteria of the phyllospheres are essentially oligonitrophils, and they may belong to different species. Their contact with the plant is extremely tenacious. Even after repeatedly washing leaves of *Cajanua cajan*. Sen *et al*, 1955 found representatives of the phyllosphere on their surface. The microflora of the phyllosphere is apparently very important in the nitrogen nutrition of the plants.

Mycorrhiza

Mycorrhizae are found on the roots of numerous plants—a close union of fungal hyphae with the cortical parenchyma of the host plant. The term 'mycorrhiza' was introduced by Frank in 1885 to describe the complex fungal-root organs of the birch and beech families.

In the past few years much has been published about mycorrhizae (for example Klechetov, 1947; Mishustin and Pushkinskaya, 1949; Kelly, 1952; Voznyakovskaya, 1954; Rubin and Obrucheva, 1954; Akhromenko, 1960; Krucheva, 1960; Gel'tser and Cherednichenko, 1960; Kryuger, 1961; Shemakhanova, 1961, 1963; Kharli, 1963; Shemakhanova and Mishustin, 1966).

Mycorrhizae are either ectotrophic or endotrophic. The cycle of development of the fungal endophyte (reaction of the host cell nucleus, penetration of the endophyte, phagocytosis and so on) is broadly similar to the corresponding process in the nodules of legumes and other plants. The action of the endophyte on the host plant is also similar. This naturally raises the question of the ability of fungi that form mycorrhiza to fix atmospheric nitrogen.

Attempts to elucidate the nitrogen-fixing powers of fungi that form ectotrophic mycorrhiza were made by Müller (1903), Möller (1906, 1947), Mehlin (1925, 1936) and Bond and Scott (1955). The results of Mehlin's experiments (1925, 1936), carried out with a nitrogen-free medium in sterile conditions, using pure cultures of mycorrhiza-forming fungi and seedlings with or without ectotrophic mycorrhiza, demonstrated no ability of the fungi to fix nitrogen. Fungi in the medium without nitrogen grew poorly for a time and then died. The amount of nitrogen was identical in the mycorrhizal and non-mycorrhizal seedlings.

The nitrogen-fixing capacity of fungi that form endotrophic mycorrhiza was investigated by Rayner (1922), Ternetz (1937) and Bond and Scott (1955). Ternetz, studying in pure cultures the pycnidial fungi she had isolated from the mycorrhiza of members of the Ericaceae, designated some of them nitrogen fixers. The quantity of nitrogen fixed depended on the fungal species and was between 10·92 and 22·14 mg per gram of dextrose. Rayner (1932), noticed that the seedlings of *Calluna vulgaris* growing without nitrogen and infected with nitrogen-fixing fungus supplied by Ternetz, looked healthier than those not infected, although these were growing in a medium containing nitrogen.

We should also like to note the findings of Richards and Voigt (1964) that in soils in the United States in which *Pinus radiata* usually grew there was an annual accumulation of about 50 kg/hectare of nitrogen. In 1965 (quoted by

Becking, 1965a) they confirmed, using nitrogen-15, the fixation of nitrogen in a joint culture of *Pinus radiata* and *Rhizopogon roseolus*. In their view the fungus can fix nitrogen only in symbiosis with the plant and does not have this ability in a pure culture.

Bond and Scott (1955), however, using labelled nitrogen in sterile conditions, found no nitrogen-fixing ability in either the fungi forming ectotrophic mycorrhiza (in *Pinus sylvestris*) or those forming endotrophic mycorrhiza (*Calluna vulgaris*). The inconsistency and paucity of information we have described shows the incompleteness of knowledge about the functions of this large group of fungal endophytes, and their role in the nutrition of plants. At the same time, as Kostychev wrote '.... the problem of the nutrition of mycotrophic plants comes within the category of the most important problems in the soil nutrition of plants ... and until this problem is solved the essence of root nutrition cannot be fully grasped' (1933).

References

Adjanohoun, E. 1957. Etude cytologique et genèse des galles bacteriennes des Rubiacees Africanes due genre *Psychotria*. *C.R. Acad. Sci.*, **245** (5), 576–578.

Akhromeiko, A. I. 1960. New data on the role of mycorrhiza in the nutrition of woody plants. *Lesnoe khozyaistvo*, (10).

Aldrich-Blake, R. N. 1932. On the fixation of atmospheric nitrogen by bacteria living symbiotically in root nodules of *Casuarina equisetifolia*. *For. Mem.*, Oxford, **14**, 1–20.

Allen, E. K. and Allen, O. N. 1949. The anatomy of the nodule growth on the roots of *Tribulus cistoides* L. *Soil Sci.*, **14**, 179–183.

Allen, E. K. and Allen, O. N. 1958. Biological aspects of symbiotic nitrogen fixation. in *Encyclopedia of Plant Physiology*, **9**, 48–118.

Allen, E. K. and Allen, O. N. 1964. Nonleguminous plant symbiosis. in *Microbiology and Soil Fertility. Proc. Twenty-fifth Annual Biol. Colloq.* (3–4) 77–106.

Allen, J. D., Silvester, W. B. and Kalin, M. 1966. Streptomyces associated with root nodules of *Coriaria* in New Zealand. *N.Z.J. Bot.*, 57–65.

Arcularius, J. J. 1928. Zytologische Untersuchungen an einigen endotrophen Mykorrhizen. *Zbl. Bakteroil. Parasitenkunde, Infektionskrankh. und Hyg.*, Abt. 2, **74**, 191–207.

Arzberger, E. A. 1910. The fungous root-tubercles of *Ceanothus americanus*, *Elaeagnus argentea* and *Myrica cerifera*. *Missouri Bot. Garden Rept.* **21**, 60–102.

Aseeva, K. B., Evstigneeva, Z. G. and Kretovich, V. L. 1966. Amino acid composition of alder, *Phaseolus* and lupin nodules. *Dokl. Akad. Nauk SSSR*, **169** (2), 463–465.

Baylis, G. T. S., McNabb, R. E. R. and Morrison, T. M. 1963. The mycorrhizal nodules of podocarps. *Brit. Mycol. Soc. Trans.*, **46** (3), 378–384.

Becking, J. H. 1961. A requirement of molybdenum for symbiotic nitrogen fixation in alder. *Plant and Soil*, **15**, 217–227; *Nature*, **192** (4808), 1204–1205.

Becking, J. H. 1965a. Nitrogen fixation and mycorrhiza in *Podocarpus* root nodules. *Plant and Soil*, **23** (2), 213–226.

Becking, J. H. 1965b. *In vitro* cultivation of alder root-nodule tissue containing the endophyte. *Nature*, **207** (4999), 885–887.

Becking, J. H. 1966. Interactions nutritionnelles plantes-actinomycetes. *Rapport général*. *Ann. Inst. Pasteur*, **111** (3) Suppl., 211–246.

Becking, J. H., Boer, W. E. and Houwink, A. L. 1964. Electron microscopy of the endophyte of *Alnus glutinosa*. *Antonie van Leeuwenhoek*. *J. Microbiol. and Serol.*, **30** (4), 343–376.

Beijerinck, M. W. 1888. Die Bacterien der Papilionaceen Knöllchen. *Bot. Ztg.*, **46**, 725–735.

Beking, D. K. (Becking, J. H.) 1966. Symbiosis of root nodules of the alder *Alnus glutinosa*: in *IX Mezhdunarodnyi Mikrobiologicheskii Kongress. Tezisy dokladov* Ninth International Microbiology Congress, Moscow). Summaries of speeches, 289.

Bergersen, F. J. and Costin, A. B. 1964. Root nodules on *Podocarpus lawrencei* and their ecological significance. *Austral. J. Biol. Sci.*, **17**, 44–48.

Bergersen, F. J., Kennedy, G. S. and Wittman, W. 1965. Nitrogen fixation in coralloid roots of *Macrozamia communis* L. Johnson. *Austral. J. Biol. Sci.*, **18** (6), 1135–1142.

Boas, F. 1911. *Ber. Dtsch. bot. Ges.*, **29** (416), (Quoted by R. Pillai and A. Sen, 1966).

Bond, G. 1952. *Ann. Bot.* New Ser., **16** (64), 467–475 (Quoted by Becking, J. 1966).

Bond, G. 1956. A feature of the root nodules of *Casuarina*. *Nature*, **177**, 191–192.

Bond, G. 1957. Isotopic studies on nitrogen fixation in nonlegume root nodules. *Ann. Bot.* New Ser., **21**, 513–521.

Bond, G. 1959. Fixation of nitrogen in nonlegume root-nodule plants: in *Symposium Soc. Exptl Biol.* 13. *Utilization of Nitrogen and its Compounds by Plants*, Cambridge University Press, 59–72.

Bond, G. 1963. The root nodules of nonleguminous angiosperms. *Thirteenth Symposium Soc. Gen. Microbiol.*, London, 72–91.

Bond, G. 1964. Isotopic investigations of nitrogen fixation in nonlegume root nodules. *Nature*, **204** (7), 600.

Bond, G., Fletcher, W. W. and Ferguson, T. P. 1954. The development and function of the root nodules of *Alnus*, *Myrica* and *Hippophaë*. *Plant and Soil*, **5**, 309–323.

Bond, G. and Hewitt, E. J. 1961. Molybdenum and the fixation of nitrogen in *Myrica* root nodules. *Nature*, **190**, 1033–1034.

Bond, G. and Hewitt, E. J. 1962. Cobalt and the fixation of nitrogen by root nodules of *Alnus* and *Casuarina*. *Nature*, **195**, 94–95.

Bond, G. and MacConnell, J. T. 1955. Nitrogen fixation in detached non-legume root nodules, *Nature*, **176**, 606.

Bond, G., MacConnell, J. T. and McCallum, A. H. 1956. The nitrogen nutrition of *Hippophaë rhamnoides* L. *Ann. Bot.* New Ser., **80**, p. 501.

Bond, G. and Scott, G. D. 1955. An examination of some symbiotic systems for the fixation of nitrogen. *Ann. Bot.*, **19** (73), 65–77.

Bose, S. R. 1947. Hereditary (seed-borne) symbiosis in *Casuarina equisetifolia* Forst. *Nature*, **159**, 512–514.

Bottomley, W. B. 1906. The cross-inoculation of Leguminosae and other root-nodule bearing plants. *Rept. Meeting Brit. Assoc.*, *Adv. Sci.*, **76**, 752–753.

Bottomley, W. B. 1907. The structure of root tubercules in leguminous and other plants. *Rept. Meeting Brit. Assoc. Adv. Sci.*, **76**, 693.

Bottomley, W. B. 1912. Root nodules of *Myrica gale*. *Ann. Bot.*, **26**, 111–117.

Bouwens, H. 1943. Investigation of the symbiont of *Alnus glutinosa*, *Alnus incena* and *Hippophaë rhamnoides*. *Antonie van Leeuwenhoek, J. Microbiol. and Serol.*, **9**, 107–114.

Bremekamp, E. B. 1933. The bacteriophilous species of *Psychotria bacteriophilia*. *J. Bot.*, **71**, 271–279.

Centifanto, J. M. 1965. Leaf nodule symbiosis in *Psychotria bacteriophila. Düss. Abstr.*, **25** (7), 3794.

Christoph, H. 1921. Untersuchungen über mykotrophen Verhältnisse der 'Ericales und die Keimung von Pirolaceen. *Beih. Bot. Zbl. Abt.*, **1**, 38, 115–157.

Cooper, W. S. 1922. The bread sclerophyll vegetation of California. An ecological study of chaparral and its related communities. *Carnegie Inst. Publ.*, Washington, **319**, 124.

Crocker, R. L. and Major, 1955. Soil development in relation to vegetation and surface age of glacier Bay. *Alaska. J. Ecol.*, **43**, 422–448.

Danilewicz, K. 1965. Symbiosis in *Alnus glutinosa* (L.) Gaertn. *Acta Microbiologica Polonica.*, **14** (3–4), 321–325 (in English).

Davenport, H. E. 1960. Haemoglobin in the root nodules of *Casuarina cunninghamiana. Nature*, **186**, 653–654.

Delwiche, C. C., Zinke, P. J. and Johnson, C. M. 1965. Nitrogen fixation by *Ceanothus. Plant Physiol.*, **40** (6), 1036–1047.

Dommergues, J. 1963. Evaluation of the nitrogen fixation in a dune soil afforested with *Casuarina equisetifolia. Agronomica*, **7**, 335–340.

Douin, R. 1953. Sur la fixation de l'azote libré par Myxophycees endophytes Cycadacees. *C.R. Acad. Sci. Paris*, **236**, 956–958.

Douin, R. 1954. Nitrogen and carbon nutrition of *Anabaena* endophytic in Cycadaceae. *Ann. Univ. Lyon. Sci. Sect. C*, **8**, 57–70.

Faber, R. C. 1912. Das erbliche Zusammenleben von Bakterien und tropischen Pflanzen. *Jahrb. wiss. Bot.*, **51**, 285–375.

Faber, F. C. 1914. Die Bakterien-Symbiosen der Rubiaceen. *Jahrb. wiss. Bot.*, **54**, 243–264.

Ferguson, T. P. and Bond, G. 1953. Observation on the formation and function of the root nodules of *Alnus glutinosa* (L) Gaertn. *Ann. Bot. New Ser.*, **17**, 175.

Fernandez, F. and Bhat, J. V. 1945. A note on the association of *Chlorococcum humicolum* in the roots of *Cycas renoluta. Current Sci.*, **14**, 235.

Ferreira dos Santos, N. 1947. Natureza dos nodules do *Podocarpus variegatus. Hort. Rev. Agron.*, **36**, 68–72.

Fiuczek, M. 1959. Fixation of atmospheric nitrogen in pure cultures of *Streptomyces alni. Acta Microbiologica Polonica*, **8**, 283–287.

Fletcher, W. W. 1955. The development and structure of the root nodules of *Myrica gale* L. with special reference to the nature of the endophyte. *Ann. Bot. New Ser.*, **19**, 501–513.

Frank, B. 1892. Über die auf den Gasaustausch bezüghichen Einrichtungen der Wurzelknöllchen der Leguminosen. *Ber. Dtsch. Bot. Ges.*, **10**, 281.

Furman, T. 1959. The structure of the root nodules of *Ceanothus sanguineus* with special reference to endophyte. *Amer. J. Bot.*, **46**, 698–703.

Gardner, I. C. 1965. Observations on the fine structure of the endophyte of the root nodules of *Alnus glutinosa* (1) Gaerten. *Arch. Mikrobiol.*, **51** (4), 365–383.

Gardner, I. C. and Bond, G. 1957. Observation on the root nodules of *Shepherdia. Canad. J. Bot.*, **35** (3), 305–314.

Gel'tser, F. and Cherednichenko, I. 1960. New information on the causes of potato decay. *Kartofel'iovoshchi*, (6).

Goldman, C. B. 1961. The contribution of alder trees (*Alnus temifolia*) to the primary productivity of Castle Lake, California. *Ecology*, **42**, 282–287.

Greenland, D. J. 1958. Nitrate fluctuations in tropical soils. *J. Agric. Sci.*, **50**, 82.

Hariet, P. 1892. Sur une algue qui vit dans les racines des Cycadées. *C.R. Acad. Sci., Paris*, 115–325.

Hawker, L. E. and Fraymouth, J. 1951. A re-investigation of the root-nodules of species of *Elaeagnus, Hippophaë, Alnus* and *Myrica*, with special reference to the morphology and life histories of the causative organisms. *J. Gen. Microbiol.*, **5**, 369–386.

Hellmers, H. and Kellcher, J. M. 1959. *Ceanothus leucodermis* and soil nitrogen in Southern California mountains. *Forest Sci.*, **5**, 275–278.

Hewitt, E. J. and Bond, G. 1966. The cobalt requirement of non-legume root nodule plants. *J. Exp. Bot.*, **17** (52), 480–492.

Hiltner, L. 1898. Über Entstehung und physiologische Bedeutung der Wurzelknöllchen. *Forest Naturwiss. Z.*, **7**, 415–423.

Hiltner, L. 1903. Über die biologische und physiologische Bedeutung der endotrophen Mycorrhiza. *Naturwiss. Z. landwirtsch. Forestwesen*, **1**, 9–25.

Hoak, A. 1964. Über den Einfluss der Knöllchenbakterien auf die Wurzelhaare von Leguminosen und Nichtleguminosen. *Zbl. Bakteriol. Parasitenkunde, Infektionskrankh. und Hyg.*, Abt. 2, **117** (4), 343–366.

Holmsgaard, E. 1960. Amount of nitrogen-fixation by alder. *Det Forstl. Forsgsv. i Danm 2b*, 253–270 (in English).

Hooker, J. D. 1854. On some remarkable spherical exostoses developed on the roots of various species of Coniferae. *Proc. Linnean Soc.*, **2**, 335–336.

Isachenko, B. L. 1913. The nodules of *Tribulus terrestris*. *Izvestiya Imp. SPb. botanicheskogo sada*, **13** (1), 23–27.

Janse, J. M. 1897. Les endophytes radicaux de quelques plants javanaise. *Ann. Jard. Bot. Buitenzorg.*, **14**, 53–201.

Jepson, W. L. 1936. in *The Flora of California*. **2**, Berkeley Univ., California Press, 460–462.

Karavaev, M. N. 1959. Nitrogen-fixing nodules on the roots of fossil alder (*Alnus* sp.). *Botanicheskii zhurnal*, **44** (7), 1000–1001.

Kämppel, M. and Wartenberg, H. 1958. Der Formenwechsel des *Actinomyces alni* Peklo in den Wurzeln von *Alnus glutinosa* (L) Gaertn. *Arch. Mikrobiol.*, **30**, 46.

Kelli, A. (Kelly, A.). 1952. *Mikotrofiya u rastenii* (Mycotrophy in plants), IL Moscow.

Kellermann, K. T. 1911. Nitrogen-gathering plants: in *USD Yearbook for 1910*, 213–218.

Kerling, L. C. P. 1958. The leaf microflora of *Beta vulgaris*, *Tijdschr. Plantenzicken*, **64**, 402.

Kharli, Dzh. L. (Kaarli, J. L.). 1963. The biology of mycorrhiza: in *Mikoriza rastenii* (Plant mycorrhiza). Izd-vo s.-kh. lit., zhurn. i plakatov, Moscow.

Khrushcheva, E. P. 1960. Conditions promoting the formation of mycorrhiza. *Agrobiologiya*, (4).

Klechetov, A. N. 1947. Mycorrhiza in the rubber-plant tau-sagyz and the root reaction to soil conditions. *Izvestiya Kuibyshevskogo sel'skokhozyaistvennogo instituta*, (9).

Klečka, A. and Vukolov, V. 1935. Comparative studies of tree mycorrhiza. *Sbor. Českosl. Akad. Zěmed. ved.*, **10**, 443–457.

Kohnke, H. 1941. The black alder as a pioneer tree on sand dunes and eroded land. *J. Forestry*, **39**, 333–334.

Kondo, T. 1931. Zur Kenntnis des N-Gehaltes des Mykorrhiza-Knöllchens von *Podocarpus macrophylla* D. Don. *Bot. Mag.* (Tokyo), **45**, 495–501.

Kostychev, S. P. 1933. *Fiziologiya rastenii*, **1**, 241.

Kozlova, E. I., Badumyan, L. S. and Vendilo, M. V. 1966. Properties of bacteria in the nodules of the sea buckthorn. *Mikrobiologiya*, **35** (4), 699–706.

Kravtsov, B. I. 1950. Properties of the nodules of *Elaeagnus* and sea buckthorn. *Les i step'*, (9), 89–90.

Krasil'nikova-Krainova, A. I. 1962. Nature of the root nodules of *Alopecurus pratensis* L. *Mikrobiologiya*, **31** (6), 1041–1047.

Krebber, O. 1932. Untersuchungen über die Wurzelknöllchen der Erby. *Arch. Mikrobiol. Jahresb.*, **3**, 588–608.

Kryuger, L. V. 1961. Endotropic mycorrhiza of herbaceous plants of certain phyto-coenoses of the Central Pre-Urals. *Botanicheskii zhurnal*, (5).

Lawrence, D. B. 1958. Glaciers and vegetation in South-Eastern Alaska. *Amer. Sci.*, **46**, 89–122.

Lawrence, D. B. and Hulbert, L. 1950. Growth stimulation to adjacent plants by lupin and alder on recent glacier deposits in South-Eastern Alaska. *Bull. Ecol. Soc. Amer.*, **31**, 58.

Leaf, G., Gardner, J. C. and Bond, G. 1959. Observation on the composition and metabolism of the nitrogen-fixing root nodules of *Myrica*. *Biochem. J.*, **72**, 662.

Lieske, R. 1921. *Morphologie und Biologie der Strahlenpitz (Actinomyceten)*. Born-traeger, Leipzig.

Life, A. C. 1901. The tuber-like rootlets of *Cycas revoluta*. *Bot. Gaz.*, **31**, 265–271.

McLuckie, J. 1922. Studies in symbiosis. II. The apogeotropic roots of *Macrozamia spiralis* and their physiological significance. *Linnean Soc. N.S. Wales Proc.*, **47**, 319–328.

McLuckie, J. 1923. Studies in symbiosis. 3. Contribution to the morphology and physiology of the root-nodules of *Podocarpus spinulosa* and *P. elata*. *Linnean Soc. N.S. Wales Proc.*, **48**, 82–93.

Maire, R. and Tison, A. 1909. La cytologie des Plasmodiophoracees et la classe des Phytomyxinae. *Ann. Mycol. (Berl.)*, **7**, 226–253.

Melin, E. 1925. *Untersuchungen über die Bedeutung der Baummykorrhiza*. Jena, 1–125.

Melin, E. 1936. Methoden experimentellen Untersuchungen mykotropher Pflanzen: in *Handbuch der Biologischen. Arbeitemethoden*, **11**, 1015–1108.

Menna, M. E. di. 1958. Two new species of yeasts from New Zealand. *J. Gen. Microbiol.*, **18**, 269.

Miehe, H. 1912. *Jahrb. wiss. Bot.*, **50**, 29 (Quoted by H. Pillai and Al Sen, 1966).

Miehe, H. 1918. Anatomische Untersuchungen der Pilzsymbiose bei *Casuarina equisetifolia* nebst einigen Bemerkungen über das Mykorrhizenproblem. *Flora*, N.F. (11–12), 431–449.

Mikola, P. 1958. Liberation of nitrogen from alder litter. *Acta forest. fennica*, **67**, 1–10 (in English).

Milovidov, P. F. 1928. *Arch. Anat. Microbiol.*, **24**, 19 (Quoted by R. Pillai and A. Sen, 1966).

Mishustin, Ye. N. and Pushkinskaya, O. I. 1949. The mycorrhiza of woody plants and its importance in windbreak forest plantations. *Mikrobiologiya*, **18** (5).

Möller, A. 1906. Mykorrhizen und Stickstoffernährung. *Ber. Dtsch. Bot. Ges.*, **24**, 230–233.

Möller, A. 1947. Mycorrhiza and nitrogen assimilation with special reference to mountain pine (*Pinus mugro urra*) and spruce (*Picea abies* L.) *Karst. Det. Forst. For i Danmark*, **19** (2), 105–208 (in English).

Montasir, A. and Sidrak, G. H. 1952. Root nodulation in *Zygophyllum coccincum* L. *Inst. Fonad du desert.*, **8**, 68–70 (in English).

Moore, A. W. 1964. Note on non-leguminous nitrogen-fixing plants in Alberta. *Canad. J. Bot.*, **42** (7), 952–955.

Morrison, T. M. 1961. Fixation of nitrogen-15 by excised nodules of *Discaria toumatou*. *Nature*, **189**, 945.

Morrison, T. M. and Harris, G. P. 1958. Root nodules in *Discaria toumatou* Raoul choix. *Nature*, **182** (4451), 1746.

Mostafa, M. A. and Mahmound, M. Z. 1951. Bacterial isolates from root nodules of Zygophyllaceae. *Nature*, **167** (4246), 446–447.

Mudrova, A. A. 1939. A technique for isolating a pure culture of 'foxtail bacterium' *Bacillus alopecuri* Nogt. *Dokl. Akad. Nauk SSSR*, **25** (2), 163–165.

Müller, P. E. 1903. Über das Verhältnis der Bergkiefer zur Fichte in den Jutlandischen Heidekulturen. *Naturwiss. Z. Land-und Forstwirtsch.*, **1**, 289–378.

Nadson, S. 1955. Forms and formation of *Eleagnus* nodules. *Trudy Kazakhskogo gosudarstvennogo sel'skokhozyaistvennogo instituta*, **5** (1).

Niewiarowska, J. 1959. Symbiosis in sand-willow. *Acta microbiologica polonica*, **8**, 289–294.

Niewiarowska, J. 1961. Morphologie et physiologie des Actinomycetes symbiotiques des *Hippophaë. Acta microbiologica polonica*, **10**, 271–286.

Nobbe, F. and Hiltner, L. 1899. Die endotrophe Mycorrhiza von *Podocarpus* und ihre physiologische Bedeutung. *Landwirtsch. Vers. Stat.*, **51**, 241–245.

Nogtev, V. P. 1938. Nodules on the roots of the meadow foxtail *Alopecurus pratensis*; their origin and physiological function. *Botanicheskii zhurnal*, **23** (2), 145–150.

Orr, M. Y. 1923. *Rept. Roy. Bot. Garden, Edinburgh*, (14), 57.

Ovegton, J. D. 1965. Studies of the development of woodland conditions under different trees. 4. The ignition loss water, carbon and nitrogen content of the mineral soil. *J. Ecol.*, **44**, 171–179.

Panosyan, A. K. 1943. *Mikrobiologicheskii sbornik Biologicheskogo Instituta Akademii Nauk SSSR, Armyanskii filial*, (1), 147 (quoted by Kozlova, E. I., Badumyan, L. S. and Vendilo, M. V. 1966).

Peklo, J. 1910. Die pflanzlichen Aktinomykosen. *Zbl. Bakteriol., Parasitenkunde, Infektionskrankh, und Hyg.*, Abt. 2, **27**, 451–579.

Petrie, E. J. 1925. Physiological studies on *Ceanothus americanus*. Dissertation Michigan Coll. Agric. Appl. Sci.

Phillips, J. 1932. Root nodules of *Podocarpus*. *Ecology*, **13**, 189–195.

Pillai, R. N. and Sen, A. 1966. Microbiology of the phyllosphere. *Sci. and Culture*, **32** (7), 383–384.

Plotho, von O. 1941. Die Synthese der Knöllchen an den Wurzeln der Erle. *Arch. Mikrobiol. Jahresb.*, **12**, 1–18.

Pommer, E. II. 1956. Beiträge zur Anatomie und Biologie der Wurzelknöllchen von *Alnus glutinosa* (L) Gaertn. *Flora*, **143**, 603–634.

Pommer, E. II. 1959. Über die Isolierung der Endophyten aus den Wurzelknöllchen *Alnus glutinosa* Gaertn, und über erfolgreiche Re-Infektionsversuche. *Ber. Dtsch. bot. Cos.*, **72**, 138–150.

Quick, C. R. 1944. Effects of snow brush on the growth of Sierra gooseberry. *J. Forestry*, **42**, 827–832.

Quispel, A. 1954. Symbiotic nitrogen fixation in non-leguminous plants. I. Preliminary experiments on the root-nodule symbiosis of *Alnus glutinosa*. *Acta bot. Neerl.*, **3**, 495–511.

Rabotnova, I. L. and Ponomarenko, N. I. 1949. Symbiotic bacteria of the meadow foxtail. *Mikrobiologiya*, **18** (1), 54–61.

Rao, K. 1923. *Agric. J. Industr.* 18, 132 (quoted by R. Pillai and A. Sen, 1966).

Rayner, M. C. 1922. Nitrogen fixation in the Ericaceae, *Bot. Gaz.*, **73**, 226–235.

Reinke, J. 1879. Zwei parasitische Algen, *Bot. Ztg.*, **37**, 473–478.

Richards, B. N. Voigt, G. K. 1964. Role of mycorrhiza in nitrogen fixation. *Nature*, **201**, 310–311.

Richards, B. N. Voigt, G. K. Nitrogen accretion in coniferous forest ecosystems. *Proc. Soc. North Amer. Forest. Soils. Conf.* Corvallis, Oregon, USA. Quoted by Becking, 1965a.

Roberg, M. 1934. *Uber den Erreger der Wurzelknöllchen von Alnus und den Elaeagnaceen: Elaeagnus und Hippophae. Jahrb. wiss Bot.*, **79**, 472.

Roberg, M. 1938. Über der Erreger der Wurzelknöllchen europäischer Erlen. *Jahrb. Wiss. Bot.*, **86**, 344.

Rossi, S. 1964. Propagation dans le sol de l'organisme causant les nodosites dans les racines d'aune (*Alnus glutinosa*). *Ann. Inst. Pasteur*, **106** (3), 503–510.

Rubin, B. A. and Obrucheva, N. V. 1954. The physiology of the mycotropic nutrition of woody plants. *Uspekhi sovremennoi biologii*, **40** (2/5).

Ruinen, J. 1956. Occurrence of *Beijerinckia* in the 'phyllosphere'. *Nature*, **177**, 220.

Ruinen, J. 1961. The phyllosphere. 1. An ecologically neglected milieu. *Plant and Soil*, **15** (2), 81–109.

Ruinen, J. 1965. The phyllosphere. 3. Nitrogen fixation in the phyllosphere. *Plant and Soil*, **22** (3), 375–394.

Sabet, J. S. 1946. Bacterial root nodules in the Zygophyllaceae. *Nature*, **157** (3994), 656–657.

Savel'ev, N. M., Gorbaleva, P. N. and Klevenskaya, I. L. 1959. Role of nodules on the roots of grasses. *Izvestiya Sibirskogo otdeleniya Akademii Nauk SSSR*, (10), 124–128.

Saxston, W. T. 1930. The root nodules of the Podocarpaceae. *S. Afric. J. Sci.*, **27**, 323–325.

Schacht, H. 1853. Beiträge zur Entwicklungsgeschichte der Wurzel. *Flora*, **36**, 257–266.

Schaede, R. 1933. Über die Symbionten in den Knöllchen der Erle und des Sanddörnes und die cytologische Verhältnissen in ibren. *Planta*, **19**, 389–416.

Schaede, R. 1943. Über die Symbiose in den Wurzelknöllchen der Podocarpeen. *Ber. Dtsch. Bot. Ges.*, **61**, 39–41.

Schaede, R. 1944. Über die Korallenwurzeln der Cycadeen und ihre Symbiose. *Planta*, **34**, 98–124.

Schaede, R. 1962. Die pflanzlichen Symbiosen. 2 Aufl. Stuttgart, G. Fischer.

Schoenke, R. E. 1957. Influence of mountain avens (*Dryas drummondii*) on growth of young cottonwoods (*Papulus trichocarpa*) at Glacier Bay, Alaska, *Minnesota Acad. Sci. Proc.*, **25**, 55–58.

Schwartz, W. 1959. Bakterien und Actinomyceten Symbiosen. *Encyclop. Plant Physiol.*, **11**, 560–572.

Sen, A., Rewari, R. B., Paul, N. P. and Sen, A. N. 1955. *Proc. Nat. Acad. Sci. USA*, **24A**, 5 (quoted by Pillai and Sen, 1966).

Shemakhanova, N. M. 1961. Effect of pure cultures of fungal mycorrhiza producers on the development of seedlings of pine and oak. *Izvestiya Akad. Nauk. SSSR, Ser. Biol.*, (3).

Shemakhanova, N. M. 1963. in *Mikoriza drevesnykh rastenii* (*Mycorrhiza of woody plants*). Izd-vo s.-kh. literatury, zhurnalov i plakatov, Moscow.

Shibata, K. 1902. Cytologische Studien über die endotrophen Mykorrhizen. *Jahrb. wiss. Bot.*, **37**, 643–684.

Silina, K. S. 1965. Biological features of the meadow foxtail in a forest-tundra zone. *Botanicheskii zhurnal*, **40** (4), 592–596.

Silver, W. S. 1964. Root nodules symbiosis. 1. Endophyte of *Myrica cerifera* L. *J. Bacteriol.*, **87** (2), 416–421.

Silver, W. S., Bendana, F. S. and Powell, R. D. 1966. Root nodules symbiosis. 2. The relation of auxin to root geotropism in root and root-nodules of non-legumes. *Physiol. Plantarum*, **19**, 207–218.

Silver, W. S., Centifanto, J. M. and Nicholas, D. J. D. 1963. Nitrogen fixation by the leaf-nodule endophyte of *Psychotria bacteriophila*. *Nature*, **199**, 396–397.

Slodger, Ch. and Sil'ver, U. S. (Sloger, C. and Silver, W. S.). 1966. Nitrogen fixation by excised root nodules and their homogenates in *Myrica cerifera*: in *IX Mezhdurnarodnyi Mikrobiologicheskii Kongress. Tezisy dokladov* (Ninth Microbiology Congress, Moscow) Summaries of speeches, 293.

Sloger, C. and Silver, W. S. 1965. Note on nitrogen fixation by excised root nodules and nodular homogenates of *Myrica cerifera* L. in Non-Heme Iron Proteins: Role in Energy Conservation, 299–300.

Snyder, R. M. 1925. Nitrogen fixation by non-leguminous plants. *Michigan State Agric. Exptl Stat. Quart.*, **8**, 34–36.

Spratt, E. R. 1912a. The formation and physiological significance of root nodules in the Podocarpineae. *Ann. Bot.*, **26**, 801–814.

Spratt, E. 1912b. The morphology of the root tubercules of *Alnus* and *Elaeagnus* and the polymorphism of the organism causing their formation. *Ann. Bot.*, **26**, 119–128.

Spratt, E. 1915. The root-nodules of Cycadaceae. *Ann. Bot.*, **29**, 619–626.

Stewart, W. D. P. 1962. A quantitative study of fixation and transfer of nitrogen in *Alnus. J. Exp. Bot.*, **13** (38), 250–256.

Stewart, W. D. P. 1963. The effect of combined nitrogen on growth and nodule development of *Myrica* and *Casuarina. Z. allg. Microbiol.*, **3** (2), 152–156.

Stevenson, G. B. 1953. Bacterial symbiosis in some New Zealand plants. *Ann. Bot.* New Ser., **17**, 343.

Steyaert. 1932. Une epiphytic bacterienne des racines de *Coffea robusta et C. klainii. Rev. Zool. et Bot. Afric.*, **22**, 133–139.

Stone, E. L. 1955. Observation on forest fertilization in Europe. *Proc. Nat. Joint Comm. Fert. Appl.*, **31**, 81–87.

Takesige, T. 1937. Die Bedeutung der Symbiose zwischen einigen endophytischen Blaualgen und ihren Wirtspflanzen. *Bot. Mag.*, **57**, 514–524.

Tarrant, K. E., Isaak, L. A. and Chandler, R. E. 1951. Observations on litter fall and foliage nutrient content of some Pacific Northwest tree species. *J. Forestry*, **49**, 914–915.

Tarrant, R. E. 1961. Stand development and soil fertility in a Douglas-fir–red alder plantation. *Forest Sci.*, **7**, 238–246.

Taubert, H. 1956. Über den Infektionsvorgang und die Entwicklung der Knöllchen bei *Alnus glutinosa* Gaertn. *Planta*, **48**, 135–156.

Ternetz, C. 1907. Über die Assimilation des atmosphärischen Stickstoffes durch Pilze. *Jahrb. Wiss. Bot.*, **44**, 353–408.

Uemura, S. 1952. Studies on the root nodules of alders (*Alnus* spp.). 4. Experiments on isolation of Actinomycetes from alder nodules. *Bull. Govt Forest Exptl. Stat.*, Tokyo, **52**, 1–18.

Uemura, S. 1961. Studies on the *Streptomyces* isolated from alder root nodules *Alnus* spp. *Sci. Rept Agron. Forest Fish. Res. Council*, Tokyo, **7**, 1–90.

Uemura, S. 1964. Isolation and properties of micro-organisms from root nodules of non-leguminous plants. A review with extensive bibliography. *Bull. Govt. Forest. Exptl. Stat.*, Tokyo, **167**, 59–91.

Verner, A. R. 1965. Relationships between microbial flora and higher green plants in conditions of their introduction and acclimatization: in *Rastitel'nye resursy Sibiri, Urala i Dal'nego Vostoka* (Plant resources of Siberia, the Urals and the Soviet Far East), Nauka, Moscow, 36–42.

Virtanen, A. J. 1962. On the fixation of molecular nitrogen in nature. *Communs Inst. Forest Fennicae*, **55** (22), 1–11.

Virtanen, A. and Saastamoinen, S. 1936. Untersuchungen über die Stickstoffbildung bei der Erle, *Biochem. Z.*, **284**, 72–85.

Vlamis, J., Schultz, A. M. and Biswell, H. H. 1958. Nitrogen fixation by deerbrush. *Calif. Agr. Jan. Issue*, 11.

Vlamis, J., Schultz, A. M. and Biswell, H. H. 1964. Nitrogen fixation by root nodules of western mountain mahogany. *J. Range Management*, **17**, 73–74.

Voznyakovskaya, Yu. M. 1954. The problem of mycorrhiza and its practical importance, *Mikrobiologiya*, **3** (2).

Wagle, R. F. and Vlamis, J. 1961. Nutrient deficiencies in two bitterbrush soils. *Ecology*, **42**, 745–752.

Wahlenberg, W. G. 1930. Effect of *Ceanothus* brush on Western yellow pine plantations in the Northern Rocky Mountains. *J. Agric. Res.*, **41**, 601–612.

Winter, G. 1935. Über die Assimilation des Luftstickstoffs durch endophytische Blaualgen. *Beitr. Biol. Pflanzen*, **23**, 295–305.

Woronin, M. 1867. Observations sur certaines excroissences que présentent les racines de l'aune et du lupin des jardins. *Ann. Sci. Nat. Bot.*, Ser., **5, 7**, 73–86.

Yeates, J. S. 1924. The root nodules of New Zealand pines. *N.Z. J. Sci. Technol.*, **1**, 121–124.

Yendo, J. and Takase, T. 1932. On the root-nodule of *Elaeagnus*. *Bull. Serie Silk. Ind.* (Uyeda), **4, 5**.

Youngken, H. W. 1919. The comparative morphology, taxonomy and distribution of the Myricaceae in the Eastern United States. *Contribs. Bot. Lal. Univ.*, **4**, 339–400.

Zimmermann, A. 1902. Über Bakterienknoten in den Blättern einigor Rubiaceae. *Jahrb. Wiss. Bot.*, **37**, 1–11.

4 Structure and Development of
Azotobacter

In 1901 Beijerinck discovered an aerobic micro-organism that bound molecular nitrogen, and called it *Azotobacter chroococcum*. Later, further species of *Azotobacter* were described, but they are not all now recognised as independent systematic groups. Many of them proved to be mere variants of other species of *Azotobacter*.

The vast amount of work carried out on *Azotobacter* has been reviewed in several monographs (Omelyanskii, 1923; Sushkina, 1947; Blinkov, 1959; Rubenchik, 1960; Zinov'eva, 1962; Zaitseva, 1965). In presenting a brief description of *Azotobacter* we shall be concerned principally with morphological and cytological features. At an early age the size of the cells varies within the limits $2 \cdot 0$–$7 \cdot 0 \times 1 \cdot 0$–$2 \cdot 5$ μm. In individual cases the length reaches 10–12 μm.

Multiplication proceeds by simple division with the formation of a transverse septum and sometimes a constriction is observed. In unfavourable conditions the cells may bud (Petersen, 1961).

Young *Azotobacter* cells are motile and have peritrichous flagella. Krasil'nikov *et al.* (1965) found outgrowths similar to fimbria, and a similar observation has been made in our laboratory (Plate 6). In the logarithmic growth phase the cells contain structures that stain with nuclear dyes (Mencl, 1911; Petschenko, 1930; Pochon *et al.*, 1944; Imshenetskii, 1946, 1962). There is considerably less DNA in the cells than in other species of bacteria, and it fluctuates between $0 \cdot 81$ and $0 \cdot 70$ per cent (Belozerskii *et al.*, 1957).

In young cells of *Azotobacter* the plasma is finely granulated. When the cells age, they lose their motility, shorten and assume an almost coccoid form and grains appear in the plasma that sharply refract the light (they are fat drops, volutin, metachromatic inclusions and so on). At this time the cell (or group of cells) is enclosed in a thick mucilaginous capsule. If the capsule becomes denser then *Azotobacter* resembles the chlamydospores of lower fungi. The capsule is made of polysaccharides and contains about $0 \cdot 023$ per cent of nitrogen. The anhydride of uronic acid accounts for as much as 75 per cent of the mucilage, which is relatively resistant to the action of micro-organisms (Martin, 1945; Quinnel *et al.*, 1957; Cohen and Johnstone, 1964; Martin *et al.*, 1965). Martin *et al.* (1966) established that the polysaccharides of *Azotobacter*, in a complex with copper and iron salts, give compounds that are poorly decomposed by micro-organisms. The cystoid cells of *Azotobacter* have been investigated by Tchan *et al.* (1962) and Parker and Socolofsky (1966).

In old cultures giant spherical cells can be found, which are sometimes equated with gonidia. Cells as small as 0·2 μm in diameter (and even smaller) are also found; they resemble microgonidia. In favourable conditions these microscopic cells begin to develop and form normal sized cells.

The presence of these and other cells in cultures of *Azotobacter* (disorganized, shapeless masses—symplasms, giant spindle-shaped cells, branched forms and even spore formations) prompted the theory of cyclogeny, that *Azotobacter* has a complex life cycle (Löhnis *et al.*, 1916, 1921).

Further investigation suggested that the diversity of forms of *Azotobacter* is largely a consequence of its pleomorphism and partly due to the appearance of involutive forms (for example giant inflated cells). One can hardly speak of any regular cycle (for example Krasil'nikov, 1931; Lewis, 1937; Bachinskaya, 1935; Bisset and Hale, 1953). As long ago as 1932 Vinogradskii (1932*a*) pointed out that the appearance of morphologically different cells in cultures of *Azotobacter* depended on the composition of the medium. This is clearly confirmed by much experimental evidence. By changing the composition of the medium and the conditions of culture it is possible to induce changes in the phases of development of *Azotobacter*, and in the morphology of these bacteria. Thus, Van Schreven (1962) noted that the 'gonidia' of *Azotobacter* are not formed in all media. Using flow culture, Málek (1952) and Macurá and Kotková (1953) arrested the development of *Azotobacter* at the juvenile cell stage. For a long time the population in the medium was dominated by rod-like cells, which then became diplococci. Morphogenesis can be regulated by synchronizing cultures of *Azotobacter* (Zaitseva *et al.*, 1961; Ierusalimskii *et al.*, 1942; Lin Liang-Ping and Wyss, 1965).

In a medium containing ethanol as the only source of carbon, *Azotobacter* retains its motility and rod shape for a long time. In many other media the pleomorphism is very sharply manifest. This was noted early (Beijerinck, 1901; Pražmowski, 1912) and was later confirmed in a series of studies (for example Lee and Burris, 1943; Gainey 1944; Alexander and Wilson, 1954; Yamagishi and Furusaka, 1964).

Some have described a definite cycle of development for *Azotobacter* (Eisenstark *et al.*, 1950; Dondero and Zelle, 1953). On solid nutrient media containing no nitrogen compounds *Azotobacter* forms large mucilaginous, sometimes wrinkled colonies. When they age these develop green-yellowish, pink or brown-black pigments. Colonies of different species of *Azotobacter* have their own particular colour (Table 4.1).

Table 4.1. Characteristic Features of the Main Species of *Azotobacter*

Character	Az. chroococcum	Az. beijerinckii	Az. vinelandii	Az. agilis (Azomonas agilis)
Size of cell	3·1 × 2·0	4·6 × 2·4	3·4 × 1·5	3·3 × 2·8
Formation of cysts	Yes	Yes	Yes (in ethanol on tenth day)	No
Motility	Motile, especially in young culture or when grown in ethanol	Non-motile	Motile, (in old cultures motility may not be noticed)	Motile
Formation of pigment	With ageing	With ageing	Already in young culture	Does not form or forms in young culture
Characterization of pigment	From dark brown to black	From yellowish to light brown	Yellowish-green fluorescing	
	Does not diffuse into water		Diffuses into water	
Utilization of starch	Yes	No	—	—
Utilization of sodium benzoate	Facultatively	Yes (grows at a concentration of even 5%)	Yes (grows at a concentration of 1%)	No
Utilization of mannitol	Yes	Yes	Yes	No
Utilization of rhamnose	Yes	Yes	Yes	No

This table is after Voets and Dedekan, 1966.

Classification of the Genus *Azotobacter*

Several classifications have been suggested for *Azotobacter*. Some (Tchan, 1953; H. Jensen, 1954*a*; Rubenchik, 1960; Norris, 1960) recognize an independent family of Azotobacteriaceae, which we consider most useful. Others, among them Krasil'nikov (1949), include *Azotobacter* in the family Bacteriaceae.

Many species of *Azotobacter* have been described, including *Az. chroococcum, Az. beijerinckii, Az. vinelandii, Az. agilis, Az. nigricans, Az. woodstownii, Az. macrocytogenes, Az. galophilum* and *Az. miscellum*. The work of H. Jensen (1954) and Voets and Dedeken (1966) has indicated that *Az. chroococcum, Az. beijerinckii* and *Az. vinelandii* are well defined species. A fourth is assigned by Voets and Dedeken to the genus *Azomonas*; this is *Azomonas agilis. Azomonas* has characteristic, almost round cells. The principal properties of *Azotobacter* are presented in Table 4.1.

On the basis of a study of the DNA base ratios (guanine + cytosine) of nitrogen-fixing bacteria, de Ley and Park (1966) decided that the genus *Azotobacter* is heterogeneous and consists of three groups—*Azotobacter, Azomonas* and *Azotococcus*. The first group is divided into three species: *Az. chroococcum, Az. beijerinckii, Az. vinelandii*, and the second into two: *Az insignis* and *Az. macrocytogenes*, and the third consists of one species—*Az. agilis*.

Some features of the structure of the cell of *Azotobacter*, in particular its nuclear apparatus (reticulation and low content of DNA) convinced several investigators that the blue-green algae are phylogenetically related to *Azotobacter* (Fischer, 1905; Heinze, 1906; Imshenetskii, 1946). But this view is not shared by all microbiologists (Pražmowski, 1912; Bisset, 1955).

It is difficult to identify species of *Azotobacter* according to their host specificity, and so other criteria have been sought. In particular, there have been attempts to establish differences in the effects of various compounds such as pyronine, saffronin and potassium tellurite on the species of *Azotobacter* (Callao *et al.*, 1961). It should be borne in mind that under the influence of varied factors the properties of a particular species may change substantially (Mandal, 1963; van Schreven, 1966). Particularly noteworthy is the possibility of converting one species of *Azotobacter* into another by the transforming action of DNA (M. and S. Sen, 1965).

Physiological Properties of *Azotobacter*

Azotobacter is characteristically able to assimilate molecular nitrogen as well as bound forms of the element. The amount of nitrogen accumulated depends primarily on the properties of the particular strain of *Azotobacter*. Active and passive cultures are distinguished according to this property. In unfavourable environmental conditions *Azotobacter* may sometimes lose the ability to assimilate molecular nitrogen.

Most cultures assimilate no more than 10 mg of molecular nitrogen per gram of carbon source consumed. Higher values have been reported, in Kluyver and Becking's experiments (1955) individual strains of *Az. chroococcum* fixed as much as 15 mg of nitrogen per gram of glucose utilized, and in Lopatina's experiments (1949) as much as 30 mg of nitrogen was fixed.

The nitrogen-fixing capacity of *Azotobacter* may vary greatly with the conditions of cultivation and with the composition and acidity of the nutrient medium (Burk *et al.*, Mishustin *et al.*, 1939*a* and *b*); temperature (Iswaran and Sen, 1958); aeration (Butkevich and Kolesnikova, 1941); the presence of bound sources of nitrogen (for example Greaves *et al.*, 1940; Ebert, 1959; Iswaran, 1960); the character of the carbon source (for example, Fedorov, 1952,

Rabotnova and Rodionova, 1953); the presence in the medium of trace elements and biologically active substances (for example Bortels, 1937; Esposito and Wilson, 1956; Krilova, 1963; Kasimova, 1965); the frequency of passage of the culture (Rao *et al.*, 1959); and other factors.

When Fedorov (1952) investigated the effect of various carbon sources on the assimilation of molecular nitrogen by *Azotobacter* he found that the productivity of nitrogen assimilation depended on the structure of the organic matter available and its store of chemical energy. The ease of oxidation during respiration is also important. The more oxidized the molecules, the less their store of chemical energy, and when they are utilized the productivity of nitrogen fixation is correspondingly low.

Azotobacter in symbiotic cultures with other micro-organisms often assimilates molecular nitrogen more vigorously than in pure culture. Maybe the other microbes supply *Azotobacter* with biologically active substances (for example Bouisset and Breulland, 1960; Panosyan *et al.*, 1962; Szegi and Timár, 1965). Some of the nitrogen assimilated is released by *Azotobacter* into the surrounding medium in the form of proteins, amino acids and, in part, ammonia (Fedorov, 1952; Sandrak, 1958; Bailly, 1965).

Azotobacter may develop by assimilating the salts of ammonia, nitric and nitrous acids and urea (for example Thompson, 1932; Fedorov, 1952; Ebert, 1959; Voinova-Raikova, 1966). There are, however, indications that not all cultures of *Azotobacter* can assimilate nitrites and nitrates (Rouqueroi, 1962; Bernard, 1963). *Azotobacter* does not develop well on all amino acids, although aspartic and glutamic acids are regarded as quite satisfactory nutrient materials (Wilson and Burris 1947; Fedorov, 1952).

How does bound nitrogen in the medium affect the nitrogen-fixing function of *Azotobacter*? A good deal of work has been devoted to this problem. Available evidence suggests that small doses of nitrogen-containing mineral compounds (in particular, ammonium salts) enhance nitrogen assimilation, while higher concentrations suppress it. Some consider that nitrogen fixation is depressed completely (Rubenchik and Roizin, 1939; Burris and Wilson, 1943), others consider the depression to be partial (Fedorov, 1952; Gushcherov, 1956; Iswaran, 1960; Ebert, 1959). Undoubtedly, the degree of depression of nitrogen fixation by mineral nitrogen compounds depends primarily on their concentration in the nutrient medium.

It is worth noting Iswaran's attempt (1960) to produce an equation for the dependence of molecular nitrogen fixed by *Azotobacter* on the amount of bound nitrogen present in the medium. The numerical indices differ for different nitrogen sources. Thus, for a medium containing ammonium salts the equation has the following form:

$$y = 4\cdot84 - 1\cdot59 \log_{10} x$$

while for a medium with the salts of nitric acid:

$$y = 3\cdot73 - 3\cdot31 \log_{10}$$

where y is the amount of nitrogen fixed per gram of mannitol (measured in mg), and x is the amount of nitrogen in the medium (in mg per 100 ml.).

The equations presented, however, must be considered approximate, for it has been shown that different cultures of *Azotobacter* react differently to the same doses of nitrogenous compound introduced into the medium. Some of them rapidly go over to assimilation of bound forms of nitrogen, but others do so more slowly (Azim and Roberts, 1956; Belozerskii *et al.*, 1957). Cultures of *Azotobacter* have also been found that are able to assimilate free and bound nitrogen simultaneously (Newton *et al.*, 1953).

It is interesting to note that Holme and Zacharias (1965) detected certain differences in the structure of the antigens of *Azotobacter* grown on mineral and molecular nitrogen.

The carbon source for *Azotobacter* can be provided by various organic compounds. This bacterium utilizes carbohydrates (monosaccharides, disaccharides and certain polysaccharides), organic acids of the fatty and aromatic series, monohydric and polyhydric alcohols (ethyl alcohol, glycerol, mannitol) and other substances. As noted, it is difficult to find another organism comparable with *Azotobacter* in terms of the diversity of the compounds available to it (for example Beijerinck, 1901; Krainskii, 1908; Löhnis and Pillai, 1908). Omelyanskii (1923) very aptly called it a polyphage. *Azotobacter* can utilize some volatile organic compounds even in the gaseous state, for example vapours of ethyl alcohol, acetone and certain volatile organic acids (Kholodnyi *et al.*, 1945; Molina and Quant, 1960). *Azotobacter* does not grow in formic acid and develops weakly in acetic acid (Macura and Stavlčík, 1953).

It must be emphasized that the individual species and even strains of the same species of *Azotobacter* differ with respect to their sources of carbon nutrition (for example Smith, 1935; H. Jensen, 1951; Schlüter and Bukatsch, 1959; V. Jenson, 1961; Petrushenko, 1961; Shethna and Bhat, 1962; Darzniek, 1962; Sasson and Daste, 1963; Prša, 1963; Babak, 1965). Thus, there are strains of *Azotobacter* that do not assimilate lactose mannitol, in benzoic acid and so on.

As Vančura and Macura noted (1961) the character of the carbon source greatly influences the phases of development of *Azotobacter*. For example, in media containing organic acids there is virtually no lag phase.

If the medium contains several carbon sources, then they are utilized successively in the order of their accessibility to *Azotobacter* (Schafler, 1959; Vančura and Macura, 1961). Sometimes several peaks of development of the microorganism can be detected, each evidently connected with the period of utilization of a definite carbon source.

Aristovskaya (1941) noted that an increase in the content of carbon dioxide in the atmosphere favourably influences reproduction in *Azotobacter*. Later, Hermann (1963), investigating *Az. chroococcum*, confirmed this with a nutrient medium continuously aerated with air containing different amounts of carbon dioxide. An increase in the concentration of carbon dioxide to 0·5 per cent enhanced the increase in the biomass and nitrogen fixation. But it has not yet been established whether carbon dioxide is directly incorporated into the cell metabolism of *Azotobacter* or whether it has some other influence.

According to existing data the greater part of carbon-containing organic compounds is oxidized by *Azotobacter* to carbon dioxide (for example Beijerinck and van Delden, 1902; Kržemieniewski, 1907; Jones, 1913; Omelyanskii, 1923). Several investigators have noted the accumulation of organic acids (formic, acetic, lactic, butyric) and ethyl alcohol (Stoklasa, 1908) in the medium when *Azotobacter* develops. Careful investigation (Macura and Stávélik, 1953) established that when *Az. chroococcum* assimilates glucose, pyruvic and lactic acids and also ethyl alcohol accumulate in the medium. Their accumulation is at a maximum in the lag phase and decreases later; evidently they are oxidized. Cohen and Johnstone (1963) also found that the medium became acid as *Azotobacter* developed. The greatest acidification is given by cultures of *Az. macrocytogenes*.

Geiko *et al.* (1966) noted the presence of keto acids (pyruvic, α-ketoglutaric, α-keto-β-methylbutyric, α-keto-β-methylvaleric, oxaloacetic and hydroxy-pyruvic) in the cells of *Az. vinelandii* and its culture fluid. The quantity of keto acids formed depends on the developmental phase reached. When a culture is exposed to ultrasonics its size almost doubles.

Voets (1963) investigated the oxidation of the benzene ring by *Azotobacter*. He established that benzoic acid is initially transformed into salicylic acid and then to catechin which undergoes further oxidation.

Hardisson *et al.* (Hardisson and Pochon, 1966; Hardisson and Robert-Gero, 1966) cultured *Azotobacter* on benzoic acid and established the formation of a humin-like substance. Synthesis of this compound was intensified in the presence of manganese. When it was fractionated fulvin, humin and haematomelanin fractions were identified. The dark pigment of colonies of several species of *Azotobacter* in various nitrogen-free media resembles the melanins. According to Proctor (1959), when starved, *Azotobacter* may assimilate the capsule mucilage as a source of carbon. In the soil, *Azotobacter* utilizes various organic compounds and products of the breakdown of animal and plant cells. In particular, its multiplication is intensified in soils fertilized with straw and straw manure, and also in various composts containing cellulose (for example Makrinov and Troitskii, 1930; Fedorov and Tepper, 1947; Darzniek, 1961; Nepomiluev and Shishov, 1962; Erofeev and Vostrov, 1964). The favourable effect of straw can be explained by the fact that *Azotobacter* assimilates substances formed in the substrate when cellulose breaks down.

In organic materials (manure, composts, garbage, poultry manure and so on) the processes of degradation of cellulose proceed weakly, and very few cells of *Azotobacter* are present (Naguib, 1963, 1965).

There are many indications that the store of mobile organic matter in soil and clays is one of the chief factors influencing the development of *Azotobacter* (for example Kononova, 1930; Geller and Kgariton, 1951; Rakhno, 1953; Rodina, 1954; Tribunskaya, 1954; Gushcherov, 1956; Semikhatova, 1958;

Daraseliya, 1963; Zhdannikova, 1963; Rovira, 1965; Di Menna, 1966; Lasting *et al.*, 1966). We shall consider the results of one of our experiments that fully confirms this proposition. Samples of the black-earth soil of the Institute of Agriculture of the Central Black-Earth Zone in the USSR (Voronezh Region) were taken from different fields under crop rotation. *Azotobacter* was not found in this soil by the usual methods of analysis. In order to activate multiplication of the cells present the moisture of the soil samples was adjusted to 60 per cent of the total water capacity. We added 0·5 per cent mannitol and 0·1 per cent K_2HPO_4 to some of the test samples. Then the samples, in Koch dishes, were placed in a thermostat at 25°C. After two weeks *Azotobacter* was estimated

Table 4.2. Activation of *Azotobacter* in Black-Earth

Crop in field before soil sampling	Initial colony count	No. of colonies after activation	
		Without additives	With addition of K_2HPO_4 and mannitol
Spring wheat, furrow not tilled	2	110	1,000
Spring wheat, furrow tilled	6	40	1,000
Sunflower	9	20	1,000
Old sod field	0	3	1,000

by the soil plate method. Table 4.2 shows that there was an appreciable increase in the number of *Azotobacter* colonies after incubation in soil from furrows that had been tilled, that is, that were rich in fresh organic residues. Heavy activation of *Azotobacter* was observed when the K_2HPO_4 and mannitol were added to test soils. Such activation of the development of *Azotobacter* was also noted by Anderson (1966) when sucrose was introduced into the soil. This indicates that *Azotobacter* actively multiplies in the soil only in the presence of mobile organic matter.

It should be noted that in H. Jensen's view (1950) a lack of organic substances is the chief factor limiting the spread of *Azotobacter* in soils with a favourable pH and a sufficiently high phosphorous content. This view is shared by Sasson and Daste (1961). Several investigators have noted an increase in the content of *Azotobacter* in soils richer in organic matter, but do not consider this to be the only factor affecting multiplication of the bacterium (Gushcherov, 1963; Daraseliya, 1963; Rovira, 1965). Of course, as has been rightly pointed out, when considering the development of *Azotobacter* in the soil it is necessary to take other factors into account (for example Pochon and Coppier, 1951; Moreau and Gams, 1960; Chandra *et al.*, 1962).

Soriano and Atlas (1963) suggested that when the energy stores in the soil decrease, *Azotobacter* develops in the deeper layers. *Azotobacter* can utilize humus in the soil only to an insignificant degree which does not ensure satisfactory development (Kurdina, 1951; Bernard and Voronkova, 1960). And so even in soils rich in humus *Azotobacter* does not multiply greatly in the absence of fresh organic residues, It is also known that small doses of humus greatly stimulate the growth of *Azotobacter* (Kržemieniewski, 1908; Dikusar, 1945; Chizhevskii and Kudakov, 1952; Khatipova, 1962), which can be explained as the effect of the trace elements and colloidal compounds present in the humus.

The multiplication of *Azotobacter* largely depends on the presence in the medium of phosphorus and potassium compounds—this organism is particularly sensitive to phosphorus. An absence or lack of phosphorus in the medium slows the development of the culture and diminishes nitrogen fixation (for example Ziemiecka, 1932; Mirotvorskii, 1945; Esposito and Wilson, 1956; Gutschy, 1959; Sabel'nikova, 1960; Galimova, 1960). The source of phosphorus for *Azotobacter* may be either inorganic or organic phosphorus-containing compounds (H. Jensen, 1954; Voinova-Raikova, 1966). The presence of active phosphatase has been established for *Azotobacter* (Surman, 1961, 1965; Kanopkaite and Pauluskaite, 1961). Fixation of nitrogen by *Azotobacter* begins when the concentration of phosphate is 4 mg per 100 ml of medium. In the presence of about 800 mg of phosphate per 100 ml of medium assimilation of nitrogen ceases (Becking, 1961).

There have been many investigations of the effect of phosphorus on the state of *Azotobacter* in the soil, (for example Ernandes, 1953; Degtyareva, 1959; Shepetina, 1959; Golovacheva, 1960; Smalii, 1961; Subramoney and Abraham, 1962). They have shown that substances containing phosphorus have a positive effect on carbon exchange, multiplication and nitrogen fixation in *Azotobacter*. Singh and Shrikhande (1953) established that a young culture of *Azotobacter* requires less phosphorus-containing compounds than an old culture.

The high sensitivity of *Azotobacter* to phosphorus enabled Uspenskii (1930) to devise a microbiological method for determining the soil requirements for phosphorous fertilizers, using *Azotobacter* as the test organism.

Potassium is also necessary for the development of *Azotobacter*, but in relatively small quantities (Krzemieniewski, 1908). Large doses of potassium salts suppress the development of *Azotobacter*. According to Chentsov (1960) and Babak (1966) the degree of toxicity is determined by the anionic part of the salts.

Calcium and magnesium are important in the metabolism of *Azotobacter*. Many have noted the high requirements of most species of *Azotobacter* for calcium (for example Dianova and Voroshilova, 1927; Esposito and Wilson, 1956; Hovenkamp, 1958; Peshakov, 1959; Zaitseva et al., 1960; Sabel'nikova, 1960). A lack of calcium leads to a lengthening of the lag phase, heavy vacuolization and inflation of the cells. Synthesis of ATP and polyphosphates is disturbed, and so is coupled oxidative phosphorylation.

The concentration of calcium salts must not exceed a certain optimum or else the viability of *Azotobacter* decreases (Iswaran and Sen, 1960b). All species of *Azotobacter* do not have identical requirements for calcium salts (Norris and Jensen, 1958; Bullock *et al.*, 1960). For *Az. chroococcum* 0·01 per cent calcium chloride is close to the optimum. An increase in the amount of calcium chlorate reduces nitrogen assimilation more than does a decrease in calcium chloride (Krishna *et al.*, 1965). Large doses of calcium carbonate, however, do not suppress nitrogen assimilation (Iswaran and Sen, 1960*b*).

The physiological importance of calcium for *Azotobacter* has still not been

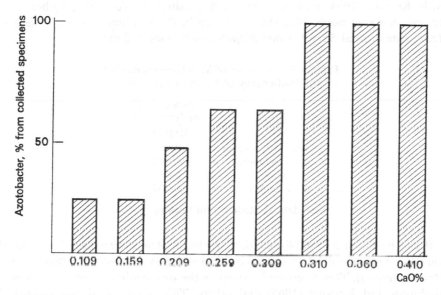

Fig. 23. Distribution of *Azotobacter* in soddy-podzolic soils in relation to their content of calcium (after Pavlovich).

established fully. Burk and Lineweaver (1931) assumed that it was necessary for the assimilation of molecular nitrogen. Zaitseva *et al.* (1960), working with *Az. vinelandii*, showed that an absence of calcium inhibited the growth of *Azotobacter* on either molecular or ammoniacal nitrogen. Thus, the effect of the calcium ion cannot be specific in relation to nitrogen fixation. On the basis of the high sensitivity of *Az. chroococcum* to calcium, Uspenskii (1930) proposed that it be used to determine soil requirements for lime. Figure 23 shows the dependence of the spread of *Azotobacter* in soddy podzolic soils on their content of calcium.

According to H. Jensen (1954*b*), Voets (1962) and Abushev (1966) the magnesium requirements of *Azotobacter* are high.

Azotobacter needs about ten times as much magnesium as iron, but it is possible that magnesium is not involved in nitrogen fixation.

A distinguishing feature of *Azotobacter*, as with other nitrogen fixers, is high sensitivity to trace elements in the medium. Molybdenum is a special need for most cultures of *Azotobacter*. It is a requirement both for the fixation of molecular nitrogen and when the bacteria are growing on nitrates (with the exception of *Az. agilis* which does not grow on salts of nitric acid). However, when growing on a medium with a source of bound nitrogen, *Azotobacter* requires less molybdenum than when growing on a nitrate-free medium (Burema and Wieringa, 1942).

Species of *Azotobacter* differ in their requirements for molybdenum (Table 4.3). N. B. Krylova (1964) arranged nitrogen fixers similarly. According to her data, *Az. chroococcum* needs 10–100 mg/l of Mo in the medium and *Az. agile* and *Az. vinelandii* need 5 mg/l, while *Beijerinckia* requires 0·5 mg/l.

Table 4.3. Quantities of Molybdenum needed
for Half-energy of Nitrogen Fixation

Species	Amount of molybdenum (p.p.m.)
Az. chroococcum	0·05
Az. vinelandii	0·004
Az. agile	0·002
Beijerinckia	0·004

These data are taken from Becking (1962).

It can be taken as established that species of *Azotobacter* found in neutral soil require more molybdenum than bacteria which live in acid soils (*Beijerinckia*, *Mycobacterium*). This assertion is based on the observations of Becking (1961), Mishustin and Krylova (1965) and others. This is apparently an ecological adaptation, for in acid soils molybdenum salts are more readily soluble, and therefore more accessible to the micro-organisms.

At the same time there is an obvious adaptation of strains of *Azotobacter* to molybdenum and vanadium as geochemical factors of the environment (Koval'skii and Letunova, 1966a and b). Strains of *Az. chroococcum* isolated from the soils with a low content of molybdenum ($2·0 \times 10^{-4}$ per cent) and vanadium ($2·2 \times 10^{-3}$ per cent) require for optimal fixation of nitrogen $0·5 - 25$ mg of $Na_2MoO_4.2H_2O$ and 0·2 mg of $NaVO_3.6H_2O$ per litre. Strains isolated from areas with a high natural content of molybdenum ($2·4 \times 10^{-3}$ per cent) and vanadium ($2·8 \times 10^{-2}$ per cent) in the soil require greater concentrations of these elements for nitrogen fixation to proceed (150 and 15 mg respectively).

As Becking noted (1962), vanadium is not a substitute for molybdenum for *Az. chroococcum* and *Az. agile*. Koval'skii *et al.* (1966) showed that when introduced into the medium molybdenum was concentrated in the cells of *Az. chroococcum* to a considerably greater extent than was vanadium. Some strains of *Az. chroococcum* do not accumulate vanadium at all. Molybdenum is concentrated in the protoplasm of *Azotobacter* (Keeler *et al.*, 1956). Molybdenum and vanadium are equivalent for most cultures of *Az. vinelandii*. Neither molybdenum nor vanadium activates respiration or the dehydrogenase activity in nitrogen-fixing micro-organisms (Krylova, 1963).

A series of results suggests that boron compounds stimulate the multiplication of *Azotobacter* and the fixation of nitrogen (for example Voicu, 1942; Herzinger, 1940; Ernandez, 1953; Gerretsen and Hoop, 1954; Mulder, 1957; Karasevich, 1962; Voinova-Raikova, 1966). There are isolated items of information of a contradictory character (Matusashvili, 1947) which can hardly be considered convincing. Boron compounds (borax) have a stimulating effect on the nitrogen-fixing activity of *Azotobacter* when in a concentration of roughly 0·005 per cent (Shchepkina, 1961).

Anderson and Jordan (1961) established that *Azotobacter* develops normally and assimilates molecular nitrogen in the absence of boron compounds, which only stimulate metabolic activity in *Azotobacter*. This indicates that boron is useful but not essential for *Azotobacter*. At different times during development different doses of boron are optimum. For young cultures three parts per million is best and for old cultures twenty parts per million (Jordan and Anderson, 1950).

Species of *Azotobacter* require very small quantities of iron compounds (Bassalik and Neugebauer, 1933). There are some indications of a stimulation of the synthesis of fluorescent pigments in *Azotobacter* in the presence of iron salts (Johnstone, Pfeffer and Blanchard, 1959). According to Keeler *et al.* (1958), iron is localized in the cytoplasmic membrane of the cell.

Many have noted a favourable effect of manganese salts on *Azotobacter* (for example Olaru, 1920; Steinberg, 1938; Kholopov, 1952; Rakhno, 1953; Iswaran and Rao, 1960; Solov'eva, 1963; Klintsare and Kreslinya, 1963; Kirsanina *et al.*, 1965). Manganese, however, is not an element that is necessary for nitrogen fixation (Voets, 1962). *Az. chroococcum* has the greatest need for manganese. It requires roughly twenty to thirty parts per million in the medium, twice the concentration that is adequate for *Az. beijerinckii* and *Az. vinelandii*. Manganese may to some extent replace magnesium.

Contradictory information has been published about copper. Several investigators say copper is toxic for *Azotobacter* even in low concentrations (Greaves, 1933; Becking, 1961; Krylova 1963). Some have noted that very small doses of copper are necessary for this species (Schröder, 1932; Abutalybov and Gazieva, 1961).

It is possible that copper acts differently during the different phases of development of *Azotobacter* (Shchepkina, 1961). According to Mulder (1939), copper is necessary for the formation of pigment. According to Kova'skii and Letunova (1966) strains of *Azotobacter* are adapted to the copper content of different soils.

Varying results have been obtained in investigation of the effect of iodine compounds on nitrogen fixation (for example Fedorov, 1952; Hovadik and

Polách, 1958; Mekhtiev and Melkumova, 1962; Kirsanina *et al.*, 1965; Voinova-Raikova, 1966). Bromine is apparently a depressant (Federov, 1952).

The aluminium ion is undoubtedly toxic to *Azotobacter* (Mickovski and Popovski, 1960; Smirnov, 1960; Gromyko, 1963; Ryys, 1963, 1966; Zinov'eva, 1966). This may be an important reason for the absence of *Azotobacter* from podzolic soils.

There is some information about the stimulating effect of arsenic (Greaves, 1913), zinc (Schroder, 1932; Shchepkina, 1961; Becking, 1961; Mekhtieva and Melkumova, 1962; Solov'eva, 1962, 1963; Samosova *et al.*, 1964; Voinova-Raikova, 1966), cobalt (Iswaran and Rao, 1960; Abushev, 1966), strontium and several other elements.

Radioactive elements (radium, thorium and uranium), even in very small concentrations, activate the development of *Azotobacter* and the process of nitrogen fixation (for example Stocklasa, 1913; Kayser *et al.*, 1920, 1925; Shtern, 1938; Krasil'nikov *et al.*, 1955a; Drobkov, 1963).

Further investigations are needed to clarify the mechanism of the stimulatory effect of trace elements. It may be that some of them form part of the molecules of the enzymes responsible for nitrogen fixation, while others have a catalytic effect on particular functions of the microbial cell.

Cultures of *Azotobacter* usually produce considerable quantities of biologically active substances. Among these are vitamins of the B group, nicotinic and pantothenic acids, biotin, heteroauxin and gibberellin. However, Hennequine and Blachère (1966) were unable to detect the production of gibberellin by *Azotobacter*. Table 4.4 gives references to investigators who have established the presence of biologically active substances in *Azotobacter*.

Although cultures of *Azotobacter* produce a valuable series of biologically active substances, the addition of vitamins, gibberellin and heteroauxin to the culture medium accelerates growth (for example Meshkov, 1959; Karasevich, 1962; Szegi and Gulyas, 1963; Žák, 1965; Dommergues and Dusausoy, 1966). According to Klyushnikova (1966) individual strains react characteristically to vitamin supplements.

Brakel and Hilger (1965) and Romanow (1965) showed that the synthesis of auxins (β-indolylacetic and β-indolyllactic acids) by *Azotobacter* is intensified sharply in the presence of tryptophan. According to Voinova-Raikova (1966) the formation of growth substances by *Azotobacter* is stimulated when nucleic acid is added to the medium.

In 1938, Berezova *et al.* made observations that led them to assume that certain cultures of *Azotobacter* produce antibiotics. This fact was firmly established by Mishustin *et al.* (1963), who showed that *Az. chroococcum* released fungistatic substances with a broad activity spectrum but inactive against bacteria and Actinomycetes.

Individual species and even races of *Azotobacter* differ greatly in their production of biologically active compounds (Strzelczyk, 1964–1965).

Table 4.4. Vitamins and Other Biologically Active Substances formed
by *Azotobacter*

References

Vitamin B_1 (thiamine)	Bonner, 1937; Detinova, 1937; Nakhimovskaya, 1941; Lee and Burris, 1943; Brundtsa, 1949; Gebgardt, 1958, 1961; Pakarskite, 1961; Gvamichava, 1962; Bershova and Kozlova, 1962; Yatskunene, 1962; Dachyulite and Pakarskite, 1962; Zinov'eva and Klyushnikova, 1962; Voznyakovskaya, 1963; Rusakova, 1965; Klyushnikova and Kvasnikov, 1966; Klyushnikova, 1966.
Vitamin B_2 (riboflavin)	Lee and Burris, 1943; Pridham, 1952; Johnstone *et al.*, 1959; Pakarskite, 1961; Gvamichava, 1962; Voznyakovskaya, 1963; Klyushnikova, 1966.
Vitamin B_6 (pyridoxine)	Gebgardt, 1958, 1961; Gvamichava, 1962; Yatskunene, 1962; Klyushnikova, 1966.
Vitamin B_{12} (cyancobalamine)	Gupta and Das, 1959; Kanopkaite *et al.*, 1959; Petrushenko, 1961; Dachyulite, 1962; Zinov'eva and Klyushnikova, 1962; Bershova and Kozlova, 1961; Shpokauskas *et al.*, 1963, 1965; Babak, 1964; Bagdasryan, 1965; Rubenchik *et al.*, 1965.
Nicotinic and pantothenic acids	Lee and Burris, 1943; Gebgardt, 1958, 1961; Bershova and Kozlova, 1962, 1965; Smalii, 1962; Zinov'eva, 1962; Gvamchina, 1962; Yatskunene, 1962; Voznyakovskaya, 1963; Rusakova, 1965; Rubenchik *et al.*, 1965; Tribunskaya, 1966; Klyushnikova, 1966.
Folic acid Biotin	Brundtsa, 1949; Shpokauskas *et al.*, 1963. Thorn and Walker, 1936; Bonner, 1937; Allison and Minor, 1940; Nakhimovskaya, 1941; Lee and Burris, 1943; Brundtsa, 1949; Shavlovskii, 1954; Gebgardt, 1958; Bershova and Kozlova, 1962, 1965; Yatskunene, 1962; Zinov'eva and Klyushnikova, 1962; Voznyakovskaya, 1963; Rusakova, 1965; Rubenchik *et al.*, 1965; Klyushnikova, 1966.
Heteroauxin (β-indolyl acetic acid)	Berezova *et al.*, 1938; Raznitsina, 1938; Bukatsch *et al.*, 1952, 1956; Smalii and Bershova, 1957; Pochon and de Barjac, 1958*a* and *b*; Burger and Bukatsch, 1958; Krupina, 1960; Rubenchik, 1960; Vančura and Macurá, 1960; Manuca and Gheorghiu, 1960; Petrushenko, 1961; Chalvignao, 1962; Naumova *et al.* 1962; Fallot, 1963; Gheorghiu and Manuke, 1964; Nutman, 1964; Nita, 1965; Bun'ko, 1965; Brakel, and Hilger, 1965; Romanow, 1965; Novikova and Irtuganov, 1966; Hennequin and Blachère, 1966.
Gibberellins	Vančura, 1961; Naumova *et al.*, 1962; Nutman, 1964.

Effect of external factors on *Azotobacter*

Azotobacter makes extreme demands on the medium. The bacteria of this genus gravitate to neutral soils and tolerate acidification but poorly.

The range of pH suitable for the development of *Azotobacter* may shift somewhat with the composition of the medium and other factors. Thus, for example, according to Burk *et al.* (1934) *Azotobacter* may grow on bound nitrogen in a more acid medium than it would if living off molecular nitrogen (Fig. 24).

Azotobacter can be considered capable of developing in media with a pH range of from 4·5 to 9, but, according to Kirakosyan *et al.* (1966), nitrogen fixation occurs only within the narrower range of 5·5 to 7·2 (sometimes up to 7·7). Many say that strains isolated from soils of different pH react differently to the pH of the nutrient medium (Fig. 25). However, Kirakosyan *et al.* (1966) found that the degree of nitrogen fixation by ecological strains isolated from soils of different pH did not correlate with the reaction of the medium.

Individual species, and probably even strains, are not equally sensitive to an

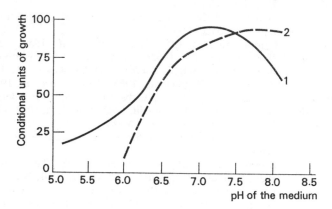

Fig. 24. Effect of pH on the development of *Azotobacter*.
1, In medium containing NO_3; 2, in the presence of gaseous nitrogen.

acid medium. Thus, the minimum pH of the medium for *Az. chroococcum* and *Az. beijerinckii* is about 5·5 and for *Az. macrocytogenes* it is about 4·6 (Rabotnova, 1953; H. Jensen, 1955). According to Blinkov (1951) the optimum pH for *Azotobacter* is in the range 7·2 to 8·2. But Rabotnova (1957), studying the effect of pH on the reproduction of *Az. chroococcum, Az. agile* and *Az. vinelandii*, noted the highest yield of cells at pH 6·5 to 6·7. At pH 5·0 and 8·0 the cells divided very slowly. Kaleshko (1961) established that acidification of the medium to pH 5·3 inhibits reproduction and the fixation of atmospheric nitrogen in *Azotobacter* much more than does alkalinization to pH 9·0.

An acid medium is unfavourable to *Azotobacter* and leads to a sharp decline in viability and nitrogen-fixing activity (for example Fedorov and Ernandez, 1953; W. and J. Boyd, 1962; Karasevich, 1962). Inactive forms of *Azotobacter* which have lost the ability to bind molecular nitrogen (Mishustin *et al.*, 1939a and b) can be isolated from acid soils. *Azotobacter* is not usually encountered in soils with a pH below 5·6 to 5·8 (Visser, 1966). This microorganism seems to be influenced not only by the active acidity, but also by the exchange acidity of the soil (for example Martin *et al.*, 1937; H. Jensen, 1952;

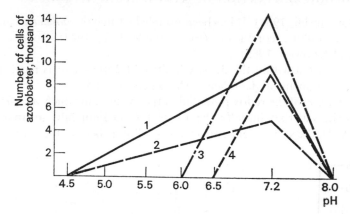

Fig. 25. Reproduction of *Azotobacter* in relation to the pH of the nutrient medium. Cultures were isolated from soil at pH:

(1) 4·6; (2) 4·8; (3) 6·0; (4) 6·5.

Rakhno, 1953; Pavlovicha, 1953; Toskov, 1962; Meiklejohn, 1962; Mizsirliu, *et al.*, 1964; Rangaswami and Sadasivan, 1964; Zawiślak, 1965; Ulyasevich, 1965; Di Menna, 1966; Ryys, 1966). The dependence of the spread of *Azotobacter* on soil pH is well illustrated in Fig. 26.

When considering the oxygen regime necessary for *Azotobacter* it should be noted that this micro-organism is an aerobe. Aeration, as many have noted promotes the multiplication of *Azotobacter* (for example Duda, 1956; Rabotnova,

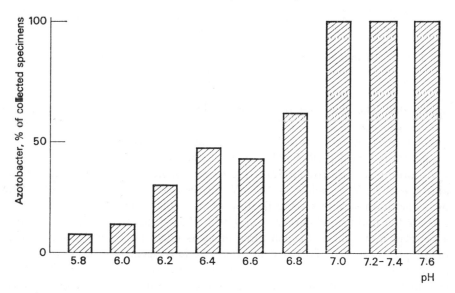

Fig. 26. Spread of *Azotobacter* in soddy-podzolic soils in relation to soil reaction (after Pavlovicha).

1957; Dorosinskii, 1958). It has been established that *Azotobacter* can multiply in micro-aerophilic conditions (for example Fife, 1943; Rouqueroi, 1962; Chang and Knowles, 1965).

Soil compacted within the density limits of 1·2–1·4 g/cm³ is most favourable for the development of *Azotobacter* (Minchenko, 1965).

Investigations of the redox potential of the medium in which *Az. chroococcum* develops showed that initially the Eh of the medium falls to approximately −200 mV. In the subsequent period when cell lysis begins, the Eh rises but falls

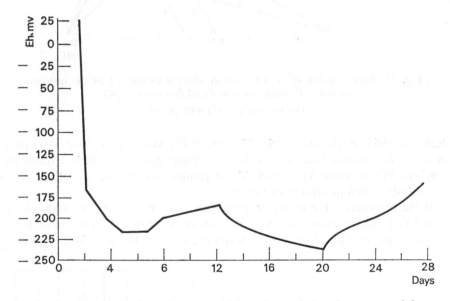

Fig. 27. Change in the redox potential in culture of *Az. chroococcum* (after Rabotnova).

again later. The character of the change in redox potential in the culture of *Az. chroococcum* is shown in Fig. 27. In a culture of *Az. agile* no cellular disintegration is observed and therefore at the end of the experiment the fall in Eh is greatly retarded (Rybalkina, 1937; Rabotnova, 1957).

According to Fedorov (1952) if the redox potential of the medium is either above + 200 mV or below − 200 mV, nitrogen assimilation diminishes. Thus heavy aeration suppresses the fixation of molecular nitrogen. Consistent with this are the findings of Parker *et al.* (1954, 1958, 1960) that in the presence of 4 per cent atmospheric oxygen three times as much nitrogen is assimilated as in the presence of 10–20 per cent atmospheric oxygen. On the basis of these results Parker decided that oxygen and nitrogen compete as hydrogen acceptors and that nitrogen fixation is a form of the respiratory process. Others have noted the inhibitory effect of oxygen on nitrogen fixation (Tschapek and Giambiagi, 1955; Bulen *et al.*, 1963). There is some evidence in the literature to suggest that when aeration is increased the utilization of carbon compounds diminishes.

It may be that with an abundant supply of air, products of the incomplete oxidation of organic substances are formed. And so, to obtain unit biomass the organism needs to oxidize a large quantity of the carbon source (Khmel' *et al.*, 1965, 1966). In Dilworth's (1962) view when *Azotobacter* is growing in a medium with sucrose, an increase in the partial pressure of oxygen inhibits the oxidation of pyruvic and α-ketoglutaric acids in the Krebs cycle.

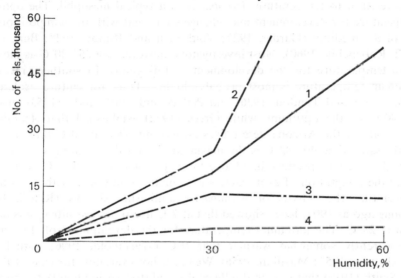

Fig. 28. Effect of moisture on the development of *Azotobacter* (in black earth from the Voronezh Region). 1, 80 per cent of total water capacity; 2, 60 per cent; 3, 50 per cent; 4, 40 per cent.

Azotobacter has heavy demands for soil moisture. According to Mishustin (1937) it has a lower osmotic pressure than fungi or Actinomycetes. Dommergue's (1962) findings were in complete accord with this view. Requirements for moisture were similar to those of higher plants. At an osmotic pressure of about 4·0 atmospheres the development of *Azotobacter* stops. Dommergues assigned this bacterium to the hygrophilic group. High hygrophilicity has been noted by others (for example Enikeeva, 1952; Karaguisheva, 1961).

This explains the frequent dependence on soil moisture of changes in the numbers of *Azotobacter* (Tanatin, 1954; Timofeev, 1958; Koshkarova, 1959; Degtyareva, 1959; Zimna, 1962; Tanatin and Pafikova, 1963; Walezyna and Przesmycka, 1964; Kopčanovà and Řehořkovà, 1961). The depth of penetration of *Azotobacter* into the soil is also largely determined by the supply of moisture in the soil (Miloševič, 1963; Lazarev, 1964).

It is worth noting the findings of Kirakosyan (1961), indicating that cultures of *Azotobacter* from soils in wet zones are more hygrophilic than those isolated from dry regions. These agree with observations by Dommergues (1964) who noted that cultures of *Azotobacter* from tropical regions tolerate drying best. Cysts of *Azotobacter* are more resistant to drying (Socolofsky and Wyss, 1961).

Clearly *Azotobacter* can develop intensively in soils that are sufficiently wetted (Fig. 28). On the basis of the hygrophilicity of the forms of *Azotobacter* found in the soil, Pshenin (1964), who found the marine species *Az. miscellum* in the Black Sea, assumed that its ancestors might also be the ancestors of the soil species of *Azotobacter*.

In relation to temperature *Azotobacter* is a typical mesophil. The optimum temperature for development may change somewhat with the composition and pH of the medium (Garder, 1927; Voitkevich and Runov, 1928; Burk *et al.*, 1932; Rubenchik, 1960). Most investigators, however, see 25°–30°C as the optimum temperature for the development of this genus. In southern races the optimum temperature is most probably higher. Thus, for cultures isolated in India, Dhar and Tandon (1936) and Pathak and Shrikhandel (1953) regarded 35°–40°C as the optimum, while Green (1932) established that *Azotobacter* from soils of the Arizona desert fixes no less nitrogen at 40°C than at 30°C. At the same time 40°–42°C is the maximum for northern cultures and even at 37°C these strains produce involutive forms (Naplekova, 1959). Garder (1927) noted the adaptation of azotobacters to climatic conditions. With a fall in the temperature the activity of *Azotobacter* diminishes sharply (Kaarli, 1966). As long ago as 1907, Koch showed that at 7°C five times less nitrogen is bound than at 24°C. The minimum temperature for the development of *Azotobacter* is apparently somewhat warmer than 0°C (Warmbold, 1906; Pathak and Shrikhandel, 1953; Mazilkin, 1956). We must, however, note the view of Brown and Smith (1919) that in a gradually cooled soil (to freezing point), the nitrogen fixing activity of *Azotobacter* increases. In the northern soils of the Kol'sk peninsular, Ezrukh (1962) observed some multiplication of *Azotobacter* at 0°C.

Azotobacter tolerates well a decrease in temperature and therefore, in the winter, even in northern latitudes, counts of the bacterium in the soil do not decline appreciably.

Vegetative cells of *Azotobacter* do not withstand high temperatures, and a prolonged period at 45°–48°C leads to degeneration and death (Iswaran and Sen, 1960*a*; van Schreven, 1962; Bychkovskaya and Shklyar, 1965). According to Omelyanskii (1915) at 50°C cells of *Azotobacter* die after 30 minutes. A temperature of 80°C kills them very rapidly (Rovira, 1965).

It should be noted that some investigators (Norkina, 1953; Garbosky and Giambiagi, 1963; Bychkovskaya and Shklyar, 1965; Abd-el-Malek and Ishac, 1966) have established higher heat tolerance in cultures of *Azotobacter*. Thus, for example, Garbosky and Giambiagi (1963) exposed individual cultures for 15 minutes at 90°C and for 3–7 minutes at 100°C. In their view this can be explained by the ability of *Azotobacter* to form intracellular heat-stable 'granules' acting as spores but not resembling them. Van Schreven (1966), however, was unable to confirm this.

Of the biological factors that effect the *Azotobacter* in the soil, we would note first the micro-organisms that can influence indirectly the metabolism of the bacterium by modifying pH, redox conditions, and so on, or directly by providing nutrient and biologically active substances.

In the soil are many micro-organisms belonging to different systematic and physiological groups which stimulate or depress the multiplication and nitrogen-fixing activity of *Azotobacter*. The activating effect of cellulolytic, butyric and other soil micro-organisms on *Azotobacter* has often been noted (for example Pringsheim, 1909; Imshenetskii and Solitseva, 1940; Vintika, 1953; Voznyakov-skaya, 1954; Van Tsy-fan, 1958; Golovacheva, 1959; Gadgil and Bhide, 1960; Maliszewska, 1961; Panosyan *et al.*, 1962; Balinschi *et al.*, 1962; Saubert and Grobbeiaar, 1962; Marendiak, 1964; Szegi *et al.*, 1963; Remacle, 1966).

Antagonism has often been demonstrated between *Azotobacter* and many soil micro-organisms (for example Nickell and Burkholder, 1947; Vintika, 1953; Afrikyan, 1954; Krasil'nikov *et al.*, 1955*b*; Stille, 1958; Savkina, 1958; Szelényi and Hélméczi, 1960; Stepanova, 1961; Katznelson and Strzelczyk; Maliszewska, 1961; van Schreven, 1963; Vedenyapina, 1963; Vardapetyan, 1965; Callao, *et al.*, 1966). The activating and inhibitory effect of micro-organisms on *Azotobacter* can be assumed to be less obvious in the soil than in artificial media. In the main this is determined by the lack of large stores of organic compounds in the soil. Of course, other factors are important in limiting the multiplication of micro-organisms that produce biologically active compounds which can also be taken up by soil colloids (van Schreven, 1963) and destroyed by diverse microflora. Szegi (1960) demonstrated experimentally that antagonism to *Azotobacter* is less strongly marked in the soil than in laboratory media.

With this in mind we consider it artificial to attempt to relate the course of development of *Azotobacter* in the soil to the activity of the microflora depressing and activating the bacterium, as others have done (Pavlovicha, 1953; Koleshko, 1960). Some antagonism between species of *Azotobacter* has been noted in culture (Kirakosyan and Melkonyan, 1964).

Phage have been described which lyse *Azotobacter* (for example Smalii and Sergienko, 1936; Monsour *et al.*, 1955). Some investigators consider that the numbers and activity of *Azotobacter* in the soil are influenced significantly by bacteriophage (El'bert and Koleshko, (1961). But the role of phage in the vital activity of *Azotobacter* has not been studied in detail.

Ecology of *Azotobacter*

The ecology of *Azotobacter* has been the subject of many investigations, the results of which are summarized in the monographs of Sushkina, Blinkov, Rubenchik and Zinov'eva. There is no need to repeat generalizations already made, and we shall give only our own ideas on the patterns of distribution of

Azotobacter; ideas formed on the basis of wide experience. They do not always coincide with views expressed by others, in particular by Sushkina.

We have shown that *Azotobacter* makes many demands on the environment. It develops in the presence of sufficient moisture in neutral soils supplied with phosphorus, potassium and a series of trace elements. Such exacting demands explain the relatively restricted distribution of *Azotobacter*.

It should be noted that the incidence of individual species in different soils has hardly been studied. Nevertheless, on the basis of available information it can be taken that the dominating species is usually *Az. chroococcum. Az. agile* is found in very wet soils (Walczyna and Przesmycka, 1964; Soriano, 1966).

In attempting to establish patterns of distribution of *Azotobacter* it is necessary to emphasize the significance of the procedures used to detect it in the soil. The common method is to spread out lumps of soil on a nutrient medium, on the assumption that if the soil contains cells of *Azotobacter* they will develop and form colonies readily distinguishable from colonies of other micro-organisms.

Another procedure worth noting consists of preparing from the test soil a paste to which are added substances that are nutritive for *Azotobacter* (sources of carbon, phosphorus and potassium). The paste is incorporated into a layer of washed charcoal placed in a small Petri dish (diameter 5 cm). To aerate the soil, a glass tube is inserted into the charcoal layer with its end above the soil plate. If the soil contains *Azotobacter*, small colonies form quite rapidly on the surface of the soil plate. Their numbers give an idea of the richness of *Azotobacter* in the soil. The principle of the method of soil plates was proposed by Vinogradskii and was improved by Uspenskii.

There are other methods for detecting *Azotobacter* in the soil, but they are used less often (for example Fenglerowa, 1965; Amor Asunciòn Manuel, 1965).

When studying the ecology of *Azotobacter* it is desirable to use both the methods described, or at any rate, to give preference to the second method. We have come to this conclusion because, in our experience the soil plate method is the more sensitive. Not infrequently when the soil lump method gives negative results the other method will establish the presence of *Azotobacter*. As a rule if less than 100 colonies of *Azotobacter* form in the soil plate, when lumps of soil are placed on nutrient medium, *Azotobacter* is not detected (Fig. 29). For analysis of soils containing a large number of *Azotobacter* cells both methods give good results. Table 4.5 confirms the higher sensitivity of the soil plate method. When investigating patterns of distribution of *Azotobacter* it is necessary also to bear in mind the possibility of seasonal development in the soil. At certain times this micro-organism may multiply vigorously in the arable layer and yet its numbers may fall to values not perceivable by ordinary methods. Obviously a single analysis of the soil cannot give a proper idea of the ecology of *Azotobacter* or of any other micro-organisms.

The need to allow for changes in numbers when studying the development of *Azotobacter* in the soil is illustrated by Table 4.6, taken from our own data.

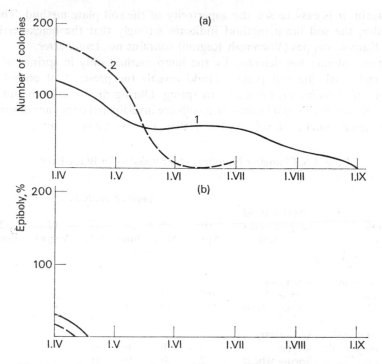

Fig. 29. Numerical pattern of *Azotobacter* in black-earth soil (Voronezh Region) *a*, Determination using soil plates; *b*, determination by the method of soil lumps. 1, Ploughed land; 2, virgin land.

Table 4.5. Presence of *Azotobacter* Established in Soils by Different Methods

Soil	Crop	No. of colonies of *Azotobacter* per plate	Soil lumps with *Azotobacter* (%)
Soddy-podzolic	Rye (NPK)	25	0
Moscow Region	Rye (manure)	10	0
Ordinary black-earth,	Grass mixture	116	2
Voronezh Region	Spring wheat	93	8
Light chestnut soil,	Virgin	135	5
Volgograd Region	Wheat	189	4
Sierozem,	Steppe, virgin	13	0
Azerbaijan SSR	Salt Steppe, virgin	140	0

Here again, it is easy to see the superiority of the soil plate method. Thus, in particular, the soil lump method indicates wrongly that the black-earth soil of the Kamen Steppes (Voronezh Region) contains no *Azotobacter*.

Solitary colonies are detected by the lump method only in spring, whereas with black-earth the soil plate method reveals the presence of considerable numbers of *Azotobacter*, especially in spring. During the summer the numbers of cells fall to zero. Thus if black-earth soils are investigated only once in summer, the erroneous conclusion will be drawn that *Azotobacter* is absent.

Table 4.6. Changing Numbers of *Azotobacter* in Black-Earth

Method	Agricultural exploitation of soil	Time of analysis					
		10 April	30 May	27 June	27 July	17 August	20 September
Soil plate count; (number of colonies)	Uncutting steppe	192	4	6	0	0	0
	Spring wheat	120	39	40	3	0	0
Soil lump count (% of lumps with *Azotobacter*	Uncutting steppe	3	0	0	0	0	0
	Spring wheat	2	0	0	0	0	0
Moisture content of soil (%)	Uncutting steppe	34·6	28·6	27·5	30·6	28·6	26·8
	Spring wheat	30·4	16·4	25·7	20·3	20	25·7

Thus, only observations of the dynamics of *Azotobacter* development can give a correct idea of its ecology. Investigators who have attempted to base the patterns of distribution on solitary seasonal analyses have made serious errors. This, in particular, applies to Sushkina whose results we shall discuss later.

To consider the problem of the spread of *Azotobacter* in different soils we shall draw heavily on the material of our laboratory. Starting from the northern soils we shall gradually move south.

Azotobacter is practically absent from virgin podzols and soddy-podzolic soils. These soils, usually poor, and with an acid reaction, are not suitable for this micro-organism. When podzols are worked, conditions improve for the development of *Azotobacter*. According to Stepanova (1930) *Azotobacter* multiplies in cultivated podzols only if the pH is not less than 6·2 and the store of phosphorus not less than 0·16 per cent. But, we found using soil plates, that *Azotobacter* is also detected in rather more acid soils.

In well-cultivated horticultural soils *Azotobacter* is usually present in considerable numbers and can be rapidly detected by any method (Table 4.7).

Many investigators note that when cultures of *Azotobacter* are isolated from acid podzolic soils they are usually weakened and sometimes possess heightened resistance to acid (Bychkovskaya and Shklyar, 1959; Vigorov, 1961; Karasevich, 1962). Active strains of *Azotobacter* are usually found in cultivated soils.

Soil aggregation stimulates the development of *Azotobacter*. No direct connection can be found, however, between the numbers of *Azotobacter* and the degree of aggregation (Hely and Bonnier, 1954).

It should be noted that the development in the soil of nitrogen-fixing microorganisms, in particular *Azotobacter*, is greatly influenced by the character of the fertilizers used. As a rule organic and phosphorus mineral fertilizers stimulate the multiplication of nitrogen-fixers while nitrogen mineral compounds

Table 4.7 Presence of *Azotobacter* in Podzolic and Soddy-Podzolic Soil

Soil investigated	pH of aqueous extract	Number of colonies on soil plates	Lumps with *Azotobacter* (%)
Virgin meadow and forest	4·4–5·3	0	0
Medium cultivated, not manured	5·5–6·1	5–40	0
Medium cultivated, manured	5·5 6·2	30–150	0–10
Horticultural soil	6·6–7·1	up to 2,000	50–100

often suppress their growth (for example H. Jensen, 1940; Delwiche and Wijler, 1956; Barrow and Jenkinson, 1962; Chang and Knowles, 1965; Pokorná-Kozová, 1965; Todorova, 1966; Kadyrova, 1966).

In soils of the hydromorphous series with an increased moisture content and a predominance of meadow vegetation *Azotobacter* is usually encountered throughout the growing season. These soils are primarily those of the alluvial plains, which have a podzolic zone with a favourable pH and a relatively high content of nutrients. *Azotobacter* is found in these soils by both the soil lump and soil plate methods. Lowland water-logged soils often contain *Azotobacter* but in the unfavourable conditions it multiplies weakly. *Azotobacter* is not encountered in the soils of high moors, and rarely in low virgin peats (for example Zimienko, 1957; Tešić and Todorovič, 1963). When peaty soil is cultivated, conditions improve for the multiplication of *Azotobacter* (Trizno and Vavulo, 1951), but certain peats are toxic (Rybalkina, 1938) to the bacterium.

Our findings on the spread of *Azotobacter* in the soils of podzolic and soddy-podzolic zones agree well with those of Sushkina, H. Jensen (1959) and others. This is not the case with other types of soil.

We shall now consider the black-earth soils of the steppe zone. According to Sushkina the process of steppe transformation, that is the transformation of meadow soils into black-earths (and also chestnut, brown and sierozem soils) has a destructive effect on *Azotobacter*. This conclusion is plainly false and, as we shall show, based on a methodological error.

Our investigations have shown that in the more northerly zone of adequately wetted deep black-earths *Azotobacter* can be found throughout the growing period by both the soil plate and soil lump methods. In periods when the soil has dried out, of course, the cell counts decrease somewhat. Table 4.8 shows the

Table 4.8. Number of *Azotobacter* in Black-Earth Soil
Throughout the Year

Date	Virgin	After wheat
	No. of colonies by plate method count	
17 April	340	410
14 May	210	320
24 June	260	320
12 July	110	270
11 August	100	1300
	Soil lumps with *Azotobacter* (%)	
17 April	80	90
14 May	80	64
24 June	85	36
12 July	30	21
11 August	75	43

results of a study of deep black-earths of the Kursk Zone Station (layer 0–10 cm) carried out in our laboratory.

We have found that in the zone of common and southern black-earth *Azotobacter* appears only ephemerally in the spring when there is no irrigation. Often, even in spring the soil lump method gives values close to zero for these black-earths, which indicates the presence of moderate numbers of *Azotobacter*.

Many investigators have also noted a spring maximum in the development of *Azotobacter* in the soils of the southern USSR (for example Kirsanina and Volkova, 1960; Kolker and Dakhnova, 1960; Gvamichava, 1962; Karaguishieva, 1963; Goguadze, 1966; Pakusin, 1966).

Why do the numbers of *Azorobacter* decrease in these black-earths in the summer; these are such fertile soils? A water shortage cannot be the reason, for at this time deep black-earths lose just as much moisture but *Azotobacter* survives in them. In our view the cause of the summer decline of *Azotobacter* in black earths is a deficit of mobile organic matter. Common and southern black-earths are not so rich in organic residues as deep black-earths. We can assume that by the end of summer the store of mobile organic compounds is depleted and *Azotobacter* begins to die off. Presumably a few cells remain in the soil but are not detected by the methods of analysis used. This point of view has been confirmed experimentally. If an energy-rich substance is introduced into samples of common black-earth taken in the summer and wetted, many *Azotobacter* cells are found in the samples after a short incubation at 25°C.

Table 4.9. Presence of *Azotobacter* in Common Black-Earths of the Kamen Steppe

Type of agricultural exploitation of soil	No. of colonies by plate method count						Soil lumps containing *Azotobacter* (%)					
Date	10/4	30/5	27/6	27/7	17/8	20/9	10/4	30/5	27/7	27/7	17/8	20/9
						Unirrigated						
Virgin steppe	192	4	6	0	0	0	3	0	0	0	0	0
Spring wheat	120	30	40	8	3	0	1	0	0	0	0	0
Forest belt	112	0	0	0	0	0	8	0	0	0	0	0
						Wetted soil						
Rice, inundation	—	2000	—	2000	120	—	—	100	—	94	9	—
Spring wheat rainfall	—	93	—	70	200	—	—	8		0	65	—

These figures refer to the top 10 cm of soil.

To illustrate this point, Table 4.9 shows the numerical pattern of *Azotobacter* in irrigated and unirrigated common black-earths of the Kamen steppe (Voronezh Region). In the unirrigated black-earths *Azotobacter* could be found only in spring while in the irrigated soil it was found throughout the summer. In the latter case more favourable conditions are apparently created for the development of *Azotobacter*, not only in relation to the water regime but also because of the greater availability of organic matter.

If unirrigated chernozems were taken for analysis only once in the season and investigated only by the lump method, the wrong conclusion that *Azotobacter* was absent would be drawn. Sushkina made such an error. It is quite obvious that we cannot speak of the 'destructive' influence of *Azotobacter* on steppe transformation. In chernozems of other localities of the southern USSR we have found a pattern of distribution similar to that already described.

Table 4.10 Presence of *Azotobacter* in the Dark Grey Soil of the Tellerman Forest

Level	Soil layer (cm)	Number of colonies by plate method							Soil lumps with *Azotobacter* (%)		
		April	May	June	July	Aug.	Sept.	Oct.	May	June	Oct.
A_0	0–3	0	0	0	0	0	0	0	0	0	0
$A_{1/2}$	3–10	1	0	0	4	90	43	15	0	0	0
A_1	10–20	0	0	93	0	200	130	50	0	0	0
A_2	20–30	0	0	205	1	178	42	85	0	36	2
B_1	30–40	0	0	150	0	9	5	90	0	76	6

There have been reports that *Azotobacter* multiplies better in calcareous chernozems than in podzolized black-earths (Mekhtiev, 1966).

It should be noted that meadow soils in river valleys are richer in *Azotobacter* in the chernozem zone.

As in common chernozems, *Azotobacter* is found only in the spring in virgin and unirrigated, cultivated chestnut soils. In the spring, both methods gave us positive results. In dry weather the soil lump method no longer detected *Azotobacter*, which, however, continued to be detected in soil plates for some

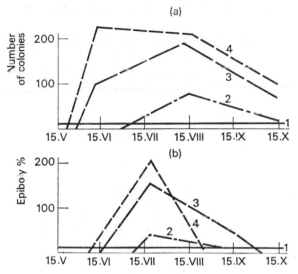

Fig. 30. Pattern of numbers of *Azotobacter* in the dark grey soil of the Tellerman Forest of the Voronezh Region. (*a*) Determined by the soil plate method and (*b*) determined by the soil lump method. 1, A_0; 2, A_1 (3–10 cm); 3, A_1 (10–20 cm); 4, A_2 (20–30 cm).

time. The state of *Azotobacter* is very distinctive in grey forest soils, which in Sushkina's view are also 'destructive' for *Azotobacter*. The pattern of development of *Azotobacter* in the dark grey soil of the Tellerman Forest of the Voronezh Region is shown in Table 4.10 and Fig. 30. The soil samples were taken from a mixed woody plantation, and as can be seen, more *Azotobacter* was detected by the soil plate method. Numbers were greatest in the summer when the temperature and water regime of the soil were favourable for the development of *Azotobacter*. Only when the soil had dried a little (for example in July) did the *Azotobacter* count decrease. In the first 10 cm of soil, where saprophytic bacteria multiply in the greatest numbers, the *Azotobacter* count was less than in the layer between 10 and 30 cm deep.

Evidently the microflora and the products of the decomposition of the fresh woody litter suppress the growth of *Azotobacter* to some extent. Thus, here again is a very specific pattern of numbers of *Azotobacter*.

It can be assumed that the summer maximum in the multiplication of *Azotobacter* holds for several soils. Thus, for example Milŏsevič (1963) found the same situation with sandy soils, and Rouqueroi (1962) with the soils of rice fields.

In the black alkali soils of steppes (solonetz), in the common black-earths and the chestnut soils *Azotobacter* develops most actively in the spring. This is quite understandable, for in these soils the most favourable air-water regime for *Azotobacter* is set up in spring. *Azotobacter* is present in level B_1 of these soils in relatively large numbers. In this case the lump method gives a clear picture

Table 4.11. Presence of *Azotobacter* in the Solonetz of Arsham-zel'mene

Soil	Level	No. of colonies counted by plate method				Soil lumps with *Azotobacter* (%)			
		17/4	20/6	30/7	1/9	17/4	20/6	30/7	1/9
Crustal solonetz, virgin	A_1	54	49	48	18	0	0	0	0
	B_1	171	130	69	15	10	0	0	0
	B_2	40	90	33	17	5	0	0	0
Crustal solonetz, cultivated	0–10	10	21	19	35	0	0	0	0
	10–20	120	64	36	84	5	0	0	0
	20–30	150	100	73	80	10	0	0	0

of the numbers. In Tanatin's view (1953) the development of *Azotobacter* in solonetz and solonchak (saline soil) is limited not by the concentration of salts but by the moisture deficit. Table 4.11 illustrates this point with data obtained from steppe solonetz in Arsham-zel'mene (Ergeni).

It should be noted that *Azotobacter* is not detected in all solonetz (Push-kinskaya, 1966; Bol'shakova, 1966). Thus for example, we did not find it in the residual solonetz of the Tellerman Forest (Voronezh Region). One cause of this is likely to be the reduction in the mobility of the organic matter as a result of the expulsion of sodium from the absorbing complex. We could not isolate *Azotobacter* from certain steppe solonetz of western Kazakhstan either. Extremely short wet periods and the low content of organic matter may be an obstacle to the development of *Azotobacter* in these soils.

In Sushkina's view *Azotobacter* is encountered only in the B_1 and B_2 levels of solonetz, while it is absent from the upper-solonetz level A. As our investiga-tions have shown (Table 4.10) this statement does not correspond with the facts. Sidorenko and Naplekova (1962) note that the oxidative enzymes of *Azotobacter* from solonetz have a reduced activity.

We shall now discuss briefly the distribution of *Azotobacter* in the sierozems and solonchaks. We investigated the sierozems of Uzbekhistan and Azerbaijan, and found that in the virgin and bogharic soils of this zone *Azotobacter* mostly develops ephemerally in the spring. It is regularly detected in irrigated, cultivated and virgin soils characterized by a high surface or ground moisture content. Thus, here too, steppe transformation has no depressant effect on *Azotobacter*. The chief restraint on the development of *Azotobacter* rests in the quantity of moisture and organic residues in the soil. Similar conclusions have been reached by several investigators studying sierozems (for example Mirotvorskii, 1945; Belyakov, 1949; Kvasnikov, 1951; Darzniek, 1960, 1962; Rankov, 1965).

Table 4.12. Effect of Soil Salinity on the Development of *Azotobacter*

Condition of cotton	No. of colonies on soil plates	No. of soil lumps containing *Azotobacter* (%)	Soil residue (%)
Soil from the State Farm, Pakha-Aral, Uzbekhistan.			
Normal development	2,000	100	0·09
Weak depression	1,500	70	0·34
Heavy depression	130	0	1·6
Sowings died		0	2·6
Soil of Mil'sk Station, Azerbaijan			
Normal development	500	100	0·5
Weak depression	20	8	1·3
Heavy depression	0	0	1·6
Sowings died	0	0	2·2

We must mention the resistance of *Azotobacter* to increased concentrations of salts, for solonchaks are quite rich in them. Sushkina established that there are races of *Azotobacter* resistant to salt. This was confirmed by Sidorenko (1961), Blinkov (1962), Babak (1966) and Mitrofanova (1963) and Iswaran *et al.* (1966). But it cannot therefore be concluded that saline soils are rich in *Azotobacter*. They usually contain very small numbers of the micro-organism. Sometimes in summer, however, *Azotobacter* is found in virgin solonchaks much more often than in typical sierozems (Verner, 1951). This can be explained by the fact that solonchaks are often located in low lying areas where the water regime is relatively favourable.

Table 4.12 compiled from our findings, shows the distribution of *Azotobacter* in sierozems with different degrees of salinity. The salinity of the soil was evaluated on the basis of the development of cotton and the content of salts ('solid residue') in the soil. As the table shows, as soil salinity increases the numbers of *Azotobacter* decrease sharply. Others have noted a poorer development of *Azotobacter* with an increase in soil salinity (Kvasnikov, 1950; Babak, 1962; Rankov, 1964; Kadyrova, 1966). Rankov established that *Azotobacter* was more sensitive to soda salinity than to sulphate chloride salinity. Nasyrova made a similar observation (1966).

In the tropical zone a considerable proportion of adequately wetted soils have a low pH (laterites). Instead of *Azotobacter* these soils contain the acid-resistant micro-organism *Beijerinckia*. But if, under the influence of certain factors, the soil becomes neutral, it can support *Azotobacter* (for example Kluyver and Becking, 1955; Meiklejohn, 1962). In a dry zone of the tropical belt the development of *Azotobacter* chiefly depends on the occurrence of periods of moisture.

We consider that the many investigations of the ecology of *Azotobacter*—carried out in different continents—amply confirm our conclusions (for example Swaby, 1939; H. Jensen, 1940; Tchan, 1962; Meiklejohn, 1962; Moore, 1963).

Effect of *Azotobacter* on higher plants

A great deal of work has revealed that *Azotobacter* develops more briskly in roots than in soil (for example Federov, 1952; Krasil'nikov, 1958; Vančura *et al.*, 1959, 1965; Döbereiner and Puppin, 1961; Zimna, 1962; Manil, 1962; Saric and Rasòvìc, 1963; Litovchenko, 1965; Malyshkin and Prynova, 1965, 1966). Fig. 31 shows the quantitative distribution of *Azotobacter* in the rhizosphere of rye and vetch and in other soil. Kotelev (1964), using an autoradiographic method, established that cells of *Azotobacter*, labelled with phosphorus and introduced into the soil with seeds, concentrate around the growing root system when the seedlings develop. A similar spread of *Azotobacter* from seed to root was noted by Jackson and Brown (1966).

In poor soils, for example deserts, *Azotobacter* is found chiefly around the roots of plants (Tanatin, 1953; Kvasnikov *et al.*, 1959; Elwan and Mahmoud, 1960; Minasyan and Nalbandyan, 1961; Kaplun, 1964).

All this strongly suggests that root secretions and root litter are a source of the various micro-organisms of the rhizosphere, which include *Azotobacter*. According to Rovira (1965), however, *Azotobacter* does not constitute more than 1 per cent of the total micropopulation of the root zone. The absolute numbers of *Azotobacter* depend on the plant species (for example Krasil'nikov, Katznelson and Strzelczyk, 1961, Milòsevič, 1963; Petkov, 1964; Vidal and Deborgne, 1964; Sadimenkova, 1965), the phase that its development has reached (Pavlovicha, 1953; Federov and Nepomiluev, 1954; Marendiak, 1964; Alieva, 1965); the type of soil and its micropopulation (Stille, 1958; Sarìc and Rašovìc, 1963; Vigorov and Tribunskaya, 1965) and several ecological and geographical factors.

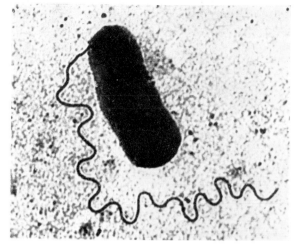

1 (*a*)

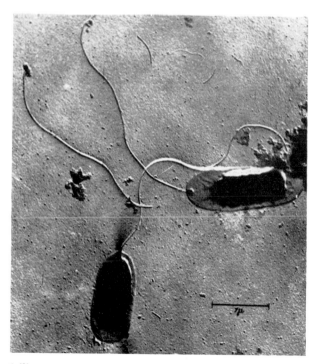

1 (*b*)

1   Electron micrograph of cells of *Rhizobium* with monotrichous flagella. (*a*) From the nodule of *Phaseolus* (after
Ziegler); (*b*) from the nodule of *Glycine hispida* (after De Ley
and Rassel).

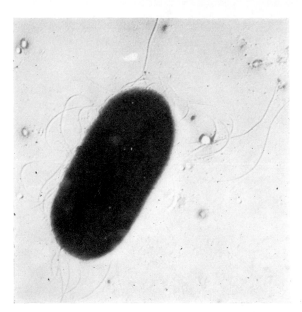

2   Electron micrograph of *Rhizobium* cell with peritrichous flagella from peanut nodule (after De Ley and Rassel).

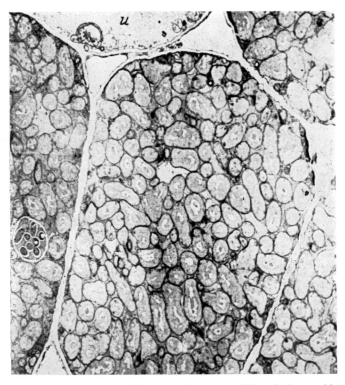

3   Cell of nodular tissue of *Trifolium subterraneum* filled with bacteroids. The cytoplasm occupies a limited volume of the cell and the mitochondria are pushed against the cell wall. Residues of the infection thread can be seen containing bacteria.

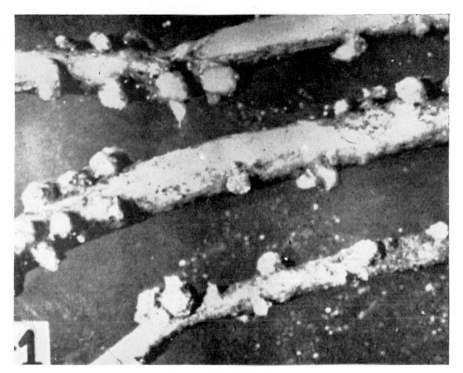

4 Nodules on the roots of *Tribulus cistiodes* (after E. and O. Allen).

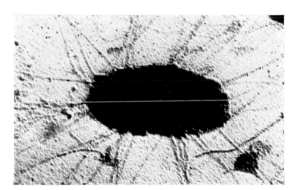

5 Fimbria in *Azotobacter* (electron microscopy preparation after Nikitin).

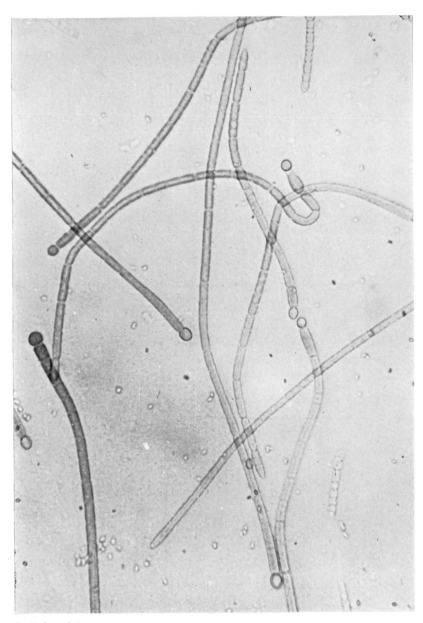

6 *Tolypothrix tenuis.*

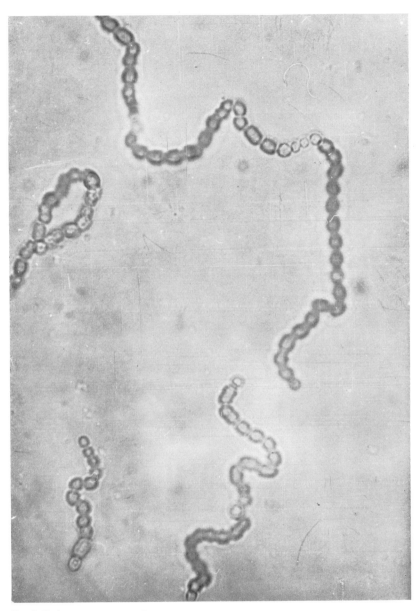

7   *Cylindrospermum constricta.*

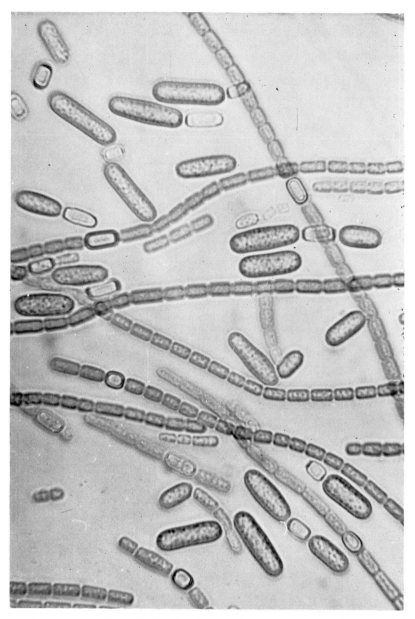

8   *Anabaena cylindrica*. The free-floating oval bodies are spores and the rounded cells within the algal filament are heterocysts, the sites of nitrogen fixation.

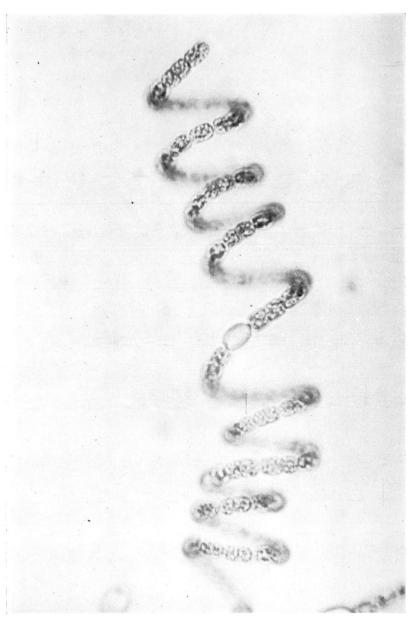

9a

9   *Anabaena flos-aquae* with a heterocyst and gas vacuoles within the other cells.
(*a*) Spiral form; (*b*) a straight filament.

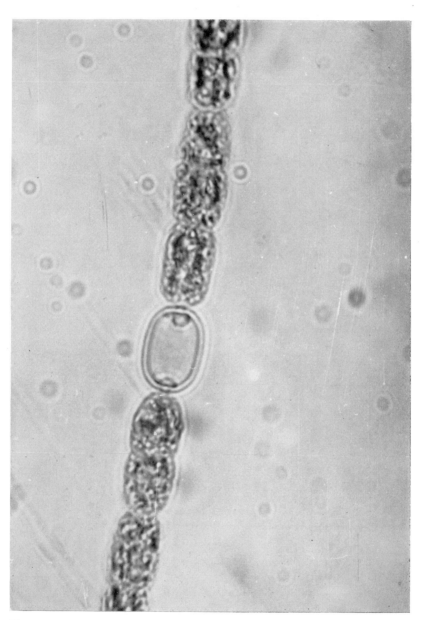

9b

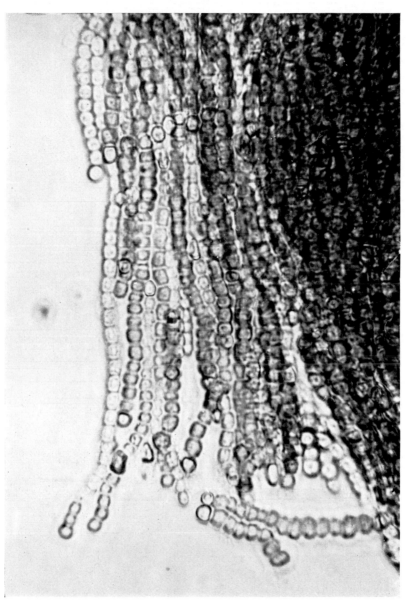

10 *Nostoc muscorum.*

11   Open system for culture of algae (after Pinevich).

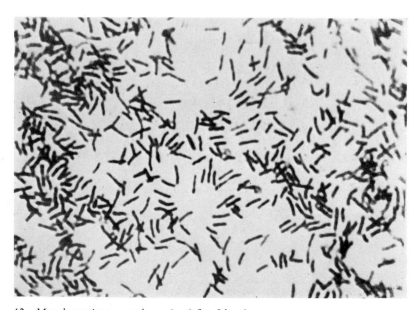

12   *Mycobacterium azot-absorption* (after L'vov).

Krasil'nikov (1958) considered the plant to be the chief factor determining the development of *Azotobacter* in the soil. For example, he found that in sterile conditions, wheat does not promote the growth of *Azotobacter*. In the rhizosphere of wheat this micro-organism either does not develop at all or does so very weakly. Under lucerne, however, it multiplies actively. Krasil'nikov's reported that the situation is the same in the field. It is difficult to agree with this. According to Krasil'nikov's own findings, one gram of rhizosphere soil under wheat in the Moscow Region contains up to ten *Azotobacter* cells, while

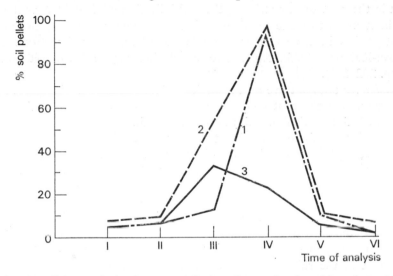

Fig. 31. Course of development of *Azotobacter* in the rhizosphere of (1) rye; (2) vetch and (3) the rest of the soil (after Sobieszczanski).

in the Moldavian SSR the number is 500. In the latter case the rhizosphere of wheat contains just as many cells of *Azotobacter* as the rhizosphere of lucerne (600 cells per gram). In the Vakhsk valley (Central Asia) 10,000 cells of *Azotobacter* are found in a gram of soil surrounding lucerne roots, seventeen times more than in Moldavia. Clearly it is not the plant alone, but a whole set of conditions including soil and climate, that determines the development of *Azotobacter* in the rhizosphere.

Of course, the products of exosmosis from the roots of different plants are not identical, and this must influence the multiplication of *Azotobacter* (Dorosinskii and Krupina, 1960; Bukatsch and Rollmann, 1965). Some root secretions may contain toxic compounds. According to Zayed (1963) root extracts of *Populus tremula* and *Fagus sylvatica* suppress the development of *Azotobacter*, and according to Rice Elroy (1965) root secretions of plants such as *Ambrosia elatior*, *Euphorbia corollata* and *Helianthus annuus* are toxic for *Azotobacter*. The products of exosmosis from the roots of *Ambrosia* and *Helianthus* include a compound similar to chlorogenic acid, while those of

8

Euphorbia contain a substance closely related to the gallotannins. Tribunskaya (1966) noted bacterial and bacteriostatic effects in root secretions and extracts of Canadian chamomile, cress and *Nasturtium*. According to Drobot'ko (1958) these roots secrete 'phytoncides'.

There are also indications that *Azotobacter* is depressed by the root systems of other plants. Clark (1948) and Ulyasheva (1964) noted this in the rhizosphere of tomatoes.

Vančura and Macurá (1961) believe that there may sometimes be an oxygen deficit in the root zone, restricting the multiplication of *Azotobacter*. Attempts have been made to relate the distribution of *Azotobacter* cells to the content of antagonists in the soil (for example Koleshko, 1960; Strzelczyk, 1961; Voinova-Raikova, 1965). No definite conclusion can be drawn yet however (see pp. 202–203).

Azotobacter does not live on the surface of the roots (rhizoplane) but in the soil surrounding the root (rhizosphere) (for example Kurdina, 1957; Karaguishieva, 1958; Mar'enko, 1954; Peshakov, 1965).

Trolldenier (1965) found, using the fluorescent microscope, that even the rhizosphere contains only a few *Azotobacter* cells. Many antagonists impede the multiplication of this micro-organism in the root zone (Hovadik *et al.*, 1965). The numbers of *Azotobacter* in the rhizosphere are also affected by the properties of the soil (Martinez, 1964). But if seeds are inoculated with a suspension of *Azotobacter* a certain amount (roughly a sixth of the amount introduced) will be in close contact with the surface of the roots (Jackson and Brown, 1966).

The data available nevertheless compel us to recognize that *Azotobacter* is frequent in the rhizosphere. It cannot, however, be considered symbiotic with plants, as Bukatsch and Heitzer (1953) and Petrenko (1961, 1966) saw it. They thought that *Azotobacter* isolated from the rhizosphere of certain plants was adapted to survive best in the root zone of that particular species. It is possible that there is selection of *Azotobacter* cultures in the rhizosphere, but this does not indicate the existence of special races peculiar to definite plant species. By passing *Azotobacter* through the rhizosphere of a particular species of plant it is possible to modify the properties of the micro-organism (Zinov'eva, 1962; Strzelczyk, 1964–1965).

The knowledge that *Azotobacter* develops in root zones encouraged the idea that this micro-organism may improve the nitrogen nutrition of plants in the soil, especially when the multiplication of *Azotobacter* is stimulated. Since the 1930s Azotobacterin, a fertilizer preparation containing *Az. chroococcum*, has been in use in the USSR at the suggestion of Kostychev *et al.* (1926). At one time it was regarded exclusively as a nitrogen fertilizer for improving the growth of different species of farm crops. Later, when the restricted nitrogen accumulating powers of *Az. chroococcum* were established, it became impossible to equate the effect of Azotobacterin with that of mineral nitrogen fertilizers. Therefore there was a growing tendency to relate its beneficial effect to the formation of various physiologically active substances (vitamins, gibberellins and so on).

Quite simple calculations indicate that the nitrogen-fixing powers of *Azotobacter* are restricted in the atmosphere. The raw materials are considered to be chiefly the root secretions of the plants concerned. Exosmosis can hardly yield more than 5 per cent of the organic material synthesized by the plant. For a wheat harvest of 2,000 kg/hectare, root exosmosis can give a maximum of 250 kg/hectare of organic compounds, for in such a case the gross dry weight of all plants is about 5 tons. The organic substances released by the root pass

Table 4.13. Effects of *Azotobacter* on Yields

Experimental variant	Yield (g per vessel)	
	Control	*Azotobacter* infection
Complete Hellriegel mixture	41·9	44·8
Hellriegel mixture with 0·1 of the nitrogen norm	7·2	7·9
	Yield (%)	
	Control	*Azotobacter* infection
Complete Hellriegel mixture	100·0	106·0
Hellriegel mixture with 0·1 of the nitrogen norm	17·2	18·9

through a biological filter (the micro-organisms on the root surface) and so it is difficult to accept that *Azotobacter* utilizes more than 10 per cent of the products of exosmosis, that is more than 25 kg/hectare. It seems that even this calculation must be considered over generous.

One gram of energy-rich material enables *Azotobacter* to assimilate 10–15 mg of molecular nitrogen. Using 25 kg/hectare of the products of exosmosis *Azotobacter* cannot assimilate more than 0·25–0·37 kg of nitrogen per hectare per year, which is equivalent to 1·3–2·0 kg of sodium nitrate. The potentialities of *Azotobacter* were estimated at roughly this figure by H. Jensen (1940, 1950) and rather higher by Meshkov and Khodakova (1966).

The role of *Azotobacter* in the nitrogen nutrition of plants can be determined more exactly only in a sterile monobacterial plant culture. This has been done by Fedorov (1945) and ourselves with the same results. Table 4.13 gives the results of one of Fedorov's experiments with maize.

In complete Hellriegel mixture *Azotobacter* increased the yield of maize by only 6 per cent; about a quarter of the nitrogen fixed remained in the medium. With less nitrogen (0·1 of the Hellriegel norm) *Azotobacter* could not provide the plant with a satisfactory supply of nitrogen; it increased the yield of maize by only 1·7 per cent. Thus the root secretions of plants cannot ensure an appreciable accumulation of nitrogen by *Azotobacter*, even when it is the only bacterium present.

Fedorov (1952) thought it desirable to intensify the activity of *Azotobacter* in the soil by adding straw: 1 g of straw fertilizer, according to Fedorov, ensures the accumulation of 6–8 mg of nitrogen in the soil. Somewhat lower figures (3–4 mg) were reported by H. Jensen (1940), which means that 1 ton of straw can enrich the soil with 3–6 kg of nitrogen. Unfortunately the introduction

Table 4.14. Favourable effects of *Azotobacter* on Growth of Seedlings

Experimental variant	Germinated seeds (%)	Mean length of seedlings (cm)
Controls	65	10·25
Seeds inoculated with *Azotobacter*		
Strain 49	87	12·71
Strain 70	78	12·40

of straw in the first year sharply reduces the grain harvest and a useful effect is obtained only in the second or third year.

Thus, the positive effect of *Azotobacter* on the plant cannot be explained as an enrichment of the medium with bound nitrogen. For example, according to Kholopov (1952) and Petrenko (1961), in several cases the nitrogen requirements of plants are better satisfied after the introduction of *Azotobacter* into the soil than after the addition of mineral nitrogen fertilizers. Such results are inexplicable, for they imply that the level of exosmosis exceeds five to ten times the weight of the plants, or that *Azotobacter* utilizing 1 g of root secretions fixes 100–400 mg of molecular nitrogen.

In certain types of media *Azotobacter* undoubtedly improves plant growth. Its stimulating effect can be demonstrated simply. If seeds which have been sterilized on the surface are treated with a dense suspension of *Azotobacter*, the seedlings that develop grow considerably better than control seedlings (grown form seeds treated with water). Table 4.14 shows the results of one such experiment with spring wheat. Test seeds were grown in glass vessels on sterile filter paper. The seedlings were measured on the eighth day.

Bacterization clearly improved germination and accelerated the growth of the young plants. It may be due to the ability of *Azotobacter* to synthesize biologically active substances which influence the seedlings (Toplyakova, 1955;

Pochon and Chalvignac, 1964; Margo, 1964; Sobieszan'ski and Niewiadoma, 1966).

Az. chroococcum synthesizes 50–100 γ of thiamine per gram of cell dry matter, 240–600 γ of nicotinic acid, up to 600 γ of pantothenic acid, about 6 γ of pyridoxine, 3–12 γ of biotin and possibly other vitamins. According to Ratner and Dobrokhotova (1956, 1958) the vitamins (group B and others) stimulate enzymatic processes in the root and accelerate the first stages of synthesis of organic nitrogenous compounds.

Growth substances of the auxin type can be demonstrated in *Az. chroococcum* by Turetskaya's method (1947). This involves establishing the formation of additional rootlets on *Phaseolus* cuttings under the influence of auxins. 'Pioneer' dwarf pea can be used to determine gibberellin-like compounds in a culture of *Azotobacter*. These compounds only stimulate germination and accelerate growth when sufficient *Azotobacter* cells are available.

As well as the stimulating effect of *Azotobacter* on the seeds of various species, Bel'tyukova (1953) found an antagonism against the agents of bacterial diseases in plants. Our own observations have shown us that *Azotobacter* also has fungistatic properties. This work, which was initially carried out with Petrova (1958), was extended by Mar'enko (1963).

Several fungi encountered in seeds and soil (including species of *Fusarium*, *Alternaria* and *Penicillium*) can depress the development of many plant species, especially in cold weather. *Azotobacter*, producing antibiotics to delay the development of fungi, is a particularly important aid to growth in early development.

Az. chroococcum can restrict the growth of various species of fungi. Antagonistic action, however, varies from species to species. For example, Table 4.15 shows that the culture of *Az. chroococcum* 70 is most active and has a wide activity spectrum. A culture of *Az. chroococcum* 82 depresses only *Alternaria*.

The suppression of fungi by *Az. chroococcum* can be demonstrated when seeds germinate on wetted, unsterile filter paper. In control seeds and the roots of control seedlings (after soaking in sterile water) fungi multiply rapidly, often containing a dark pigment. When seeds are treated with a culture of *Az. chroococcum* the development of the fungi is suppressed.

In some experiments we also inoculated seeds with a culture of species of *Alternaria*. This appreciably reduced germination and depressed development of seedlings. *Alternaria* produces several toxic compounds (Pares and Afanchao, 1960), although some species do not have toxic properties (Khudyakov and Marshchunova, 1966).

When seeds contained a fungal infection, bacterization with a culture of *Az. chroococcum* had a good restorative effect. The results of one such experiment with seeds of Moskova spring wheat are presented in Table 4.16.

Control and bacterized seeds were grown in unsterile sand. We used two cultures of *Az. chroococcum*: strain 49 with weakly marked antagonistic properties, and strain 70 which actively depressed the development of the fungi.

Table 4.15. Fungistatic Effect of Cultures of *Az. Chroococcum*

Culture of Az. chroococcum	Aspergillus glaucum	Penicillium (section Biverticillata)	Penicillium (section Monoverticillata)	Fusarium sp.	Alternaria sp.
			Test Fungus		
70	2·0	3·4	3·4	11·4	10·2
79	2·0	4·2	3·6	5·1	5·8
80	0	1·0	0	0	2·8
82	0	0	0	0	5·0
49	2·0	0	0·7	0	4·2

Numbers indicate the radius of the sterile zone of wort-agar (mm) (where there are no fungi) around a block of Ashby's agar carrying a culture of *Az. chroococcum*.

Strain 70 had a particularly marked effect on seed germination. Bacterization counteracted the harmful effect of *Alternaria* and speeded up the growth of the seedlings. This is evidently the result of fungistatic and growth effects.

Several of Mar'enko's experiments with maize and spring wheat revealed a favourable effect of *Az. chroococcum*. We shall discuss only the case of maize, for the results with wheat were identical. The work was carried out on the soddy-podzolic soil of the Timiryazev Agricultural Academy (Moscow). Phosphorous fertilizers were applied to the extent of 150 kg P_2O_5 per hectare, and lime was

Table 4.16. Effects of *Alternaria* and *Azotobacter* on Germination and Growth of Seedlings

Experimental variant	Germinated seeds (%	Mean length of 9-day old seedlings (cm)
Controls (wetted with water)	75	12·6
Inoculation:		
Alternaria	63	10·6
Alternaria + Azotobacter (strain 49)	64	14·5
Alternaria + Azotobacter (strain 70)	90	15·6

added in an amount adjusted to the acidity of the soil together with 75 tons/
hectare of manure.

As Table 4.17 shows, maize seed injected with *Alternaria* gave a yield reduced
by about 30 per cent. When the soil (soddy-podzolic) was not manured bacterization reversed the harmful effect of *Alternaria* but did not increase the yield over
the controls. When the manure was added *Azotobacter* increased the yield by
15–24 per cent compared with the controls and by 48–57 per cent compared

Table 4.17. Effect of *Az. chroococcum* on the Yield of Maize

Experimental variant	Soil fertilized with P_2O_5		Soil fertilized with P_2O_5 and Ca		Soil fertilized with P_2O_5, Ca and manure	
	g	%	g	%	g	%
Controls (wetted with water)	$22 \cdot 86 \pm 0 \cdot 4$	100	$23 \cdot 0 \pm 0 \cdot 75$	100	$56 \cdot 6 \pm 1 \cdot 01$	100
Inoculation: *Alternaria* sp.	$15 \cdot 29 \pm 0 \cdot 8$	66·9	$15 \cdot 3 \pm 0 \cdot 33$	66·5	$38 \cdot 5 \pm 1 \cdot 04$	66·8
Alternaria + *Azotobacter* (strain 70)	$22 \cdot 31 \pm 1 \cdot 2$	97·6	$23 \cdot 0 \pm 1 \cdot 24$	100	$66 \cdot 3 \pm 0 \cdot 67$	115·2
Azotobacter (strain 70)	$23 \cdot 15 \pm 0 \cdot 3$	101·3	$24 \cdot 2 \pm 0 \cdot 55$	105·0	$73 \cdot 7 \pm 0 \cdot 66$	128·0
Azotobacter (strain 53)	$23 \cdot 8 \pm 0 \cdot 3$	104·3	$24 \cdot 7 \pm 0 \cdot 59$	107·2	$66 \cdot 2 \pm 0 \cdot 69$	115·0

The weight of the air-dried mass is given in grams per vessel for the flowering phase.

with *Alternaria* injection. The reason may be that *Azotobacter* could only
multiply actively and have its maximum effect in the substrate rich in organic
matter. Different cultures of *Azotobacter* differed in their effectiveness.

The stimulating effect of *Az. chroococcum* is easy to understand when there
is sufficient organic matter in the soil, for this bacterium develops actively only
in fertile soil When mobile organic substances are not provided an agronomi-
cally significant effect is unlikely to be obtained from *Azotobacter*. We and others
have confirmed this.

It is not possible to outline here the results of the many experiments carried
out with *Azotobacter* preparations in the USSR. We think it desirable, however,
to relate the results of Russian work carried out between 1958 and 1960. Only in
eight cases out of twenty-three (34·4 per cent) did the Azotobacterin significantly
increase the yield. Field experiments carried out in Russian institutes are summarized in Table 4.18. Between eight and sixty were carried out with each crop.

Table 4.18. Effect of Azotobacterin on crops
(our own results)

Crop	No. of tests	Gain in yield given by Azotobacterin (%)
Winter wheat	4	6·0
Spring wheat	15	7·6
Maize	13	5·1
Potato	14	10·4

The table takes account only of experiments with positive results and enables us to represent graphically the possible effectiveness of Azotobacterin.

The only evidence that Azotobacterin has a positive effect is that statistical treatment often shows the differences to be significant, and the gains in yield from *Azotobacter*, though small, were regular in several places. Such an effect, of course, cannot be compared with the effect of mineral nitrogen. Many years of experimentation by Tyurin and Sokolov (1958) showed them that 120 kg/ hectare of mineral nitrogen gives an increase in yield of between 65 and 80 per cent.

More than fifty experiments with Azotobacterin have been carried out in soddy-podzolic and black-earth soils. Bacterization of field crops gave results confirming the conclusion we had drawn from published material. The gain in yield, as a rule, did not exceed 10 per cent.

We now return to an analysis of the data which indicate that *Azotobacter* has a positive effect on the plant. Azotobacterin undoubtedly has a greater effect in manured soils. Table 4.19, compiled from material published by the Institute of Agricultural Microbiology of the VASKhNIL, shows that organic fertilizers increase the efficacy of Azotobacterin to a greater or lesser degree.

Table 4.19. Dependence of the Effect of Azotobacterin on Manuring

Crop	Fertilizer	Harvest (100 kg/hectare)	Gain given by Azotobacterin (100 kg/hectare)	(%)
Maize (seed)	Manure	54·7	18·4	33·4
	Mineral fertilizers	55·0	7·0	12·7
Potato	Manure	170·0	42·0	25·3
	Without fertilizer	109·0	14·0	12·9
Sugar beet	Manure	402·0	31·0	7·7
	Without fertilizer	389·0	9·0	2·3

We have also detected the favourable influence of manure on the effectiveness of Azotobacterin. We have described our experiments (Table 4.18) and in Table 4.20 we give the results of experiments with Azotobacterin and a vegetable. These results are significantly better than those for field crops (Tables 4.17 and 4.18) which is understandable since the soil under vegetables is usually manured.

Table 4.20. Effect of Azotobacterin on
Yield of Cucumber

Yield per frame, kg	Controls	Azobacterin introduced
1st 2nd pickings	9·33	16·25
3rd–6th pickings	22·99	26·39
Total 100 kg/hectare	558	565

According to the Institute of Agricultural Microbiology of VASKhNIL, the yield of cabbage is increased by 19 per cent and of tomatoes by 28 per cent. In an experiment with cucumber the bacterial preparation was introduced into the soil with the seedlings, which were grown in propagating frames. Bacterization had a favourable effect on the yield at the first picking, but the total yield of the season was practically identical in control and test frames. We obtained similar results in experiments with tomatoes. Fruit ripened much faster after bacterization by 100 kg/hectare of Azotobacterin.

Interesting results were obtained when Azotobacterin was used to treat vegetable seedlings grown in blocks of peat and manure prepared according to the formula of Edel'shtein (1954). The blocks measuring 7 cm³ were prepared from screened peat, filings and cow dung in a ratio of $3:1:0.5$. The mixture was enriched with mineral salts. Where tomatoes were concerned, we added to each kilogram of the mixture 3 g of NH_3NO_3, 24 g of superphosphate, 6 g of KCl and 1 g of lime. For cauliflower we added to each kilogram of the mixture 1 g of NH_4NO_3, 12 g of superphosphate, 3 g of KCl and 2 g of lime. The control mixture was wetted with water and the test mixture was treated with a suspension of *Az. chroococcum*. The seedlings were first grown in a propagating frame and then planted out in the field. The bacterized plants developed better and ripened earlier. The gross yield of tomatoes from the test plot was 33·8 per cent higher than for the controls. Cauliflower, when bacterized, gave an increased yield of almost 40 per cent. We also noted that the plants in the test plots were less subject to disease than the others.

Table 4.21 gives results from several of our experiments with cauliflower grown on blocks.

Thus, the favourable effect of *Az. chroococcum* on plants cultivated on a rich substrate may be explained by the good growth of *Azotobacter* in the soil and the root zone of the plant. In poor soil *Azotobacter* rapidly dies off and its effect on the plants is therefore ephemeral.

It is known, however, that in fertile soils *Azotobacter* usually begins to grow spontaneously. How can we then explain the favourable effect of further infection? We believe that this depends on the small numbers of *Azotobacter* cells present, even in fertile soils. Bacterization greatly increases these numbers and this creates favourable conditions for the development of the root systems,

Table 4.21. Effect of Bacterization on
Yield of Cauliflower

Blocks	Controls	Bacterial preparation introduced
Transitional peat		
Yield in 100 kg/hectare	94·6	131·0
Yield in %	100	138·0
Lowland peat		
Yield in 100 kg/hectare	158·1	240·3
	100	142·9
Manure and filings		
Yield in 100 kg/hectare	98·9	135·7
	100	139·0

especially in their initial period of growth. Here, both the stimulating effect of growth substances and the suppression of fungal flora become manifest.

Our ideas about the mechanism of action of *Azotobacter* have been confirmed by Shende (1965), who used a culture of *Az. chroococcum* to bacterize rice seeds in a two year experiment in the Kuban district where the soil is quite fertile. As Table 4.22 shows, *Azotobacter* was ineffective on an unfertilized soil, whereas mineral nitrogen increased the rice harvest appreciably. Bacterization yielded results only when fertilizer was applied too. This indicates that *Azotobacter* acts not as a nitrogen fertilizer but as a growth stimulant.

Experiments in the use of *Azotobacter* for inoculating seeds of farm plants have not only been carried out in the USSR, (for example Allison *et al.*, 1947; Clark, 1948; Timonin, 1948; Fähraeus *et al.*, 1948; Stapp, 1951; Maliczewska, 1953; Kaš and Žák, 1955; Rovira, 1960 and 1965; Georghiu, 1960; Me Mao-Ying, 1961; Sandra, 1963; Marszewska-Ziemiezka and Maliszewska, 1963; Manil and Brakel, 1965; Cacciari, 1966). In most cases *Azotobacter* either did not increase the harvest or had only a slight effect. Close scrutiny of the literature, however, reveals that in rich soil *Azotobacter* acts in a more regular manner (Maliszewska, 1961; Me Mao-Ying, 1961; Brown *et al.*, 1964).

Table 4.22. Effect of *Azotobacter* on the Rice Harvest

Introduction of	Harvest without bacterization (100 kg/hectare)		Increased yield given by bacterization
	1961	1963	(%)
Without fertilizer	41·4	39·1	4·5
N_{60}	52·1	57·8	12·1
$N_{60}P_{60}$	58·6	64·9	18·6
$N_{60}K_{60}$	52·2	56·7	16·6
$N_{60}P_{60}K_{60}$	59·5	69·3	19·6
N_{120}	53·3	65·7	15·9
$N_{120}P_{90}$	58·6	73·4	27·2
$N_{120}K_{90}$	56·9	64·9	21·4
$N_{120}P_{90}K_{90}$	59·8	77·3	30·5

Taken together, these findings confirm that *Azotobacter* cannot be regarded as analogous to nitrogenous fertilizers. Its favourable effect, in certain conditions is connected with the production of biologically active substances. It can now be considered demonstrated that the supplementary use of these compounds favourably influences plant growth (for example Ovcharov, 1953, 1955; Ratner and Dobrokhotova, 1958; Voznyakovskaya, 1963; Sobieszczan'ski, 1961, 1965).

References

Abd-el-Malek, Y. and Ishac, Y. Z. 1966. Longevity of *Azotobacter*. *Plant and Soil*, **24** (2), 325 327.

Abutalybov, M. G. and Gazieva, N. I. 1961. Effect of trace elements on the development of azotobacters and the activity of nitrogen fixing micro-organisms in the soil. *Uch. zapiski Azerb, un-ta Ser. Biol. Nauki*, (1), 45–53.

Abushev, R. A. 1966. Effect of certain trace elements on the development of azotobacters in the grey-brown soil of Apsheron. *Uch. zapiski Azerb, un-ta, Biol. Seriya*, (1), 63–66.

Afrikyan, E. K. 1954. Antagonistic effect of sporulating bacteria on a culture of *Azotobacter*. *Trudy In-ta mikrobiologii, Akad. Nauk SSSR*, (3), 154–165.

Afrikyan, E. K. and Artyunyan, R. Sh. 1953. Effect of antibiotics on azotobacters. *Mikrobiol. sb. Akad. Nauk ArmSSR*, **1** (7), 21–35.

Alexander, M. and Wilson, P. W. 1954. Large-scale production of the *Azotobacter* enzymes. *Appl. Microbiol.*, **2** (3), 135.

Alieva, N. Sh. 1965. The microflora of the rhizosphere of certain leguminous crops. *Uch. zapiski Azerb. un-ta, Ser. Biol. Nauki*, (4), 41–45.

Allison, F. E. 1947. *Azotobacter* inoculation of crops. 1. Historical. *Soil Sci.*, **64**, 413–427.

Allison, F. E. and Minor, F. 1940. Synthesis of coenzyme R by certain rhizobia and *Azotobacter chroococcum*. *J. Bacteriol*, **39**, 273.

Allison, F. E., Caddy, V. L., Pinck, L. A. and Armiger, W. H. 1947. *Azotobacter* inoculation of crops. 2. Effect on crops under greenhouse conditions. *Soil Sci.*, **64**, 489–497.

Amor Asunción Manuel J. 1965. Determinacion cuantitativa de azotobacter, en placas de silico-gel, sin secado, *Cienc. e investig.*, **21** (8), 368–370.

Anderson, G. R. 1966. Field plots and microbial studies. Biologie du sol., *Bull. Internat. inform.*, *Assoc. Internat. Sci. sol, Commiss.* III (5), 6–7, UNESCO, Paris.

Anderson, G. R. and Jordan, J. V. 1961. Boron: a non-essential growth factor for *Azotobacter chroococcum*. *Soil Sci.*, **98** (2), 113–116.

Aristovskaya, T. V. 1941. Use of CO_2 and the possibility of reducing carboxyl by heterotrophs. *Mikrobiologiya*, **10** (6), 701–715.

Azim, M. A. and Roberts, E. R. 1956. Studies in the biological fixation of nitrogen. VI. *Biochim. Biophys. Acta*, **21**, 308.

Babak, N. M. 1962. Distribution and species composition of azotobacters in the saline soils of Moldavia. *Trudy Mold. n-i. in-ta oroshaemogo zemled. i ovoshchevodstva* **4** (1), 163–168.

Babak, N. M. 1964. Synthesis of vitamin B_{12} by various *Azotobacter* cultures. *Trudy Mold n-i. in-ta oroshaemogo zemled. i ovoschevodstva*, **5** (2), 40–45.

Babak, N. M. 1965. Importance of carbon sources in the medium when establishing azotobacters in the saline soils of Moldavia. *Mikrobiologiya*, **34** (5), 868–870.

Babak, N. M. 1966. Effect of various salts on *Azotobacter chroococcum* and *Azotobacter galophilum*. *Mikrobiologiya*, **35** (1), 162–167.

Bachinskaya, A. A. 1935. Structure and development of azotobacters. *Trudy Vses. n-i. in-ta s-kh. mikrobiologii*, **6** (1), 3.

Bagdasaryan, E. G. 1965. Formation of vitamin B_{12} by soil bacteria *Mikrobiologiya*, **34**, (3), 502–505.

Baillie, A., Hodgkiss, W. and Norris, J. 1962. Flagellation of *Azotobacter* spp. as demonstrated by electron microscopy. *J. Appl. Bacteriol.*, **25** (1), 116–119.

Bailly, J. 1965. Renation entre le variation des substances azote du milieu et le modification morphologique d'*Azotobacter chroococcum*. *Rev. gen. bot.*, **72** (852), 257–284.

Balinschi, I., Dragușanu, V. and Văleski, C. 1962. Research on the stable relationship between *Azotobacter chroococcum* and *Bacillus megatherium* var. phosphaticus. *Lucrări ştiintifice*, 145–154.

Barrow, N. J. and Jenkinson, D. S. 1962. The effect of water-logging on fixation of nitrogen by soil incubated with straw. *Plant and Soil*, **16**, 258–262.

Bassalik, K. and Neugebauer, I. 1933. Über die Stimulation von *Azotobacter chroococcum* durch Eisen. *Acta Soc. bot. polon.*, **10** (4), 5, 481.

Becking, J. H. 1961. Studies on nitrogen-fixing bacteria of the genus *Beijerinckia*. *Plant and Soil*. **14** (1), 49–81, (4), 297–322.

Becking, J. H. 1962. Species differences in Mo and V requirements and combined nitrogen utilization by the Azotobacteriaceae. *Plant and Soil*, **2**, 171–202.

Beijerinck, M. W. 1901. Über Oligonitrophile Mikroben. *Zbl. Bakteriol, Parasitenkunde, Infektionskrankh und Hyg.*, Abt. 2, **7**, 561–582.

Beijerinck, M. W. and Delden A. van 1902. Über die Assimilation des freien Stickstoffs durch Bakterien. *Zbl. Bakteriol., Parasitenkunde, Infektionskrankh und Hyg.*, Abt. 2, **9**, 3–43.

Belyakov, E. V. 1949. Distribution of azotobacters in the virgin and cultivated soils of the desert and steppe zones of Central Kazakhstan. *Izv. Akad. Nauk KazSSR.*: in *Osvoenie pustyn'*. (Desert reclamation), **1** (75), 49–87.

Belozerskii, A. N. et al. 1957. The chemistry of azotobacters. I. Mikrobiologiya, 24, (409), 26 (4).

Bel'tyukova, K. I. 1953. Effect of Azotobacterin on the vulnerability of farm plants to bacterial diseases: in Voprosy primeneniya baktericelnykh udrobrenii (Problems of the use of artificial fertilizers). Izd-vo Akad, Nauk UkrSSR, Kiev, 123–134.

Berezova, E. F., Naumova, A. N. and Raznitsina, E. A. 1938. Nature of the effect of azotogen. Dokl. Akad. Nauk SSSR, 18 (6), 357–362.

Bernard, V. V. 1961. Conditions of acclimatization of Azotobacter in soddy-podzolic soil. Trudy In-ta mikrobiologii SSSR (2), 102–110.

Bernard, V. V. 1963. Evolution of ammonia during fixation of atmospheric nitrogen by azotobacters. Agrobiologiya, (3), 430–433.

Bernard, V. V. and Voronkova, E. A. 1960. Multiplication of azotobacters and fixation of nitrogen in presence of organic fertilizers and other sources of organic matter. Agrobiologiya, (1), 103–108.

Bershova, O. I. and Kozlova, I. A. 1962. Formation of vitamins by certain rhizospheric micro-organisms and the effect of trace elements on them. Part 1, 2, Mikrobiol. zh., 24 (2), 30–34; (5), 15–20 (in Ukrainian).

Bershova, O. I. and Kozlova, I. A. 1965. Effect of various combinations of trace elements on the synthesis of vitamins by soil micro-organisms: in Primenenie mikroelementov v sel'skom khozyaistue. (Use of trace elements in agriculture), Naukova dumka, Kiev, 87–90.

Bisset, K. A. 1950. Differentiation between certain genera of Bacteriaceae by the morphology of the microcyst stage. J. Gen. Microbiol., 4, 414.

Bisset, K. A. 1955. The value of cytological studies in elucidating natural relationships among bacteria. J. Gen. Microbiol., 12 (2), 325–329.

Bisset, K. A. and Hale, C. M. F. 1953. The cytology and life-cycle of Azotobacter chroococcum. J. Gen. Microbiol., 8, 442–448.

Blinkov, N. G. 1951. Connection between azotobacters and the pH of the medium. Uch. zapiski Tomskogo ped. in-ta, 8.

Blinkov, G. N. 1959. Azotobacter i ego znachenie dlya vysshykh rastenii (Azotobacter and its importance for higher plants. Izd. Tomskogo un-ta.

Blinkov, G. N. 1962. Salt-resistant azotobacters. Mikrobiologiya, 31 (5), 880–883.

Bol'shakova, V. S. 1966. The microflora of the soils of the saline complex of the Dzhanybensk district of the West Kazakhstan region: in Mikroflora pochv severa i srednei chasti SSSR. (The microflora of the soils of the north and central parts of the USSR), Nauka, Moscow, 352–364.

Bonner, I. 1937. The role of vitamins in plant development. Bot. Rev., 3 (12), 616.

Bortels, H. 1937. Über die Wirkung von Molybdan und Vanadiumdungunger auf Azotobacter-Zahl and Stickstoffbindung in Erde. Arch. Mikrobiol., 8 (1), 13–26.

Bouisset, L. and Breulland, J. 1958. Étude bacteriologique de germes oligonitrophiles isolés de sol. Bull. Soc. histoire natur. Toulouse, 93 (3–4), 479–482.

Boyd, W. L. and Boyd, J. W. 1962. Presence of Azotobacter species in polar regions. J. Bacteriol. 83 (2), 429–430.

Brakel, J. and Hilger, F. 1965. Etude qualitative et quantitave de la synthese de substances de nature auxinique par Azotobacter chroococcum in vitro. Bull. Inst. Agron et stat. rech. Gembloux, 33 (4), 469–487.

Brown, J. W. and Smith, N. M. 1912. Bacterial activities in frozen soils. Zbl. Bakteriol., Parasitenkunde, Infektionskrankh und Hyg., 34, 369–374.

Brown, M. E., Burlingham, S. K. and Jackson, R. M. 1964. Studies on Azotobacter species in soil. III. Effects of artificial inoculation on crop yields. Plant and Soil, 20 (2), 194–214.

Brundtsa, K. 1949. Mikrobiologiya (Microbiology), Kaunas (in Lithuanian).

Bukatsch, F., Burger, K. and Schluter, M. 1956. Untersuchungen uber Eiweiss und

Eiweistoffwechsel bei *Azotobacter* mit besonderer Berücksichtigung der Indolkorpcr. *Zbl. Bakteriol., Parasitenkunde, Infektionskrankh und Hyg.*, **11** (109), 225–232.

Bukatsch, F. and Heitzer, J., 1952. Beitrage zur Kenntnis der Physiologie von *Azotobacter. Arch. Mikrobiol.*, **17** (1), 79–96.

Bukatsch, F. and Rollmann, W. 1965. Zur Frage der Anpassung von *Azotobacter chroococcum* an die Rhizosphäre. *Zbl. Bakteriol Parasitenkunde Infektionskrankh und Hyg.*, Abt. 2, **119** (6), 555–565.

Bulen, W. A., LeComte, J. R. and Bales, H. E. 1963. Short-term $^{15}N_2$ incorporation by *Azotobacter. J. Bacteriol*, **85** (3), 666.

Bullock, G. L., Bush, J. A. and Wilson, P. W. 1960. Calcium requirements of various species of *Azotobacter. Proc. Soc. Exptl. Biol.*, **105** (1), 26–30.

Bun'ko, I. P. 1965. The question of the activity of *Azotobacter* as a producer of physiologically active substances. *Vieści Akad. Năuka BSSR, Sierija selskahaspad. navuki, Izvestiya Akad. Nauk BSSR, Seriya s-'kh nauki*, (2), 54–59 (in Byelorussian).

Burema, S. J. and Wieringa, K. L. 1942. Molybdenum as a growth requirement of *Azotobacter chroococcum. Antonie v. Leeuwenhoek. J. Microbiol and Serol.* (8), 123–133.

Burger, K. and Bukatach, F. 1958. Uber die Wuchsstoffsynthese in Bodenfreilebender stickstoffbindender Bakterien. *Zbl. Bakteriol Parasitenkunde Infektionskrankh und Hyg.*, **11** (111), 1–28.

Burk, D., Horner, C. and Lineweaver, H. 1932. Injury and recovery of respiration in *Azotobacter. J. Cell Comp. Physiol.*, **1**, 435–439.

Burk, D. and Lineweaver, H. 1931. The influence of calcium and strontium upon the catalysis of nitrogen fixation by *Azotobacter. Arch Mikrobiol.*, **2**, 155–186 (in English).

Burk, D., Lineweaver, H. and Horner, G. 1934. The specific influence of acidity on the mechanism of nitrogen fixation by *Azotobacter. J. Bacteriol.*, **27**, 335.

Burris, R. H. and Wilson, P. W. 1943. Utilization of combined nitrogen by *Azotobacter. J. Bacteriol.*, **45**, 17.

Butkevich, V. S. and Kolesnikova, N. A. 1941. Formation of ammonia on fixation of molecular nitrogen by azotobacters. *Dokl. Akad. Nauk SSR*, **33** (1), 66–69.

Bychkovskaya, A. L. and Shklyar, M. Z. 1959. Acid tolerant variant of *Azotobacter. Mikrobiologii*, **28** (3), 336–342.

Bychkovskaya, A. L. and Shklyar, M. Z. 1965. Resistance of cysts of azotobacters to unfavourable factors. *Trudy VNII s-kh. mikrobiologii*, (19), 9–19.

Cacciari, L. 1966. Modificazional dell attivita biologica di suoli trattati con anidride carbonica. *Agronimica*, **10** (2), 142–148.

Callao, V., Alvarado, R., Sedano, A., Olivares, J. and Montoya, E. 1966. Efecto antagonico del *Myxococcus xanthus* sobre los *Azotobacter. Microbiol esp.*, **19** (1), 45–51.

Callao, V., Hernandez, E. and Montoya, E. 1961. In metodo rapido para la caracterizacion des los *Azotobacter* des suelo. *Agrochimica*, **6** (1), 51–55.

Chalvignao, A. 1962. Incidence des *Azotobacter* sur la nutrition et la croissance du lin (*Linum usitatissimum*). *Agrochimica*, **6** (3), 286–292.

Chandra, Purna, Bollen, W. B. and Kadry, L. T. 1962. Microbiological studies of two Iraqi soils representative of an ancient site. *Soil Sci.*, **94**, 251–257.

Chang, P. C. and Knowles, R. 1965. Non-symbiotic nitrogen fixation in some Quebec soils. *Canad. J. Microbiol.*, **11** (1), 29–38.

Chentsov, B. V. 1960. Development of microflora in peat manure phosphorite compost with different forms of potassium fertilizers. *Nauch. trudy Sev-Zap. n-i. in-ta sel'sk kh-va*, (1), 32–39.

Chien Tse-Shu, Ho Fu-Heng Feng Xao-Shan *et al.* 1964. The microbiological characteristics of red soils. *Acta pedol. sinica*, **12** (4), 390–400.

Chizhevskii, M. G. and Dikusar, M. M. 1959. Effect of various forms of natural organic matter on the development of *Azotobacter* and the productivity of nitrogen fixation by it. *Izvestiya TSKhA*, (3), 69–80.

Chizhevskii, M. G. and Kulakov, E. V. 1952. Joint introduction of organic and mineral fertilizers. *Pochvovedenie*, (5).

Clark, F. E. 1948. *Soil Sci.*, **65**, 193 (quoted by Jackson R. and Brown,, M. 1966)

Cohen, G. H., Johnstone, D. B. 1963. Acid production by *Azotobacter vinelandii*. *Nature*, **198**, 211.

Dachyulite, Ya. A. 1962. Effect of certain antibiotics on the biosynthesis of vitamin B_{12} in an *Azotobacter* culture: in *Voprosy fiziologii i biokhimii* (problems of physiology and biochemistry), Vilnyoci, 169–175 (in Lithuanian).

Daraseliya, N. A. 1963. The microflora of the red-earth and brown forest soils of Georgia. *Trudy In-ta pochvovedeniya GruzSSR*, **11**, 55–64.

Darzniek, Yu. O. 1960. Conditions for take of *Azotobacter* in newly irrigated soils on bacterization. *Izvestiya Akad. Nauk Turkm SSR, ser. biol.* (6), 51–58.

Darzniek, Yu. O. 1961. Rate of multiplication of the cells of *Azotobacter* in the soil. *Mikrobiologiya*, **30** (6), 1042–1044.

Darzniek, Yu. O. 1962. *Usloviya zhiznedeyatel'nosti azotobactera v pochvakh Murgabskogo oazisa pod khlopchatnikom*. (Conditions for the vital activity of *Azotobacter* in the soils of the Murgab oasis under cotton). Author's abstract of M.Sc. Thesis. Ashkhabad.

Delwiche, C. C. and Wijler, J. 1956. Non-symbiotic nitrogen fixation in soil. *Plant and Soil*, **7**, 113–129.

Denarie, J. and Blachere, H. 1966. Inoculation de graines de végétaux cultives a l'aide de souches bacteriennes. *Ann. Inst. Pasteur*, **111** (3), Suppl., 57–74.

Degtyareva, M. G. 1959. Effect of irrigation and mineral fertilizers on the spread and nitrogen-fixing activity of azotobacters in the chestnut soil of the Kulundinsk steppe. *Trudy Biol. in-ta Sib. otd. Akad. Nauk SSSR*, (4), 61–67.

Detinova, L. V. 1937. Synthesis of vitamin B by bacteria. 2. Biological testing of bacteria for vitamin B. *Uspekhi zootekhn nauk*, **3** (2), 175.

Dhar, N. K. and Tandon, S. P. 1936. Influence of temperatures on nitrogen fixation by *Azotobacter*. *Proc. Nat. Acad. Sci., Allahabad.*, **6**, 35–39.

Dianova, F. V. and Voroshilova, A. A. 1927. Study of the reasons for the absence of azotobacters in cultivated soils. *Nauchno-agronom. zh.*, (7–8), 483–496.

Dikusar, M. M. 1945. Bactericidal effect of humus. *Dokl. TSKhA*, (2).

Dilworth, M. J. 1962. Oxygen inhibition in *Az. vinelandii* pyruvate oxidation. *Biochim. Biophys. Acta.*, **56** (1), 127–138.

Dilworth, M. J. and Parker, C. A. 1961. Oxygen inhibition of respiration in *Azotobacter*. *Nature*, **191**, 520–521.

Di Menna, M. E. 1966. The incidence of *Azotobacter* and of nitrogen-fixing clostridia in some New Zealand soils. *N.Z. Agric. Res.*, **9** (2), 218–226.

Döbereiner, J. and Puppin, R. A. 1961. Inoculacao do arroz com bacterias fixadoras de nitrogenio, do genero *Beijerinckia* Derx. *Rev. brasil. biol.*, **21** (4), 397–407 (summary in English).

Dommergues, J. 1962. Contribution a l'étude de la dynamique microbienne des sols en zone semiaride et en zone tropical seche. *Ann. agron.*, **13** (4), 262–324; (5), 391–468.

Dommergues, K. 1964. Étude de quelques facteurs influant sur le comportement de la microflore du sol au cours de la dessiccation. *Sci. sol.* (2a), 141–155.

Dommergues, J. and Dusausoy, M., 1966. Evaluation de la densite des microorganismes vivants dans les cultures d'*Azotobacter*. Biologie du sol. *Bull. internat. inform Assoc., Internat. Sci. sol. Commiss.* III (5), UNESCO, Paris, 23.

Dondero, N. C. and Zelle, M. R. 1953. Observations on the formation and behaviour of conjugation cells and large bodies in *Azotobacter agile*. *Science*, **118**, 34–36.

Dorosinskii, L. M. 1958. Depth cultivation of azotobacters: in *Poluchenie i primenenie bakterialnykh udobrenii* (The preparation and use of bacterial fertilizers), Akad. Nauk UkrSSR, Kiev, 209–215.

Dorosinskii, L. M. and Krupina, L. I. 1960. The adaptation of azotobacters in the rhizosphere of various plants. *Byull. nauchno-tekh. inf. po s-kh. mikrobiologii*, (8), 14–17.

Drobkov, A. A. 1963. Importance of natural radioactive elements in the life of soil micro-organisms and higher plants: in *Mikroorganizmy v sel'skom khozyaistve* (micro-organisms in Agriculture), Moscow State University, 59–80.

Duda, J. 1956. Use of the pedology method in *Azotobacter* culture, *Acta microbiologica polonica*, **5** (1–2), 113–115.

Ebert, K. 1959 Über die Beeinflussung des Wachstums und der Stickstoffbindung von *Azotobacter* durch Amid-Ammoniak und Nitratstickstoff. *Z. Pflanzenernähr. Düng, Bodenkunde* **87** (2), 118–134.

Edel'shtein, V. I. 1954. Nutrient feeding blocks by the hydropeat method. *Kartofel'i ovoshchi*, (3), 18–19; *Kolkhoznoe proizvodstvo*, (3), 20–22.

Eisenstark, A. McMahon, K. J. and Eisenstark, R., 1950. A cytological study of a pleomorphic strain of *Azotobacter* with the electron and phase microscopes and the Robinow nuclear staining technique. *J. Bacteriol.*, **59** (1), 75–81.

El'bert, B.Ya. and Koleshko, O. I. 1961. Certain little studied biological properties of the micro-organisms of the genus *Azotobacter*: in *Mikro-organizmy pochvy i ikh rol 'v urozhainosti rasteni* (Soil micro-organisms and their role in plant harvest). Moscow State University, 50–51.

Elwan, S. H. and Mahmoud, S. 1960. Note on the bacterial flora of the Egyptian desert in summer. *Arch. Mikrobiol.*, **36** (4), 360–364.

Enikeeva, M. G. 1952. Moisture content of soil and the activity of micro-organisms. *Trudy In-ta mikrobiol., Akad. Nauk SSSR*, **2**, 130.

Ernandes, A. 1953. *Vliyanie dlitel'nogo primeneniya razlichnykh udobrenii i dlitel'noi monokul'tury nekotorykh sel'skokhozyaistvennykh rastenii na rasprostranenie i azotfiksiruyushchuyu aktivnost' azotobaktera v podzolistoi pochve* (Effect of prolonged use of different fertilizers and prolonged monoculture of certain farm plants on the distribution and nitrogen-fixing activity of *Azotobacter* in podzolic soil) Author's abstract of M.Sc. Thesis, Moscow.

Erofeev, N. S. and Vostrov, I. S. 1964. Use of straw as a direct fertilizer. *Izvestiya Akad, Nauk SSSR, ser. biol.*, (5), 668–676.

Esposito, R. and Wilson, P. W. 1956. Trace metal requirements of *Azotobacter*, *Proc. Soc. Exptl. Biol.*, **3**, 93.

Ezrukh, E. N. 1962. The distribution and adaptation of azotobacters in the soils of the Kolsk peninsula: in *Voprosy botaniki i pochvovedeni v Murmanskoi oblasti* (Problems of Botany and Pedology in the Murmansk Province). *Izd-vo Akad. Nauk SSSR*, 148–154.

Fallot, J. 1963. Bacteries et tissus végétaux en culture. *Ann. Inst. Pasteur*, **1051** (2), 188–194.

Fedorov, M. V. 1940. Effect of azotobacters on the nitrogen balance of the soil and yield of farm plants when the soil is fertilized with straw. *Mikrobiologiya*, **9** (6), 541–557.

Fedorov, M. V. 1945. Effect of *Az. chroococcum* on the nitrogen balance and maize harvest in a monobacterial culture with added organic substances. *Izvestiya Akad. Nauk SSSR, Seriya Biol.*, (6), 703–708.

Fedorov, M. V. 1950. Effect of individual nutrient elements on the fixation of atmospheric nitrogen by *Azotobacter*. *Dokl. Akad. Nauk SSSR*, **72** (1), 157.

Fedorov, M. V. 1952. *Biologicheskaya fiksatsiya azota atmosfery*. (Biological fixation of atmospheric nitrogen). Sel'khozgiz. Moscow.

Fedorov, M. V. and Nepomiluev, V. F. 1954. Distribution and nitrogen-fixing activity of *Azotobacter* in the rhizosphere of perennial grasses. *Mikrobiologiya*, **23** (3), 275–282.

Fedorov, M. V. and Tepper, E. Z. 1947. Conditions determining the take of *Azotobacter* in the rhizosphere of farm plants and in the soil. *Mikrobiologiya*, **16** (6), 498–507.

Fenglerowa, W. 1965. Simple method for counting *Azotobacter* in soil samples. *Acta Microbiol. Polonica*, **14**, 203–206 (in English).

Fife, J. M. 1943. The effect of different oxygen concentrations on the rate of respiration of *Azotobacter* in relation to the energy involved in nitrogen fixing and assimilation. *J. Agric. Res.*, **66**, 421–440.

Fischer, H. 1905. Ein Beitrag zur Kenntnis des Lebensbedingungen von stickstoffsammelnden Bakterien. *Zbl. Bakteriol., Parasitenkunde, Infektionskrankh. und Hyg.*, Abt. II (14), 33.

Gadgil, P. D. and Bhide, V. P. 1960. Nitrogen fixation by *A zotobacter* in association with some associated soil micro-organisms. *Prod. Nat. Inst. Sci. India*, B **26** (2), 60–63.

Gainey, P. L. 1944. Measuring the growth of *Azotobacter*. *J. Bacteriol.*, **48** (3), 285–294.

Gainey, P. L. and Fowler, E. 1945. Growth curves of *Azotobacter* at different pH levels. *J. Agric. Res.*, **70**, 219–236.

Galimova, R. A. 1960. Effect of fertilizers on the vital activity of azotobacters in the near-root zone of maize in irrigated soils. *Trudy Kazakhsk. s-kh. in-ta*, **8**, 138–141.

Garder, L. A. 1927. Effect of temperature on the development of azotobacters and their assimilation of atmospheric nitrogen. *Trudy Otd. s-kh. mikrobiologii Gos. in-ta opytn. agronomii*, **2** (4), 134–138.

Garbosky, A. J. and Giambiagi, N. 1963. Thermoresistance chez les Azotobacteriaceae. *Ann. Inst. Pasteur*, **10–5** (2), 202–208.

Gebgardt, A. C. 1958. Effect of introducing certain vitamins and micro-organisms on the uptake of nitrogen by plant seedlings. *Nauchn. dokl. vyssh. shkoly ser. biol.*, (3), 160–163.

Gebgardt, A. G. 1961. *O sushchnosti deistviya azotobakterina i putyakh povyssheniya ego effektivnosti.* (Essential aspects of the action of azotobacterin and ways of improving its effectiveness). Author's abstract of doctoral thesis. Moscow.

Geiko, N. S. Lyubimov, V. I. and Kretovich, V. L. 1965. Free keto acids in *Azotobacter vinelandii. Dokl. Akad. Nauk SSSR*, **160** (4), 944–945.

Geller, I. A. and Khgariton, E. G. 1951. *Azotobacter* in the soil and under the grass-arable system. *Mikrobiologiya*, **20** (2), 113.

Georgiu, V. and Menuke, L. (Georghiu, V. and Manuca, L.). 1964. *Azotobacter chroococcum* (strain 100) as a producer of auxin-like substances: in *Bakterial'nye udobreniya* (Bacterial Fertilizers), Moscow, 166–176.

Gerretsen, F. C. and Hoop, A. 1954. Boron as an essential micro-element for *Azotobacter chroococcum. Plant and Soil*, (4), 5.

Giambiagi, N. and Sedeño, A. R. de. 1966. Fertilizacion bacteriana en suelos de pradera. *IDIA*, (217), 42–50.

Goguadze, V. D. 1966. Microbiological characterization of the red-earth soils of Georgia: in *Mikroflora poch vyuzhnoi chasti SSSR*. (The Microflora of the Soils of the Southern Part of the USSR), Nauka, Moscow, 246–263.

Golovacheva, R. S. 1959. Relationships of *Azotobacter* and *Clostridium pasteuranium* in soil conditions. *Izv. EstSSr, Ser. Biol.*, **8** (3), 223–236.

Golovacheva, R. S. 1960. Some findings on the effect of fertilizers on the development of nitrogen-fixing bacteria in the soil. *Byull. nauchno-tekhn inf. Est. n-i. in-ta zemled. i melioratsii*, (5), 46–51.

Greaves, J. E. 1913. The occurrence of arsenic in soils. *Biochem. Bull.*, **2**, 519–523.

Greaves, J. E. 1933. Some factors influencing nitrogen fixation. *Soil Sci.*, **36**, 267.

Greaves, J., Jones, L. and Anderson, A. 1940. The influence of amino acids and protein on nitrogen fixation by *Azotobacter chroococcum Soil Sci.*, **49** (1), 9–19.

Greene, R. A. 1932. The effect of temperature upon fixation by *Azotobacter*. *Soil Sci.*, **33**, 153–161.

Gromyko, E. P. 1963. Aluminium as a factor determining the toxicity of soddy-podzolic soil for soil micro-organisms: in *Pochvennaya i sel'sko- khozyaistvennaya mikrobiologiya* (Soil and Agricultural Microbiology), Tashkent, Izd-vo akad. Nauk UzbSSR, 61–67.

Gupta, Y. P. and Das, N. B. 1959. Preliminary studies on the synthesis of vitamin B_{12} by *Azotobacter chroococcum*. *Sci. and Culture*, **25** (2), 149–150.

Gushterov, G. 1956. *Razprostranenie na azotobacter v rizosferata na tyutyne ot stankedimitrovsko* (Distribution of *Azotobacter* in the rhizosphere of Stanke Dimitrov tobacco). *Izd Bălgarskata Akademiya no naukite, Bull de l'Institut Botanique*, **5**, 433–451. (In Bulgarian.)

Gutschy, I. 1959. New method of quantitative estimation of nitrogen assimilation in soil by microbiological means. *Savremena Pojlopr.*, **7** (12), 1086–1091 (in Serbocroat).

Gvamichava, N. E. 1962. Micro-organisms as a source of accumulation of vitamins in the soil. *Soobsh. Akad. Nauk GruzSSR*, **28** (2), 223–226.

Hardisson, C. and Pochon, J. 1966. Synthèse de substances parahumiques par *Azotobacter chroococcum* I. Conditions de culture et extraction de la substance. *Ann. Inst. Pasteur*, **111** (1), 66–75.

Hardisson, C. and Robert-Gero, M. 1966. Synthèse de substances parahumiques par *Azotobacter chroococcum* 2. Etude chimique comparative avec des extraits humiques de sols. *Ann. Inst. Pasteur*, **111** (4), 486–496.

Heinse, B. 1906. Über die Stickstoffassimilation durch niedere Organismen. *Landwirtsch. Jahresb*, **35**, 889.

Hely, F. W. and Bonnier Ch. 1954. The effects of induced soil aggregation on the occurrence of *Azotobacter* and *Streptomyces*. *Antonic v. Leeuwenhoek. J. Microbiol and Serol.*, **20** (4), 359–373.

Hennequine, J. R. and Blachere, H. 1966. Recherches sur la synthèse de phytohormones et de composes phénoliques par *Azotobacter* et des bacteries de la rhizosphere. *Ann. Inst. Pasteur.*, **111** (3), Suppl., 89–102.

Hermann, D. 1963. Der Einfluss von CO_2 Partialdruck und Glucose-Konzentration auf Wachstum und Stickstoffbindung von *Azotobacter chroococcum* Beijerinckii. *Arch Mikrobiol.*, **45** (4), 373–397.

Herzinger, F. 1940. Beiträge zum Wirkungskreislauf des Bors. *Bodenkunde und Pflanzenernähr*, **16**, 141–168.

Holme, T. and Zacharias, B. 1965. Differences in the antigenic pattern of *Azotobacter* growth in different nitrogen sources. *Nature*, **208**, 1235–1236.

Hovadík, A. and Polách, J. 1958. Iodine and soil micro-organisms. *Bull Výzkumn Ústav Zelinir Olomouci*, **2**, 41–60.

Hovadík, A. Vanćura, V., Vlček, F. and Mácura, J. 1965. Bacteria in the rhizosphere of red pepper: in *Plant Microbes Relationship*. Czechosl. Acad. Sci. Prague, 109–119 (in English).

Hovenkamp, H. G. 1958. Stability of the enzymes for respiratory chain phosphorylation in *Azotobacter vinelandii* 4 *Internat. Congr. Biochem., Vienna. Suppl. Internat. Abstr. Biol. Sci.*, London, 66.

Ierusalimskii, N. D., Zaitzava, G. N. and Khmel', I. A. 1962. Study of the physiology of *Azotobacter vinelandii* in flow culture. *Mikrobiologiya*, **31** (3), 417–423.

Imshenetskii, A. A. 1946. Phylogeny of *Azotobacter chroococcum*. *Mikrobiologiya*, **15** (5), 459–466.

Imshenetskii, A. A. 1962. Evolution of the biological fixation of nitrogen: in *Trudy V Mezhdunarodnogo biokhimicheskogo Kongressa. Simpozium III. 'Evolutsionnaya biokhimiya* (*Proceedings of the Fifth International Biochemical Congress*, Symposium III, Evolutionary biochemistry), 141–151.

Imshenetskii, A. A. and Solntseva, L. I. 1940. Symbiosis of cellulose-, and nitrogen-fixing bacteria. *Mikrobiologiya*, 9 (9–10), 783–803.

Iswaran, V. 1958. Effect of humus of legume and non-legume origin of the nitrogen fixation in *Azotobacter chroococcum*. *Current Sci.*, 27 (12), 489–490.

Iswaran, V. 1960. Influence of combined nitrogen on fixation of nitrogen by *Azotobacter*. *Proc. Indian Acad. Sci.*, B, 57 (1), 32–36.

Iswaran, V. and Rao, W. V. B. 1960. The effect of trace elements on nitrogen fixation by *Azotobacter*. *Proc. Indian Acad. Sci.*, B, 57 (3), 103–115.

Iswaran, V. and Sen, A. 1960a. Inactivation of *Azotobacter* by heat. *Current Sci.*, 27 (9), 341–342.

Iswaran, V. and Sen. A. 1960b. Effect of calcium salts on nitrogen fixation by *Azotobacter* sp. *Ann. Biochem and Exptl. Med.*, 20 (8), 197–204.

Iswaran, V., Subba-Rao, N. S. and Sundara Rao, W. V. B. 1966. Sodium chloride tolerance by *Azotobacter chroococcum*. *Current Sci.*, 35 (5), 126–127.

Jackson, R. M., Brown, M. E. 1966. Behaviour of *Azotobacter chroococcum* introduced into the plant rhizospheres. *Ann. Inst. Pasteur*, 111 (3), Suppl., 103–112.

Jensen, H. L. 1940. Contributions to the nitrogen economy of Australian wheat soils, with particular reference to New South Wales. *Proc. Linnean Soc. NS Wales*, 65, 1–122.

Jensen, H. L. 1950. *Om forekomst af Azotobacter i dyrkede jorder* (Occurrence of *Azotobacter* in cultivated soils). Trykt nos Nielsen and Lydrche, Copenhagen, 622–649.

Jensen, H. L. 1951. Notes on the biology of *Azotobacter*. *Proc. Soc. Appl. Bacteriol.*, 74 (1), 89–94.

Jensen, H. L. 1954a. The Azotobacteriaceae. *Bacteriol. Revs.*, 18 (4), 195.

Jensen, H. L. 1954b. The magnesium requirements of *Azotobacter* and *Beijerinckia*, with some additional notes on the latter genus. *Acta Agric. Scand.*, 4 (2), 224–236 (in English).

Jensen, H. L. 1955. *Azotobacter macrocytogenes* n. sp. a nitrogen-fixing bacteria resistant to acid reactions. *Acta agric. Scand.*, 5, 278–294 (in English).

Jensen, V. 1955. The *Azotobacter* flora of some Danish watercourses. *Saertryk af Bot. Tidsskr.*, 52, 143 (in English).

Jensen, V. 1961. Rhamnose for detection and isolation of *Azotobacter vinelandii* Lipman. *Nature*, 190, 832–833.

Johnstone, D. B. Pfeffer, M. and Blanchard, G. C. 1959. Fluorescence of *Azotobacter*. 1. A comparison of the fluorescent pigments with riboflavin. *Canad. J. Microbiol.*, 5, 299–304.

Jones, D. H. 1913. A morphological and cultural study of some azotobacters. *Zbl. Bakteriol., Parasitenkunde Infektionskrankh und Hyg.*, Abt 2, 38, 14.

Jordan, J. V. and Anderson, G. R. 1950. Effect of boron on nitrogen fixation by *Azotobacter*. *Soil Sci.*, 69, 311–319.

Kaarli, L. 1966. Microbiological processes of the natural breakdown of organic soil matter and fertilizers in different temperature conditions. *Tead tööde kogumik. Eesti Maaviljel. ja Maaparand. Tead. Uurimise. Inst.*, (8), 97–142.

Kadyrova, T. M. 1966. Distribution of azotobacters in the grey-earth meadow soils of the Shirvan stepps. *Uch. zapiski Azerb. un-ta, ser. biol.*, (1), 53–62.

Kanopkaite, S. I. and Pauluskaite, L. V. 1961. Phosphatases of *Azotobacter chroococcum*. *Trudy Akad. Nauk LitSSR, Ser. Biol.*, (3), (26), 3.

Kanopkaite, S. I. *et al.* 1959. A new producer of vitamin B_{12} in sapropel. *Trudy Akad. Nauk LitSSR Ser. Biol.*, (1), (17), 125–133.

Kaplun, S. A. 1964. *Azotobacter* in the virgin soils of Uzbekhistan. *Trudy n-i. in-ta pochvoved.*, (4), 105–107.

Karaguishieva, D. 1958. Effect of bacteria fertilizers on the number of azotobacters and the yield of farm crops in the virgin and cultivated soils of Kazakhstan: in *Izuchenie i primenenie bakterial'nykh udobrenii* (Study and use of bacterial fertilizers). Akad. Nauk UkrSSR, Kiev., 112–121.

Karaguishieva, D. 1961. Effect of moisture content of soils in the development of azotobacters. *Trudy In-ta pochvoved.*, *Akad. Nauk KazSSR*, 12, 69–75.

Karaguishieva, D. 1963. Ecology of *Azotobacter* in the soils of Kazakhstan in connection with the use of its preparations. *Trudy In-ta pochvoved.*, *Akad. Nauk KazSSR*, 14, 11–38.

Karasevich, E. K. 1962. *Fiziologicheskie osobennosti neaktivnykh ras azotobaktera* (Physiological aspects of inactive *Azotobacter* races) Author's abstract of M.Sc. Thesis, Moscow.

Kaś, V., Žak, I. 1955. Development and activity of *Azotobacter* in relation to concurrent microflora in soil and in the rhizosphere of various plants. *Sbor. Českosl. Akad. zeměd. věd. Rostl. vyroba*, 228 (33).

Kasimova, G. S. 1965. Effect of various stimulators on the fixation of atmospheric nitrogen. *Uch. zapiski. Azerb. Gos. Un-ta, Ser. Biol.*, (2).

Katznelson, H. and Strzelczyk, E. 1961. Studies on the interaction of plants and free-living nitrogen fixing micro-organisms. *Canad. J. Microbiol.*, 7, 437–446, 507–513.

Kayser, E. 1920. Influence des radiations lumineuses sur un fixateur d'azote. *C.R. Acad. Sci.*, 171, 969.

Kayser, E. and Delavel, H. 1925. Radioactive et fixateurs d'azote et levures alcooliques. *C.R. Acad. Sci.*, 189 (3), 151–153.

Keeler, R. F., Bulen, W. A. and Varner, J. E. 1956. Distribution of molybdenum-99 in cell free preparations of *Azotobacter vinelandii J. Bacteriol.*, 721, (3).

Keeler, R. F., Carr, L. B. and Varner, J. E. 1958. Intracellular localization of iron, calcium, molybdenum and tungsten in *Azotobacter vinelandii. Exptl. Cell. Res.*, 15.

Khatipova, Kh.M. 1962. Effect of humoammophis on the development of micro-organisms in the soil: in *Guminovye udobrenii, Teoriya i praktika ikh primeneniya* (Humus fertilizers. The theory and practice of their application). Part 2, Gossel'-khozizdat, Kiev, 233–237.

Khmel', I. A. 1966. Effect of environmental factors on the growth and metabolism of *Azotobacter vinelandii* in conditions of continuous cultivation: in *IX Mezhdunarodnyi Mikrobiologicheskii Kongress. Tezisy dokladov* (Ninth International Microbiology Congress. Summaries of speeches), Moscow, 300.

Khmel', I. A., Gabinskaya, K. N. and Ierusalimskii, N. D. 1965. Growth and nitrogen fixation of *Azotobacter vinelandii* in conditions of variable aeration. *Mikrobiologiya*, 34 (4), 689–694.

Kholodnyi, N. G., Rozhdestvenskii, V. S., Kil'chevskaya, A. A. 1945. Assimilation of volatile organic substances by soil bacteria. *Pochvovedenie*, (7).

Kholopov, V. D. 1952. *Effektivnost' azotobakterina v usloviyakh Zapadnoi Sibiri*. (Effectiveness of Azotobacterin in the conditions of western Siberia). Author's abstract of M.Sc. Thesis, Tomsk.

Khudyakov, Ya.B. and Marshchunova, G. N. 1966. Effect of fungi of the genus *Alternaria* on barley yield: in *Ispol'zovanie mikro-organizmov dlya povyssheniya urozhaya sel'sko-khozyaistvennykh kul'tur*. (Use of micro-organisms for raising the yield of farm crops), Kolos, Moscow, 82–85.

Khudyakova, Yu.A. 1950. *Sravnitel'naya kharakteristika kul'tur azotobaktera* (Comparative characterization of *Azotobacter* culture). Author's Abstract of M.Sc. Thesis, Moscow.

Kirakosyan, A. V. Effect of moisture on the fixation of nitrogen by the ecological forms of *Azotobacter*: in *Mikrobiologicheskiis bornik Akad. Nauk ArmSSR*, (Microbiology handbook of the Armenian Akademy of Sciences), (11), 249–260.

Kirakosyan, A. V., Ananyan, L. G. and Melkonyan, Zh.S. 1966. Effect of *p*H of medium on nitrogen fixation of the ecological forms of *Azotobacter chroococcum*. *Vopr. mikrobiologii*, (3), Erevan, Izd-vo Akad. Nauk ArmSSR, 13–23.

Kirakosyan, A. V. and Melkonyan, Zh.S. 1964. Intraspecies relationships of *Azotobacter*: in *Mikrobiologickeskiis bornik Akad. Nauk. ArmSSR*, (Microbiology Handbook of the Armenian Academy of Sciences), (12), 73–86.

Kirsanina, E. F. and Volkova, V. A. 1960. Some information on the distribution of azotobacters in the soils of the Gorno-Altai Autonomous Province. *Mikrobiologiya*, **29** (4), 551–554.

Kirsanina, E. F., Volkova, V. A. and Izotova, A. N. 1965. Effect of trace elements on the development of azotobacters in the meadow black-earth soils of the Gorno-Altai Autonomous Province. *Izvestiya Alt. otd. Vses. geogr. ob-va*, (5), 141.

Klintsare, A. Ya. and Kreslinya, D.Ya. 1963. Effect of trace elements on the development of nitrogen-fixing bacteria in the rhizosphere of farm plants. in *Pochvennaya i sel'sko-khozyaistvennaya Mikrobiologiya* (Soil and agricultural microbiology), Izd-vo Akad. Nauk UzbSSR, Tashkent, 155–162.

Klyushinkova, T. M. and Kvasnikov, E. I. 1966. Formation of vitamins of the B group by azotobacters in conditions of depth cultivation. *Mikrobiol. zhurn.*, **28** (2), 13–17.

Klyushnikova, T. M. 1966. *Sintez vitamino gruppy B azotobakterom v usloviyakh glubinnogo kul'tivirovaniya.* (Synthesis of vitamins of group B by *Azotobacter* in conditions of depth cultivation). Author's abstract of M.Sc. Thesis, Kiev.

Kluyver, A. U. and Becking, J. H. 1955. Some observations on the nitrogen-fixing bacteria of the genus *Beijerinckia* Derx. *Suomaluis. tiedeakat Toimituks.* Ser. A, **11** (60), 367–380.

Koch, A. 1907. Ernahrung der Pflanzen durch frei im Boden lebende stickstoffsammelnde Bakterien. *Mitt. landwirtsch. Ges.*, **12**, 110.

Koleshko, O. I. 1960. Seasonal changes in the development of azotobacters as a function of the presence of microbial antagonists. *Dokl. Akad. Nauk BSSR*, **4** (4), 176–178.

Koleshko, O. I. 1961. Effect of external factors on the morphology of *Azotobacter*: in *Voprosy fiziologii rastenii i mikrobiologii*. (*Problems of Plant Physiology and Microbiology*). 2nd Ed. Minsk, 146–154.

Kolker, I. I. and Dakhnova, E. N. 1960. Distribution of azotobacters in the soils of the Crimea. *Mikrobiologya*, **29** (4), 555–562.

Kononova, M. M. 1930. Microbiological characterization of the soils of certain districts of Central Asia. *Trudy Sr-Az. opytn. issled. in-ta vodnogo khozyaistva*, **3** (15), 3.

Kopčanová, L., Řehořková, V. Effect of drought on the vitality and nitrifying activity of *Azotobacter chroococcum*. *Sbor. Vysokei školy pol'nohospod. Nitre Agron. fak*, (9), 99–104 (in Slovak).

Koshkarova, G. M. 1959. Development of nitrogen-fixing bacteria and the nitrogen regime in cotton-lucerne crop rotation. *Byull. Azerb. n-i. in-ta klopkovodstva.*, (3–4), 41–46.

Kostychev, S. P., Sheloumova, A. M. and Shul'gina, O. G. 1926. Investigations on soil biodynamics. *Trudy Otd. S-kh. mikrobiologii*, **1**, 5–46.

Kotelev, V. V. 1964. Use of ^{32}P for establishing the distribution of azotobacters around the root system of the plant when seeds are bacterized: in *Ispol'zov. mikroIspol'zovanie mikroorganizmov v khozgaistve nardodnom.* (Use of micro-organisms in the national economy). 1, Kishinev, Karta Moldovenyaske, 18–23.

Koval'skii, V. V. and Letunova, S. V. 1966a. Adaptation of *Azotobacter chroococcum* to different contents and rations of molybdenum, vanadium and copper in the soils: in *IX Mezdunarodnyi Mikrobiologicheskii Kongress. Tezisy dokladov* (Ninth International Microbiology Congress. Summaries of speeches) Moscow, 206, *Agrokhimiya*, (7), 73–92.

Kovalskii, V. V. and Letunova, S. V. 1966b. Adaptation of *Azotobacter chroococcum* to the geochemical factors of the habitat. *Dokl. VASKhNIL* (4), 2–6.

Koval'skii, V. V., Letunova, S. V. and Gribovskaya, I. F. 1966. Accumulation of molybdenum, vanadium, and copper by strains of *Azotobacter chroococcum*. *Agrokhimiya*, (9), 56–62.

Krainskii, V. 1908. Assimilation of free nitrogen in the soil by *Azotobacter chroococcum*, its physiological properties and activity in the soil. *Zhurn. opytn. agronomii*, **9**, 689.

Krasil'nikov, H. A. 1931. History of development of *Azotobacter* in connection with the problem of polymorphism. *Mikrobiol. zh.*, **12** (1), 16.

Krasil'nikov, N. A. 1949. *Opredelitel Ibacterii i aktinomitsetov.* (Classification of bacteria and actinomycetes), Izd. Akad. Nauk SSSR, Moscow and Leningrad.

Krasil'nikov, N. A. 1958. *Mikroorganizmy pochvy i vysshie rosterit* (Soil micro-organisms and higher plants). Izd-vo Akad. Nauk SSSR, Moscow.

Krasil'nikov, N. A., Boltyanskaya, E. V., Sokolov, A. A., Melkonyan. Zh. 1965. Flagellate-like outgrowths in *Azotobacter*. *Dokl. Akad. Nauk SSSR*, **164** (4), 931–933.

Krasil'nikov, N. A., Drobkov, A. A., Shirokov, O. G. and Shevyakova, N. I. 1955a. *O deistvii radioaktivnykh elementov na razvitie kluben'kovykh bakterii i azotobaktera. Mikroelementy. Tezisy dokladov Vsesoyuznogo soveshchanii po mikroelementam.* (Action of radioactive elements on the development of nodule bacteria and *Azotobacter*. Trace elements. Summaries of speeches of the All-Union Conference on Trace-Elements). Riga.

Krasil'nikov, N. A., Drobkov, A. A., Shirokov, O. G. and Shevyakova, N. I. 1955b. Toxicosis of podzolic soils. *Izv. Akad. Nauk SSSR, ser. biol.*, (3), 33–48.

Kreslin', D. Ya. 1961. Change in the reaction (pH) of the nutrient medium in relation to the intensity of growth of local *Azotobacter* strains. *Trudy In-ta mikrobiologii Akad. Nauk LatvSSR*, (14), 49–63.

Krishna Bahadur, Krishna, A. and Murari Lal. 1965. Influence of calcium ions on growth and nitrogen fixation in cultures of the nitrogen-fixing micro-organism *Azotobacter chroococcum* strain no. 120/21/2B/B, at pH values 6·5, 7·0 and 7·5 *Zbl. Bakteriol., Parasitenkunde, Infektionskrankh und Hyg.*, Abt 2., **119** (3), 245–255.

Krongauz, R. A. 1959. Ways of improving the adaptation and effectiveness of *Azotobacter* in northern black earths. *Mikrobiologiya*, **28** (2), 242–245.

Krupina, L. I. 1960. Formation of heteroauxin by various strains of *Azotobacter*. *Byull. nauchno-tekhn. po s-kh. mikrobiologii. Leningrad*, (8), 11–13.

Krylova, N. B. 1962. Role of micro-organisms in nitrogen fixation. *Izv. Akad. Nauk SSSR, Ser. Biol.*, (5), 718–731.

Krylova, N. B. 1963. Effect of molybdenum and vanadium on nitrogen fixation. *Dokl. TSKhA*, (84), 293–299.

Krylova, N. B. 1964. *Vliyanie molibdena, vanadiya i drugikh mikro-elementov na fiksatsiyu atmosfernogo azota nekotorymi svobodno-zhivushchimi mikro-organizmov.* (Effect of molybdenum, vanadium and other trace elements on the fixation of atmospheric nitrogen by certain free-living micro-organisms) Author's abstract of M.Sc. Thesis, Moscow.

Krzemieniewski, S. 1907. Physiologische Untersuchungen über *Azotobacter chroococcum* *Bull. internat. Acad. sci. Cracowie*, 746.

Krzemieniewski, S. 1908. Physiologische Untersuchungen über *Azotobacter chroococcum* *Bull. Internat. Acad. sci. Cracowie*, 333, 929.

Kurdina, E. S. 1951. Effect of humic acid on the growth of groups of soil micro-organisms and its importance for these micro-organisms as a source of nutrition. *Trudy In-ta mikrobiologii Akad. Nauk SSSR*, 3.

Kurdina, R. M. 1955. *Azotobacter i ego vzaimootnosheniya s vysshimi rasteniyami* (*Azotobacter* and its relationships with higher plants). Author's abstract of M.Sc. Thesis. Alma Ata.

Kvasnikov, E. I. 1951. *Azotobacter* in the wetted soils of Uzbekistan with the grass-arable system of crop rotation. *Dokl. Akad. Nauk UzbSSR*, (5), 35–40.

Kvasnikov, E. I. Petrushenko, O. P. and Zhvachkina, A. A. 1959. Bacterial nitrogen fixers in the soils of Uzbekistan: in *Mikrobiologiya na sluzhbe sel'skogo khozyaistva.* (Microbiology in the service of agriculture). Sel'khozgiz, Moscow, 115–120.

Lasting, V., Kaarli, L. and Gurfel, D. 1966. Microflora of various soils of North-Estonian provenance and related soil cultures. *Teaduslike tööde kogumik*, 8, *Microbroloogia*, Tallin, 7–59 (in Estonian).

Lazarev, S. F. 1964 Microbiological processes in the soils of the natural zones of Central Asia. *Agrobiologiya*, (1), 29–36.

Lee, S. B., Burris, R. H. 1943. Large-scale production of *Azotobacter*. *Industr. and Engng. Chem.*, 35, 354–357, 381.

Letunova, S. V. 1958. Formation of vitamin B_{12} by various species of Actinomycetes and bacteria isolated from the soils of geochemical cobalt provinces. *Mikrobiologiya*, 27 (4).

Lewis, J. H. 1937. Cell inclusions and the life cycle of *Azotobacter*. *J. Bacteriol.*, 34, 191–205.

Ley, J. de and Park, I. W. 1966. Molecular biological taxonomy of some free-living nitrogen-fixing bacteria. *Antonie van Leeuwenhoek. J. Microbiol and Serol.*, 32, 6–16.

Lin Liang Ping and Wyss, O. 1965. Synchronization of *Azotobacter* cells. *Texas Repts. Biol. and Med.*, 23 (2), 474–480.

Litovchenko, E. R. 1965. The soil microflora of an apple orchard in the Crimea: in *Sadovodstvo* (Fruit farming), Urozhai, Kiev, 114–125.

Löhnis, F. 1921. Studies of the life cycles of the bacteria. *Mem. Nat. Acad. Sci.*, 16 (2), 1.

Löhnis, F. and Pillai, N. K. 1908. Über stickstoffixirende Bakterien *Zbl. Bakteriol., Parasitenkunde. Infektionskrankh und Hyg.* Abt. 2, 20, 781.

Löhnis, F. and Smith, L. 1916. Life cycle of the bacteria. *J. Agric. Res.*, 6, 675.

Lopatina, G. V. 1949. Effectiveness of various *Azotobacter* cultures: in *Trudy Vses. n-i. in-ta s-kh. mikrobiologii 1941–1945 gg.* (Proceedings of the All-Union Scientific Research Institute of Agricultural Microbiology for 1941–1945). Sel'khozgiz, 108–112.

Macurá, J. 1955. Cultivation of *Azotobacter* by different methods. *Československ. Biol.*, 4, 274 (in English).

Macurá, J. and Kotkova, M. 1958. Cultivation of *Azotobacter* in current media. *Československ. Biol.*, 2 (1), 41–49.

Macurá, J. and Stavělii, Z. 1953. *Azotobacter* glucose exchange. *Československ. biol.*, 2, 154–161.

Makrinov, M. A. and Troitskii, V. B. 1931, *Ispol'zovanie mikro-organizmov dlya podnyatiya proizvodstel'nosti pochv.* (Use of micro-organisms for increasing soil productivity) Sel'khozgiz, Moscow.

Malek, J. 1962. Cultivation of *Azotobacter* in current media. *Československ. biol.*, 1 (2), 91–92.

Maliszewska, W. 1963. Inoculation of plants by *Azotobacter*. *Roczn. nauk roln.* A, 68, (1), 33. *Acta Microbiologica Polonica*, (2).

Maliszewska, W. 1961. Inoculation of plants by *Azotobacter*. *Roczn. nauk roln.* D, 91, 103–235.

Malyshkin, P. E. and Prynova, M. M. 1965 (1966). The microflora of the rhizosphere of tomatoes, cabbage and of meadow-soddy soil of the Volgo-Akhtubinsk alluvial plain. *Trudy Volgogradsk. opytn. stantsii Vses. n-i. in-ta rastenievodstva*, (4), 131–139.

Mandel, R. K. 1963. Effects of some antibiotics and antimetabolites on protein and nucleic acid synthesis in *Azotobacter vinelandii* and *Lactobacillus arabinosus*. *Trans. Rose. Res. Inst.*, 26 (2), 43–54.

Manil, P. 1962. Vegetaux et microbes du soleation des *Azotobacter*. *Nature Sci. Progr.* (3331), 473–474.

Manil, P. and Brakel, J. 1965. *Azotobacter* et rhizosphere. A propos de quelques essais effectués en Belgique. *Compt. Rend. Soc. Biol.*, 159 (4), 1025–1027.

Manuca, L. and Georghiu, V. 1960. Auxinic effect of cultures of *Azotobacter chroococcum*. *Probleme Agric.*, 12, 23–30.

Marendiak, D. 1964. Questions of the dynamics and quantitative relationships of certain strains of micro-organisms in the rhizosphere of maize. *Shor. Vysokei školy pol'nohospod. Nitre Agron. fak*, (71), 161–169.

Mar'enko, V. G. 1963. Fungistatic action of *Azotobacter chroococcum*. *Dokl. TSKhA*, (74), 301–306; (89), 328–333.

Mar'enko, V. G. 1964. Dependence of the maize harvest and nitrogen balance on *Azotobacter chroococcum* in a monobacterial culture. *Dokl TSkhA*, (99), 399–406.

Margo, A. A. 1964. Physiologically active substances of azotobacters. in *Issledovanie pochv i primenenie udobrenii*. (Investigation of soils and use of fertilizers). Urozhai, Minsk, 40–50.

Marszewska-Ziemiecka, J. and Maliszewska, W. 1963. Mutual distribution of *Azotobacter* and soil amoebae. *Pamiętnik Palawski*, 2, 35–48.

Martin, J. P. 1945. Some observations on the synthesis of polysaccharides by soil bacteria. *J. Bacteriol.*, 50, 349–360.

Martin, J. P., Ervin, J. O. and Shepherd, R. A. 1965. Decomposition and binding action of polysaccharides from *Azotobacter indicus* (*Beijerinckia*) and other bacteria in soil. *Soil Sci. Soc. America Proc.*, 29 (4), 397–399.

Martin, J. P., Ervin, J. O. and Shepherd, R. A. 1966. Decomposition of the iron, aluminium, zinc, and copper salts or complexes of some microbial and plant polysaccharides in soil. *Soil Sci. Soc. America Proc.*, 30 (2), 196–200.

Martin, W. P., Walker, R. H. and Brown, P. E. 1937. The occurrence of *Azotobacter* in Iowa soils and factors affecting distribution. *Res. Bull.*, 217, 227–256.

Martinez, V. R. 1964. Influencia de la rizosfera del maiz en la microflora. *Poeyana. Inst. biol.*, A, (1), 1–11.

Matuashvili, S. I. 1947. Effect of boron and molybdenum on the morphological and physiological properties of *Azotobacter chroococcum*. *Mikrobiologiya*, 16 (1), 19–31.

Mazilkin, I. A. 1956. Microbiological characterization of the soils of the Olekminsk district. *Trudy Yakutsk. fil. Akad. Nauk SSSR*, (1), 135–175.

Me Mao-Ying, 1961. Effectiveness of azotobacterin, *Tuzhang*, (7), 55–77 (in Chinese).

Meiklejohn, J. 1962. Microbiology of the nitrogen cycle in some Ghana soils. *Empire J. Exptl. Agric.*, 30 (11b), 115–126.

Mekhtiev, S. Ya. 1966. The microflora of the soils of Moldavia: in *Mikroflory pochv severnoi i srednei chasti SSSR*. (The microflora of the soils of the northern and central parts of the USSR), Nauka, Moscow, 274–296.

Mekhtiev, N. A. and Melkumova, T. A. 1962. Effect of trace elements on the development of azotobacters in certain soils of Azerbaijan. *Dokl. Akad. Nauk AzerbSSR*, 18 (10), 59–63.

Mencl. 1911. Die Kernäquivalente und Kerne bei *Azotobacter chroococcum* und seine Sporenbildung. *Arch. Protistenkunde*, 22, 1.

Meshkov, N. V. 1959. Effect of thiamine and biotin on the development of certain soil microbes. *Mikrobiologiya*, 28 (16), 895–899.

Meshkov, N. V. and Khodakova, R. N. 1966. Biological fixation of atmospheric nitrogen in a vegetation experiment on soddy podzolic soil: in *Tezisy dokladov III Vsesoyuznogo s"ezda pochvovedov*. (Summaries of speeches of the third All-union conference of pedologists) Tartu, 88–89.

Mickovski, M., Popovski, D. 1960. Microbiological features of red-earths of the Macedonian People's Republic. *Arh. poljopr. nauke*, **13** (40), 23–26 (in Serbo-Croat).

Milošević, R. 1963. Distribution of *Azotobacter* in some sandy soils with various vegetation. *Zemljište i biljka*, **12** (1–3), 357–366 (in Serbo-Croat).

Minasyan, A. I. and Nalbandyan, A. D. 1961. The microflora of the semi-arid stony soils—'kirov' and their modifications during reclamation *Agrobiologiya*, (6), 842–848.

Minchenko, A. Ya. 1965. Effect of certain agrotechnical procedures on the development of *Azotobacter*. *Trudy Vses. n-i. in-ta khlopkovodstva UzbSSR*, (6), 166–173.

Mirotvorskii, K. A. 1945. Azotobacters of the soils of Turkmenia. *Izv. Turkm. fil. Akad. Nauk SSSR*, (5–6), 118–123.

Mishustin, E. N. 1937. Effect of intracellular pressure in soil bacteria. *Mikrobiologiya pochvy* (Soil microbiology) *Trudy VIUA*, **2**, 103–118.

Mishustin, E. N. 1953. Principle of regionalization of *Azotobacter*: in *Rol' mikroorganizmov v pitanii rastenii*. (The role of micro-organisms in plant nutrition). Sel'khozgiz, Moscow.

Mishustin, E. N. 1954. Ecologogeographic distribution of azotobacters in the soils of the USSR. *Trudy in-ta mikrobiologii Akad. Nauk SSSR*, (III), 81–97.

Mishustin, E. N. and Bakhareva, Z. I. 1939. Soil acidity as a factor determining the appearance in the soil of inactive azotobacters. *Mikrobiologiya*, **8** (9–10), 1063–1072.

Mishustin, E. N. and Krylova, N. B. 1965. Molybdenum requirements of free-living, nitrogen-fixing bacteria. *Mikrobiologiya*, **34** (4), 683–688.

Mishustin, E. N., Naumova, A. N. and Mar'enko, V. G. 1963. *Azotobacter* and its effectiveness. *Izv. TSKhA*, **4** (53), 42–54.

Mishustin, E. N. and Petrova, A. N. 1958. Antagonist relations of the micro-organisms of the rhizosphere and the effectiveness of azotobacterin: in *Poluchenie i primenenie bakterial'nykh udobrenii*. (The preparation and application of bacterial fertilizers). Izd-vo Akad. Nauk UkrSSR, Kiev.

Mishustin, E. N. and Semenovich, M. I. 1939. Soil acidity as a factor determining the appearance in the soil of inactive azotobacters. *Mikrobiologiya*, **8** (1), 19–32.

Missirliu, E., Papacostea, P., Preda, C., Manolescu, V., Popa E. and Cămírzan, C. 1964. Contribution to the study of Alpine acoperite soil microflora in association with Nardetum strictae from the Bucegi massif. *Studii tehn. şi econ. C.*, (12), 67–93.

Mitrofanova, N. S. 1963. Microbiological characterization of saline and alkaline soils. *Trudy In-ta pochvoved. Akad. Nauk KazSSR*, **15**, 85–104.

Molina, J. S. and Quant, J. 1960. Fuentes de carbono utilizadas por el *Azotobacter* en places de tierra moldeada. III. Substancias volatiles producidas en fermentaciones anaerobias *et al*. *Suele Cienc e invest.*, **16** (12), 474–476.

Monsour, V., Wyss, O. and Kellogg, D. S. 1955. A bacteriophage for *Azotobacter*. *J. Bacteriol*, **70** (7), 486–487.

Moore, A. W. 1963. Occurrence of non-symbiotic nitrogen-fixing micro-organisms in Nigerian soils. *Plant and Soil*, **19**, 385–395.

Moreau, R. and Gams, W. 1960. Note préliminaire sur la microflore de quelques sols Alpins. *C.R. Acad. Sci.*, **251** (15), 1560–1562.

Mulder, E. G. 1939 (1940). On the use of micro-organisms in measuring a deficiency of copper, magnesium and molybdenum in soils. *Antonie van Leeuwenhoek, J. Microbiol. and Serol.*, **6**, 99–109.

Mulder, E. C. 1957. Spore-elements and micro-organisms. *Landbouwkund. tijdschr.*, **69** (7–8), 575–591.

Naguib, A. I. 1963 (1965). Note on the comparative study of the microflora content of some locally available organic products of manurial value. *UAR. J. Bot.*, **6**, 101–103.

Nakhimovskaya, M. I. 1941. Synthesis of growth substances of the 'bios' type by soil bacteria. *Mikrobiologiya*, **10** (6), 688–700.

Naplekova, N. N. 1959. Effect of temperature on the development of azotobacters. *Izv. Sibirsk. otd. Akad. Nauk SSSR*, (11), 69–72.

Nasyrova, Z. A. 1966. Effect of bacterization of jougbara seeds on the development of plants in saline soils: in *Voprosy mikrobiologii.* (Problems of microbiology), Nauka, Tashkent, 38–42.

Naumova, A. N., Mishustin, E. N. and Mar'enko, V. G. 1962. Nature of the action of bacterial fertilizers. *Izv. Akad. Nauk SSSR, Ser. Biol.*, (5), 709–717.

Nepomiluev, V. F. and Shishov, L. L. 1962. Distribution and nitrogen-fixing activity of azotobacters in certain types of soil. *Mikrobiologiya*, **31** (2), 294–300.

Newton, J., Wilson, P. and Burris, R. 1953. Direct demonstration of ammonia as an intermediate in nitrogen fixation by *Azotobacter. J. Biol. Chem.*, **204**, 445.

Nickell, L. G. and Burkholder, P. R. 1947. Inhibition of *Azotobacter* by soil Actinomycetes. *J. Amer. Soc. Agron.*, **39**, 9.

Niţă, L. 1965. Activating micro-organisms. *Natura, Ser. biol.*, **17** (2), 14–20.

Norkina, S. P. 1953. Search for a rational method of drying *Azotobacter* cultures. *Trudy Vses. n-i. in-ta s-kh. mikrobiologii*, **8**, 131–140.

Norris, J. R. 1960. *Notes on the classification of the family Azotobacteria ceae.* Glasgow. Dept. Bacteriol.

Norris, J. and Jensen, H. L. 1958. Calcium requirement of *Azotobacter. Arch. Mikrobiol.*, **31**, 198.

Novikova, A. T. and Irtuganova, L. D. 1966. Formation of heteroauxin by soil micro-organisms. *Mikrobiologya*, **35** (4), 701–711.

Novogrudskii, D. M. and Naumova, A. N. 1932. Causes of the inactivity of *Azotobacter. Mikrobiologiya*, **1** (2), 181–191.

Nutman, P. S. 1964. *Soil Microbiology Department. Rept. Rotnamsted. Exptl. Stat. for 1964.*, 85–97.

Okuda, A. and Yamuguchi, M. 1960. Nitrogen-fixing micro-organisms in paddy soils., 6. *Soil and Plant Food.*, **6** (2), 76–85.

Olaru, D. A. 1920. *Contribution a l'étude du role du manganese en agriculture.* Theses Fac. sci. Univ. Paris.

Omelyanskii, V. L. 1915. The physiology and biology of nitrogen-fixing bacteria. *Atkhiv biol. nauk*, (19).

Omelyanskii, V. L. 1923. *Svyazvanie atmoyfernogo azota pochvennymi mikroorganizmov.* (Fixing of atmospheric nitrogen by soil micro-organisms). Selected works, Izd-vd. Akad. Nauk SSR, Moscow, 1951, ch 6.

Omelyanskii, V. L. 1926. *Osnovy mikrobiologii.* (Fundamentals of microbiology), Moscow-Leningrad.

Ovcharov, K. E. 1953. Importance of vitamins in the vital activity of plants. *Uspekhi sovr. biologii*, **36** (3).

Ovcharov, K. E. *Vitaminy v zhizni rastenii.* (Vitamins in the life of plants). Izd-vo Akad. Nauk SSR, Moscow.

Pakarskite, K. Yu. 1961. Vitamins B_1, B_2 and B_{12} and folic acid in certain natural sources: in *Trudy I biokhimicheskoi konferentsii Pribaltiiskikh respublik i Belorussii*, 1960. (Proceedings of the first biochemical conference of the Baltic Republics and Byelorussia, 1960), Tartu, 248–257.

Pakusin, A. G. 1966. The microflora of the soils of Azerbaijan. in *Mikroflory pochv yuzhoi chasti SSSR.* (The soil microflora of the southern part of the USSR), Moscow, Nauka, 49–77.

Panosyan, A. K. *et al.* 1962. Relationships of azotobacters and other soil micro-organisms. *Izv. Akad. Nauk ArmSSR, Ser. Biol.*, **15** (2), 13–24.

Pares, Y. and Afanchao, A. M. 1960. Pouvoir antibiotique d'*Alternaria humicola* Oudemans. *C.R. Acad. sci.*, **250** (14), 2601–2602.

Parker, C. A. 1954. Effect of oxygen on the fixation of nitrogen by *Azotobacter*. *Nature*, **173**, 780.

Parker, C. A. and Scutt, P. B. 1958. Competitive inhibition of nitrogen fixation by oxygen. *Biochim. Biophys. Acta*, **29** (3), 662.

Parker, C. A. and Scutt, P. B. 1960. The effect of oxygen on nitrogen fixation by *Azotobacter*. *Biochim. Biophys. Acta*, **38** (2), 230–238.

Parker, L. T. and Socolofsky, M. D. 1966. Central body of the *Azotobacter* cyst. *J. Bacteriol.*, **91** (1), 297–303.

Pashakov, G. Khr. Distribution of *Azotobacter* in the rhizosphere of vines for different soils and its nitrogen-fixing capacity: in *Nauchni trudy n-i in-ta lozarstvo i vinarstvo*. (Scientific proceedings of the Scientific Research Institute of Viticulture and Wine Production), Vol. 2, Pleven, 342–352.

Pathak, K. N. and Shrikhandel, I. G. 1953. Optimum temperature for nitrification and nitrogen fixation. *J. Indian Soc. Sci.*, (1).

Pavlovicha, D. Ya. 1953. *Rasprostranenie azotobaktera v pochvakh Latuiiskoi SSR* (The Distribution of azotobacters in the soils of the Latvian SSR), Author's abstract of M.Sc Thesis Riga.

Per, F. L. 1957. *Usloviya fiksatsii svobodnogo azota bakteriami grup 'Azotobacter' i 'Clostridium' v ozerakh Latviiskoi SSR.* (Conditions for the fixation of free nitrogen by bacteria of the *Azotobacter* and *Clostridium* groups in the lakes of the Latvian SSR), Authors abstract of M.Sc. Thesis. Riga.

Peshakov, G. 1965. Development of micro-organisms on the root surface of the grape vine and the effect of fertilizers on the rhizosphere microflora: in *Plant Microbes Relationship*. Czechosl. Acad. Sci., Prague, 120–125.

Petersen, E. J. 1961. Studies on the morphology and cytology of *Azotobacter chroococcum*. *Kgl. Vet. Landbohogskol. Arsskr.*, 115–155 (in English).

Petkov, P. D. 1964. Effect of non-rotation of wheat, barley and oats regarding the quantity of certain types of micro-organisms on it. *Rastenievădni nauki*, **1** (8), 89–92 (in Bulgarian).

Petrenko, G. Ya. 1961. Factors influencing the symbiotic relationships of azotobacters and higher plants: in *Mikro-organizmy i effektivnoe plodorodie pochvy*. (Micro-organisms and effective soil fertility). (11), 11–129.

Petrenko, G. Ya. 1966. Use of gaseous atmospheric nitrogen by non-leguminous crops with the aid of azotobacters: in *IX Mezhdunarodnyi Mikrobiologischeskii Kongress. Tezisy dokladov.* (Ninth International Microbiology Congress. Summaries of speeches), Moscow, 301.

Petrushenko, O. P. 1961. *Azotobakter v pochvakh Uzbekistana i nekotorye osobennosti ego biologii.* (Azotobacters in the soils of Uzbekistan and some special features of its biology), Author's abstract of M.Sc. Thesis. Tashkent.

Petschenko, P. 1930. Über die Biologie und Morphologie un den Entwicklungszyclus von micro-organismen aus der *Azotobacter*—Gruppe. *Zbl. Bakteriol. Parasitenkunde. Infektionskrankh. und. Hyg.*, Abt. 2, **80**, 161.

Pochon, J. and Barjac, H. de 1958*b*. *Traité de microbiologie des soils*. Dunod., Paris.

Pochon, J. Chalaust, R. and Tchan, Y.T. 1948. Inoculation d'*Azotobacter* dans le sol. Fixation d'azote et croissance des végétaux. *Acad. d'Agriculture de France. Extrait du proces verbal de la sceance du* 30 *juin.*

Pochon, J. and Chalvignac, M. A. 1964. Bacterisation des graines et physiologie de la plante. I. Action de quelques éspeces bacteriennes sur la croissance du lin. *Ann. Inst. Pasteur*, (3), Suppl., 242–249.

Pochon, J. and Coppier, O. 1951. *Ann. Agron*, (4), 425. (Quoted by Pochon and Barjac, 1960, Moscow, 514).

Pochon, J., Tchan, Y. T. and Wang, T. L. 1948. Recherches sur le cycle morphologique et l'appareil nucleare des *Azotobacter*. *Ann. Inst. Pasteur*, 14 (3), 182.

Pokorná-Kozová, J. 1965. *Azotobacter* in the presence of various organic and mineral fertilizers. *Sbor. Českosl. akad. zeměd. věd. Rostl. vyroba*, 11 (9), 985–992.

Pražmowski, A. 1912. Die Entwicklungsgeschichte, Morphologie und Cytologie des *Azotobacter chroococcum* Beijer. *Zbl. Bakteriol., Parasitenkunde, Infektionskrankh. und. Hyg.*, Abt. 2, 33, 392.

Pridham, T. G. 1952. Microbial synthesis of riboflavin. *Econ. Botany*, 6, 185–206.

Pringsheim, H. 1909. Über die Verwendung von Zellulose als Energie quelle zur Assimilation des Luftstickstoffes. *Zbl. Bakteriol., Parasitenkunde, Infektionskrankh. und Hyg.*, Abt. 2, 23, 300.

Proctor, M. H. 1959. A function for the extracellular polysaccharide of *Azotobacter vinelandii*. *Nature*, 184, (4703) suppl. 25, 1934–1935.

Prsă, M. 1963. The cyclical development of biotypes of azotobacters isolated from terra rossa of Instria. *Bull. Sci. Conseil Acad. RSFY*, 8 (3-4), 89 (in English).

Pshenin, L. N. 1964. *Azotobacter miscellum* nov. sp. inhabiting the Black Sea. *Mikrobiologiya*, 33 (4), 684–691.

Pushkinskaya, O. I. 1966. The microflora of solonetz: in *Mikroflora pochv severnoi i srednei chasti SSSR*. (Microflora of the northern and central part of the USSR), Nauka, Moscow, 297–351.

Quinell, M. C. Knight, S. C. and Wilson, P. W. 1957. Polysaccharide from *Azotobacter indicum*. *Canad. J. Microbiol.*, 3, 277–288.

Rabotnova, I. L. 1957. *Rol'fiziko-khrmicheskikh uslovii* (pH *i* rH$_2$) *v zhiznedeyatel'nosti mikroorganizmov*. (The role of physiochemical conditions (pH and rH$_2$) in the vital activity of micro-organisms), Izd-vo Akad. Nauk SSSR, Moscow.

Rabotnova, I. L. Zaitseva, G. N. and Mineeva, L. A. 1959. Study of the lag-phase in the micro-organisms. III. Changes in the cells of azotobacters growing on molecular and ammoniacal nitrogen. *Mikrobiologiya*, 28 (5), 683–689.

Rabotnova, I. L. and Rodionova, G. S. 1953. Enrichment of the substrate with nitrogenous substances during development of *Azotobacter beijerinckii*. *Mikrobiologiya*, 22 (4), 414–422.

Rakhno, N. Kh. 1953. *Rolazotobaktera v povysshenii urozhainosti polevykh kul'tur v pochvennykh i klimaticheskikh usloviyakh Estonskoi SSR*. (Role of *Azotobacter* in increasing the yield of farm crops in the soil and climatic conditions of the Estonian SSR), Author's abstract of M.Sc. Thesis, Leningrad.

Rangaswami, G. and Sadasivam, K. V. 1964. Studies on the occurrence of *Azotobacter* in some soil types. *J. Indian Soc. Soil Sci.*, 12 (1), 43–49.

Rankov, V. 1964. The salination of the soil and the development of the nitrogen fixation micro-organisms. *Agrochimica*, 2 (4), 330–342.

Rankov, V. 1965. Effect of salinization of the soil on the mutual relationship of *Azotobacter* with other types of soil micro-organisms. *Rastenievădni nauki*, 2 (3), 107–114 (in Bulgarian).

Rao, W. V. B. and Sundara and Iswaran, V. 1959. Nitrogen fixation by *Azotobacter chroococcum* in culture. *J. Indian Soc. Soil Sci.*, 7, 91–95.

Ratner, E. I. and Dobrokhotova, I. N. 1956. Possible role of vitamins produced by soil micro-organisms in the root nutrition of plants. *Fiziol. rast.*, 3 (2), 101–109.

Ratner, E. I. and Dobrokhotova, I. N. 1958. Identification of the nature of the effect of vitamins on the synthetic activity of the roots on assimilation by the plant of mineral nitrogen. *Dokl. Akad. Nauk SSSR*, 122 (5), 944–947.

Raznitsyna, E. A. 1958. Formation by bacteria of growth substances of the auxin group. *Dokl. Akad. Nauk. SSSR*, 18 (6), 353–356.

Regel, S. de. 1932. Entwicklungsgang von *Azotobacter chroococcum*. *Zbl. Bakteriol., Parasitenkunde, Infektionskrankh. und Hyg.*, Abt. 2, **86**, 44.

Remacle, J. 1966. Evolution d'une population d'*Azotobacter chroococcum* dans la rhizosphere de l'orge en présence d'un bacillus pectinolytique. *Ann. Inst. Pasteur*, **111** (3) Suppl., 149–154.

Rice Elroy, L. 1965. Inhibition of nitrogen-fixing and nitrifying bacteria by seed plants. Characterization and identification of inhibitors. *Physiol. plantarum*, **18** (1), 255–268.

Rodina, A. G. 1954. Effect of plant fertilizers on the process of nitrogen fixation and the results of use of azotogen in fish ponds. *Mikrobiologiya*, **23** (6), 684–692.

Romanow, J. 1965. Recherches sur la synthèse de l'acide indole-3-acetique par les Azotobacter. *Ann. Inst. Pasteur Supp.*, (3), 280–292.

Rouqueroi, Th. 1962. Sur le phenomene de fixation de l'azote dans les rizieres de camargue. *Ann. agron.*, **13** (4), 325–346.

Rovira, A. D. 1965. Effects of *Azotobacter*, *Bacillus* and *Clostridium* on the growth of wheat. in *Plant Microbes Relationship*, Prague, Czechosl. Acad. Sci., 193–200 (in English).

Rubenchik, L. I. 1960. *Azotobakter i ego primenenie v sel'skom khozyaistve*, (*Azotobacter* and its use in agriculture), Izd-vo Akad. Nauk UkrSSR, Kiev.

Rubenchyk, L. I. and Royzyn, M. B. (Rubenchik, L. I. and Royzin, M. B.). 1960. Effect of bound nitrogen on *Azotobacter*: in *Minlyvist' mikrobiv i bakteriofahiya* (Variability of microbes and bacteriophage), Vyd-vo Akad. Nauk UkrSSR). (in Ukranian).

Rubenchik, L. I., Smalii, V. T. and Zinov'eva, Kh. G. Formation of vitamins by micro-organisms of the rhizosphere of farm plants: in *Rol' mikroorganizmov v pitanii rastenii i povysshenii effektionosti udobrenii* (Role of micro-organisms in plant nutrition and efficiency of fertilizers). Leningrad, 14–21.

Rusakova, O. C. 1965. Accumulation of vitamins in the soil under the action of various strains of *Azotobacter*: in *Vykorystannya zroshuvanykh zemel'*. (Use of irrigated land), Urozhai, Kiev, 139–142. (in Ukranian).

Rybalkina, A. V. 1937. Connection between the redox potential in laboratory cultures of *Azotobacter chroococcum* Beijer and its stages of development. *Mikrobiologiya*, **7** (3).

Rybalkina, A. V. 1938. Survival of cultures of *Azotobacter chroococcum* Beijer. in peat. *Mikrobiologiya*, **7** (8), 933–936.

Ryys, O. O. 1963. Distribution of free-living, nitrogen-fixing bacteria in the soddy podzolic soils of the Estonian SSR. *Izv. Akad. EstSSR, Ser. Biol.*, **12** (4), 274–281.

Ryys, O. O. 1966. Correlation between the number of bacteria and the properties of soils: in *Tezisy dokladov III Vsesoyuznogo s"ezda pochvovedov*. (Third All-Union Congress of pedologists. Summaries of speeches). Tartu, 102–103.

Sabel'nikova, V. I. 1960. Dependence of the development of azotobacters on the presence of $CaCO_2$ in the soils of Moldavia. *Izv. Mold. fil. Akad. Nauk SSSR*, (6), 75–80.

Sabel'nikova, V. I. 1965. Patterns of distribution of azotobacters in the black earth of Moldavia (in a cross section of sub-types): in *Mikrobiologicheskie protsessy v pochvakh Moldavii*, (Microbiological processes in the soils of Moldavia). (2), Kishinev, Kartya Moldovenskyaske, 29–41.

Sabinin, D. A. and Vilovskaya, A. E. 1931. Guza-pai as a fertilizer in cotton growing. *Za khlopkovuyu nezavisimost'*, (2), 55–62.

Sadimenkova, L. V. 1965. Effect of windbreak forest plantations on the microflora of the soil: in *Novosti agrolesomelioratsii*, (New developments in agro-forest improvement). *Vses. n-i. inta. agrolesomelior.*, (49), Volgograd, 121–124.

Samosova, S. M., Munina, A. A. and Kiprova, R. R. 1964. Effect of trace elements introduced into the soil for raising the frost hardiness of winter wheat on the microflora of the rhizosphere. *Uch. zapiski Kazansk. un-ta*, **124** (8), 44–59.

Sandrak, N. A. 1958. Chemical composition and accessibility to plants of nitrogenous secretions. *Dokl. TSKhA*, (34), 101.

Sarić, Z. and Raśović, B. 1963. Influence of maize on the dynamics of *Azotobacter* in the soil. *Zemljište i biljka*, 12 (1–3), 273–277.

Sasson, A. and Daste, Ph. 1961 (1962). Observations concernant l'écologie de l'*Azotobacter* dans les sols arides du Maroc. *C.R. Soc. Biol.*, 155 (10), 1997–2002.

Sasson, A. and Daste, Ph. 1963. Observations nouvelles concernant l'écologie del '*Azotobacter* dans certains sols arides du Maroc. *C.R. Acad. sci.*, 257, 3516–3518.

Saubert, S. and Grobbeiaar, N. 1962. The identification and nitrogen fixation of some free-living micro-organisms from the Northern Transvaal. *S. Afrik. tydskr. landbouwetensk.*, 5 (2), 283–292 (in English).

Savel'eva, L. V. 1954. *Vliyanie sposobov vneseniya granulirovannogo superfosfata i azotobakterina na fiziologicheskie svoistva i urozhai ozimnoi pshenitsy* (Effect of methods of introducing granulated superphosphate and azotobacterin on the physiological properties and yield of winter wheat). Author's abstract of M.Sc. Thesis, Moscow.

Savkina, E. A. 1958. *Rizosfernye bakterii kukurizy, ikh fiziologicheskire osobennosti i uzaimootnosheniya s rasteniem i azotobakterom* (Rhizosphere bacteria of maize, their biological features and relationship with plants and *Azotobacter*). Author's abstract of M.Sc. Thesis, Moscow.

Schäfler, S., Voiculescu, R., Toma, E. and Nas, L. Experimental variability of *Azotobacter chroococcum*. *Bul. ştiint. Acad. R.P.R. Soc. biol., geol., si geogr.*, 6, 571.

Schlüter, M. and Bukatsch, F. 1959. Physiologische Untersuchungen zur Rassenfrage von *Azotobacter*. *Zbl. Bakteriol. Parasitenkunde, Infektionskrankh. und Hyg.*, Abt. 2 (112), 16–20, 509–543.

Schmidt-Lorenz, W. and Rippel-Baldes, A. 1958. Wirkung des Sauerstoffpartialdruckes auf Wachstum und Stickstoffbindung von *Azotobacter chroococcum* Beij. *Arch. Mikrobiol.*, 28, 45.

Schreven, D. A. van. 1962. Effect of the composition of the growth medium on morphology and reproduction of *Azotobacter chroococcum*. *Antonie van Leeuwenhoek. J. Microbiol and Serol.*, 28 (2), 97–120.

Schreven, D. A. van. 1963. The effect of some Actinomycetes on the growth and nitrogen-fixing power of *Azotobacter chroococcum*. *Plant and Soil*, 19 (1), 1–18.

Schreven, D. A. van. 1966. Effect of penicillin on the morphology and reproduction *Azotobacter chroococcum*. *Antonie van Leeuwenhoek. J. microbiol. and Serol.*, 32 (1), 67–93.

Schroder, M. 1932. Die Assimilation des Luftstickstoffs durch einige Bakterien. *Zbl. Bakteriol., Parasitenkunde, Infektionskrankh und Hyg.*, Abt. 2, 25, 177–211.

Semikhatova, O. A. 1958. *Mestnyi shtamm azotobaktera, ego effektivnost' i usloviya primeneniya* (A local strain of *Azotobacter*, its effectiveness and conditions of use). Author's abstract of M.Sc. Thesis, Voronezh.

Sen, M. and Sen, S. 1965. Interspecific transformation in *Azotobacter*. *J. Gen. Microbiol.*, 41 (1), 1–6.

Sergienko, F. 1936. Question of the causes of bacteriophage occurrence in the soil. *Mikrobiol. zhurn. Akad. Nauk UkrSSR*, 3 (2).

Sergunina, L. A. 1963. Distribution of azotobacters in the alluvial soils of the Moskva and Klyaz'ma rivers. in *Poimennye pochvy Russkoi ravniny*, (Alluvial soils of the Russian plain). Moscow State University, (2), 141–145.

Shavlovskii, G. M. 1954. *Uchastie mikroorganizmov rizosfery v vitaminnom i aminokislotnom pitanii rastenii* (Participation of micro-organisms in the vitamin and amino acid nutrition of plants). Author's abstract of M.Sc. Thesis, L'vov.

Shchepkina, O. I. 1961. Effect of trace elements on the microflora of the Priazov black earth; in *Rol' mikroelementov v sel'skom khozyaistve* (Role of trace elements in agriculture). Moscow State University, 229–238.

Shende, Sh. T. 1965. *Primenenie azotobakterina pod ris v sochetanii s mineral'nymi udobreniyami* (Application of azotobacterin under rice in conjunction with mineral fertilizers). Moscow.

Shepetina, F. A. (1959). *Vliyanie fosforobakterina i azotobakterina na rost i razvitie i urozhainost' podsolnechnika i kleshcheviny* (Effect of phosphorobacterin and azotobacterin on the growth, development and yield of the sunflower and the castor oil plant). Author's abstract of M.Sc. Thesis, Voronezh.

Shethna, Y. I. and Bhat, J. V. 1962. Studies on soil bacteria decomposing glycerol. *J. Indian Inst. Sci.*, **44** (4), 141–147.

Shklyar, M. Z. 1951. Mutual influence of *Azotobacter* and *Clostridium pasteurianum* in a mixed culture. *Trudy Vses. n-i in-ta s-kh. mikrobiol.*, **11** (2).

Shpokauskas, A. K. and Dachyulite, Ya. A. 1965. Distribution of *Azotobacter* in the soils of the Lithuanian SSR, its nitrogen-fixing activity and biosynthesis of vitamin B. in *Rol' mikroorganizmov v pitanii rastenii i povyssheniya effektivnosti udobrenii* (Role of micro-organisms in plant nutrition and improving the effectiveness of fertilizers). Kolos, Moscow, 61–63.

Shpokauskas, A. K., Pakarskite, K. I. and Dachyulite, Ya. A. 1963. Distribution of *Azotobacter* in the soils of the Lithuanian SSR, its nitrogen-fixing activity, biosynthesis of vitamins of group B (B_1, B_2, B_{12}, folic acid). *Tezisy dokladov Vsesoyuznoi Konferentsii po sel'sko-khozyastvennoi mikrobiologii*. (All-Union Conference on Agricultural Microbiology. Summaries of speeches), Leningrad, 131–132.

Shtern, E. A. 1938. Effect of radium emanation on soil microbe. *Azotobacter chroococcum*. in *Trudy godichnoi sessii Gosudarstvennogo rentgenologicheskogo i rakovogo instituta* (Proceedings of the annual session of the State Institute of Radiology and cancer), 44.3.

Sidorenko, A. I. 1962. The microflora of virgin and reclaimed solonetz of Baraba. *Trudy biol. in-ta Sib. otd. Akad. Nauk SSSR*, (9), 127–147.

Sidorenko, A. I. and Naplekova, N. N. 1962. Enzymatic activity of cultures of azotobacters isolated from the solonetz of Baraba. *Trudy biol. in-ta Sib. otd. Akad. Nauk SSSR*, (9), 157–162.

Singh, G. and Shrikhande, J. G. 1953. Non-symbiotic nitrogen fixation by *Azotobacter*. *J. Indian Soc. Soil Sci.*, **1**, 47–54.

Smalii, V. T. 1936. Bacteriophage of *Azotobacter*. *Mikrobiol. zh. Akad. Nauk UkrSSR*, **3** (3).

Smalii, V. T. and Bershova, O. I. 1957. Formation of heteroauxin in *Azotobacter* cultures. *Mikrobiologiya*, **26** (5), 526–532.

Smalyy, V. T. (Smalii, V. T.). 1962. Effect of rhizosphere bacteria on the content of nicotinic and pantothenic acids in wheat plants. *Mikrobiol. zhurn.*, **24** (1), 15–19. (In Ukranian).

Smirnova, V. I. 1960. Study of the effectiveness of azotobacterin and adaptation of azotobacters in the rhizosphere of maize. *Dokl. TSKhA*, (52), 357–365.

Smith, N. R. 1935. Strain variation of *Azotobacter* and the utilization of carbon compounds. *J. Bacteriol.*, **30**, 323–328.

Sobeshan'skii, E. and Nevyadoma, T. (Sobieszański, J. and Niewiadoma, T.). 1966. Effect of metabolites of *Azotobacter* on the growth and development of plants): in *IX Mezhdunarodnyi Mikrobiologischeskii Kongress. Tezisy dokladov* (Ninth International Microbiology Congress. Summaries of speeches). Moscow, 285–286.

Sobieszczan'ski, J. 1961. Methods of investigating the effect of micro-organisms on higher plants. *Zesz. nauk WSR*, **38**, 1.

Sobieszczanski, J. 1965. Role of micro-organisms in the life of cultivated plants. 1, 2. *Acta Microbiologica Polonica*, **14** (2), 161–202.

Socolofsky, M. D. and Wyss, O. 1961. Cysts of *Azotobacter J. Bacteriol.*, **81** (6), 946–964.

Solov'eva, E. P. 1962–1963. Effect of trace elements on soil microflora: in *Khimizatsiya sel'skogo khozyaistva Bashkirii* (Chemical transformation of the agriculture of Bashkiria). (4–5), Ufa, 211–214.

Soriano, S. 1966. Nitrifying and nitrogen-fixing bacteria of the soils of the Argentine: in *IX Mezdunarodnyi Mikrobiologicheskie Kongress. Tezisy dokladov* (Ninth International Microbiology Congress. Summaries of speeches). Moscow, 275–276.

Soriano, S. and Atlas, E. 1963. Distribution del *Azotobacter chroococcum* en prot undidad en el. *Suelo Cienc e invest.*, **19** (5), 128–132.

Stapp, C. 1951. The value of Az. inoculating preparations for German agriculture. *Landwirtsch. Forsch.*, **3**, 176–205.

Steinberg, R. A. 1938. Applicability of nutrient solution purification of the study of trace element requirements of *Rhizobium* and *Azotobacter*. *J. Agric. Res.*, **57** (6), 461–476.

Stepanova, M. K. 1930. Distribution of azotobacters in some soils of the north-west region. *Trudy In-ta s-kh. mikrobiologii*, **4** (2), 53–67.

Stepanova, L. N. 1961. Soil bacteria—antagonists to *Azotobacter*: in *Povysshenie plodorodiya pochv nechernozemnoi polosy* (Increasing the soil fertility of the non-black earth belt), (1), Moscow, 80–88.

Stille, B. 1958. Beobachtung über das verhalten von *Azotobacter* im Wurzelbereich öherer Pflanzen. *Arch. Mikrobiol.*, **1** (31), 255–261.

Stoklasa, J. 1908. Beiträge zur kenntnis der chemischen Vergange bei der Assimilation des elementären Stickstoffs durch *Azotobacter* und *Radiobacter*. *Zbl. Bakteriol. Parasitenkunde. Infektionskrankh. und Hyg.*, Abt. 2, **21**, 484.

Stokiasa, J. 1913. De l'influence de la radioactivite sur les micro-organismes fixateurs d'azote en transformateurs des matières oazotées: *CR. Acad. sci.*, **157**, 879–882.

Strzelczyk, E. 1961. Studies on the interaction of plants and free-living nitrogen-fixing micro-organisms. II. Development of antagonists of *Azotobacter* in the rhisosphere of plants at different stages of growth in two soils. *Canad. J. Microbiol.*, 7 (4), 507–513.

Strzelczyk, A. 1964–1965. Investigations of the effect of cultivated crops on azotobacterin, 1, 2. *Ann. Univ. M. Curie-Sklodowska*, E, **19**, 115–146.

Subramoney, N. and Abraham, A. 1962. A note on the non-symbiotic nitrogen fixation in red loam soils. *Sci. and Culture*, **28** (7), 339–340.

Surman, K. I. 1961. Splitting of phosphoric acid from sodium nucleate by *Azotobacter chroococcum, Bac. mucilaginosus* and some other soil bacteria. *Byull. nauchno-tekhn inf. po s-kh. mikrobiologii*, **10** (2), 22–24.

Surman, K. I. 1965. Phosphatase activity of soil micro-organisms as one of the factors of their positive influence on plants. *Trudy Vses. n-i. in-ta s-kh. mikrobiologii*, (19), 27–33.

Sushkina, N. N. 1949. *Ekologo-geograficheskoe rasprostranenie azotobaktera v pochvakh SSSR* (Ecologo-geographical distribution of *Azotobacter* in the soils of the USSR). Izd-vo Akad. Nauk SSSR, Moscow-Leningrad.

Swaby, R. S. 1939. The occurrence and activities of *Azotobacter* and *Clostridium butyricum* in Victorian soils. *Austral. J. Exptl. Biol. Med. Sci.*, **17**, 4.

Szegi, J. 1960. Effect of antibiotics on a system of enzymes participating in *Azotobacter* nitrogen fixation. *Agrokém. és talaj.*, **9** (1), 1–10; **2**, 227–236.

Szegi, J. and Gulyas, F. 1963. Effect of the metabolic products of some cellulose-decomposing micro-organisms on the respiration of *Azotobacter* and the germination of lucerne seeds. *Agrokém. és talaj*, **12** (1), 99–106.

Szegi, J. and Timár, E. 1965. The effect of the metabolic products of cellulose decomposing organisms on the growth of some other micro-organisms. *Acta agron. Acad. sci. hung.*, **13** (3–4), 337–348 (in English).

Szelényi, F. and Hélméczi, B. 1960 (1961). Effect of antibiotics on a system of enzymes participating in *Azotobacter* nitrogen fixation. *Debreceni mezőgazd. akad. tud. évk.*, Debrecen, 41–47.

Tanatin, B. Ya. 1953. *Rasprostranenie azotobaktera v pochvakh i rizosfere sel'skokhozyaistvennykh rastenii Leninabadskoi oblasti Tadzhikskoi SSR* (Distribution of *Azotobacter* in the soils and rhizosphere of farm plants in the Leninabad province of the Tadzhik SSR). Author's abstract of M.Sc. Thesis, Leninabad.

Tanatin, B. Ya. 1954. Presence of azotobacters in certain unworked soils of North Tadzhikistan. *Mikrobiologiya*, **23** (1), 37–42.

Tanatin, B. Ya. and Pafikova, M. A. 1963. Distribution of azotobacters in the soils of the Isfarinsk district of North Tadzhikistan. *Uch. zapiski Leninabad. gos. ped. in-ta im. Kirova*, (8), 112–115.

Tchan, Y. T. 1953. Studies on N-fixing bacteria. IV. Taxonomy of genus *Azotobacter*. *Proc. Linnean Soc. N.S. Wales*, **78**, 85.

Tchan, Y. T. Birch-Andersen, A. and Jensen, H. L. 1962. The ultrastructure of vegetative cells and cysts of *Azotobacter chroococcum*. *Arch. Mikrobiol.*, **43**, 50–56.

Teplyakova, Z. F. 1955. Effectiveness of bacterial fertilizers in Kazakhstan. *Vestn. Akad. Nauk KazSSR*, (10), (127), 33–45.

Tešić, Ž. and Todorović, M. 1963. Microbiological features of peat from the neighbourhood of Lake Ohridsko. *Zemljište i biljka*, **12** (1–3), 333–338.

Thompson, L. G. 1932. The periodism of nitrogen fixation in soil and the influence of inoculation with *Azotobacter*. *J. Amer. Soc. Agron.*, **24**, 442.

Thorne, D. and Walker, R. 1936. Physiological studies on *Rhizobium* 4. Accessory factors. *Soil Sci.*, **42** (3), 301.

Timofeev, V. A. 1958. Distribution of azotobacters in soils of Kirgiz Ala-Tau. *Byull. Kirg. n-i. in-ta zemledelenie*, (3), 51–55.

Timonim, M. I. 1948. Az. preparation (azotogen) as a fertilizer for cultivated plants. *Soil Sci. Soc. Amer. Proc.*, **13**, 246–250.

Todorova, B. 1966. Effect of animal manure and green manure on the yield and development of certain groups of micro-organisms. *Pochvoznanie i agrokhimiya*, **1** (1), 95–101 (in Bulgarian).

Toskov, N. 1962. Some physiological features of various strains of *Azotobacter* of the soil of Bulgaria. *Nauchni Trudy Vissh. selskostop in-ta 'Kolarov-Plovdiv'*, **12** (1), 207–217.

Tribunskaya, A. Ya. 1954. Adaptation of azotobacters on the roots of the spring wheat in grey forest soils. *Mikrobiologiya*, **23** (3), 283–290.

Tribunskaya, A. Ya. 1966. Role of azotobacters in the nutrition of plants in peat cans, lime burrows and in urban conditions. in *Okhrana prirody na Urale* (Nature conservation in the Urals). (5), Sverdlovsk, 65–68.

Trizno, S. I. and Vavulo, F. P. 1951. Effectiveness of bacterial fertilizers on peat and boggy soils: in *Sbornik nauchnykh trudov Akad. Nauk BSSR*, **1**, Minsk, 132–153.

Trolldenier, G. 1965. Fluoreszenzmikroskopische Untersuchung von Mikroorganismenreinkulturen in der Rhizosphere. *Zbl. Bakteriol Parsitenlande Infektionskrankh. und Hyg.*, Abt. 2, **119** (3), 256–258.

Turetskaya, R. Kh. 1947. Method for determining the activity of growth substances on root formation. *Dokl. Akad. Nauk SSSR*, **57** (3), 295–297.

Tyurin, I. V. and Sokolov, A. V. 1958. Types of soils and the effectiveness of fertilizers. *Izv. Akad. Nauk SSSR, Ser. Biol.*, (6), 651–660.

Ulyasevich, E. I. *Deistvie izvestkovanie na mikroflory dernovo-podzolistoi pochvy, pochvy i rizosfery kukuruzy v usloviyakh Poles'ya USSR* (Action of liming on the

9

microflora of soddy-podzolic soils, soils and the rhizosphere of maize in the conditions of Polessya (UkrSSR).) Author's abstract of M.Sc. Thesis, Kiev.

Ulyashova, R. M. 1965. The microflora of a gravel tomato culture. *Priroda*, (3), 120–121.

Uspenskii, E. E. 1930. Microbiological evaluation of the soil requirements for lime, phosphates and other fertilizers. *Udobrenie i urozhai*, (7–8), 581–599.

Vančura, J. 1961. Detection of gibberellic acid in *Azotobacter* cultures. *Nature*, 192 (4797), 88.

Vančura, V., Abd-El-Malek, Y. and Zayed, M. N. 1965. *Azotobacter* and *Beijerinckia* in the soils and rhizosphere of plants in Egypt. *Folia microbiol.*, 10 (4), 224–229.

Vančura, V. and Macurá, J. 1960. Indole derivatives in *Azotobacter* cultures. *Folia microbiol.*, 5, 293–297.

Vančura, V. and Macurá, J. 1961. The effect of root excretions on *Azotobacter*. *Folia microbiol.*, 6 (4), 250–259.

Vančura, V., Macurá, J., Fischer, O. and Vondracek, J. 1959. The relation of *Azotobacter* to the root system of barley. *Folia microbiol.*, 4 (2), 119–129.

Van Tsy-fan (Wang Tzu-fang). 1958. *Vzaimootnosheniya mezhdu azotobakterom i rizosfernymi bakteriyami pshenitsy* (Interaction between *Azotobacter* and the rhizosphere bacteria of wheat). Author's abstract of M.Sc. Thesis, Moscow.

Vardapetyan, Sh. S. *Aktinomitsety pochve Armyanskoi SSR i ikIvliyanie na Azotobaktery i ammonifikatory* (Actinomycetes of the soil of the Armenian SSR and their effect on *Azotobacter* and ammonium fixers). Author's abstract of M.Sc. Thesis, Erevan.

Vasil'eva, T. A. 1959. Azotobacters in soils of some fields of the grass-arable system. *Trudy Pochv. in-ta Mold. fil. Akad. Nauk SSSR*, (3), 45–48.

Vedenyapina, N. S. 1963. Ecology of *Azotobacter* and means of increasing the effectiveness of azotobacterin in the Volgograd province. *Tezisy dokladov Vsesoyuznoi konferentsii po sel'sko-khozyaistvennoi mikrobiologii* (All-Union Conference on Agricultural Microbiology. Summaries of speeches), Supplement, Leningrad.

Verner, A. R. 1951. Further remarks on the salt resistance of *Azotobacter*. *Mikrobiologiya*, 20 (5), 400–405.

Vidal, G. and Deborgne, L. 1964. Nouvelles recherches sur la rhizosphere de la vigne. *Ann. Inst. Pasteur*, 106 (4), 651–653.

Vigileva, A. I. 1954. *Sovmestnye kul'tury azotobaktera i kluken'kovykh bakterii i ikh ispol'zovanie v sel'skom khozyaistve* (Combined cultures of *Azotobacter* and nodule bacteria and their use in agriculture). Author's abstract of M.Sc. Thesis, Moscow.

Vigorov, L. I. 1961. Acclimatization of various *Azotobacter* races on the roots and rhizosphere of spring wheat grown in podzolic soil. *Mikrobiologiya*, (3), 478–483.

Vigorov, L. I. and Tribunskaya, A. Ya. 1965. Spectrum of bacteria effecting the turnover of nitrogen in podzolic soils and the rhizosphere of wheat grown in this soil. *Trudy Ural'skogo n.-i. in-ta sel'skogo khozyaistva*, 6, 307–321.

Vinogradskii, S. N. (Winogradskii, S. N.). 1932. Sur la synthese de l'ammoniac par les azotobacter du sol. *Ann. Inst. Pasteur*, 48.

Vintika, J. 1953. Questions of the study of *Azotobacter* symbiosis. *Českosl. biol.*, 2 (2), 89–92.

Visser, S. A. 1966. Annual variation in the distribution and activity of different groups of micro-organisms in a Nigerian ground-nut soil. *W. Afric. J. Biol. Appl. Chem.*, 9 (1), 20–24.

Voets, J. P. Effect of manganese and magnesium on nitrogen fixation by *Azotobacter*. *Medel landbouwhogenschooan opzoakingsstaat*. Staat. Gent, 27 (4), 1441–1454 (in Flemish).

Voets, J. P. 1963. Le metabolisme du benzoate par *A. vinelandii*. *Ann. Inst. Pasteur*, 105 (2), 382–391.

Voets, J. P. and Dedeken, M. 1966. A physiological approach to the classification of the genus *Azotobacter*: in *Mezhdunarodnyi Mikrobiologicheskii Kongress* (International Microbiology Congress), Moscow, 73–77 (in English).

Voicu, J. 1924. Influence de humus sur la sensibilité de l'*Azotobacter chroococcum* vis à vis de bore. *C.R. Acad. Sci.*, **175**, 317–319.

Voinova-Raikova, Zh. (Vojnova-Rajkova, J.). 1966. Sources of nutrition for the development of *Azotobacter*. in IX *Mezhdunarodnyi Mikrobiologicheskii Kongress*, *Tezisy dokladov* (Ninth International Microbiology Congress. Summaries of speeches), Moscow, 285.

Voitkevich, A. F. and Runov, E. V. 1928. Distribution of azotobacters in the soil. *Vestn. bakteriol. agron. stantsii*, (25).

Vojnova-Rajkova, J. 1965. On the specificity of rhizosphere microflora: in *Plant Microb Relationship*. *Prague* Czechosl. Acad. Sci., 126–133 (in English).

Voznyakovskaya, Yu. M. 1954. Relationship between cellulose-decomposing bacteria and *Azotobacter*. *Agrobiologiya*, (4), 81–87.

Voznyakovskaya, Yu. M. 1963. Choice of micro-organisms for use in bacterial fertilizers. *Mikrobiologiya*, **32** (1), 168–174.

Walczyna, J. and Przesmycka, W. 1964. Occurrence of *Azotobacter chroococcum* and *Azotobacter beijerinckii* in lowland meadow soils. *Roczn. nauk roln*, F, **76** (1), 163–182.

Wang Yü-ts'ing. (1960). Survival of *Azotobacter* in the soil of rice fields. *Nung-ye xuebao*, *Acta agric. sinica*, **2** (1), 83–89.

Warmbold, H. 1906. Untersuchungen über die Biologie stickstoffbindender Bakterien. *Landwirtsch. Jahresb.*, **35** (1).

Wilson, P. and Burris, R. 1947. The mechanism of biological nitrogen fixation. *Bacteriol. Revs.*, **11**, 41.

Yamagishi, H. and Furusaka, Ch. 1964. Growth of *Azotobacter chroococcum* in the medium with colloidally dispersed calcium phosphate gel. *Sci. Repts. Res. Inst. Tohoku Univ.*, D, **15** (1), 1–5.

Yatskunene, A. V. 1962. Increasing the content of vitamins in the carrot under the action of *Azotobacter*. *Byull. nauchn-tekhn. inf. Litovsk n-i. in-ta zhivotnovodstva*, (3), 46–50.

Yuzhina, Z. I. 1958. Relation between toxic properties of soil of the Kolskii peninsula and the number of microbe-antagonists to *Azotobacter*. *Mikrobiologiya*, **7** (2), 201.

Zaitseva, G. 1965. *Biokhimiya azotobaktera*. The biochemistry of *Azotobacter*. Nauka, Moscow.

Zaitseva, G. N. and Belozerskii, G. N. and Novozhilova, L. P. 1960. Effect of the calcium ion on nitrogen and phosphorus metabolism of *Azotobacter vinelandii*. *Mikrobiologiya*, **29** (3), 343–350.

Zaitseva, G. N., Khmel', I. A. and Belozerskii, A. N. 1961. Biochemical changes in a synchronized culture of *Azotobacter vinelandii*. *Dokl. Akad. Nauk SSSR*, **141**, 740–745.

Žák, J. 1965. The effect of growth stimulators on the plant rhizosphere: in *Plant Microbes Relationships*. *Prague. Czechosl. Acad. Sci.*, 91–98.

Zawiślak, K. 1965. Occurrence of *Azotobacter chroococcum* in slope soils of the Warmińsko-Mazurski lake district. *Zesz. nauk. Wyzsza Szkola Roln. Szczecinie*, (18), 161–171.

Zayed, M. N. 1963. The antibacterial effect of some plant residues on soil micro-organisms. *J. Soil Sci. UAR*, **3** (2), 257–271.

Zhdannikova, E. N. 1963. Microbiological characterization of the peat bog soils of the Tomsk region: in *Zabolochennye lesa i bolota Sibiri* (The boggy forests and peat bogs of Siberia). Izd-vo Akad. Nauk SSSR, Moscow, 170–182.

Zimiecka, J. 1932. Microbiological tests of the soils' fertility, nitrification and nitrogen fixation. *Proc. Second Internat. Congr. Soil Sci. London, 1930*, Com. 3, 53–55.

Zimienko, T. H. (Zimenko, T. G.). 1957. Toxicity of certain peatbog soils to *Azotobacter chroococcum. Vieśći Akad. navuk BSSR, sierija bijal. navuk*, (1), 49–53 (in Byelorussian).

Zimna, J. 1962. Influence of certain species of meadow plants on the abundance of *Clostridium* and *Azotobacter* in their rhizospheres. *Ekol. polska*, **8** (2), 165–171 (in English).

Zinov'eva, Kh. G. 1962. *Azotobakter i sel'skokhozyaistvennye rasteniya (Azotobacter* and farm plants). Sel'khozgiz, Kiev.

Zinov'eva, Kh. G. and Klyushnikova, T. M. 1962. Synthesis of vitamins by cultures of *Azotobacter chroococcum* adapted to certain farm plants. *Mikrobiol. zh.*, **24** (6), 26–31.

Zinov'eva, Kh. H. (Zinov'eva, Kh. G.). 1966. Toxic properties of soddy-podzolic soils of Polessya (UkrSSR) towards *Azotobacter. Mikrobiol. zh.*, **28** (1), 21–25 (in Ukrainian).

5 Nitrogen-Fixing Bacteria of the Genus 'Beijerinckia'

Bacteria of the genus *Beijerinckia* are similar to *Azotobacter* but differ in their considerable resistance to acid (they can even grow at pH 3·0), calcium phobicity and certain other properties. These bacteria were first isolated by Starkey and De (1939) from the acid soils of the paddy fields of India, and were given the name *Azotobacter indicum*. In 1950 Derx, working in the Bogor Botanic Gardens (Java), also discovered this micro-organism in the soil and proposed that it be classified as a new genus—*Beijerinckia*.

Cells of *Beijerinckia* may be round, oval or rod-shaped. Sometimes the rods are bent. As they age the cells may become deformed. Cell size is very irregular: young cells are on average $0·5 — 2·0 \times 1·0 — 4·5 \mu$m, old cells $1·2 \times 4·8 — 7·0 \mu$m. They are Gram-negative. A characteristic feature is the presence at the ends of the cells of granules that stain with Sudan III and increase in size with age. These may be mobile or immobile. Some strains form capsules enclosing a group of cells but not cysts or endospores. Cultures are characterized by slow growth with a latent period of between three and fifteen days. Typical colonies usually form after three weeks at 30°C. On nitrogen-free agar with glucose (pH 5·0), cultures of *Beijerinckia* form highly convex, often folded, glossy mucilaginous colonies of a viscous consistency. When they age cultures normally develop pigment.

On the basis both of his own and other published findings, Hilger (1965) included the genus *Beijerinckia* in the family Azotobacteracae. He considered that *Beijerinckia* can be divided into three groups: *B. indica, B. fluminensis* and *B. derxii* (Table 5.1).

He included *B. indica* (Starkey and De) Derx, *B. lacticogenes* (Kauffmann and Toussaint) Tchan and *B. mobilis* Derx in his first group. To confirm the validity of this grouping he referred to Jensen (1954) and Petersen (1959) who questioned the legitimacy of the existence of the species *B. mobilis*, and Dommergues (1953) who doubted the existence of the species *B. lacticogenes*. He was unconvinced by the objections that the cultures of these three species differed in outward appearance; he has isolated and described various intermediate forms from soils in the Congo. As for the immobility of *B. lacticogenes*, distinguishing it from *B. indica* and *B. mobilis*, Hilger thought that this difference could be disregarded, being very difficult to discern.

251

Table 5.1. Characteristic Features of the Principal Groups of *Beijerinckia*

Feature	B. indica	B. fluminensis	B. derxii
Size of cells (μm)	0·5–1·5 × 1·7–3·0	1·1–1·5 × 3·0–3·5	1·5–2·0 × 3·5–4·5
Mobility	Mobile or immobile	Yes	No
Formation of pigment	Possible in old cultures	Yes	Yes
Characterization of pigment	From red to brown; does not diffuse into medium	Dark brown	Usually fluoresce yellow-green; in media with salts of organic acids pink lilac; does not diffuse into medium
Character of colonies on nitrogen-free medium with readily utilizable sugars	Round with regular edge, slightly convex, smooth, relatively mucilaginous, vitreous or non-transparent sometimes folded	Convex with irregular contours, rough, dry, granular, hard and not transparent	Round, with regular edge, convex, smooth mucilaginous, dense, not transparent, often iridescent and folded
Formation of sticky mucilage (gum type) in liquid nitrogen-free medium	Relatively abundant	Very weak or absent	Abundant
Rate of nitrogen fixation	Fast	Relatively slow	Slow

This table is taken from Hilger (1965).

In this second group Hilger included only *B. fluminensis*, Döbereiner and Ruschel. The third group, *B. derxii* Tchan, *B. acida* (Roy) Petersen and *B. congensis*, he characterized by considerable similarity in morphological and cultural properties.

Jensen (1954), Norris (1959, 1960) and Rubenchik (1960) also included *Beijerinckia* in the family Azotobacteracae. The systematic study of *Beijerinckia*, however, cannot be considered complete yet.

For each gram of energy-rich material used, these bacteria fix between 16 and 20 mg of molecular nitrogen (Jensen 1954; Kluyver and Becking 1950). The nitrogen-fixing activity of cultures of *B. indica* and *B. fluminensis* has been reported to increase considerably in the presence of the oligonitrophilic yeast *Lipomyces starkey* isolated from the same tropical soil as *Beijerinckia* (Dommergues and Mutaftschiev, 1965).

Bacteria of the genus *Beijerinckia* are strict aerobes. When cultured in media containing carbohydrates they produce compounds of an acid or alkaline nature according to the composition and pH of the media.

In neutral media the products are acid, including acetic acid and a certain, so far unidentified, non-volatile acid (Quinnel, 1957; Becking, 1961). Lactic acid has been found in a culture of *B. lacticogenes* (Kauffmann and Toussaint, 1951).

Bacteria of the genus *Beijerinckia* assimilate efficiently monosaccharides, disaccharides and starch as sources of carbon. They assimilate organic acids less well, utilizing little or no acetic, citric or tartaric acids. Nor do they assimilate aromatic compounds (for example, benzoic acid) such as those utilized by *Azotobacter*.

Beijerinckia makes better use of ammoniacal and nitrate nitrogen and many amino acids than atmospheric nitrogen (Jensen, 1948). As we have noted, this group of bacteria resembles *Azotobacter* with respect to vigour of nitrogen fixation.

The mucilage of *Beijerinckia*, like that of *Azotobacter*, is of a polysaccharide nature (Martin *et al.*, 1965). It decomposes very slowly in the soil. Eight weeks after introduction into acid and neutral soils, 19 per cent of this mucilage has decomposed, whereas in the same conditions up to 67 per cent of the mucilage of *Azotobacter* is destroyed. The polysaccharides of *Beijerinckia* are highly toxic to several soil micro-organisms (Döbereiner and Ruschel, 1964). In pure cultures of *Beijerinckia* the formation of mucilage is correlated with the intensity of nitrogen assimilation (Barooah and Sen, 1964).

Beijerinckia is affected by the concentration of phosphorous compounds in the medium. According to Becking (1961) the bacterium begins to develop in the presence of about 0·2 mg of phosphate per 100 ml. of medium, while *Azotobacter* requires about 4·0 mg in the same volume of substrate. *Beijerinckia* withstands considerably greater concentrations of phosphates than does *Azotobacter*.

For neither the development nor the nitrogen assimilating activities of *Beijerinckia* is calcium necessary. Small doses of calcium (22·5 p.p.m. of $CaCl_2$) inhibit its growth (Hilger, 1964). This depressant effect is most apparent at pH 5·0–6·0.

Beijerinckia requires much smaller amounts of magnesium compounds than does *Azotobacter* (Jensen, 1954). *Beijerinckia* begins to develop in a medium containing approximately twenty times less magnesium than required by *Azotobacter*. But when *Beijerinckia* is isolated from soils formed on magnesium-rich parent rocks the cultures are characterized by the same magnesium requirements as *Azotobacter* (Becking, 1960).

Becking (1961) showed that *Beijerinckia* has a high resistance to iron salts (up to 50 mg in 25 ml. of medium). Dommergues (1963) considered that the presence of *Beijerinckia* in the soil may serve as an indicator of processes of ferrugiation (lateritization).

Beijerinckia also tolerates high concentrations of aluminium (up to 80 p.p.m.) and titanium; the soils which it inhabits usually contain large quantities of these elements (Becking, 1961; Dommergues, 1963).

Beijerinckia is less sensitive to iron and aluminium salts than is *Azotobacter*. The two genera also differ in their requirements for molybdenum (Becking, 1961; Mekhtieva and Melkumova, 1962; Krylova, 1963), probably reflecting their ecological adaptation. Species of *Beijerinckia* which require least molybdenum inhabit acid soils containing insignificant amounts of available molybdenum. For vigorous nitrogen fixation *Azotobacter* requires considerably more molybdenum. The optimum concentration for *Beijerinckia* is between 0·004 and 0·034 p.p.m. Vanadium does not stimulate the fixation of atmospheric nitrogen by cultures of *Beijerinckia* and is no substitute for molybdenum.

Beijerinckia is more resistant to high concentrations of manganese than is *Azotobacter* (Becking, 1961); copper stimulates nitrogen fixation by *Beijerinckia* (Becking, 1961; Krylova, 1963).

As we have noted, *Beijerinckia* is an acidophilic genus. Its optimum pH is between 4·5 and 6·0. At pH 7·0 development is depressed (Hilger, 1964). These bacteria usually continue to grow at pH 3·0, however.

The literature records that certain species of *Azotobacter* can live in an acid medium. Thus Jensen (1955) reported a variant of *Az. macrocytogenes* developing at pH 4·5–5·0. This property is also possessed by *Az. beijerinckia acido-tolerans* described by Tchan (1953); nevertheless, this is an exception, for *Azotobacter* does not usually grow at a pH of less than 5·5.

Beijerinckia is resistant to soil drought in the tropical summer (Becking, 1961). The range of temperatures suitable for the development of *Beijerinckia* is somewhat narrower than that for *Azotobacter*. Cultures of *Beijerinckia* develop well on laboratory media at temperatures between 16° and 37°C, while for *Azotobacter* the best temperatures are 16° to 45°C (Becking, 1959, 1961). Cooler temperatures inhibit the growth in both cases. At freezing and sub-zero temperatures, however, *Beijerinckia* cultures remain viable for a long time (Tchan, 1953). The shortest lag phase is observed in cultures of *Beijerinckia* growing at 30°–35°C and pH 4·5–5·3 (Strydom, 1966).

During the 1960's various investigations have considerably extended knowledge of the characteristics and ecology of this genus (Becking, 1961; Voets, 1962; Hilger, 1963; Dommergues, 1963).

Bacteria of the genus *Beijerinckia* are widespread in the soils of southern and tropical zones but are less often encountered in temperate climates (Anderson, 1966). They have been found in the soils of India (Starkey and De, 1939), Malaya (Altson, 1936), Indonesia (Derx, 1950; Kluyver and Becking, 1955), Japan (Suto, 1957; Becking, 1971), the Ivory Coast (Kauffman and Toussaint, 1951), Madagascar (Dommergues, 1963), Tanganyika (Meiklejohn, 1954),

Northern Australia (Tchan, 1953), Brazil (Döbereiner and De Castro, 1955), North and South Africa (Becking, 1961) and elsewhere.

Beijerinckia has also been found in certain countries of Europe and America. Becking (1961) found it in Yugoslavia, Goguadze (1963, 1966), and Emtsev and

Table 5.2. Presence of Bacteria of *Beijerinckia* and *Azotobacter* in Different Soils

Soil taken from	Number of samples studied	Number of samples in which were found	
		Beijerinckia	*Azotobacter*
(I) Non-tropical soils			
Europe	155	2	75
North Africa (Tunis, Algiers, Egypt)	8	0	6
South Africa	40	15	5
Asia (Japan)	8	5	0
Australia (South Australia, Victoria, New South Wales)	30	0	5
North America (Florida)	3	0	3
Total samples	244	22	94
Percentage	100	9·0	38·5
(II) Tropical soils			
Africa	53	39	11
Asia	43	22	16
South America	52	19	10
Total	148	71	37
Percentage	100	48·0	25·0

These data are taken from Becking (1961).

Gogorikidze (1966) in the red soils of Georgia, and Anderson (1966) on the soils of the United States (Idaho) and Canada. Ruinen (1956) reported finding *Beijerinckia* on the surface of leaves of many tropical plants in Indonesia.

Becking (1961) examined many soils (about 400 samples collected in different parts of the world) for the presence of *Beijerinckia*. His findings are given in Table 5.2, which shows that *Beijerinckia* is encountered much more frequently in tropical soils than in temperate climates. Thus, out of 155 soil samples collected in Europe, *Beijerinckia* was found in only two, whereas thirty out of fifty-five African soils contained this micro-organism. Other soils from the tropical zone were also rich in *Beijerinckia*. It is distributed in the red-yellow soils of China with a pH of 5·4 (Wu Chang Chih, 1965). Vančura *et al.* (1965) found *Beijerinckia* in the soil of the Harga oasis (Egypt).

According to Becking (1961) and Rangaswami and Sadasivam (1964) *Beijerinckia* is most often encountered in tropical acid soils with a pH between 5·0 and 5·9. It is not usually found in more acid or more alkaline soils. In neutral and more alkaline soils *Azotobacter* is the dominant nitrogen fixer. The absence of *Beijerinckia* from these soils can be explained by competition with *Azotobacter*, which is the more probable because *Beijerinckia* can grow in nutrient media with a fairly high pH (9·0–10·0). The only exception is *B. mobile*—which grows appreciably less well at pH 9·0. However, Anderson (1966) has established the presence of *Beijerinckia* in alkaline soils.

The bulk of *Beijerinckia* cells are concentrated in the upper layer of soil (0–20 cm) but development is possible down to 100 cm (Emtsev and Gogorikidze, 1966). The available evidence suggests that there are more cells of *Beijerinckia* in cultivated acid soils than in virgin soils, while virgin meadow soils contain more cells than forest soils.

The distribution of *Beijerinckia* in soils is regulated not only by the solidity and moisture content, but also by a whole series of factors peculiar to the given soil (Strydom, 1966). As a rule, *Beijerinckia* is associated with tropical lateritic leeched soils and not with tropical alluvial soils (Kluyver and Becking, 1955). In lateritic soils, iron, aluminium, titanium and manganese ions accumulate in the upper layers as a result of the process of lateritization. Iron is usually in the form of sesquioxides or concretions of brown haematite. Phosphorus is firmly fixed in such an iron complex. The content of calcium, magnesium, potassium and sodium in these soils is negligible. The physiological features of *Beijerinckia* reflect these conditions.

Attempts have been made to relate the adaptation of *Beijerinckia* to the distribution of the tropical leguminous plants of the family Caesalpiniaeae, such as *Cassia fora*, which are possibly facultative symbionts (Derx, 1953). It is assumed that unlike nodule bacteria, *Beijerinckia* has not lost its nitrogen-fixing powers in the absence of the host plant.

Ruinen (1956), who isolated *Beijerinckia* from the leaves of many tropical plants, thought that the bacteria may be epiphytes, receiving sugars and nutrient salts in the cuticular secretions of the host plants and supplying the plants with bound nitrogen. Meiklejohn (1962) found *Beijerinckia* on the leaves of woody and shrub vegetation in Ghana.

There are some indications of a link between *Beijerinckia* and the root systems. Thus, according to Döbereiner and Alvahydo (1959) the increased content of these nitrogen-fixing bacteria in the proximity of sugar cane roots in Brazil can be explained by the enrichment of the soil with the sucrose excreted from the root. Further away from the roots there are fewer cells of *Beijerinckia*. In the flooded soils of rice fields a higher content of *Beijerinckia* has been noted in the rhizosphere, especially the rhizoplane, than in the surrounding soil (Döbereiner and Puppin, 1961). Vančura *et al.* (1965) also found that the root system of plants had a favourable effect on *Beijerinckia*. The rhizosphere of lucerne, for example, contains far more *Beijerinckia* than soil without plants.

The influence of *Beijerinckia* on the plants has rarely been considered. Of course *Beijerinckia* may play a role as a nitrogen accumulator; there are indications that it forms biologically active substances. Thus Fallot (1960) noted that a culture of *Beijerinckia* and its extracts induced germination in tubers of topinambour and resting tissues of other plants (Fallot, 1966). A similar effect is produced by a culture of *Azotobacter chroococcum*.

The reaction is evidently due to growth factors of the auxin type, and to compounds of the purine derivative type. The failure of cultures of *Beijerinckia* to synthesize auxin (Burger and Bukatsch, 1958) is apparently connected with the absence of certain necessary substances. Brakel and Hilger (1965) found that tryptophan sharply stimulates the formation of β-indolylacetic and β-indolylbutyric acids in cultures of *Beijerinckia*. It is to be hoped that future investigation will help to establish the significance of *Beijerinckia* in the nitrogen balance of soils and the life of plants.

References

Altson, R. A. 1936. Studies on *Azotobacter* in Malayan soils. *J. Agric. Sci.*, **26**, 268–280.

Anderson, G. R. 1966. Identification of *Beijerinckia* from Pacific north-west soils. *J. Bacteriol*, **91** (5), 2105–2106.

Barooah, P. and Sen, A. 1964. Nitrogen fixation by *Beijerinckia* in relation to slime formation. *Arch. Mikrobiol.*, **48** (4), 381–385.

Becking, J. H. 1959. Nitrogen-fixing bacteria of the genus *Beijerinckia* in South African soils. *Plant and Soil*, **11** (3), 193–206.

Becking, J. H. 1961. Studies on nitrogen-fixing bacteria of the genus *Beijerinckia*. *Plant and Soil*, **14** (1), 49–81; (4), 297–332.

Brakel, J. and Hilger, F. 1965. Étude qualitative et quantitative de la synthèse de substances de nature auxinique par *Azotobacter chroococcum in vitro*. *Bull. Inst. agron. stat. rech. Gembloux*, **33** (4), 469–487.

Burger, K. and Bukatsch, F. 1958. Über die Wuchsstoffsynthese in Boden freileben der stickstoffbindender Bakterien. *Zbl. Bakteriol., Parasitenkunde, Infektionskrankh und Hyg.*, **2** (III), 1–28.

Derx, H. G. 1950. *Beijerinckia*, a new genus of nitrogen-fixing bacteria occurring in tropical soils. *Proc. Koninkl nederl Akad. wet.*, (53), 140–147.

Derx, H. G. 1953. Sur la cause de la distribution geographique limitée des *Beijerinckia*. *Proc. 6th Internat. Congr. Microbiol. Roma*, **6**, 353–355.

Döbereiner, J. 1961. Nitrogen-fixing bacteria of the genus *Beijerinckia* Derx in the rhizosphere of sugar cane. *Plant and Soil*, **15** (3), 211–216.

Döbereiner, J. and Alvahydo, R. 1959. Sobre a influencia da cana de acuar na ccorrencia de *Beijerinckia* no solo. *Rev. brasil. biol.*, **19**, 401–412.

Döbereiner, J. and De Castro, A. F. 1955. Occorencia e capacidade de fixacao de nitrogenio de bacterias do genero *Beijerinckia* nas series de soles da area territorial do Centro Nacional de Ensino e Pesquisas Agronomicas. *Bol. Inst. Ecol. Exptl Agric.*, **16**.

Döbereiner, J. and Puppin, R. A. 1961. Inoculacao do arrox com bacterias fixadoras de nitrogenio do genero *Beijerinckia* Derx. *Rev. Brasil. biol.*, **21** (4), 397–407.

Döbereiner. J. and Ruschel, A. P. 1964. Methods for the study of *Beijerinckia*. *Soil Biol. Internat. News Bull.*, New Ser., **1**, 3–5.

Dommergues, J. 1953. *Mem. Inst. Sci. Madah.*, **5**, 327, 353 (quoted by Hilger, F. 1965).

Dommergues, J. 1963. Distribution des *Azotobacter* et des *Beijerinckia* dans les principaux types de sol de l'Ouest Africain. *Ann. Inst. Pasteur*, **105** (2), 179–187.

Dommergues, Y. and Mutaftschiev, S. 1965. Fixation synergique de l'azote atmospherique dans les sols tropicaux. *Ann. Inst. Pasteur.* Suppl., (3), 112–120.

Emtsev, V. T. and Gogorikidze, N. I. 1966. Distribution of free-living nitrogen fixers in the red earths of Georgia. *Izv. TSKhA*, (5), 31–40.

Fallot, J. 1960. Action positive de quelques micro-organismes sur la proliferation *in vitro* de tissus de topinambour. *Bull. Soc. franc. physiol veget.*, **6** (2), 90–91, Discuss 91.

Fallot, J. (Fallo, Dz.) 1966. Stimulation by saprophytic bacteria of the growth of stem tissues cultivated *in vitro*: in *IX Mezdunarodnyi Mikrobiologicheskii Kongress Tezisy dokladov* (Ninth International Microbiology Congress. Summaries of speeches). Moscow, 227–228.

Goguadze, V. D. 1963. The microflora of the red earth of Georgia. *Izv. Akad. Nauk SSSR, Ser. Biol.*, (1), 40–48.

Goguadze, V. D. 1966. Microbiological characterization of the red earth of Georgia: in *Mikroflora pochv. yuzhnoi chasti SSSR* (The microflora of the soils of the southern part of the USSR), Nauka, Moscow, 246–263.

Hilger, F. 1964. Comportement des bacteries fixatrices d'azote du genre *Beijerinckia* a l'égard du pH et du calcium. *Ann. Inst. Pasteur*, **106** (2), 279–291.

Hilger, F. 1965. Études sur la systematique du genre *Beijerinckia* Derx. *Ann. Inst. Pasteur*, **109** (3), 406–423.

Jensen, H. L. 1948. The influence of molybdenum, calcium and agar on nitrogen fixation by *Azotobacter indicum*. *Proc. Linnean Soc. N.S. Wales*, **72**, 299–310.

Jensen, H. L. 1954. *Bacteriol. Rev.*, **18**, 195 (quoted by Hilger, F., 1965).

Jensen, H. L. 1954. The magnesium requirements of *Azotobacter* and *Beijerinckia*, with some additional notes on the latter genus. *Acta agric. scand.*, **4** (2), 224–236 (in English).

Jensen, H. L. 1955. *Azotobacter macrocytogenes* n. sp. a nitrogen-fixing bacteria resistant to acid reactions. *Acta agric. scand.*, **5**, 278–294 (in English).

Kauffman, M. and Toussaint, P. 1951. Un nouveau germe fixateur de l'azote atmospherique: *Azotobacter lacticogenes*. *Rev. gen. bot.*, **58**, 553–561.

Kluyver, A. J. and Becking, J. H. 1950. Some observations on the nitrogen-fixing bacteria of the genus *Beijerinckia* Derx. *Ann. Acad. Sci. fennicae*, ser. A. II, **60**, 367–380 (in English).

Krylova, N. B. 1963. Effect of molybdenum and vanadium on nitrogen fixation. *Dokl. TSKhA*, (84), 293–299.

Martin, J. P., Ervin, J. O. and Shepherd, R. A. 1965. Decomposition and binding action of polysaccharides from *Azotobacter indicum* (*Beijerinckia*) and other bacteria in soil. *Soil. Sci. Proc.*, **29** (4), 397–399.

Martin, J. P., Ervin, J. O. and Shepherd, R. A. 1966. Decomposition of iron, aluminium, zinc and copper salts or complexes of some microbial and plant polysaccharides in soil. *Soil Sci. America Proc.*, **30** (2), 196–200.

Meiklejohn, J. 1954. Notes on nitrogen-fixing bacteria from East African soils. *Trans. Fifth Internat. Congr. Soil Sci. Leopoldville*, **3**, 123–125.

Meiklejohn, J. 1962. Microbiology of the nitrogen cycle in some Ghana soils. *Empire J. Exptl. Agric.*, **30** (118), 115–126.

Mekhtieva, N. A. and Melkumova, T. A. 1962. Effect of trace elements on the development of azotobacters in some soils of Azerbaijan, *Dokl. Akad. Nauk AzSSR*, **18**, (10) 59–63.

Norris, J. R. 1959. The isolation and identification of azotobacters. *Lab. Pract.*, **8** (7), 239–243.

Norris, J. R. 1960. Notes on the classification of the family Azotobacteriaceae. Glasgow Dept. Bacteriol.

Petersen, E. J. 1959. *Roy. Veterin. Agric. Coll. Copenhagen*, 70 (quoted by Hilger, F., 1965).

Quinnel, C. M. 1957. Observations on the physiology of *Azotobacter indicum*. *J. Bacteriol.*, **73**, 688–689.

Rangaswami, G. and Sadasivam, K. V. 1964. Studies of the occurrence of *Azotobacter* in some soil types. *J. Indian Soc. Soil Sci.*, **12**, 43–49.

Rubenchik, L. I. 1960. *Azotobacter i ego primenenie v sel'skom khozyaistve* (*Azotobacter* and its use in agriculture). Izd-vo Akad. Nauk UkrSSR, Kiev.

Ruinen, J. 1956. Occurrence of *Beijerinckia* species in the phyllosphere. *Nature*, **177**, 220–221.

Ruinen, J. 1966. The phyllosphere. 3. Nitrogen fixation in the phyllosphere. *Plant and Soil*, **22** (3), 375–394.

Starkey, R. J. and De P. K. 1939. A new species of *Azotobacter*. *Soil Sci.*, **47**, 326.

Strydom, B. W. 1966. The effect of soil pH soil type on the occurrence of *Beijerinckia* species in non-lateritic soils. *Biol. Abstr.*, **47** (13), No. 63729, *S. Afric. J. Agric. Sci.*, 1965, **8** (3), 853–861.

Suto, T. 1957. Some properties of an acid tolerant *Azotobacter*, *Azotobacter indicum*. *Tohoku J. Agric. Res.*, **7**, 369–382.

Tchan, J. T. 1953. Studies of N-fixing bacteria. III. *Azotobacter beijerinckii* (Limpan, 1903) var. acido-tolerans (Tchan, 1952). *Proc. Linnean Soc. N.S. Wales*, **78**, 83–84.

Vančura, V. Ald-El-Malek, J. and Zayed, M. N. 1965. *Azotobacter* and *Beijerinckia* in the soils and rhizosphere of plants in Egypt. *Folia microbiol.*, **10** (4), 224–229.

Voets, J. P. 1962. Effect of manganese and magnesium on nitrogen fixation by *Azotobacter*. *Meded. Landbouwhogenschooen opzoekingsstaat, staat Gent.*, **27** (4), 1441–1454 (in Flemish).

Wu Chang Chih. 1965. Eco-geographical distribution of the free-living aerobic nitrogen-fixing bacteria in several agricultural soils of Hupeh province. *Acta pedol. sinica*, **13** (2), 216–222.

6 Anaerobic Nitrogen Fixers (Genus *Clostridium*)

Classification

The first anerobic micro-organism assimilating molecular nitrogen was isolated and described by Winogradskii in 1893; it was a sporulating bacterium called *Clostridium pasteurianum*. Cells of this bacterium measure about 2·5 — 7·5 × 0·7 — 1·3 μm. In culture they are solitary, paired or in short chains. Young cells are motile and their plasma is homogeneous but with age it sometimes becomes granulated. The cells are Gram-positive. Spores are oval or oblong, arranged centrally, and when they are formed the cell becomes spindle-shaped. The spores are 1·3 × 1·6 μm across. Krasil'nikov and Duda (1964, 1966) found that the nuclear apparatus of *Clostridium* changed during spore formation.

Several investigators have described anaerobic species of the genus *Clostridium*. Bredemann (1909) considered it possible to group them into a single entity, which he called *Bacillus amylobacter*. Many, however, including Pringsheim (1909) and Omelyanskii (1913), did not agree. MacCoy *et al.* (1930) proposed that the anaerobic nitrogen fixers be assigned to the genus *Clostridium*. A characteristic feature of these bacteria is thought to be a granulase reaction with iodine and the absence of catalase. Strong powers of fermentation distinguish them from the pathenogenic anerobes. Saprophytic bacteria of the genus *Clostridium* are divided into two groups—agents of butyric acid and aceto-butyl fermentation. They are apparently divided according to whether they give butyric acid or acetone plus butanol.

The following nitrogen fixers of the genus *Clostridium* have been described: *Cl. pasteurianum*, *Cl. butyricum*, *Cl. butylicum*, *Cl. beijerincki*, *Cl. multifermentans*, *Cl. pectinovorum*, *Cl. acetobutylicum*, *Cl. aceticum*, *Cl. felsineum*, *Cl. kluyveri*, *Cl. lactoacetophilum* and *Cl. madisoni* (Sjolander and MacCoy, 1937; Rosenblum and Wilson, 1949). The most developed classifications of anaerobes are those of Bergey (1957) and Prevôt (1957).

Bergey brought all the sporulating, rod-shaped bacteria into the family Bacillaceae, which includes the genus *Bacillus*, uniting the obligate and facultative rod-shaped aerobes and the genus *Clostridium*, to which belong anaerobes that do not form catalase.

Table 6.1 is a compilation of the eleven species of nitrogen-fixing bacteria of the genus *Clostridium* that are described in Bergey's classification. The table includes the principal properties of the anaerobic fixers of molecular nitrogen, but not features common to all species.

Prevôt's classification is more detailed. The 133 nitrogen-fixing species of *Clostridium* belong, according to his classification, to the family Clostridiaceae in the order Clostridiales. The latter is included in the class Sporulales.

Physiology of *Clostridium*

The nitrogen nutrition of the various bacteria of the genus *Clostridium* can be very diverse. They are capable of assimilating many organic nitrogen compounds and also nitric acid and ammonium salts. According to Emtsev (1962), some cultures of *Clostridium* do not fix molecular nitrogen in media not containing bound nitrogen. Nitrogen is fixed only in substrates containing small quantities of organic nitrogen compounds.

Several sets of results indicate that the most vigorous nitrogen accumulator among these micro-organisms is *Cl. pasteurianum*, which usually binds about 5–10 mg of nitrogen per gram of carbon source consumed. But there are indications that other representatives of the genus can bind considerably larger quantities of free nitrogen (Parker, 1954).

It should be noted that the level of nitrogen accumulation depends not only on the species (or strain) of bacteria, but also on the conditions of cultivation. This may explain the considerable variations in reports of nitrogen fixed by anaerobes. Thus 1–4 mg of nitrogen has been said to be assimilated per gram of sugar utilized (Winogradskii, 1893, 1895; Willis, 1934; Sjolander and MacCoy, 1937), while there is also information that this value is 4–12 mg and more (Nersesyan, 1930; Bodily, 1939; Jensen and Spencer, 1947; Alexander, 1961). In Rybkina's experiments (1960) the amount of nitrogen assimilated reached 16 mg, while Parker (1954) found that it was 27 mg per gram of energy source consumed.

It should be noted that Parker (1954) used a nutrient medium rich in growth substances and vitamins. Others also noted that if elective media for nitrogen-fixers were replaced by optimum media a considerable increase in the amount of nitrogen fixed could be achieved (Rosenblum and Wilson, 1949; Hino, 1955; Emtsev and L'vov, 1965). In several cases nitrogen assimilation is no less vigorous than it is by *Azotobacter*.

According to Emtsev (1960) *Cl. pasteurianum* fixes nitrogen much more actively in the presence of *Bac. closteroides*—its constant companion. Evidently, this micro-organism produces a series of biologically active compounds that improve the development of *Cl. pasteurianum*.

All these factors make it impossible to measure with any precision the level of nitrogen accumulation by anaerobic micro-organisms in the soil. Indirect pointers however indicate that it cannot be substantial and hardly exceeds a few kilograms of nitrogen per hectare per year (Tyurin, 1957; Ross, 1960). When

Table 6.1. Principal Properties of Nitrogen-Fixing Species of *Clostridium*

Property	*Cl. butyricum*	*Cl. butylicum*	*Cl. beijerinckia*	*Cl. multi-fermentans*	*Cl. pasteurianum*
Shape of cell	Straight or slightly bent rods: solitary, paired, chains	Rods; solitary	Rods; solitary sometimes chains	Rods; solitary or short chains	Rods; solitary or chains
Size of cell (microns)	0·7 × 5·0–7·0	—	—	—	0·9–1·7 × 3·5–4·7
Shape of spore	Oval	Ovoid	Oval	Ovoid	Characteristic presence of 'capsule'
Size of spore (microns)	—	—	1·5 × 2·0	—	1·5–2·0
Position of spore	Non-central to subterminal	Central	Subterminal or terminal	Subterminal or terminal	—
Gram staining	Positive in young cultures	Positive	Positive	Positive in young cultures	Positive
Formation of granulase	In stage of sporulation	In young cultures	At start of sporulation	Yes	Yes
Character of colonies in medium with glucose (surface growth in anaerobic conditions)	Round and slightly irregular, weakly convex, moist, cream–white colonies	Moist, round or irregular, convex, cream colonies	Moist, round or slightly irregular, convex, white or cream	Grey, non-transparent, convex with sharp edges with irregular outlines. With age the colonies turn white and viscous	Round, slightly convex moist, cream granular structure with dense centre and even margins
Distinguishing features	Utilizes potato and maize starch	Utilizes potato starch not maize starch	Does not utilize potato or maize starch	Haemolysis blood agar	Does not utilize calcium lactate or starch. Presence of spore in 'capsule' characteristic
Optimum temperature (°C)	30–37	30	30	37	25

Cl. madisonii	Cl. acetobutylicum	Cl. lactoacetophilum	Cl. kluyverii	Cl. felsineum	Cl. pectinovorum
Rods; solitary or short chains	Rods; solitary or paired but not in chains	Rods; solitary or paired, short chains	Straight or slightly bent rods, solitary paired, rare chains	Rods; single, paired, chains	Rods; some-times bent, solitary less often short chains
0·5–1·0 × 3·0–5·8	0·6–0·7 × 2·6–4·7	0·7–0·9 × 3·0–8·0	0·9–1·5 × 3·0–11	0·3–0·4 × 3·5–5·0	0·5–0·6 × 3·9–4·2
Ellipsoid	Ovoid	Ovoid	Ovoid	Ovoid	Ovoid
0·7–1·3 × 1·3–2·4	—	1·1 × 1·5	1·3 × 1·8	—	1·4 × 2·3
Subterminal or terminal	Non-central	Subterminal	Terminal	Subterminal	Terminal
Positive in young cultures	Positive, pas-sing into negative	Positive, pas-sing into negative	Usually nega-tive, some weakly positive in young culture	Positive, pas-sing into negative	Positive
In young cultures	In stage of sporulation	—	—	In stage of sporulation	In stage of sporulation
Dark, cream round, convex mucilaginous—viscous	Dense, convex, quite regular, non-transparent, cream or white	Dense, dark grey with filamentous outgrowths, large lobular margin, jagged, rubbery, dia-meter 1–2 mm	No growth	Convex, smooth, slightly irregular shape, yellowish orange to brown	Round or irregular, smooth or pectinate, often star shaped, sometimes com-pact, sticky, waxy. Comes away from agar completely.
Forms hydrogen sulphide	Liquefies coag-ulated albumin, vigorously ferments or actively coagu-lates milk. Does not ferment pectin	Ferments lactate with formation of butyric acid	Ferments ethanol with formation of caproic acid	Forms yellowish-orange pigment in media. Liquefies gelatine	Liquefies gelatine
29–33	37	39	34	37	27

the medium contains considerable quantities of bound nitrogen the assimilation of molecular nitrogen is suppressed. Individual investigators have approached differently the question of the doses of nitrogen compounds that depress nitrogen fixation by anaerobes. Thus Winogradskii (1893), MacCoy et al. (1928) and Willis (1934) considered that nitrogen assimilation ceases when the ratio of ammoniacal nitrogen to carbon source is 6:1,000; Omeyanlskii (1913) and Fedorov (1952) gave a ratio of 15:1,000, while Truffaut and Bezssonoff (1921) gave one of 62:1,000. Possibly these discrepancies arose because the experiments were conducted with different media containing different microbial cultures. It is interesting that Rabotnova et al. (1952) showed that the level of nitrogen fixation does not decrease in freshly isolated cultures of Clostridium when the medium is enriched with bound nitrogen.

Reports of the effect of mineral nitrogen fertilizers on Clostridium are highly conflicting. In some cases the introduction of mineral forms of nitrogen had no effect on the numbers of anaerobic nitrogen fixers, while in other cases their development has been suppressed (Willis, 1934; Bylinkina and Fishkova, 1953; Danilevich and Tsyba, 1959; Golovacheva, 1960). Organic fertilizers usually stimulate the multiplication of Clostridium in the soil (Jensen and Swaby, 1940; Voznyakovskaya and Ryzhkova, 1954; Golovacheva 1960; Emtsev, 1961; Obozov, 1962; Rankov, 1964; Rovira, 1965; Skalon, 1965). As a rule soils with a high content of organic matter are richer in clostridia (Vandecaveye and Katznelson, 1933; Zimny, 1961; Zhdannikova, 1963).

For their carbon nutrition bacteria of the genus Clostridium may utilize various compounds which usually serve also as a source of energy. The absence of a narrow selective capacity for sources of nutrition and energy explains the almost universal distribution of Clostridium (Omelyanskii, 1923).

According to some findings clostridia can consume monosaccharides, disaccharides, certain polysaccharides (dextrin, starch), polyhydric alcohols and other compounds. The range of assimilable carbon sources differs for different species, and this is used as a criterion in classification.

When clostridia decompose carbohydrates anaerobically, organic acids, butanol, acetone and other substances (ethanol, CO_2, H_2) accumulate in the medium. The proportions of the various products of fermentation vary in different cultures and also depend on the stage of fermentation that has been reached. In 1910, Kirov noticed changes in the proportions of acetic and butyric acids during fermentation. Lehmberg (1956) pointed out that more acetic acid accumulates at the beginning of fermentation, produced by Cl. butyricum, while more butyric acid accumulates later.

Shaposhnikov (1948) substantiated the idea of a biphasic anaerobic fermentation of carbohydrates, according to which, in the first phase the bacteria accumulate more oxidized products and in the second they accumulate more reduced products. Thus, the agent of acetone-butyl fermentation initially forms acetic and butyric acids and then butanol, acetone and ethanol. During this fermentation carbon dioxide and hydrogen also evolve.

As with other soil micro-organisms the bacteria of the genus *Clostridium* obtain phosphorus and potassium from mineral compounds. Phosphorus may also be assimilated from several organic substances.

Fertilizing soils with phosphorus-potassium salts stimulates to a greater or lesser degree the development of the clostridia (Waksman, 1927). In some cases there is no effect (Bylinkina and Fishkova, 1953) or even a negative effect (Zimny, 1961). This may be due to the fact that there is already sufficient phosphorus or potassium in the test soils.

Golovacheva (1960) considered *Clostridium* unable to utilize the phosphorus of compounds inaccessible to *Azotobacter*, for in her experiments, clostridia did not respond to phosphorus fertilizers added to the soil, whereas *Azotobacter* responded by beginning to develop vigorously.

Clostridia are not considered to be very calcium-sensitive. Nevertheless, some work has indicated that liming or composting the soil leads to a substantial increase in the multiplication of *Clostridium* (Willis, 1934; Abyzov, 1961; Ulasevich, 1965; Komarova, 1965). Slavnina and Potekhina's results (1964) are in agreement; decalcination of dark grey and light grey soils led to an appreciable decrease in numbers of *Clostridium*. Halverson (1958) demonstrated the need for calcium for spore formation in *Clostridium*. A lack of calcium leads to a decrease in the heat stability of the spores.

Several investigators have noted a positive effect of trace elements on the multiplication and fixation of nitrogen by clostridia (Bortels, 1930; Burk and Horner, 1935; Jensen and Spencer, 1946; Turchin *et al.*, 1957; Zimny, 1961). Thus, for example, small doses of Na_2MoO_4 (5–10 γ per litre of medium) significantly enhanced the fixation of nitrogen by anaerobes. In the view of S. and H. Krzemieniewski (1907) the favourable effect of humic acid on *Clostridium* is determined by its content of molybdenum. For some strains of *Clostridium* molybdenum can be replaced by vanadium (Jensen and Spencer, 1946).

There are indications that the multiplication of *Clostridium* is stimulated by salts of nickel and cobalt (Faguet and Goudot, 1961), boron (Abutalybov and Gazieva, 1961), manganese and copper (Abutalybov and Gazieva, 1961). The requirements of *Clostridium* for iron salts have also been demonstrated in media containing no bound forms of nitrogen (Fedorov, 1952; Carnahan and Castle, 1958; Pulay *et al.*, 1959; Dinevich, 1961).

Effect of External Factors on *Clostridium*

Clostridia are relatively resistant to acid and alkaline reactions in the medium and can develop within quite a wide pH range. The minimum pH for these micro-organisms is about 4·7 (Swaby 1939,) and the maximum is above 8·5. According to Emtsev (1959) the optimum conditions for the development of

Clostridium are within the limits of pH 5·9 to 8·3. According to Norris (1962) these limits are 6·9 to 7·3. A decrease in pH unfavourably affects the effectiveness of nitrogen fixation (Hino, 1955). According to Turk (1935) *Clostridium* binds considerably more nitrogen at pH 4·95 than at pH 8·03. In the first case accumulation of nitrogen was found to be 4·21 mg per gram of carbon utilized, in the second only 1 mg. Mossel and Ingram (1955) showed that strains of *Clostridium* with sharply marked proteolytic properties are more sensitive to an acid medium than strains without high proteolytic activity.

Table 6.2. Content of Nitrogen-Fixing Bacteria of the Genus *Clostridium* in the Profile of Soddy-Podzols

Depth (cm)	Virgin Soil				Cultivated Soil			
	Humus (%)	Total No. of cells	Vegeta-tive cells	Spores	Humus (%)	Total No. of cells	Vegeta-tive cells	Spores
0–10	1·95	69·6	66·7	2·9	1·88	27·25	24·53	2·72
10–20	1·73	74·9	72·23	2·67	1·93	14·17	6·54	7·63
20–30	0·62	2·7	0	2·7	1·19	6·54	3·82	2·72
30–40	0·61	0·31	0	0·31	0·75	2·82	2·54	0·28
40–50	0·61	0·32	0	0·32	0·58	0·28	0	0·28
50–60	0·50	0·32	0	0·32	0·47	0·29	0	0·29

Numbers are expressed in thousands per gram of absolute dry weight.

The relation of *Clostridium* to the pH of the medium explains its widespread distribution in different soils, including those with a low pH. This has been established in many investigations (for example, Dorner, 1924; de Barjac, 1955; Corke, 1958; Daraseliya, 1961; Meiklejohn, 1962; Těsić and Todorovic, 1963; Di Menna, 1966; Ryys, 1963, 1966).

The effect of the air and water regime on clostridia has been studied in some detail. Being anaerobic they tolerate a high degree of soil moisture. But the optimum degree of wetting is determined by the type of soil and its content of organic matter (Jensen, 1940; Zimny, 1961).

There is considerable evidence that clostridia develop when moisture is 60–80 per cent of the total capacity (Kvasnikov and Pervushina-Grosheva, 1952; Nabiev, 1955). Usually, the upper layers of the soil, richer in organic substances, contain more clostridia than the deeper layers. As an example, Table 6.2 shows Emtsev's findings of the distribution of nitrogen fixers of the genus *Clostridium* in the soddy-podzols of the Moscow region.

As Table 6.2 shows, *Clostridium* penetrates the deeper layers of the cultivated soil. This has to do with the greater humus content of the soils used for farming.

Per (1957) also noted that in silts the multiplication of *Clostridium* depends on the content of organic compounds in the soil.

It should be noted that many bacteria inhabit the upper layers of soil as vegetative cells. These layers are quite well aerated, but it can be assumed that *Clostridium* inhabits microzones where the access of air is impeded (Parker, 1954). The onset of anaerobiosis within soil aggregates may be promoted by aerobic micro-organisms developing on their surface (Omelyanskii, 1923; Krishna, 1928; Shklyar, 1958). *Clostridium* itself may cause the redox potential of the medium to be lowered, and this would facilitate its development in a particular microzone of soil (Rabotnova, 1955). It should also be noted that some cultures of *Clostridium* are quite indifferent to the presence of oxygen in the medium (Pringsheim, 1909; Hart, 1955; Pochon and de Barjac, 1958; Popova, 1961).

The moisture content of the arable layer may have significant influence on the distribution of clostridia in the soil profile (Rybkina, 1960; Zakharchenko, 1966).

According to Omelyanskii (1926), Nabiev (1955) and others, clostridia readily tolerate a moisture deficit and even complete drying of the soil. This may be due to the fact that in a time of drought many clostridial cells form spores.

Temperature affects bacteria species of the genus *Clostridium* differently— they include both mesophilic and thermophilic bacteria. The mesophils fix molecular nitrogen. The thermophilic race of *Cl. pasteurianum* with a temperature optimum of 53°–55°C, described by Imshenetskii and Solntseva (1945), does not assimilate molecular nitrogen.

The optimum temperature for development of the mesophilic forms of *Clostridium* is usually within the range 25°–30°C, sometimes with a higher upper limit. The maximum temperature is between 37° and 45°C (Emtsev, 1965). Above 35°C the bacterial cells usually take on an abnormal shape (Vinogradskii, 1902; Omelyanskii, 1923; Fedorov, 1952).

Di Menna (1966) thought that the spread of clostridia in soils was limited more by low temperatures than by pH. Thus their content is insignificant in the soils of northern Greenland with a pH of 5·6 to 8·2 (Jensen, 1951). Psychrophilic forms of *Clostridium*, developing well at 0°C, have been observed. Their temperature maximum is below 35°C and their optimum is between 25° and 30°C (Sinclair and Stokes, 1964; Beerens *et al.*, 1965). Nitrogen fixation has not been established in these cases.

The spores of *Clostridium* are very resistant to high temperatures. They withstand 75°C for 5 hours and 1 hour at 80°C (Omelyanskii, 1923). The spores of the thermophilic race of *Clostridium* die when boiled for 30 minutes. A higher temperature (110°C) rapidly kills them.

Much work has been done on the factors that influence the metabolism of *Clostridium* in the soil. It may be in a symbiotic or metabiotic relation with soil bacteria. For example, there is the symbiosis of *Clostridium* and *Azotobacter* (Pringsheim, 1908; Omelyanskii, 1923; Shklyar, 1956). Evidently, *Azotobacter* improves conditions for *Clostridium* by taking up oxygen, while *Clostridium* converts organic compounds that are inaccessible to *Azotobacter* to assimilable organic acids.

Many investigators have noted that clostridia multiply more intensively in the rhizosphere than in the rest of the soil (Fig. 32)—the rhizosphere contains ten to a hundred times more cells. This is understandable, for many products of root exomosis provide valuable food for anaerobes (for example Katznelson, 1946; Gvozdyak, 1959; Ross, 1960; Pertseva and Vasyuk, 1962; Skalon, 1965;

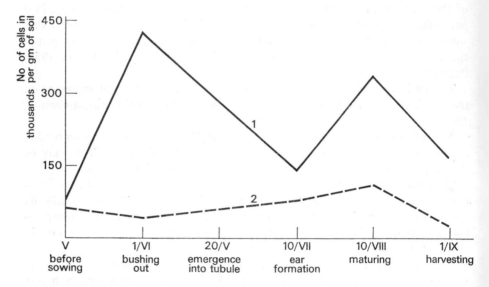

Fig. 32. Variation in the numbers of butyric acid bacteria in the rhizosphere of spring wheat (soddy-podzols) (Emtsev) in (1) rhizospheric soil; (2) soil without plants.

Nagrebetskaya, 1965; Vigorov and Tribunskaya, 1965; Til'ba and Golodyaev, 1966).

Not all the plants promote the multiplication of *Clostridium* equally well (Emtsev, 1959, 1962; Tarvis, 1960; Kolesnik, 1965; Gurfel and Lasting, 1966). It multiplies more vigorously in the rhizosphere of leguminous plants than in that of grasses. According to Komarova (1965) the rhizosphere of perennial grasses is richer in clostridia than the root zone of annual farm plants. *Cl. amylobacter* is abundant in the rhizosphere of flax (Berezova, 1940). The literature strongly suggests, however, that the type of soil has a greater effect on the numbers and composition of populations of clostridia than does the plant cover (Emtsev, 1959; Zimny, 1960, 1961; Vojnova-Rajkova, 1965 and Table 6.3).

In the rhizosphere most members of the genus *Clostridium* are found in the vegetative form, which means that they are active. According to Emtsev (1962) the rhizosphere is the site of multiplication of *Cl. pasteurianum, Cl. butyricum,*

Table 6.3. Effect of Type of Soil-Forming Process on the Numbers of Butyric Acid
Bacteria and *Clostridium pasteurianum*

Soil	State of soil	Total No. of cells of butyric acid bacteria	No. of cells of *Clostridium pasteurianum*	Literature source
Soddy-podzolic	Forest	600–1,000	6—25	Nikitina and Abenova (1960)
	Forest	—	70	Emtsev (1958)
	Meadow	250	—	Abduzhalalova (1962)
	Meadow	1,000	1–100	Zhukova (1956)
	Meadow	—	60	Emtsev (1958)
	Meadow	—	190	Miroshnichenko (1960)
	Cultivated	18,000	—	Berseneva and Morozova (1958)
	Cultivated	1,000	—	Pertseva and Golikov (1958)
	Cultivated	—	50	Emtsev (1958)
Black-earth	Virgin	10	—	Bylinkina (1953)
	Cultivated	10–1,000	1–10	Bylinkina (1953) Bylinkina and Pertseva (1953)
Chestnut	Virgin	0·6	0·01–0·1	Obozov (1962)
	Virgin	14·2	—	Chulakov (1961)
	Cultivated	30	22	Tarvis (1960)
	Cultivated	—	0·1–1	Sidorenko (1959)
Red earth	Cultivated	—	0·1	Daneliya (1954)
Yellow earth	Virgin	—	1–10	Pakusin (1960) Malyugin, Lazarev and Bylinkina (1958)
Arid and semi-arid	Virgin	0·1–1	0·01	
		0·01–0·1	0·01–0·1	Genusov, Drabkina and Stibman (1956)
		0·6	0·01–0·1	Obozov (1962)
	Cultivated	—	0·1–1	Kvasnikov and Pervushina-Grosheva (1953)
	Cultivated	0·25–11	0·06	Vukhrer (1962)
	Cultivated	25	—	Buyanovskii (1959)

Numbers are expressed in thousands per gram of soil. —, No observations made.

Cl. felsineum and *Cl. pectinovorum.* But no appreciable increase in the numbers of *Cl. acetobutylicum* is observed there (Fig. 33), evidently because the development of acetone-butyl bacteria requires protein substances which are present in insignificant amounts in the root secretions. Numbers of *Cl. acetobutylicum* increase considerably when horse manure is added to the soil.

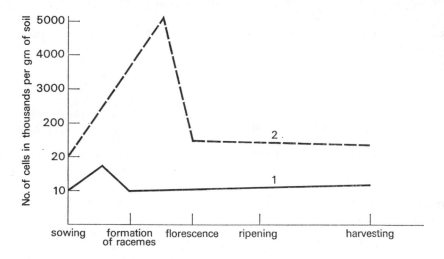

Fig. 33. Variation in the numbers of anaerobic nitrogen-fixing bacteria in the rhizosphere of buckwheat (sand culture). (1) *Cl. pasteurianum.* (2) *Cl. butylicum.*

There are indications in the literature that the multiplication of butyric acid bacteria depends on the phase of development of the plants concerned (for example Pertseva and Vasyuk, 1962; Lupinovich and Yanushkevich, 1959; Geller, 1960; Emtsev, 1961; Filippova *et al.*, 1966). The maximum number of cells of *Cl. pasteurianum* is usually observed early in the growing season of the plants. The numbers of other anaerobes, for example *Cl. butyricum*, change little in the rhizosphere during the summer.

Ecology of *Clostridium*

Clostridia, as we have noted, are extremely widespread in different soils. They are found in the soil of the Arctic (Isachenko, 1914) and Greenland (Barthel, 1918) and in all the soils investigated in any of the continents (for example, Freudenreich, 1903; Bredemann, 1909; Truffaut and Bezssonoff, 1925; Swaby, 1939; Parker, 1954; Bulgadaeva, 1961; Ilyakhina, 1962; Meiklejohn, 1956,

1965; Chang and Knowles, 1965; Dudareva, 1966; Potekhina, 1966; Zimenko, 1966). They can be considered to be universal in occurrence.

Emtsev (1965) summarized investigations of *Clostridium* in the soils of the USSR. His generalization allows us to draw several fundamental conclusions. It seems that the numbers of *Clostridium* may vary considerably in the soil zone. Usually cultivated soils, especially those fertilized with organic substances, are richest in representatives of the genus *Clostridium*. There is also no doubt that northern soils contain more clostridia than do southern soils. This is evidently connected with features of the organic matter in the soil. Table 6.4 show Emtsev's analyses (1965) of butyric acid, bacteria and *Cl. pasteurianum* in different soils of the USSR.

Until recently practically no differential counts had been made of *Clostridium* in different soils. Such an attempt was made by Emtsev (1966) who, using elective media for each species, was able to monitor the presence and numbers of *Cl. pasteurianum, Cl. butyricum, Cl. pectinovorum* and *Cl. acetobutylicum*. His results (1965) are presented in Table 6.4. They show that counts of *Cl. pasteurianum, Cl. butyricum* and *Cl. pectinovorum* are less in northern than southern soils. In the northern zone the upper layers of virgin soils are often considerably richer in clostridia than cultivated soils. This can be explained by the accumulation on the surface of virgin soils of considerable amounts of organic material (litter) which promotes the multiplication of anaerobic nitrogen fixers.

There are considerably more cells of *Cl. acetobutylicum* in southern than in northern soils, possibly because this species has an enzyme mechanism that enables it to assimilate compounds that are unavailable to other anaerobes. It can be assumed that in southern soils the readily mobile organic compounds rapidly decompose, so that the multiplication of micro-organisms assimilating more complex substances is promoted. Another important factor is that *Cl. acetobutylicum* develops best on substrates rich in protein. In southern soils with rich populations of microbes the protein store is more substantial than in the soils of the north.

Deeper into the soil the numbers of clostridia decrease (for example, Nabiev, 1955; Manteifel and Kozlova, 1955; Tarvis, 1960; Lazarev, 1960; Ryys, 1963; Visser, 1964; Vavulo, 1966; Gurfel and Lasting, 1966). The depth to which these bacteria penetrate depends on the profile of the soil and the status of the organic matter in it (Samtsevich, 1966). Some idea of the distribution of clostridia in the soil is given by Table 6.4.

Many investigators have noted little change in the number of clostridia during the year (for example Zimny, 1960; Daraseliya, 1961; Rouquerol, 1962). The seasonal fluctuations sometimes observed (Loub, 1965; Enikeeva and Gar, 1966) may be due to various factors—changes in the degree of aeration of the soil (Koshkarova, 1959), or the rate of entry of organic matter to the soil (Sheloumova *et al.*, 1927; Vavulo, 1966) and so on.

Table 6.4. Species of butyric acid and acetone-butyl bacteria in different soils of the USSR

Thousands of bacteria per gram of absolutely dry soil

Republic and region	Soil	Cl. pasteuranium			Cl. butyricum			Cl. pectinovorum			Cl. acetobutylicum		
		Forest	Virgin meadow	Culti-vated	Forest	Virgin meadow	Culti-vated	Forest	Virgin meadow	Culti-vated	Forest	Virgin meadow	Culti-vated
RSFSR, Arkhangel	Soddy-podzolic	—	672·0	300·0	—	672·0	300·0	—	784·0	720·0	—	0·7	0·6
RSFSR, Moscow	Soddy-podzolic	14·3	78·0	5·1	0·27	325·0	5·1	0·77	32·5	6·18	0·99	—	2·58
RSFSR Voronezh	Black earth of Kamen steppe	202·8	14·0	8·3	10·9	5·0	34·5	3·9	5·0	96·6	0·93	0·05	—
RSFSR, Saratov	Chestnut	6·3	2·6	—	2·6	21·0	20·8	6·3	—	26·2	7·35	13·65	26·0
Kaz. SSR, Karagand	Chestnut	—	2·5	0·7	—	6·12	0·63	—	0·25	0·06	—	—	—
Uzb. SSR, Tashkent	Typical grey earth	—	—	—	—	—	—	—	—	—	—	—	—
Tashkent	Light grey earth, State farm	—	—	1·14	—	—	—	—	—	45·5	—	—	—
	'Paknta-Aral'	—	0·25	0·7	—	0·3	1·11	—	7·07	6·06	13·13	—	30·3
Georgia SSR	Red earth	—	0·8	—	—	14·3	67·2	—	6·6	63·6	27·0	2·75	28·0
SSR	Yellow earth	0·6	6·5	2·7	0·64	27·0	2·65	26·7	—	—	0·06	27·0	74·2

—, No observations made. The analyses were made on level A_o of virgin soils and the arable layer.

Effect of *Clostridium* on Plant Growth

The influence of clostridia on the growth and productivity of farm plants has been the subject of a great deal of work, but the problem is not yet solved conclusively.

Among the first to tackle this problem were Truffaut and Bezssonoff (1925). Using Shulov's method, they infected maize with pure cultures of nitrogen-fixing bacteria (*Azobacter*, *Cl. pasteurianum* and *Bacillus truffauti*). Sterilized maize seed was sown in calcined sand in a porous vessel surrounded by sterile solution containing all necessary nutrient elements except nitrogen, which was supplied entirely by nitrogen-fixing bacteria. No carbon-containing organic

Table 6.5. Effect of Culture of *Clostridium* on the Yield of Flax

| Experiment | Yield per vessel | | | |
| | Sand | | Soil | |
	g	%	g	%
Controls	2·31	100	4·49	100
Clostridia introduced, thousands of cells per kg				
100	2·61	113	4·51	101
200	2·38	103	4·72	105
400	2·32	100	4·75	106

compounds were introduced into the medium and the nitrogen-fixing bacteria were able to use as nutrients only the products of root exosmosis.

Control plants with no source of bound nitrogen died. Inoculation with a mixture of cultures of nitrogen-fixers ensured the normal growth of maize, which gave a yield of 40 g dry weight per vessel. A nitrogen increment of 585 mg was recorded in the system. It was calculated that the assimilation of that much nitrogen by the bacteria would require about 60 grams of organic compounds, since for each gram of organic substance consumed, about 10 mg of molecular nitrogen is assimilated. Thus, the results of the experiment described are difficult to explain, and have not been confirmed by anyone else.

Berezova (1940) carried out some experiments with flax in non-sterile conditions. Vessels were filled with either sand fertilized with K and P or soddy-podzolic soil, which was then inoculated with different numbers of clostridia. As Table 6.5 shows, no significant gains in the harvest were recorded. Nevertheless, Berezova considered that the clostridial culture gave a positive result in sand. It is difficult to agree with this conclusion, for the results of the experiments were not treated statistically. We consider them to be within the limits of experimental error. In another experiment carried out on heavier soil, the introduction of clostridia resulted in a reduction of flax yield.

The paper published by Shkylar in 1958 provided information about the effectiveness of bacterization of farm plants by nitrogen-fixing bacteria, including *Clostridium*. The experiment was carried out in little cultivated, medium podzolized, loamy, sandy soil. The soil was not sterilized and 2,000 cells of nitrogen-fixing bacteria were introduced per gram. Table 6.6 shows that the introduction

Table 6.6. Effect of nitrogen-fixers on the yield of farm plants

Bacteria introduced	Millet (1954)		Maize (1955)		Barley (1955)		
	Straw	Grain	Straw	Grain	Green mass	Straw	Grain
Azotobacter	106·0	109·7	101·8	98·7	117·0	103·0	100·0
Clostridium	127·4	148·0	81·5	93·7	116·0	104·0	100·0
Both	112·6	126·4	127·3	117·4	139	115·0	112·0

Figures are expressed as percentages of controls.

of neither *Clostridium* nor *Azotobacter* had a constant effect. Only in one case, in 1954, was the yield of millet increased significantly by the introduction of *Clostridium*. This may have been a random result. A mixture of *Clostridium* and *Azotobacter* was more effective, but no scientific explanation is given for this.

In Emtsev's experiments (1961), the injection of buckwheat with a culture of *Clostridium* (various strains) resulted in an increase in the harvest yield of 21–28 per cent, while the yield of oats was increased by 32–43 per cent. This result with oats was repeated in a small plot experiment (harvest increased by 28 per cent).

Finally, we must mention the work of Rovira (1965) who bacterized wheat grown in quartz sand and silty loam with cultures of nitrogen-fixing bacteria. *Cl. pasteurianum*, *Az. chroococcum* and *Bact. polymyxa* were used; the best effect was obtained with the first of the three. *Cl. pasteurianum* produced an increase of 57 per cent in the yield on sand. On silty loam it produced an increase of 46 per cent.

Thus, in the experiments of both Emtsev and Rovira, cultures of *Clostridium* appreciably improved the growth of the plants into which they were inoculated. It is not possible, however, to say whether this effect was connected with nitrogen fixation or with the action of biologically active substances formed by the micro-organisms that were used (Raibaud *et al.*, 1965).

On the whole, the effects of bacterization of farm plants with *Clostridium* seems to be very unreliable. The practical potentialities of this micro-organism need to be investigated further.

References

Abduzhalalova, M. U. 1962. *Pochvenno-mikrobiologicheskieusloviya proizrastaniya rastenii na vnov' osvaivaemykh pochvakh Leningradskoi oblasti i priemy ikh uluchsheniya* (Soil-microbiological conditions of the germination of plants on soils of the Leningrad province which have been newly brought under cultivation and means of improving them). Author's abstract of M.Sc. thesis, Leningrad.

Abutalybov, M. G. and Gazieva, N. I. 1961. Effect of trace elements on the development of *Azotobacter* and on the activity of nitrifying micro-organisms of the soil. *Uch. zapiski Azerb. un-ta, seriya biol.*, (1), 45–53.

Abyzov, S. S. 1961. Multiplication of fatty-acid bacteria in the preparation of peat composts with the addition of various doses of lime. *Byull. nauchno-tekhn. inf. po s.-kh. mikrobiologii*, 9 (1), 6–9.

Akbroit, E. Ya. 1960. Possibility of transition of anaerobes into aerobes. *Sb. trudov Odessk. med. In-ta*, (13), 230–237.

Alexander, M. 1961. *Introduction to Soil Microbiology*. L4 N.J., USA.

Barjac, H. de. 1955. *Thèses Sciences*. Paris.

Barthel, Chr. 1916–1918. Recherches bactériologiques sur le sol du Gröenland Septentrional: in *Den II. Thule Expedition til Grönlands Nordkyst København*, 1.

Beerens, H., Sugama, S. and Tahon-Castel. 1965. Psychrotrophic clostridia. *J. Appl. Bacteriol.*, 28 (1), 36–48.

Berezova, E. F. 1940. Role of *Bac. amylobacter* in the development of flax. *Mikrobiologiya*, 9 (9–10), 813–823.

Beresneva, V. N. and Morozova, N. F. 1958. Microflora of soddy-podzolic soil with different methods of cultivation. *Trudy Vses. n-i. in-ta s-kh. mikrobiologii.*, 14, 63–74.

Bergey. 1957. *Manual of Determinative Bacteriology*, 1957, Baltimore, Williams and Wilkins Comp.

Blinkov, G. N. and Romanova, A. A. 1964. Distribution of *Clostridium pasteurianum* in the USSR: in *Sbornik statei biologo-khimicheskago i fiziko-matematichoskogo fakul'tetov Tomskogo gosudarstvennogo pedagogicheskogo instituta* (Collection of papers of the biological-chemical and physico-mathematical faculties of the Tomsk State Pedagogic Institute). *Uch. zapiski*, 21, 114–123.

Bodily, H. L. 1939. The morphological, biochemical and serological properties of *Bacillus pasteurianum* and its ability to fix atmospheric nitrogen in comparison with other anaerobic bacilli. *J. Bacteriol*, 38 (1), 120.

Bortels, H. 1930. Molybdän als Katalisator bei der biologischen Stickstoffbildung. *Arch. Mikrobiol.*, Abt. 1, 333–342.

Bortels, H. 1936. Weitere Untersuchungen über die Bedeutung von Molybdän, Vanadium, Wolfram und anderen Erdschenstoff für stickstoffbindende und andere Mikroorganismen. *Zentralbl. Bacteriol., Parasitenkunde, Infektionskrankh und Hyg.*, Abt. 2, 951, 193–218.

Bredemann, G. 1909. *Bac. amylobacter*. A. M. et Bredemann. *Zbl. Bacteriol., Parasitenkunde, Infektionskrankh und Hyg.* Abt. 2, 23, 385–568.

Bulgadaeva, R. V. 1961. Microflora of the soddy-calcareous brown soils and black earth of the Angar region. *Trudy Vost. Sib. fil. Sib. otdel. Akad. Nauk SSSR, Ser. Biol.*, (27), 114–123.

Burk, D. and Horner, C. 1935. The specific catalytic role of molybdenum and vanadium in nitrogen fixation and amide utilisation by *Azotobacter. Trans. Third Internat. Congr. of Soil Sci.*, **1**, 152–155.

Buyanovskii, G. A. 1959. The microflora of the meadow grey earth soils of the Karabakh steppe. *Dokl. Akad. Nauk AzSSR*, **15** (2), 159–162.

Bylinkina, V. N. 1953. Effect of forest belts and irrigation on the microbiological status of black earth soils of the Kamen steppe. *Trudy Vses. n-i. in-ta s-kh. mikrobiologii*, **73**, 22–27.

Bylinkina, V. N. and Pertseva, A. N. 1953. The microflora of black earth soils and its importance in the root nutrition of plants. *Trudy Vses. n-i. in-ta s-kh. mikrobiologii*, **13**, 5–13.

Bylinkina, V. N. and Fishkova, E. S. 1953. Effect of prolonged use of fertilizers on the microflora of black earth soils. *Trudy Vses. n-i. in-ta s-kh. mikrobiologii*, **13**, 22–27.

Carnahan, J. E. and Castle, J. E. 1958. Some requirement of biological nitrogen fixation. *J. Bacteriol.*, **75** (2).

Chang, P. C. and Knowles, R. 1965. Non-symbiotic nitrogen fixation in some Quebec soils. *Canad. J. Microbiol.*, **11** (1), 29–38.

Chulakov, Sh. A. 1961. Microbiological characterization of virgin and worked darkchestnut soils of the Akmolinsk region. *Trudy in-ta mikrobiologii i virusaologii. Akad. Nauk KazSSR*, **4**, 146–165.

Corke, C. T. 1958. Nitrogen transformations in Ontario forest podzols. *Forest Soils Conf.*, 116–121.

Daneliya, M. K. 1954. Microbiological processes in red earth soil and the effect of various fertilizers on the species composition of bacteria. *Byull. Vses. n.-i. in-ta chaya i subtropicheskikh kultur.*, (2), 123–131, Makharadze-Anaseuli.

Danilevich, V. M. and Tsyba, M. M. 1959. Effect of various fertilizers on the microflora of newly reclaimed peat soil of the transitional type from the marshes of the Karelsk ASSR, *Izv. Karelsk. i Kolsk. fil. Akad. Nauk SSSR*, (2), 93–99.

Daraseliya, N. A. 1961. Nitrogen fixers in the red earth soils of Georgia.*Pochvovedenie*, (4), 111–114.

Daraseliya, N. A. 1963. Microflora of the red earth and brown forest soils of Georgia. *Trudy in-ta pochvoved. Gruz SSR*, **11**, 55–64.

Di Menna, M. E. 1966. The incidence of *Azotobacter* and of nitrogen-fixing *Clostridium* in some New Zealand soils. *N.Z. Agric. Res.*, **9** (2).

Dinevich, L. S. 1961. Certain growth requirements of *Cl. perfringens. Trudy Mold n-i. in-ta epidemiol. mikrobiol. gigieny*, (5), Kishinev.

Dorner, S. 1924. Beobachtgungen über das Verhalten der Sporen und vegatativen, Formen von *B. amylobacter* A. M. et Bredemann by Nachweiszuchtversuchen. *Landwirtsch. Jahrb. Schweiz.*, **36**, 175–201.

Duda, V. I. 1966. Certain new soil anaerobic bacteria and their role in soil toxicosis: in *Tezisy dokladov III Vsesoguznogo s"ezda pochvovedov* (Third all-Union Congress of soil specialists. Summaries of speeches), Tartu, 102–103.

Dudareva, T. E. 1966. The microflora of the alkaline soils of Siberia: in *Mikroflora pochv severnoi i srednei chasti SSSR* (The microflora of the soils of the northern and central parts of the USSR), Nauka, Moscow, 365–389.

Emtsev, V. T. 1958. Distribution of *Cl. pasteurianum* in the soils and its role in plant nutrition *Nauchn trudy in-ta zemled. Tsentr. r-nov nechernozemn. polosy*, (18), 147–171.

Emtsev, V. T. 1959. Certain problems of the morphology and physiology of nitrogenfixing *Clostridium. N-i. in-t zemled. tsentr. r-nov nechernozemn. polosy*, Moscow.

Emtsev, V. T. 1960. Symbiotic relationships of *Cl. pasteurianum* and *Bac. closteroides*. *Mikrobiologiya*, **29** (4), 529–535.

Emtsev, V. T. 1961. Fixation of atmospheric molecular nitrogen by butyric acid bacteria of the genus *Clostridium*. *Agrobiologiya*, (5), 749–761.

Emtsev, V. T. 1962. Sources of carbon nutrition for the nitrogen-fixing micro-organisms of the genus *Clostridium*. *Mikrobiologiya*, **31** (1), 18–23.

Emtsev, V. T. 1965. Distribution of anaerobic bacteria of the genus *Clostridium* in various soil zones. *Dokl. TSKhA*, (115), Part 2, 133–138.

Emtsev, V. T. 1966. Anaerobic fixers of molecular nitrogen: in *Biologicheskaya azotfiksatiya i ee rol' v sel'skom khozyaistve* (Biological nitrogen fixation and its role in agriculture). *Izd-vo Akad. Nauk SSSR*, Moscow.

Emtsev, V. T. 1966. Anaerobic fixers of atmospheric nitrogen, their geographical distribution and activity: in *IX Mezdunarodnyi Mikrobiologicheskii Kongress. Tezisy dokladov* (Ninth International Microbiology Congress. Summaries of speeches), Moscow, 286–187.

Emtsev, V. T. and L'vov, N. P. 1965. Use of elective and optimal media in soil microbiology. *Izv. TSKhN*, **3**, 117–125.

Enikeeva, M. G. and Gar, K. A. 1966. Excess wetting and the activity of micro-organisms in the soils of certain types of spruce forests of the southern Taiga: in *Vliyanie izbytochnogo uvlazhneniya pochv na produktivnos, lesov* (Effect of excess wetting of the soils on the productivity of forests), Nauka, Moscow, 76–95.

Faguet, M. and Goudot, A. 1961. Sur quelques aspects électroniques de l'action de certains cations metalliques sur la croissance des bactéries aérobies (*Escherichia coli*) et des bactéries anaérobies (*Clostridium sporogenes*). *Ann. Inst. Pasteur*, **101** (6), 860–868.

Fedorov, M. V. 1952. *Biologicheskaya fiksatsiya Azota atmosfery*. (Biological Fixation of atmospheric nitrogen), Sel'khozgiz, Moscow.

Filippova, K. F., Kolotova, S. S. and Ovcharov, K. E. 1966. Trace elements as mediators of the relationships between soil micro-organisms and higher plants: in *Fiziologo-biokhimicheskie osnovy vzaimnogo uliyaniya rastenii v fitotsenoze*. (The physiologico-biochemical foundations of the mutual influence of plants in phyto-coenoses), 291–295.

Freudenreich, B. E. 1903. Über stickstoffbindende Bakterien. *Zbl. Bakteroil., Parasit-enkunde, Infektionskrankh und Hyg.*, Abt. 2, **10**, 16–17.

Geller, I. A. 1960. Microbiological foundations of a rational system of soil treatment in the period of vegetation of sugar beet. *Trudy In-ta mikrobiologi. Akad. Nauk SSSR*, (7), 124–132.

Genusov, A. Z., Drabkina, A. V. and Stibman, B. Sh. 1956. The microflora of the takyr soils of the Kun-Darin plain. *Trudy In-ta pochvovedeniya, Akad. Nauk Uzb SSR*, (2).

Golovacheva, R. S. 1960. Some findings on the effect of fertilizers on the development of nitrogen-fixing bacteria in the soil. *Byull. nauchno-tekhn. inf. Est. in-ta zemled. i melioratsii*, (5), 46–51.

Gurfel, D. and Lasting, V. 1966. Change in the microflora of high bog on cultivation: in *Sbornik nauchnykh trudov Est. n-i. in-ta zemledeleniya i melioratsii. Mikrobiologiya*. (Collection of scientific works of the Estonian Scientific Research Institute of Agriculture and Soil Improvement, Microbiology), **8**, Tallin, 60–71.

Gvozdyak, R. I. 1959. *Vliyanie predposeunoi polimikrobnoi obrabotki semyan pshenitsy i yachmenya na rizosfernuyu mikrofloru i urzohainost'*. (Effect of pre-sowing polymicrobial treatment of wheat and barley seeds on the rhizospheric microflora and yield) Author's abstract of M.Sc. Thesis, Kiev.

Halverson, H. O. 1958. *The Physiology of the Bacterial Spore*. Trondheim (in English).

Hart, M. G. R. 1955. *A study of spore-forming bacteria in soil.* Thesis Univ. London, (quoted by Chang, P. C. and Knowles, R. 1964).

Herzinger, F. 1940. Beitrage zum Wirkungskreislauf des Bors. *Bodenkunde und Pflansernähr.*, (46), 141.

Hino, S. 1955. Studies on the inhibition by certain monoxide and nitrogen oxide of anaerobic nitrogen fixation. *J. Biochem.*, **42** (6), 775–784.

Ilyakhina, Z. V. 1962. Material for the microbiological characterization of the soils of the Krasnodarsk woody-steppe and the contiguous sub-taiga: in *Trudy I Sibirskoi konferentsii pochvovedov* (Proceedings of the First Siberian Conference of Soil Scientists), Krasnoyarsk, 430–438.

Imshenetskii, A. A. and Solntseva, L. I. 1945. Thermophilic variants of butyric acid bacteria. *Dokl. Akad. Nauk SSSR*, **47** (5), 385–392.

Isachenko, B. L. 1914. Investigations on the bacteria of the North Arctic Ocean. *Trudy Murmanskoi ekspeditsii* 1906. (Proceedings of the Murmansk Expedition, 1906), Petrograd.

Jensen, H. L. 1940. Contribution to the nitrogen economy of Australian wheat soils. *Proc. Linnean Soc. N.S. Wales*, **65** (1–2), 1–22.

Jensen, H. L. 1941. Nitrogen fixation and cellulose decomposition by soil micro-organisms. *Clostridium butyricum* in association with aerobic cellulose-decomposers. *Proc. Linnean Soc. N.S. Wales*, **66** (3–4), 239–249.

Jensen, H. L. 1951. Notes on the microbiology of soil from Northern Greenland. *Medd. Grønland*, **142**, 23–29 (in English).

Jensen, H. L. and Spencer, D. 1946. Effect of molybdenum on nitrogen fixation by *Clostridium butyricum*. *Austral. J. Sci.*, **9**, 28.

Jensen, H. L. and Spencer, D. 1947. The influence of molybdenum and vanadium on nitrogen fixation by *Clostridium butyricum* and related organisms. *Proc. Linnean Soc. N.S. Wales*, **72** (1–2), 73–86, 329–330.

Jensen, H. L. and Swaby, K. J. 1940. Further investigation on nitrogen-fixing bacteria in soils. *Proc. Linnean Soc. N.S. Wales*, **65** (291–292), Pt. 5–6, 557–564.

Katznelson, H. 1946. The rhizosphere effect of mangels on certain groups of soil micro-organisms. *Soil Sci.*, **62** (5), 343–354.

Kirov, A. 1910. Investigations on butyric acid fermentation. *Ezhegodnik Kievsk politekhn in-ta*, **1**.

Kolesnik, I. I. 1965. Microbiological characterization of the forest soils of the Donets stations. *Lesovodstvo i agrolesomelior. Resp. mezhved. temat. nauch. sborn.*, (7) 7–13.

Komarova, N. A. 1965. *Rasprostranenie mikroorganizmov, fiksiruyushchikh azot v anaerobnykh usloviyakh pod razlichnymi kul'turami, i ikh znachenie i obogashchenii pochvy azotom* (Distribution of nitrogen-fixing micro-organisms in the anaerobic conditions under various crops and their importance in the nitrogen enrichment of the soil). Author's abstract of M.Sc. Thesis, Saratov.

Koshkarova, G. M. 1959. Development of nitrogen-fixing bacteria and the nitrogen regime in cotton—lucerne crop rotation. *Byull. Azerb. n-i. khlopkovodstva*, (3–4), 41–46.

Krasil'nikov, N. A. and Duda, V. I. 1966. The ultrastructure of outgrowths on the spores of anaerobic bacteria: in *IX Mezdunarodnyi Mikrobiologicheskii Kongress. Tezisy Dokladov* (Ninth International Microbiology Congress. Summaries of speeches), Moscow, 78.

Krasil'nikov, N. A. and Duda, V. I.). (Krasilnikov, N. A. and Duda, V. I. 1964. The fate of nuclear material during sporulation in anaerobic bacteria. *Z. allgem. Mikrobiol.*, **4** (3), 242–248 (in English).

Krishna, P. C. 1928a. Cellulose decomposition products as sources of energy for *Azotobacter* and *B. amlyobacter*. *J. Amer. Soc. Agron.*, **20** (5), 511–514.

Krishna, P. C. 1928b. Nitrogen fixation by soil micro-organisms. *J. Agric. Sci.*, **18** (3), 432–438.

Krzemieniewski, S. and Krzemieniewska, H. 1907. Zur Biologie stickstoffbindender Micro-organismen. *Zbl. Bakteriol., Parasitenkunde, Infektionskrankh und Hyg.*, Abt. 2, **18**, 521–522.

Kvasnikov, E. I. and Pervushina-Grosheva, A. V. 1952. Some aspects of the effect of the moisture content on microbiological processes in the grey earth soils of a cotton field. *Dokl. Akad. Nauk UzbSSR*, (3), 30–34.

Lazarev, S. F. 1960. Formation of a biologically active arable layer of soil in conditions of irrigated cotton growing. *Trudy in-ta mikrobiologii Akad. Nauk SSSR*, (7), 319–327.

Lehmberg, Ch. 1956. Untersuchungen uber die Wirkung von Ascorbinsäure und andere von *Clostridium butyricum* Prazm. *Arch. Mikrobiol.*, **24** (4), 323.

Loub, W. von. 1965. Zur Mikrobiologie mittel und nordeuropaischer Podsole. *Z. Pflanzenernähr., Düng., Bodenkunde*, **111** (3), 157–167, VIII.

Lupinovich, I. S. and Yanushkevich, K. N. 1959. The distribution of micro-organisms in peaty-boggy soil in connection with the development of winter rye: in *Pochvennye issledovaniya BSSR*. (Soil research in the BSSR), Minsk, 5–18.

MacCoy, E., Higby, W. M. and Fred, E. B. 1928. The assimilation of nitrogen by pure cultures of *Clostridium pasteurianum* and related organisms. *Zbl. Bakteriol., Parasitenkunde, Infektionskrankh und Hyg.*, Abt. 2, **76**, 314–320 (in English).

MacCoy, E., Fred, E. B., Peterson, W. H. and Hastings, E. G. 1930. A cultural study of certain anaerobic butyric acid bacteria. *J. Infect. Diseases*, **46**, 118.

Malyugin, E. A., Lazarev, N. M. and Bylinkina, V. N. 1958. Soil-microbiological characterization of the main types of soil of the sandy territories of the North Aral region in connection with plant growing reclamation. *Trudy Vses. n-i. in-ta s-kh. mikrobiologii*, **14**, 5–33.

Manteifel, A. Ya. and Kozlova, E. I. 1955. Free-living nitrogen-fixers *Azotobacter* and *Clostridium* in soils under steppe and forest. *Mikrobiologiya*, **24** (1), 36–42.

Meiklejohn, J. 1956. Preliminary notes on numbers of nitrogen fixers on Broadbalk field. *Proc. Sixth Internat. Congr. Soil Sci.*, Paris, 243.

Meiklejohn, J. 1962. Microbiology of the nitrogen cycle in some Ghana soils. *Empire Exptl Agric.*, **30** (118), 115–126.

Meiklejohn, J. 1965. *Azotobacter* numbers on Broadbalk field, Rothamsted. *Plant and Soil*, **23** (2), 227–235.

Mikhailenko, L. E. 1966. Ecology of anaerobic nitrogen-fixers of the genus *Clostridium* in water stretches of Central Dnepr. *Gidrobiol. Zh.*, **2** (3), 31–40.

Miroshnichenko, L. A. 1960. Effect on microbiological processes of prolonged treatment of the grey forest soils of the Irkutsk region. *Trudy In-ta mikrobiol. Akad. Nauk SSSR*, (7), 239–248.

Mossel, D. A. and Ingram, M. 1955. The physiology of the microbial spoilage of foods. *J. Appl. Bacteriol.*, **72**, 232.

Nabiev, G. N. 1955. *Clostridium* in natural and cultivated meadow-sierozem soils. *Izv. Akad Nauk UzbSSR*, (3), 43–51.

Nagrebetskaya, V. V. 1965. Effect of sowings of forage crops on microbiological processes in sierozem soil. *Trudy Vses. n-i in-ta khlopkovodstva*, **6**, 143–155.

Nersesyan, A. 1930. Anaerobic nitrogen-fixing bacteria of the Caspian Sea. *Izv. Azerb. gos. un-ta.*, **9**, 39–41.

Nikitina, E. A. and Abenova, M. U. 1960. Study of soil-microbiological conditions of the growth of plants on newly reclaimed soil of the soddy-podzolic zone. *Trudy Vses. n-i in-ta s-kh. mikrobiologii*, **16**, 5–14.

Norris, D. O. 1962. The biology of nitrogen fixation. A review of nitrogen in the tropics with particular reference to pastures. *Bull. Commonwealth Bur. Past. Eld. Crops*, **36**, 113–119.

Obozov, A. 1962. *Izmeneniya mikroflory pochv syrtov Tsentral'nogo Tyan' Shanya v svyazi s ikh osvoeniem*. (Changes in the microflora of the soils of the Syrts of Central Tyan Shan in connection with their reclamation). Author's abstract of M.Sc. Thesis, Frunze.

Omelyanskii, V. L. 1913. *O mikrobakh, svyazivayushchikh svobodnyi azot atmosfery* (Microbes binding the free nitrogen of the atmosphere), St. Petersburg.

Omelyanskii, V. L. 1923. *Svyazyvanie atmosfernogo azota pochvennymi mikroorganizmamic* (Binding of atmospheric nitrogen by soil micro-organisms), Petrograd.

Omelyanskii, V. L. 1926. *Osnovy mikrobiologii* (Fundamentals of microbiology), Leningrad–Moscow.

Pakusin, A. G. 1960. The microflora of the soils of the Lenkoran subtropical zone. *Izv. Akad. Nauk AzerbSSR, ser. biol. i med. nauk*, (3), 149–159.

Parker, C. A. 1954. Non-symbiotic nitrogen-fixing bacteria in soil (1. Studies in *Clostridium butyricum*). *Austral. J. Agric. Res.*, **5** (1), 90–97.

Per, F. L. 1957. *Usloviya fiksatsii svobodnogo azota bakteriyami grupp 'Azotobacter' i 'Clostridium' v ozerakh Latviiskoi SSR* (Conditions for the fixation of free nitrogen by the bacteria of the groups *Azotobacter* and *Clostridium* in the lakes of the Latvian SSR), Riga.

Pertseva, A. N. and Vasyuk, L. F. 1962. The development of the microflora on plant roots in relation to plant species and soil. *Byull. nauchotekn. inf. po s-kh mikrobiologii*, (1), 3–7.

Pertseva, A. N. and Golikov, V. G. 1958. Effect of the various components of peat composts on the course of the microbiological processes on composting. *Trudy Vses. n-i. in-ta s-kh. mikrobiologii*, **14**, 133–141.

Pochon, J. and Barjac, H. de 1958. *Traité de microbiologie des sols*, Dunod, Paris.

Popova, L. S. 1961. Aerobic forms of *Clostridium polymyxa* from peat bog and soddy-podzolic soils: in *Mikroorganizmy i organicheskoe veshchestvo pochvy* (Micro-organisms and organic matter of the soil), Izd-vo Akad. Nauk SSSR, Moscow, 98–118.

Potekhina, L. I. 1966. The microflora of the grey forest soils of Siberia: in *Mikroflora pochv severnoi i srednei chasti SSSR* (The microflora of the soils of the northern and central parts of the USSR), Nauka, Moscow, 166–185.

Prevôt, A. R. 1957. *Manuel de classification et de determination des bactéries anaerobies*, Paris.

Pringsheim, H. 1908. Über die Verwendbarkeit verschiedener Energiequellen zur Assimilation des Luftstickstoffes und die Verbreitung stickstoffbindender Backterien auf der Erde. 2. Über Stickstoffassimilierende *Clostridium*. *Zbl. Bakteriol Parasitenkunde, Infektionskrankh und Hyg.*, Abt. 2, **20**, 248–256.

Pringsheim, H. 1909. Über die Verwndung von Zellulose als Energiequell zur Assimilation des Luftstickstoffes. *Zbl. Bakteriol., Parasitenkunde, Infektionskrankh, und Hyg.*, Abt. 2, **23**, 300.

Pulay, C., Toth, S. N. and Bakos, A. R. 1959. Die Bedeutung des Eisen im Stoffwechsel einiger in der Milchindustrie vorkommended Mikroorganismen *15 Internat Dairy Congr.*, **2**, 775.

Rabotnova, I. L. 1955. *Znachenie plH i okislitel'no-vosstanovitel'nykh uslovii dlya razvitiya i obmena veshchestv mikroorganismov* (Importance of pH and redox conditions in the development and metabolism of micro-organisms). Author's abstract of Ph.D. Thesis, Moscow.

Rabotnova, I. L., Egorova, V. K., Ozolina, T. K. and Eletskii, K. K. 1952. Certain aspects of the physiology of *Cl. pasteurianum*. *Mikrobiologiya*, **21** (4).

Raibaud, P., Valencia, R., Dickinson, A. B. and Han, H. Guyen Congr. 1965. Bio-synthèse *in vitro* de la vitamine B_{12} et autres cobalamines par des souches de bactéries isolées du tube digestif du rat. *C.R. Acad. sci.*, **260** (22), 5952–5955.

Rankov, V. 1964. Effect of time and fertilizer in salinated soil on the rhizosphere microflora of millet. *Rastenirǎrdni nauki*, **1** (12), 31–38 (in Bulgarian).

Rosenblum, E. D. and Wilson, P. W. 1949. Fixation of isotopic nitrogen by *Clostridium*. *J. Bacteriol.*, **57** (4), 413–414.

Ross, D. J. 1960a. Biological studies of some tussock grassland soils. 16. Non-symbiotic nitrogen-fixing bacteria of two cultivated soils. *N.Z.J. Agric. Res.*, **3** (2), 224–229.

Ross, D. J. 1960b. A note on the occurrence of non-symbolic nitrogen-fixing bacteria in some introduced pasture soils. *N.Z.J. Agric. Res.*, **3** (2), 245–249.

Rouquerol, Th. 1962. Sur le phénomène de fixation de l'azote dans les rizières de camargue. *Ann. agron.*, **13** (4), 325–346.

Rovira, A. D. 1965. Effects of *Azotobacter*, *Bacillus* and *Clostridium* on the growth of wheat: in *Plant and Microbes Relationships*. Prague, Czechosl., Acad. Sci., 193–200 (in English).

Rybkina, N. A. 1960. Distribution of anaerobic nitrogen-fixing bacteria over crop rotation fields. *Vestn. s-kh. nauki*, (7), 137–139.

Ryys, O. 1963. Distribution of free-living nitrogen-fixing bacteria in the soddy-podzolic soils of the Estonian SSR. *Izv. Akad. Nauk Est SSR, Ser. Biol.*, **12** (4), 274–281.

Ryys, O. O. 1966. *Ob ekologo-geograficheskom rasprostranenii nekotorykh svobodno-zhivuyushchikh azotofiksiruyushchikh bakterii v pochvakh Estonskoi SSR* (The eco-geographical distribution of certain free-living nitrogen-fixing bacteria in the soils of the Estonian SSR). Author's Abstract of M.Sc. Thesis, Tallin.

Samtsevich, S. A. 1966. The microflora of the southern black earth under forest plantations and on the steppes: in *Mikroflora pochv severnoi i srednei chasti SSSR*. (Microflora of the soils of the northern and central parts of the USSR), Nauka, Moscow, 186–214.

Shaposhnikov, V. N. 1948. *Tekhnicheskaya mikrobiologiya* (Industrial microbiology), Sov. nauka, Moscow.

Sheloumova, A. Protod'yakonov, O. and Faerman, V. 1927. Investigation of the biodynamics of soils. 4. *Trudy otd. s-kh. mikrobiologii*, **2**, 3–28.

Shklyar, M. Z. 1956. Effect of aerobes on the vital activity of *Clostridium pasteurianum* in mixed cultures. *Dokl. VASKhNIL*, (8).

Shklyar, M. Z. 1958. Relationships of aerobic and anaerobic nitrogen fixers. *Azotobacter* and *Clostridium pasteurianum* in a mixed culture: in *Dostizheniya michurinskoi nauki v mikrobiologii* (Achievements of Michurin science in microbiology), Sel'khozgiz, Moscow.

Sidorenko, A. I. 1959. Microbiological characterization of the chestnut soils of Central Kulunda, *Izv. Sib. otd. Akad. Nauk SSSR*, (9), 103–110.

Sinclair, N. A. and Stokes, J. L. 1964. Isolation of obligately anaerobic psychrophilic bacteria. *J. Bacteriol.*, **87** (3), 562–565.

Sjolander, N. O. and MacCoy, E. 1937–1938. Studies on anaerobic bacteria, 13. A cultural study of some butyric anaerobes previously described in the literature. *Zbl. Bakteroil., Parasitenkunde, Infektionshrankh und Hyg.*, **197**, 314. (in English)

Skalon, I. S. 1965. Microbiological characterization of the rhizosphere of plants of the natural communities of dry-steppe and desert-steppe subzones of Central Kazakhstan. *Trudy Bot. in-ta Akad. Nauk SSSR*, Ser 3, (17), 151–169.

Slavnina, T. P. and Potekhina, L. I. 1964. Effect of decalcination of the soil on the microflora and the intensity of biochemical processes. *Trudy Tomsk. un-ta*, **172**, 73–80.

Swaby, K. S. 1939. The occurrence and activities of *Azotobacter* and *Clostridium butyricum* in Victorian soils. *Austral. J. Exp. Biol. and Med. Sci.*, **17** (4), 401–423.

Tarvis, T. V. 1960. The basic pattern in the development of the microflora of the chestnut soil of the Trans-Volga and its agronomic importance. *Byull. nauchno-tekhn inf. pos-kh. mikrobiologii*, **8** (2), 37–40.

Tešić, Z. and Todorović, M. 1963. Microbiological features of peat from the neighbourhood of Lake Ohridsko. *Zemljište i biljka*, **12** (1–3), 333–338 (in Serbocroat).

Til'ba, T. V. and Golodyaev, G. P. 1966. Nitrogen-fixing bacteria of the arable soils of the Primor'e. in *Problemy brologii na Dal'nem Vostoka*. (Problems of biology in the Soviet Far East), Vladivostok, 152–154.

Truffaut, C. and Bezssonoff, N. 1921. Sur les variations d'energie du *Clostridium pasteurianum*, comme fixateur d'azote. *C.R. Acad. Sci.*, **173**, 868–870.

Truffaut, C. and Bezssonoff, N. 1925. Sur la predominance de l'activité des fixateurs anaerobiques d'azote dans le sol. *C.R. Acad. sci.*, **181**, 165–167.

Turchin, F. V., Berseneva, Z. N., Plyshevskaya, E. G. and Zertsalov, V. V. 1957: in *Trudy Vsesoyuznoi konferentsii po primeneniyu radioaktivnykh i stabil'nykh izotopav i izluchenii v narodnom khozyaistve i nauke* (Proceedings of the All-Union Conference on the uses of radioactive and stable isotopes and radiations in the national economy and science), Izd-vo Akad. Nauk SSSR, Moscow.

Turk, L. M. 1935. Studies of nitrogen fixation in some Michigan soils. *Agric. Exptl. Stat. Michigan, State Coll. Agric. and Appl. Sci. Sect. Soils, Techn. Bull.*, (143), 1–36.

Tyurin, I. V. 1957. Soil fertility and the problem of nitrogen in pedology and agriculture: in *Soveshchanie po voprosam effektivnykh sposobov ispol'zovaniya udobrenii* (Conference on questions of effective ways of using fertilizers), Min-vo sel'sk. khoz-va SSSR.

Ulasevich, E. I. 1965. *Deistvie izvestkovanie na mikroflora dernovopodzolistoi pochvy i rizosferu kukurizy v usloviyakh Poles'ya USSR* (Effect of liming on the microflora of soddy-podzolic soil and the rhizosphere of maize in the conditions of Polessya (UkrSSR)). Author's abstract of M.Sc. Thesis, Kiev.

Vandecaveye, S. C. and Katznelson, H. 1938. Microbial activities in soil. 4. Microflora of different zonal soil types developed under similar climatic conditions. 5. Microbial activity and organic matter transformation in Palouse and Helmer soils. *Soil Sci.*, **46** (1), 2.

Vavulo, F. P. 1966. The microflora of the soils of the Byelorussian SSR: in*Mikroflora pochv severnoi i srednei chasti SSSR* (The microflora of the soils of the northern and central parts of the USSR), Nauka, Moscow, 114–135.

Vigorov, L. I. and Tribunskaya, A. Ya. 1965. Spectrum of bacteria effecting the turnover of nitrogen in podzolic soil and the rhizosphere of wheat grown in this soil. *Trudy Ural'sk n-i, in-ta sel'skogo khozyaistva*, **6**, 307–321.

Visser, S. A. 1964. The presence of micro-organisms in various strata of deep tropical peat deposits. *Life Sci.*, **3** (9), 1061–1065.

Vojnova-Rajkova, J. 1965. On the specifity of rhizosphere microflora: in *Plant Microbes Relationships*. Prague, Czechosl. Acad. Sci., 126–133 (in English).

Voznyakovskaya, Yu. M. and Ryshkova, A. S. 1954. Effect of a mixture of organic and mineral fertilizers on the soil microflora. *Dokl. VASKhNIL*, (6), 30–33.

Vukhrer, E. G. 1962. Comparative microbiological characterization of the main soils of Central Tyan-Shan. *Izv. Akad. Nauk KirgSSR, Ser. Biol.*, **4** (2), 47–75.

Waksman, S. A. 1927. *Principles of Soil Microbiology*. Baltimore, USA.

Willis, W. H. 1934. The metabolism of some nitrogen-fixing *Clostridium*. *Agric. Exptl Stat Iowa, State Coll. Agric. and Mech. Arts., Res. Bull.*, (173), Ames.

Winogradski, S. N. (Vinogradskii). 1893. Sur l'assimilation de l'azote gazeux de l'atmosphere par les microbes. *C.R. Acad. Sci.*, **116**.

Winogradski, S. N. (Vinogradskii). 1895. Recherches sur l'assimilation de l'azote libre de l'atmosphere par les microbes. *Arkhiv. biol. nauk*, **4**, St. Petersbourg.

Winogradski, S. N. (Vinogradskii). 1902. *Clostridium pasteurianium* seine Morphologie und seine Eigenschaften als Buttersaureferment. *Zentralbl. Bacteriol. Parasitenkunde, Infektionskrankh und Hygiene*, Abt., 243.

Zakharchenko, A. F. 1968. Microflora of the soils of Tadjikstan: in *Mikroflora pochv yuzhnoi chasti SSSR* (Microflora of the soils of the southern part of the USSR), Nauka, Moscow, 149–188.

Zhdannikova, E. N. 1963. Microbiological characterization of the peat-boggy soils of the Tomsk region: in *Zabolochennye lesa i bolota sibir* (Boggy forests and the bogs of Siberia), Izd-vo, Akad. Nauk SSSR, Moscow, 170–182.

Zhukova, R. A. 1956. Microbiological investigations of the virgin soils of the Kolskii peninsula, *Mikrobiologiya*, **25** (5), 569–576.

Zimenko, T. G. The microflora of peat soils: in *Mikroflora pochv severnoi i srednei chasti SSSR* (Microflora of the soils of the northern and central parts of the USSR), Nauka, Moscow, 136–165.

Zimny, H. 1960. Investigations of the occurrence of *Clostridium* in forest soils. *Ekol. polska*, **6** (4), 311–321.

Zimny, H. 1961. Investigations of the occurrence of *Clostridium* in meadow soils. *Ekol. polska*, **7** (4), 303–314.

7 Nitrogen Fixation by Blue-Green Algae

The microscopic algae constitute one of the larger groups of micro-organisms in the soil. Many of them—green algae (Chlorophyceae), diatoms (Diatomaceae) and blue-green algae (Cyanophyceae)—are normal inhabitants of the soil.

A distinction must be made between terrestrial algae and soil algae proper. When the microscopic algae that live in the soil multiply in large numbers, the well-known phenomenon known as soil 'flowering' is observed. On the surface of arid and Arctic soils, the terrestial algae form crust-like excrescences when there is sufficient moisture. A similar phenomenon is observed on primary soil formations on eruptive and other types of rock.

Soil algae develop in the soil. Maybe the rays of the red part of the spectrum (or other invisible rays) penetrate to a certain depth and are utilized for photosynthesis by the algae (Feher, 1939). But it is more likely that deep in the soil algae exist as saprophytes, utilizing organic nutrients (Rabotnova and Konova, 1950).

Systematic study of soil algae began comparatively recently. The first work was published by Bristol in 1927, and much subsequent work has established that there are many algae in the soil. According to Shtina (1961) each gram of acid podzolic soil contains 5,000–30,000 algae; a gram of virgin soddy podzolic soil contains 220,000, and a gram of cultivated soddy-podzol contains 300,000, with up to 5 million in the surface films. Roughly the same numbers were reported by other investigators (for example, Bristol, 1927; Feher, 1939; Kondrat'eva, 1958; Umarova, 1962; Kulikova, 1965). The species composition of the algal populations in the soil is extremely varied. In the soils of the USSR about 1,200 species and forms of microscopic algae have been recorded.

There have been a considerable number of studies of microscopic soil algae. We shall indicate briefly the importance of these organisms in soil processes, dealing chiefly with the ability of the blue-green algae to bind molecular nitrogen.

The first observations of the ability of blue-green algae, or Cyanophyceae, to assimilate molecular nitrogen were made by Frank (1889). The results of his experiments were questioned because he did not have bacteriologically pure cultures of algae, and nitrogen assimilation could have been due to associated bacteria. Similar doubts were expressed about the experiments of Beijerinck (1901) and Heinze (1906). Later, Pringsheim (1914) and Maertens (1914) obtained pure cultures of algae but were unable to detect nitrogen accumulation.

Only Drewes (1928) isolated pure cultures of three blue-green algae and convincingly demonstrated their ability to assimilate molecular nitrogen. This conclusion was later confirmed by a series of experiments (for example, Allison et al., 1937; De, 1939; Fogg, 1942; Watanabe, 1951a, 1959a; Cameron and Fuller, 1960).

After it became possible to demonstrate nitrogen assimilation by microorganisms, including algae, using heavy nitrogen ($^{15}N_2$) Burris et al. (1946), Watanabe (1952, 1965) and Mayland and McIntosh (1966a) again confirmed the ability of blue-green algae to assimilate molecular nitrogen.

Blue-green algae may be unicellular, colonial or multicellular. They contain chlorophyll, and have other pigments of different colours. The flagellate stage and the sexual process are absent. Blue-green algae are divided into three main orders. First, the Chroococcales—as a rule unicellular or colonial forms and non-filamentous. The cells are not usually differentiated into a base and apex. Multiplication is by division. Endospores and exospores are absent and so also are heterocysts. Individual forms have thick-walled spores. Second, the Chamaesiphonales—filamentous, unicellular and multicellular organisms. Base and apex are differentiated, and the filaments attach to a substrate and are often interlocked by the side walls forming pseudoparenchyma. Multiplication is by exospores and endospores. Third, the Hormogonales—always multicellular filamentous organisms. The filaments are closely arranged in rows and are joined by plasmodesmata. Endospores and exospores are absent. In most cases there are hormogones—multicellular portions of the trichome that serve for reproduction.

Several physiological features of blue-green algae (the ability to assimilate molecular nitrogen, produce toxins, release organic compounds including vitamin B_{12}, and so on), distinguish them sharply from other algae. The cells of blue-green algae resemble bacterial cells more than those of other algae and higher plants. Therefore, Shaposhinkov and Gusev (1964) were inclined to see a phylogenetic connection between blue-green algae and bacteria.

Nitrogen assimilation is quite widespread among blue-green algae (for example Watanabe, 1951–1963; Allen and Arnon, 1955; Singh, 1961; Fogg, 1962; Land, 1942; Takha, 1963; Laloraya and Mitra, 1964; Cox et al., 1964; Fogg and Stewart, 1965. The ability to utilize molecular nitrogen has now been recorded in almost eighty species. With the exception of *Chlorogloea*, these species all belong to the order Hormogonales. They are different species of *Amorphonostoc, Anabaena, Anabaenopsis, Aulosira, Calothrix, Cylindrospermum, Fischerella, Gleotrichia, Hapalosiphon, Mastigocladus, Nostoc, Scytonema, Scytonematopsis, Stigonema* and *Tolypothrix. Chlorogloea fritschii* (synonym *Nostoc* sp. according to data of Stewart et al.) (Mitra, 1961) belongs to the order Chroococcales. The list of nitrogen-fixing algae is growing very rapidly.

Blue-green algae are widespread in different types of soil and water. According to Muzafarov (1953) in the south of Uzbekistan there are as many as 3,000 species of microscopic algae living in the water and soil. Of these, 590 species are blue-green algae. The proportions of the algal species may change substantially, reflecting changes in soil and climatic conditions.

Different species of algae seem to be the dominant nitrogen fixers in various environments. Singh (1961) considers that in the rice fields of India the most active nitrogen fixer is *Aulosira fertilissima*, while according to Watanabe (1959*b*) in the rice fields of Japan it is *Tolypothrix tenuis* (Plate 7). Muzafarov (1953) found that in the rice fields of Central Asia the most important nitrogen fixer was *Cylindrospermum* (Plate 8). In the future it may be possible to cultivate *Tolypothrix*. In unirrigated soils *Anabaena* seems to play a major role. Plates 9 and 10 show the cell structure of two species of *Anabaena*. According to Goryunova and Orleanskii (1966) in the soils of the southern part of the European USSR (Krasnodar Region) *Gleotrichia natans*, with a mass of 9,200 kg/hectare, makes a very important contribution to nitrogen fixation.

Most algal nitrogen fixers are free-living. However, many are symbiotic with plants (Echlin, 1966). Thus, for example, many lichens are symbiotic associations of blue-green algae with fungi (for example Henriksson, 1951; Bond and Scott, 1955. Algae are found in the cavities of liverworts where they are endophytes (Lhotzky, 1946; Bond, 1963) and also in ferns and species of *Cycas* (for example, Winter, 1935; Allison *et al.*, 1937; Bortels, 1940; Douin, 1954; Watanabe and Kiyohara, 1963).

Several observers have noted symbiotic relations between the aquatic fern *Azolla* (*A. pinarta* and *A. carolina*) and the nitrogen-fixing, blue-green alga *Anabaena azolla*. This fern is used in some countries as a green fertilizer for rice fields (for example Bortels, 1940; Sung Kong Hien, 1957). Some blue-green algae such as *Nostoc punctiforme* are found in the nodules of clover, *Trifolium alexandrinum* (Bhaskaran and Venkataraman, 1958).

It should be noted that individual species of micro-organisms can exert a positive influence on free-living, nitrogen-fixing algae. Thus, for example, Bunt (1961*b*) found that species of *Caulobacter* living in the slime of blue-green algae almost double the vigour of nitrogen assimilation by *Nostoc*. A pure culture of *Caulobacter* does not bind molecular nitrogen. According to Bjälfve (1962) the nitrogen-fixing powers of *Nostoc calcicola* are also enhanced in culture with nodule bacteria. Several other bacteria (including *Bac. megaterium*, *Agrobacterium radiobacter*) can also stimulate nitrogen fixation by a culture of *Nostoc*.

Not all bacteria, however, have a positive effect on the assimilation of nitrogen by blue-green algae; several investigators have obtained negative results (Drewes, 1928; Fritsch and De, 1939). On the other hand, there is evidence that blue-green algae have a positive effect on various saprophytic bacteria. These algae intensify the multiplication and biochemical activity of many bacteria, including *Azotobacter*, *Clostridium pasteurianum* and *Rhizobium* (for example, Drewes, 1928; Verner, 1935; Shtina, 1963; Perminova, 1964; Rougieux, 1966). The relationships between blue-green algae and bacteria have been considered by Postolitsa (1965) and the phytogenetic links between blue-green algae and the bacteria were reviewed by Echlin and Morris (1965).

The reasons for the development of numerous and diverse bacterial flora in symbiosis with blue-green algae can easily be seen. On the one hand, living blue-green algae form abundant mucilage of a polysaccharide nature (Davis and Fischer, 1966; Davis et al., 1966) and on the other hand, they continuously release into their surroundings, through exosmosis, considerable quantities of various organic compounds and apparently also biologically active substances (Verner, 1935; De, 1939; Watanabe, 1951; Fogg, 1956; Watanabe and Kiyohara, 1960; Okuda and Yamaguchi, 1960; Subrahmanyan et al., 1965). Polypeptides and amino acids, and also ammonia are often the products of exosmosis (Mayland et al., 1966). Ammonia usually accumulates in old cultures of algae and is evidently a product of autolysis. In some cases, for example in the excretions of *Calothrix brevissima* only amino acids are found (Watanabe, 1951a). The quantities excreted may be very great. Cells have been known to release as much as 40 per cent of the nitrogen they have bound in the form of organic compounds. Goryunova and Rzhanova (1964) considered all blue-green algae capable of releasing many nitrogenous substances. The dead cells of blue-green algae mineralize quite rapidly.

Among the amino acids that are products of exosmosis are serine, threonine, rather less glutamic acid and tyrosine, traces of alanine, and leucine and valine (Fogg, 1952; Dhar and Bhat, 1965). Different species of algae apparently release different groups of substances. Thus, according to Watanabe (1951b), alanine and aspartic and glutamic acids accumulate in *Calothrix brevissima*.

Magee and Burris (1954) found that in cells of *Nostoc muscorum* (Plate 11) nitrogen-15 was incorporated mostly into glutamic acid and ammonia after a short exposure. This indicates that the nitrogen is first reduced to ammonia, and then attaches to α-ketoglutaric acid. A considerable part of the nitrogen-15 is also rapidly incorporated into the purines and pyrimidines.

Geographical and ecological patterns of soil algae have been examined insufficiently. Nevertheless, there is no doubt that the microscopic algae, in the same way as other micro-organisms, are very widely distributed. Some forms are practically ubiquitous. The species most often encountered in the soil belong to the genera *Tolypothrix*, *Cylindrospermum*, *Anabaena*, *Anabaenopsis* and *Calothrix*. *Anabaenopsis* is found in the water of paddy fields. But the nature of the soil and the climate influence the state of the soil algae, and they have been responsible for the emergence of some microscopic algae that are most closely associated with particular zones. Thus, for example, according to Mitra (1961), *Porphyrosiphon notarisii*—a common soil alga of the central zone—is not found in India. *Nostoc commune* and *Microcoleus vaginatus*, widespread in a temperate climate, give way in India to other species—*Nostoc sphericum* and *Microcoleus chthonoplastes*. *Plectonema battersii* is also not found in India (Singh, 1961). There are many other examples.

Watanabe (1959a), who investigated more than 800 soil samples in Asia, found more microscopic algae in southern than in northern soils. Others have taken a similar view, though based on more limited material.

In the USSR many have worked on soil algae (for example Gollerbakh, 1936; Bolyshev and Manucharova, 1947; Keller, 1948; Gromov, 1956; Muzafarov, 1953; Kondrat'eva, 1958; Vaulina, 1958; Musaev, 1960; Umarova, 1962; Kuchkarova, 1965). Shtina's work (1956–1965) is particularly worth noting. Shtina noted that each algal community contains a complex of ubiquitous species and species strictly associated with definite ecological conditions. Thus, for example during the formation of podzol there is a comparatively simple community of algae dominated by filamentous forms—Ulotrichales. At this stage blue-green algae are very insignificant. With the development of soddy soils the algae develop more abundantly. Together with green algae, considerable numbers of blue-green and small diatomous forms appear. The steppe process is associated with the predominant development of blue-green algae (*Nostoc* and *Scytonema*) and, to a lesser degree, with diatoms, while the part played by the green algae diminishes. When desert soils develop, blue-green algae (Oscillatoriales) predominate.

Blue-green algae are thus neither quantitatively nor qualitatively uniformly represented in different soils. But they can be found everywhere—both in the Arctic and the sub-tropics. They can be considered to form a substantial part of the population of micro-algae (for example, Kondrat'eva, 1958; Gollerbakh and Syroechkovskii, 1960; Umarova, 1962; Kogan, 1966).

According to Shtina the soils of the USSR contain about thirty species of nitrogen-fixing, blue-green algae, most of which are found in all zones from tundra to deserts but with a definite connection with the location (Shtina and Roizin, 1966). Since the properties of the soil and its mode of cultivation exert a considerable influence on the distribution of algae, the presence of particular algae may serve as a good indicator of the type and condition of the soil. For example, weakly wetted virgin land is widely inhabited by *Fischerella muscicola*, *Nostoc commune*, *Nostoc sphaericum*, *Tolypothrix tenuis* and *Scytonema hofmanii*. Wetter virgin soils often contain *Nostoc muscorum*, *Anabaena variabilis*, *Calothrix brevissima* and *elenkinii*. Cultivated soils mostly contain *Cylindrospermum licheniforme*, *Cylindrospermum majus*, *Cylindrospermum muscicola* and *Anabaena cylindrica*.

There are some indications that the ecological races of the same species of algae differ somewhat in their properties. We also note that the blue-green algae develop better in the rhizosphere of plants than in the rest of the soil.

It is to be hoped that further data will enable us to establish more precisely the effect of geographical and ecological factors on the distribution of microscopic algae in the soil, including those that assimilate molecular nitrogen. It is known, however, that most blue-green algae develop best in a neutral or weakly alkaline medium. Below pH 6·5 their growth is seriously weakened. For example, Allison *et al*, (1957) found that *Nostoc muscorum* grows in the pH range 5·7–9·0 with an optimum between pH 7·0 and pH 8·5. The range varies appreciably for individual species (Table 7.1). According to Trukhin (1960) *Anabaena flos-aquae* develops best at pH 7·0–7·5 and *Microcystis aeruginosa* at pH 9·0.

Uspenskaya (1939) found different intracellular pH values in different species of algae. According to Brooks (1926) the intracellular pH of the algal cell readily shifts either way when the cells are placed in acidic or basic solutions.

Of course other factors influence the response of algae to the pH of the medium too. Thus, according to Uspenskii (1926), in stretches of water with an alkaline pH, small quantities of dissolved iron can be found, since it precipitates in the form of the insoluble hydroxide and so acidic water and soils chiefly contain species of algae which are relatively resistant not only to pH but also to high concentrations of iron. According to De and Sulaiman (1950) blue-green algae under rice assimilate molecular nitrogen at a lower pH than when the plants are absent. Many have written about the response of blue-green algae to the pH of the medium (Trukhin, 1960; Fujuwara and Okutsu, 1960; Land, 1942;

Table 7.1. pH Values Suitable for the Growth of
Different Algae

Species	Growth range pH	Optimum pH
Hapalosiphon fontinalis	6·0–8·5	7·0–8·0
Anabaena variabilis	4·5–8·0	7·0–8·0
Calothrix elenkinii	6·0–9·0	7·0–8·0

These data were obtained by Takha (1963).

Takha, 1963). In general, although blue-green algae are able to develop within quite a wide pH range, a neutral or weakly alkaline zone is optimal for their growth.

Blue-green algae have a very distinctive response to oxygen. They should probably be considered aerobes. There are indications that optimum Eh values of the medium for blue-green algae are above 400 mV. Conditions in unflooded soils are usually even more aerobic. In rice soils, 10 days after flooding, the Eh index falls to approximately 280–300 mV and lower. In these conditions the algae accumulate in the upper layers which are most strongly illuminated and most saturated with oxygen. However, Shaposhnikov and Gusev (1964) noted that oxygen, even at atmospheric concentrations, was harmful to *Anabaena variabilis*. They found that the oxygen formed during photosynthesis was harmful 'waste'. This led them to postulate that blue-green algae appeared on the earth when there was no oxygen in its atmosphere and that nitrogen acted as a hydrogen acceptor for them. It is interesting to note that when fixing nitrogen in the light, blue-green algae release a surplus of oxygen. The origin of this 'extra' oxygen is not clear.

Mesophilic blue-green algae apparently cease to develop at temperatures above 40°C (Kratz and Myers, 1955). Allen (1958) tested forty different cultures of blue-green algae, and all developed normally at 40°C. The optimum temperature for the blue-green algae is close to 28°–30°C. It should be noted that the blue-green algae include some thermophilic forms—*Mastigocladus laminosus* (Allen, 1958), *Anabaena flos-aquae* (Tischer, 1965; Davis and Fischer, 1966) and others. Fertilizers, in particular the organic type, have a favourable effect on blue-green algae (Mitra, 1951; Dhar and Bhat, 1965). Table 7.2 shows the

Table 7.2. Effect of Mineral Fertilizers and Straw on Nitrogen Fixation
(in light and dark)

	Light		Dark	
Conditions	Initial amount of nitrogen (mg)	Nitrogen increment (mg)	Initial amount of nitrogen (mg)	Nitrogen increment (mg)
No fertilizer	16	4	16	6
Straw added	53	42	53	10
$P_2O_5 + K_2O$	13	131	13	0
$P_2O_5 + K_2O$ + straw	50	176	50	4

Nitrogen was estimated 12 months after the start of the experiment.

quantity of nitrogen fixed when Bjälfve (1955) added various fertilizers, including straw (1 per cent) to sand and kept some of it in the dark and some in the light. As can be seen, nitrogen was fixed only in the light, that is, it was fixed by algae, while the introduction of mineral fertilizers (P_2O_5 and K_5O) and straw greatly increased the level of nitrogen accumulation.

It should be noted that according to Dhar (1965) the introduction of *Anabaena* or *Tolypothrix* into the soil, especially together with organic substances such as straw, considerably reduces (by 50 per cent) nitrogen losses from the soil.

The stimulatory effect of phosphorous fertilizers on nitrogen fixation by blue-green algae has also been noted by De and Mandal (1956), Dhar and Deo (1965) and others. Blue-green algae also assimilate molecular nitrogen more briskly when calcium fertilizers are used, and especially with added nitrates (for example Allison *et al.*, 1937; Allen and Arnon 1955a; Balezina, 1965). Most of the algae we have considered also require small doses of sodium, which may be replaced by lithium, potassium, rubidium and caesium.

According to L. S. Balezina (1965) the constitution of an algal population becomes more diverse when fertilizers are added to the soil (Table 7.3).

The relationship between nitrogen-fixing, blue-green algae and various nitrogen, the algae can utilize ammonium salts, nitrites and nitrates, but not hydroxylamine. Certain species, for example *Calothrix elenkinii*, cannot assimilate the nitrates.

There is information to indicate that blue-green algae develop more rapidly on bound than on molecular nitrogen (Kratz and Myers, 1955). But Allen (1958) showed that in complete medium and in favourable conditions *Anabaena cylindrica* develops with equal vigour in the presence of bound and molecular nitrogen. Intense growth of nitrogen-fixing algae (especially on N_2) requires

Table 7.3. Effect of Fertilizers on the Constitution of the
Algal Population in Soddy-Podzolic Soil

Experimental variant	Group of algae				
	Blue-green	Green	Yellow	Diatoms	Total
Controls	7	11	1	1	25
Introduced, N, P and K	6	10	4	2	22
Introduced, N, P, K and Ca	10	12	6	2	30*

Figures are numbers of species.

* *Cylindrospermum* and *Anabaena* appeared.

the presence of molybdenum salts (for example, Bortels, 1940; Kratz and Myers, 1955), and evidently several other elements (including Co, Mn, B).

As is well known, the introduction of nitrogen-containing mineral substances into the medium usually suppresses the assimilation of nitrogen by both free-living and symbiotic nitrogen-fixing bacteria. In relation to nitrogen-fixing algae similar information is very conflicting. According to Allen (1958) *Anabaena cylindrica* binds molecular nitrogen when there are relatively high concentrations of nitrates in the medium. Ammonium and molybdenum salts suppress nitrogen fixation. *Cylindrospermum sphaerica* also assimilates molecular nitrogen in the presence of nitrates Venkataraman, 1961c. But there are indications to the contrary (for example, De, 1939; Hérriset, 1946). Evidently the contradictions can be explained by individual differences between species of nitrogen-fixing algae, and also by differences in experimental design (composition of medium, presence of trace elements and so on). According to Takha (1963), ammonium salts and nitrates depress the fixation of nitrogen only if they can be assimilated by the algae. For example, *Calothrix elenkinii* does not assimilate salts of nitric acid and therefore does not influence the assimilation of molecular nitrogen. The opposite situation obtains for *Anabaena variabilis* which utilizes nitrates.

We have noted the favourable effect of molybdenum on nitrogen-fixation by blue-green algae. This element usually also enhances their development on nitrates, but with a few exceptions. Thus, according to Venkataraman (1961*c*) *Cylindrospermum sphaerica* developing on nitrates does not react to molybdenum. Molybdenum is usually introduced into the medium in a concentration of 0·1–0·2 parts per million. In the absence of molybdenum, blue-green algae turn orange-yellow, because of a decrease in their content of phycocyanin—a blue proteinaceous pigment.

There are indications that other trace elements such as cobalt and boron have a positive effect on blue-green algae. Further information on the effect of the trace elements on blue-green algae can be found in articles by Holm-Hansen *et al.* (1954), Wolfe (1954), Okuda *et al.* (1962) and Takha (1963).

An example from the experiments of Allen and Arnon (1955*b*) illustrates the effect of trace elements on the growth of algae. *Anabaena cylindrica* in a medium without vanadium and molybdenum yielded a biomass of 0·36 g per litre; within the same period in a medium containing these trace elements, the weight of the biomass became 5·8 g per litre. This of course in turn influences nitrogen fixation, for it usually correlates with algal growth. According to some findings algae accumulate several trace elements in their cells (Sr, Ba, Ti, Mn, Cu and Co).

Sunlight is very important in the development of blue-green algae. A considerable number, if not most of these micro-organisms seem to be obligate phototrophs (for example, Kratz and Myers, 1955; Fogg, 1953). Most of the blue-green algae studied by Allen (1956) multiplied well and bound molecular nitrogen only in the light. Even cell-free preparations from blue-green algae fix gaseous nitrogen better when illuminated (Schneider *et al.*, 1960). According to Feoktistova (1961) maximum nitrogen-fixation is observed in *Stratonostoc* and *Amorphonostoc* when there are 8 hours of daylight with an intensity of 12,000–24,000 lux. Not all species of blue-green algae seem to require the same illumination (for example, Allen, 1956; Feoktistova, 1959; Tret'yakova, 1965, 1966). According to Cox *et al.* (1964) in *Anabaena cylindrica* nitrogen is fixed in the pigmented photosynthesizing lamellae.

Some species of blue-green algae—in particular *Tolypothrix tenuis*—can live without utilizing solar energy, for in the dark they become heterotrophs (Watanabe, 1960). Feeding on organic compounds, *T. tenuis* can form chlorophyll, carotenoids and phycocyanin in the dark. But phycoerythrin does not accumulate. According to Fay (1965) and Watanabe (1966) other species of algae can also exist in the dark, for example, *Chlorogloea fritschi* and *Anabaenopsis circularis*. It is interesting that each algal species vigorously fixes molecular nitrogen only when growing on organic compounds specific to it. Thus, Winter (1935) showed that the endophytic symbiont of cycads, *Anabaena cycadeae*, can grow in the dark in media containing sugars, but fixes molecular nitrogen only in the presence of fructose. Allison *et al.* (1957) and Tret'yakova (1965) found that *Nostoc muscorum* can assimilate molecular nitrogen in the dark in the presence of glucose and sucrose.

Thus, the degree of myxotrophicity differs in different species of blue-green algae, but its presence explains the active existence of certain species deep in the soil.

Different algae differ considerably in their tolerance of light and shade. Thus, for example, blue-green algae are less exacting in their requirements than green algae. *Cylindrospermum licheniforme* can exist in more shady conditions than other blue-green algae. *Nostoc* develops vigorously in soil shaded by forest vegetation (Kolesnik, 1965).

There is a correlation between photosynthesis and nitrogen fixation, but low and high temperatures depress nitrogen fixation more than they do the assimilation of carbon dioxide (Fogg and Than Tun, 1960) and the effects of substances inhibiting the two processes also differ (Cox, 1966).

The vigour of nitrogen assimilation is influenced significantly by the concentration of carbon dioxide in the medium. Fogg *et al.* (1960) found that *Anabaena cylindrica* weakly fixed gaseous nitrogen in the presence of 0·03 per cent CO_2. De and Sulaiman (1950) noted that when air with an increased concentration of CO_2 was blown through a soil culture nitrogen assimilation was enhanced. But concentrations of about 0·5 per cent and more CO_2 in the air have a toxic effect. The optimum concentration of CO_2 for blue-green algae depends on the temperature. According to Fogg *et al.* for *Anabaena cylindrica* the optimum concentration of CO_2 at 15°C is 0·1 per cent, and at 20°C and higher 0·25 per cent.

There are very conflicting reports about the actual level of nitrogen accumulated by blue-green algae. This is easy to explain; experiments have not always been carried out in the same conditions and have not all lasted the same length of time. For example, in calculations of accumulation per unit area, different principles of conversion have been applied. Nevertheless, we feel that some generalizations are possible. We shall use the material obtained in laboratory conditions by several investigators (including Allison *et al.*, 1937; Stokes, 1940; Muzafarov, 1953; Allen, 1958; Watanabe, 1959–1963; Venkataraman, 1961), who used different cultures of blue-green algae (*Amorphonostoc paludusum, Nostoc muscorum, Anabaena cylindrica, Anabaenopsis circularis, Cylindrospermum sphaerica* and others). Disregarding findings that differ sharply from the mean or are not readily explicable, we can say that in 1·5–2 months algae in laboratory conditions, in the light, usually fix 2–5 mg of nitrogen per 100 ml of medium. In rare cases this value is more than 10 mg.

Individual investigators have not agreed as to which are the most vigorous nitrogen fixers. Singh (1961) considers that in India the most active form is *Aulorisa fertilissima*; Watanabe (1950) said *Tolypothrix tenuis* for Japan and Ley Shang hao (1959) said *Anabaena azotica* for China.

There have been many investigations of the amount of nitrogen accumulated by blue-green algae in the soil especially in rice fields (for example De, 1939; Singh, 1942; Prasnad, 1949; Willis and Green, 1948; Watanabe, 1951b; Watanabe and Kiyohara, 1963; Tamiya, 1957; Venkataraman, 1961a; Iha et al., 1965). In rice fields, conditions are optimum for blue-green algae during a considerable part of the growing season. Most investigators found that in these conditions the annual nitrogen increment due to the activity of blue-green algae varies between 15 and 50 kg/hectare, although sometimes it is 80 kg/hectare. Watanabe (1962), who summarized the findings of many Japanese investigators, put the mean quantity of nitrogen accumulated by blue-green algae at approximately 20–25 kg/hectare. Watanabe considered that in countries where nitrogenous mineral fertilizers are used in small quantities blue-green algae are very important in the maintenance of soil fertility.

Table 7.4. Nitrogen Fixed in Rice Fields

Nitrogen accumulation (kg/hectare)	With rice	Without rice
Controls	48·8	38·6
+ P_2O_5	69·0	61·0
+ P_2O_5 + Mo	77·9	74·9

There are also indications that blue-green algae can accumulate very large quantities of nitrogen in the soil. Thus, Allen (1956) believes, on the basis of laboratory experiments, that after conversion nitrogen accumulation by blue-green algae may reach in one month 53 kg per hectare. Keller (1948) calculating the nitrogen in the excrescences of algae, obtained a maximum figure of 205 kg of nitrogen per hectare per season. There have been references to even more nitrogen accumulation by algae, but we consider them to be overestimates.

The level of nitrogen fixation in the soils of rice fields is largely determined by the soil cover, the chemical composition of the water and the fertilizers used, as indicated by De and Mandal (1956) (Table 7.4). Nitrogen accumulation was much more effective when phosphates and molybdenum were added, and under the rice, nitrogen fixation was more vigorous than in the plains; this is possibly a consequence of the enrichment of the medium with carbon dioxide through the activity of the root system of the plants.

Thus, blue-green algae can maintain the nitrogen balance of soil at a definite level, which is too low to support high yields of rice. This is particularly true if no phosphoric acid, organic fertilizers or trace elements are used on the fields, as practice has confirmed. Where mineral nitrogenous fertilizers are not used mean rice yields are considerably lower than in countries where intensive farming is practised.

In upland and even irrigated farming land, where the development of algae is episodic, they are not very important as accumulators of nitrogen. Shtina (1963), however, noted that even in northern soddy-podzolic soils it is possible to find many nitrogen-fixing, blue-green algae such as *Amorphonostoc*, *Anabaena* and *Cylindrospermum*. They can be assumed to play a definite role in supplying nitrogen to the soil. This requires further investigation.

Clearly blue-green algae that accumulate nitrogen can be used to full advantage in rice fields, and it is not surprising that the algae of rice fields have been the subject of a tremendous amount of work (for example De *et al.*, 1956; Zhurkina, 1956; Romanenko, 1956; Watanabe, 1960; Florenzano *et al.*, 1960; Singh, 1961; Obukhova, 1961; Bunt, 1961a; Venkataraman, 1962; Subramanyan *et al.*, 1964, 1965; Relwani and Manna, 1964).

In the Far East, farmers are endeavouring to stimulate the multiplication of the natural algal microflora in the fields, including those that assimilate atmospheric nitrogen. In India at certain times of the year the arable zones are often flooded with water, and a dense layer of algae rapidly forms. This is later placed in the soil and is a good organic fertilizer. In China (for example in the province of Suchan Wan) wide use is made of 'green moss'. In the winter, organic residues (bamboo, branches of trees and so on) are taken to the rice fields. During the subsequent growing period, a mass of algae (the green moss) which serves as a fertilizer, develops in the soil in the presence of organic fertilizers.

In the Central Asian regions of the USSR, the slime from irrigation ditches, which is very rich in algae, is often used as fertilizer. In the USSR and other countries good results have been obtained by using as fertilizers masses of algae that cause blooms on stretches of open water. According to Kuchkarova (1962) algal humus and fresh algae increased the cotton harvest in Uzbekistan by 15 16 per cent.

All the facts we have gathered about the beneficial effect of algae as fertilizers are not at all surprising, for algae contain from 6 to 11 per cent nitrogen and 0·8–4·2 per cent phosphorus (converted to dry weight). Nevertheless, larger masses of algae are required to give an appreciable increase in the harvests of rice and other crops. Therefore it might be a good idea to stimulate directly the multiplication of nitrogen-fixing, blue-green algae in the rice fields when the plants are growing.

According to Obukhova (1959) a sequential change can be seen in the algal flora of rice fields. Initially, when the rice plants begin to grow, green algae such as *Spirogyra*, *Zygonema* and *Oedogonium* develop. When they multiply intensively the growth of the rice may be depressed. Therefore it is sometimes necessary to remove them from the water (by introducing $CuSO_4$, or herbicides, by mechanical cleaning or other means). Later, when the rice plants emerge above the water and shade the water stretch, the development of green algae gives way to blue-green algae which form slime colonies, films and such like. Blue-green algae are also dominant as the rice develops further, in spite of the almost complete shading of the water. According to Muzafarov (1953) as much as 16 tons per hectare fresh weight of algae may accumulate, containing as much as 100 kg of nitrogen.

During the growing period there is also a sequential change in the species of blue-green algae present. Thus, for example, according to Singh (1961) in Indian paddy fields species of *Anabaena* develop first, followed by *Cylindrospermum*, *Tolypothrix* and *Fischerella*. Later *Aulosira* begins to multiply abundantly. Changes in the algal flora of paddy fields are dependent on the type of soil and various other factors.

At one time it was suggested that multiplication of blue-green algae in rice fields could be intensified by introducing cultures into the soil. Many people took up this problem (including De and Sulaiman, 1950; Shioiri *et al.*, 1943,

Table 7.5. Effect of *Tolypothrix* and Fertilizers on
Yield of Rice

Experimental variant	Dry weight	
	Straw (g/hectare)	Grain (g/hectare)
Controls	8·2	6
+ P_2O_5	12·5	10·8
+ P_2O_5 + *Tolypothrix*	19·2	17·4
+ *Tolypothrix*	16·3	14·6
+ P_2O_5 + $(NH_4)_2 SO_4$	20·2	18·0

1944; Mitra, 1961; Watanabe, 1962; Okuda and Yamaguchi, 1952; Singh, 1961). It is worth noting the view of Watanabe who, on the basis of the experiences of Okuda and Yamaguchi, noted rapid natural multiplication of algae in certain soils. Clearly not all rice fields need added algae.

Rice plants have been treated with pure cultures of algae in laboratory and field conditions (Hosoda and Takata, 1955; Hirano *et al.*, 1955; Rao *et al.*, 1963). In many cases the algae had a beneficial effect, but the increase in yield fluctuated greatly. Sometimes it was not more than 20 per cent, but in other cases it was very considerable (Iha *et al.*, 1965). Thus, for example, in the experiments of Nawawy *et al.* (1958), a culture of *Tolypothrix* had the same effect as $(NH_4)_2SO_4$ (discussed later). In field experiments carried out by Hosoda and Takata (1955) and Subrahmanyan *et al.* (1965) the effectiveness of adding blue-green algae (*Tolypothrix tenuis* in the first case and in the second case a species of algae which was not recorded), was equivalent to introducing 72 and 60 kg/hectare of $(NH_4)_2SO_4$ respectively. The quantitative effect of *Tolypothrix* on the yield of rice plants is shown in Table 7·5 (Nawawy *et al.*, 1958).

Table 7.6 shows the results of a two-year field experiment carried out by Iha *et al.* (1965), who treated rice with a culture of *Tolypothrix tenuis*. It is interesting that in the first year when the culture was introduced under the plants at the time of sowing, the effect was slight, but when the algae were added 20 days before planting the yield increased considerably, especially when phosphorus

Table 7.6. Effect of *Tolypothrix* and Other Treatments on Yield of Rice
(Iha *et al.*)

Experiment	Yield (kg/hectare)	
	1957–1958	1958–1959
Controls	2,194	3,072
+ P_2O_5*	1,973	2,914
+ *Tolypothrix* (20 days before sowing)	3,070	3,540
+ P_2O_5 + *Tolypothrix* (20 days before sowing)	3,614	3,430
+ *Tolypothrix* (at time of sowing)	2,508	3,780
+ P_2O_5 + *Tolypothrix* (at time of sowing)	2,554	3,540

* P_2O_5 added as 45 kg/hectare.

was added with the algae. We think this is possibly connected with the release of nitrogen-containing compounds into the soil from the algae. In succeeding years there was an effect irrespective of the time of adding the culture.

Table 7.7. Effect of *Tolypothrix tenuis* Culture on Rice Harvest over
Several Years

Experiment	Laboratory experiments, (g per vessel)			Field experiments (kg/hectare)	
	1959–1960	1960–1961	1962–1963	1962–1963	1963–1964
Controls	119·6	274·1	95·8	2,191	1,826
+ P_2O_5	104·9	320·8	97·8	2,265	1,827
+ *Tolypothrix*	178·5	390·7	120·4	3,012	2,490
+ *Tolypothrix* + P_2O_5	180·8	361·6	143·8	3,186	2,859

Iha *et al.* (1965) obtained steady increases in the rice harvest in both laboratory and field (Table 7.7) for several years. All rice plants were inoculated 20 days before sowing.

In one of Allen's experiments (1958), rice plants in control pots weighed 0·19 g while those in pots with added *Anabaena cylindrica* weighed 2·0 g. The nitrogen content of the plants in the treated soil was twenty times higher than in the controls.

In field experiments, the increase in the rice harvest when algae are added

Table 7.8. Effects of *Tolypothrix* and Other Treatments on Yields of Rice
(Subrahamanyan *et al.*)

Experiment	No *Tolypothrix tenuis* added	*Tolypothrix tenuis* added
Controls	100	130
+ Manure (20 kg of N per hectare)	130	137
+ Green fertilizer (*Seslaria*)	149	149
+ (NH$_4$)$_2$SO$_4$ (20 kg of N per hectare)	131	132
+ Urea (20 kg of N per hectare)	129	137

The results are expressed in percentages of controls.

usually varies between 15 and 30 per cent. The variations depend both on the conditions of cultivation and the fertilizer used. As an example, Table 7.8 shows a field experiment carried out by Subrahmanyan *et al.* (1964). In the control plot, the rice harvest (no nitrogenous substance or algae added) was 2,615 kg/

Table 7.9. Effects of *Tolypothrix tenuis* on Satisfactory
Rice Harvests

Experiment	Harvest 100 kg/hectare	Gain after algae added (%)
	Well aerated soils	
Controls	30·30	
+ *Tolypothrix*	34·85	15
	Poorly aerated soils	
Controls	20·45	
+ *Tolypothrix*	25·72	25

hectare. In a corresponding plot with algae added the harvest increased by 30 per cent. Against a background of fertilizer the algae had no effect.

Watanabe's experiment (1951*b*) is another example (Table 7.9). In this case added algae increased quite satisfactory harvests by between 15 and 25 per cent.

In a field experiment carried out by Relwani and Subrahmanyan (1963) with a mixture of blue-green algae added to the soil there was an increase of 8·3 per cent in the harvest of rice grain, whereas 20 kg of nitrogen in the form of $(NH_4)_2SO_4$ brought about an increase of 26·2 per cent. Nishigaki *et al.* (1951, 1953, quoted by Watanabe, 1966) using an isotopic method, confirmed that a culture of *Tolypothrix tenuis* can fix nitrogen in field conditions.

It should be noted that in several cases there has been an appreciable increase in the rice harvest in the third or fourth years after the introduction of nitrogen-fixing, blue-green algae. A similar phenomenon was noted in experiments in the laboratory by De and Sulaiman (1950) and in the field by Watanabe (1965). Table 7.10 shows the increases in the harvest of crude rice after the addition of

Table 7.10. Effects on Crude Rice of Adding *Tolypothrix tenuis* to the Soil

Change in yield, (%) in relation to each subsequent year after algae added*		
Year after algae	Test station No. 1	Mean data for nine other test stations
1	− 4·0	+ 2·0
2	+ 17	+ 8·0
3	+ 56	+ 15·1
4	+ 44	+ 19·5
5	+ 11	+ 10·6

* Yields varied between 2,510 and 3,740 kg/hectare.

a culture of *Tolypothrix tenuis* to the soil at ten experimental stations. In one of them (station No. 1), the soil was limed with pH 8. Naturally here the algae developed more briskly and gave a higher increase in harvest than in the other cases. 'Algolization' on the test stations was carried out for five years. The yearly increase in the effectiveness of the algae is explained by Watanabe as due to the gradual increase in their mass in the soil.

Cultures of *Anabaena* and *Nostoc* have been shown to affect rice plants favourably, especially when lime, superphosphate or molybdenum were also added (Subrahmanyan and Manna, 1966).

There is no doubt that algae in rice fields can have a positive effect. Results have been inconsistent, probably because the techniques are unrefined and experimental conditions have varied. It can be assumed that in fields fertilized with mineral and organic fertilizers the effectiveness of added algae declines.

If soil is to be supplemented successfully with algae it is vital that cultures be chosen carefully for different conditions. The success of the process, however, sometimes depends on other factors. Thus, for example the growth of rice plants is often depressed by rapidly developing green algae. Copper sulphate is often used to control the algae, but it also depresses blue-green algae. Therefore, when pure cultures of blue-green algae are used in rice fields other methods of controlling green algae are needed and are now being developed (see p. 303).

Sometimes the algae are eaten by daphnia, rotifers and other small organisms. In such cases chemicals (such as pholidol and parathione) are used which act

Table 7.11. Effect of Algae and *Azotobacter* on Grain Yields

Addition to soil	Percentage increase over control	
	1962	1963
Amorphonostoc punctiforme	11·7	24·8
Amorphonostoc paludosum	27·7	28·6
Anabaena cylindrica	3·1	32·0
Tolypothrix tenuis	15·7	20·0
Algal mixture	10·4	17·3
Algal mixture + azotobacters	25·8	14·9
Azotobacter	23·0	15·9

Control yields were 13·0/vessel in 1962 and 4·09/vessel in 1963.

selectively on the protozoa, causing no harm to the blue-green algae (Hirano et al., 1955).

The question of using algae on farm crops growing on upland or wet ground has rarely been considered. A start has been made with some trials of algae in ordinary farming conditions (for example, Shtina, 1956; Aiyers et al., 1964). We feel that the prospects of developing the technique in this direction are less hopeful than in the case of rice cultivation. Intensive development of algae in the soil and on the surface is possible only in periods of sufficiently heavy rainfall, that is at certain times during the growing season. For example, according to Umarova (1962), soil algae increase 3–4 days after cotton crops have been wetted, but after 7 days they decrease. The short period of intensive development of soil algae would obviously prevent them from accumulating a great deal of nitrogen.

Perminova (1964) working in Shtina's laboratory, found that in conditions of optimum soil moisture, added algae could increase the harvest of the Winner variety of barley by 25 to 30 per cent. Different cultures of the same species of alga did not all have the same effect on the plant.

Shtina's group also carried out a field experiment with barley growing on soddy-podzolic soil, adding a culture of *Stratonostoc linckia* (130 g dry weight per hectare of seeds). As Table 7.12 shows, where it was difficult to regulate soil moisture the algae had no appreciable effect, but there was a significant gain in percentage yield when algae were added with azotobacters.

In other field experiments cultures of blue-green algae (mostly *Nostoc muscorum*) increased the harvest of barley by 4·5 to 10·0 per cent in 25 per cent of cases. Such a degree of effectiveness is within the limits of experimental error. The best result—a positive effect in 75 per cent of cases—was obtained with a culture of algae and *Azotobacter chroococcum*; then the gain in yield was 10·3 to 15 per cent, but still low.

Table 7.12. Effect of Added Algae on the Yield of Barley in a Field Experiment

Experiment	Grain yield (100 kg/hectare)	Increase in yield	
		(100 kg/hectare)	(%)
Controls	30·9	—	—
+ azotobacters	34·2	3·3	10·7
+ azotobacters + algae	35·5	4·6	14·8
+ Algae	32·3	1·4	4·5

Field experiments involving enrichment with algal cultures have been rare, however. More experience must be gained before we know whether algal cultures can be used in practical farming. Conditions of soil and climate will doubtless influence the solving of this problem.

Inoculation of fields requires a large quantity of algae. In Watanabe's view it is necessary to introduce about 5 kg of dry culture of blue-green algae per acre of soil in rice fields. Obviously only a dry culture can be transported any distance, and the techniques involved in its preparation are especially important. These are being worked out on the basis of field investigation of the physiology of blue-green algae, and available knowledge about the multiplication of protococcal algae. For practical purposes it is desirable to consider the multiplication not of completely pure cultures, but of those that are pure from the point of view of the algae.

There are various ways of encouraging the multiplication of blue-green algae. The simplest method is used in China where algae multiply in cuvettes containing a mineral medium. Part of the culture is periodically decanted into another vessel for use, and nutrient medium is added to the other vessel. This method can, of course, be used directly in the soil. Watanabe (1959c and d) devised a very refined apparatus. In this, the algae multiply in a tank of mineral medium containing nitrates and lit by electricity (Fig. 34), and air enriched with carbon dioxide (up to 2–3 per cent) is blown through the tank. It is economically advantageous to heat the apparatus by making use of hot springs and to use natural gases as a source of carbon dioxide. One litre of culture medium gives up to 2 g of algal mass each day. Watanabe did not exclude the possible use of

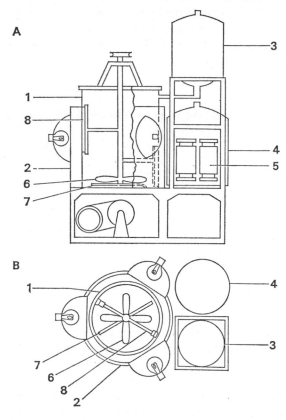

Fig. 34. Diagram of tank for culturing algae by Watanabe's method
A, Longitudinal section; *B*, transverse section. 1, Tank; 2, thermostat; 3, reservoir for medium; 4, heater; 5, cotton wool filter; 6, ventilator; 7, aeration pipe; 8, stirrer.

an open system (Plate 12). Vessels of capacity 250 l are used, and for each square metre of apparatus, 7·9 g of algae is obtained daily. From each apparatus it is possible to obtain as much as 7 tons of algae.

Pinevich and Verzilin (1963) designed a column type apparatus with natural lighting, The upper part of the column is made of transparent plastic and the lower part of stainless steel placed in a thermostat. The culture is aerated. Various other devices for culturing algae have been described (Florenzano, 1958).

For agricultural purposes most of the culture obtained can be multiplied further in open systems, and then spun down and dried. Blue-green algae rapidly settle and are easily centrifuged at 2,500–3,000 r.p.m.

Many ways of drying algae have been proposed. Lyophilization is expensive, and so Venkataraman (1961b) recommended sunlight and Tseng Chi-Mien (1959) a current of hot air. Watanabe *et al.* lyophilized algae and then dried them adding serum albumin. In dried preparations of algae 50 per cent of the cells are lost in two years. Watanabe *et al.* also cultivated algae on the surface of volcanic gravel prepared as a fine porous mass (Fig. 35). The algae that developed

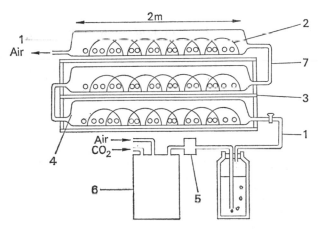

Fig. 35. Diagram of apparatus for cultivating blue-green algae on gravel (after Watanabe). 1, Polyvinyl reservoir for cultivating the algae; 2, wire frame; 3, glass plate; 4, gravel inoculated with algae; 5, cotton wool filter; 6, reservoir for gas; 7, pipe for aeration.

were kept in chlorovinyl bags or dry glass bottles, retaining their activity for 2–5 years, although some of them died.

There is no well formulated technique for the mass production of blue-green algae, although some important work has been carried out. It should be noted that with mass cultivation in open systems the development of blue-green algae may be affected adversely by various organisms. Thus the culture may be contaminated with green and diatomous algae. These can be controlled by introducing anisomycin, actidione or nystatin into the medium. These antibiotics do not act on the blue-green algae but depress other groups of algae. According to Ishizawa (1964) the development of algae can be prevented with pentachlorophenol (100 parts per million) to which blue-green algae are insensitive. To suppress daphnia and certain other organisms, pholidol and parathione (o, o diethyl-s-(4-nitrophenyl)-thiophosphate, or thiophos) are recommended. The latter, which is very effective, can be used in a dose of 0·05 parts per million.

The aquatic fern *Azolla* (*A. pinnata*, *A. carolina*) is very important in the agriculture of eastern countries. This plant, living in symbiosis with the blue-green alga, *Anabaena azolla*, has often been studied (for example, Oes, 1913; Sung-Hong Hin, 1957; Le Van Kan and Sobachkin, 1963; Shen *et al*., 1963). *Azolla* is small (0.8×1.5 cm) and multiplies rapidly, doubling its mass in 5–6 days in favourable conditions.

Azolla was used first on the soil in Vietnam in the province of Thae Binh by a peasant called Ba Heng. *Azolla* was so effective that after her death Ba Heng was immortalized; in the village where she lived a pagoda was built in honour of the 'goddess of *Azolla*' and each autumn solemn prayers are offered with sacrifices.

Table 7.13. Effect of *Azolla* on Rice Harvest

Province		Controls	*Azolla* introduced
Hai Duong	1958	26·7	31·3
Ha Tinh	1960	26·9	34·7
Haiphong	1961	34·1	36·1
Nam Dinh:	1960	16·4	17·5
	1961	23·3	26·6

Figures are in 100 kg/hectare.

For agricultural use, *Azolla* is multiplied in small artificial or natural stretches of water. When it has grown, the green mass is transferred to thoroughly wetted fields of young rice. Fifteen to twenty kilograms of *Azolla* are added per hectare. After some time on the surface of the water it forms a continuous green cover of multiplying fern. When the hot weather begins at approximately the tillering phase of the rice, the fern dies off and is rapidly mineralized, releasing nutrient elements which are used by the rice plants.

Table 7.13 shows Le Van Kan and Sobachkin's findings (1963) that *Azolla* has different effects on the rice harvest in different provinces of Vietnam. In some cases the harvest was increased by only 6–7 per cent, in others it improved by almost 30 per cent. Clearly this depends on the conditions in which *Azolla* is used. These are still not sufficiently understood.

When *Azolla* is cultivated in water, each hectare can yield a ton of green mass of the alga daily. This ton contains 2–2·5 kg of nitrogen, 1 kg of P_2O_5 and 1·5 kg of K_2O. It is recommended that 10–20 tons of *Azolla* be introduced per hectare of soil in rice fields.

Concluding this section, we should like to emphasize that algae prevent soil erosion as well as helping nitrogen accumulation in the soil. Among the nitrogen fixers, *Nostoc* should be mentioned particularly in this context.

Singh (1961) noted the considerable importance of algae in the improvement of alkaline soils. When rain falls on alkaline soils, which are often devoid of vegetation, algae begin to develop. As a result of their activity the surface layer of the soil is enriched with organic matter, the physical properties of the soil improve, the pH is lowered and in the absorbing complex sodium is replaced by calcium with an increase in the store of P_2O_5 and nitrogen. To intensify algal development, irrigation is recommended for soils of salt marshes which are undergoing improvement. Several authoritative scientists consider such biological improvement of the soil to be quite effective.

We should also note the importance of algae in the initial stages of soil formation. Autotrophic blue-green algae are the pioneer species in the reclamation of rock. Treub (1888) was the first to point out the development of blue-green algae in the igneous rocks of Krakatoa. The many species involved included *Anabaena*, *Lyngbya*, *Tolypothrix* and *Symploca*. Later, many investigators (including Elenkin, 1936; Fogg, 1947; Odintsova, 1953) confirmed the role of blue-green algae as pioneers in the formation of soil.

Algae form part of the lichens that develop on rocks, mediating the primary stages of soil formation. Tikhomirov (1957) noted the importance of algae in the biological reclamation of tundra soils. There are indications that blue-green algae destroy muscovite, biotite and other minerals involved in the formation of secondary calcite, hydrates of iron oxide and iron-manganese concretions of the high mountain varnish. The algae promote the redistribution of the elements in the soil layers by accumulating them inside their cells.

Because blue-green algae contain vitamins, in particular B_{12}, they are a good food for fish. Some blue-green algae produce toxic substances and destroy the larvae of the Anopheles mosquito (Singh, 1961). Among these algae are *Aulosira fertilissima*, *Plectonema nostocorum*, and species of *Anabaena*. Thus algae are very valuable in public health measures against malaria. They are also involved in the nitrogen balance of the sea. Stewart (1964, 1965) detected in sea water *Calothrix scopulorum* and *Nostoc entophytum*, and using nitrogen-15 showed that they could bind molecular nitrogen.

References

Aiyers, R. 1964. Effect of nitrogen-fixing blue-green algae and *Azotobacter chroococcum* on vitamin C contents of tomato plants. *Sci. and Culture*, 30 (11), 556–557, India.

Allen, M. 1956. Photosynthetic nitrogen fixation by blue-green algae. *Scient. Monthly*, 83 (2), 100.

Allen, M. 1958. Photosynthetic nitrogen fixation by blue-green algae. *Trans. Internat. Conf. Use Solar Energy Sci. Basis*, 4 Tucson, Arizona, Univ. Arizona Press, 27.

Allen, M. and Arnon, D. 1955a. Studies on nitrogen-fixing blue-green algae. I. Growth and nitrogen fixation by *Anabaena cylindrica*. *Lemm. Plant Physiol.*, 30, 366–372.

Allen, M. and Arnon, D. 1955b. Studies on nitrogen-fixing blue-green algae. 2. The sodium requirement of *Anabaena cylindrica*. *Physiol. plantarum*, 8, 653.

Allison, F. E., Hoover, S. R. and Morris, H. J. 1937. Physiological studies with the nitrogen-fixing algae *Nostoc muscorum*. *Bot. Gaz.*, **82** (3), 433.

Balezina, L. S. 1965. Effect of fertilizers and lime on the development of soil algae. *Trudy Kirovsk. s-kh. in-ta*, **18** (30), 160–166.

Beijerinck, M. 1901. Fixation of free atmospheric nitrogen by *Azotobacter* in pure culture. *Acad. wet. Amsterdam*, **2**, 67.

Bhaskaran, S. and Venkataraman, G. 1958. Occurrence of blue-green algae in *Trifolium alexandrinum* nodule. *Nature*, **181**, 277–278.

Bjälfve, G. 1955. Fixation of atmospheric nitrogen. I. Experiments in sand with and without addition of straw or starch and in straw alone in sterile and unsterile conditions in light and darkness. *Kgl. lantbrukshögskolans ann.*, **22**, 193–217 (in English).

Bjälfve, G. 1962. Nitrogen fixation in cultures of algae and other micro-organisms. *Physiol. plantarum*, **15**, 122–129.

Bolyshev, N. N. and Manucharova, E. A. 1947. Distribution of algae in the profile of certain soils of an arid zone. *Vesr. MGU*, (8).

Bond, G. 1963. The root nodules of non-leguminous angiosperms. 13 *Sympos. Soc. Gen. Microbiol. and Symbiotic Assoc.*, 72–93.

Bond, G. and Scott, G. 1955. An examination of some symbiotic system for fixation of nitrogen. *Ann. Bot.*, **19** (73), 67–77.

Bortels, H. 1940. Über die Bedeutung des Molybdäns für Stickstoffbindende Nostocaceen. *Arch. Mikrobiol.*, **11**, 155–186.

Bristol, Roach, B. M. 1927. On the algae of some normal English soils. *J. Agric. Sci.*, **17** (4), 563–588.

Brooks, M. 1926. Studies on the permeability of living cells. The penetration of certain oxidation-reduction indicators as influenced by pH. Examination of the rU of *Valonia*. *Amer. J. Physiol.*, **76**.

Bunt, J. 1961*a*. Nitrogen-fixing blue-green algae in Australian rice soil. *Nature*, **192** (4,801), 479–480.

Bunt, J. 1961*b*. Blue-green algae growth. *Nature*, **192**, 1274–1275.

Burris, R. and Wilson, P. 1946. Characteristics of the nitrogen-fixing enzyme system in *Nostoc muscorum*. *Bot. gaz.*, **108**, 254.

Cameron, R. and Fuller, W. 1960. Nitrogen fixation by some algae in Arizona soils. *Soil Sci. Soc. America Proc.*, **24** (5), 353–356.

Cox, R. 1966. Physiological studies on nitrogen fixation in the blue-green algae. *Arch. Mikrobiol.*, **53** (3), 263–276.

Cox, R., Fay, P. and Fogg, G. 1964. Nitrogen fixation and photosynthesis in a sub-cellular fraction of the blue-green algae *Anabaena cylindrica*. *Biochim. Biophys. Acta*, **88**, 208.

Danilewicz, K. 1965. Symbiosis in *Alnus glutinosa* (L) Gaertn. *Acta Microbiologica Polonica*, **14** (3–4), 321–325 (in English).

Davis, E. B. and Fischer, R. G. 1966. Photochemical reduction of elemental nitrogen by *Anabaena flos-aquae* A-37. *Nature*, **212** (5059), 302–303.

Davis, E. B., Fischer, R. G. and Brown, L. 1966. Nitrogen fixation by the blue-green algae. *Physiol. plantarum*, **19** (3), 823–827.

De Ley, J. and Park, I. W. 1966. Molecular biological taxonomy of some free-living nitrogen-fixing bacteria. *Antonie van Leeuwenhoek. J. Microbiol. and Serol.*, **32**, 6–16.

De, P. 1939. The role of blue-green algae in nitrogen fixation in rice fields. *Proc. Roy. Soc.*, B, **127**, 121–139.

De, P. 1956. Fixation of nitrogen by algae in rice soils. *Soil Sci.*, **81** (6), 453–458.

De, P. and Mandal, L. 1956. Fixation of nitrogen by algae in rice soils. *Soil Sci.*, **81**, 453–458.

De, P. and Sulaiman, M. 1950. Fixation of nitrogen in rice soils by algae as influenced by crops, CO_2 and inorganic substances. *Soil Sci.*, **70**, 137.

Dhar, N. R. 1965. Land fertility improvement by fixing atmospheric nitrogen on applying organic matter and phosphates with and without algae. *Proc. Nat. Acad. Sci. India. Sympos. on Land Fertility Improvement by Blue-Green Algae.* Sect. A, **35**, Part 3, 259–280.

Dhar, N. R. and Bhat, G. N. 1965. Influence of light intensity, organic matter and phosphate on: (A) nitrogen fixation and (B) availability of P_2O_5 in the presence and absence of *Anabaena naviculoides* and *Chlorella pyrenoidosa*. *Proc. Nat. Acad. Sci. India, Sympos. on Land Fertility Improvement by Blue-Green Algae. Sect. A,* **35**, Pt. 3, 309–327.

Dhar, N. R. and Deo, P. G. 1965. Influence of algae on nitrogen fixation in black cotton soils. *Proc. Nat. Acad. Sci. India, Sympos. on Land Fertility Improvement by Blue-Green Algae.*, Sect. A, **35**, Pt 3, 281–293.

Douin, R. 1954. Sur la fixation de l'azote libre par les myxophycces endophytes des Cycadacees. *Ann. Univ. Lyon*, Sect. C, **8**, 57–63.

Drewes, von K. 1928. Über die Assimilation der Luftstickstoffs durch Blaualgen. *Zbl. Bakteroil., Parasitenkunde, Infektionskrankh. und Hyg.*, Abt. 2, **76**, 88–121.

Echlin, P. 1966. The blue-green algae. *Scient. Amer.*, **214** (6), 74–81.

Echlin, P. and Morris, J. 1965. The relationship between blue-green algae and bacteria. *Rev. Biol.*, **40** (2), 143–187.

Elenkin, A. A. 1936. *Sinezelenye vodrosli SSSR.* (Blue-green algae of the USSR). Izd-vo Akad. Nauk SSSR, Moscow-Leningrad.

Fay, P. 1965. Heterotrophy and nitrogen fixation in *Chlorogloea-fritschii*. *J. Gen. Microbiol.*, **39** (1), 11–20.

Feher, D. 1939. Untersuchungen über die Lichtökologie der Bodenalgen. *Arch. Mikrobiol.*, **10** (2), 245–265.

Feoktistova, O. I. 1959. Effect of length of daylight on formation of the organic matter and multiplication of algae. *Trudy In-ta biologii vodokhranilishch Akad. Nauk SSSR*, (1), 110–117.

Feoktistova, O. I. 1961. Effect of length of daylight on the formation of organic matter and multiplication of algae. *Trudy In-ta biologii vodokrhanilishch Akad. Nauk SSSR*, (1), 110–114.

Florenzano, G. 1958. Prime ricerche in Italia nell imprianto sperimentale di Firenze sulla colture massiva non sterile di alghe. *Nuovo giorn. bot. ital.*, New Scr., **65** (1–2), 1–15.

Florenzano, S., Balloni, W. and Materassi, R. 1960. Le microalghe verdiazzurre azotofissatrici e la fertilita del terreno delle risaie. *3 Sympos. Internat. Agrochem.*, 8–14.

Fogg, G. E. 1942. Studies in nitrogen fixation by blue-green algae. I. Nitrogen fixation by *Anabaena cylindrica*. Lemm. *J. Exptl. Biol.*, **19** (1), 78–87.

Fogg, G. E. 1947. Nitrogen fixation by blue-green algae. *Endeavour*, **6**, 172–175.

Fogg, G. E. 1952. The production of extracellular nitrogenous substance by a blue-green algae. *Proc. Roy. Soc.*, B, **139**, 372–397.

Fogg, G. E. 1953. *The Metabolism of Algae.* Methuen, London.

Fogg, G. E. 1956. The nitrogen fixation by photosynthetic organisms. *Ann. Rev. Plant Physiol.*, **7**, 51.

Fogg, C. E. and Stewart, W. D. P. 1965. Nitrogen fixation in blue-green algae. *Sci. Progr.*, **53** (210), 191–201.

Fogg, G. E. and Than-Tun, 1960. Interrelations of photosynthesis and assimilation of elementary nitrogen in a blue-green algae. *Proc. Roy. Soc.*, B, **153**, 111–127.

Fogg, G. 1962: in *Nitrogen fixation. Physiology and Biochemistry of Algae.* Levin, R. (Ed.). Acad. Press, N.Y., 161.

Frank, B. 1889. Über den gegenwartigen Stand unserer Kenntnis der Assimilation elementaren Stickstoffs durch die Pflanzen. *Ber. Dtsch. bot. Ces.*, **7**, 34–42.

Fritsh, F. and De, P. 1939. Nitrogen fixation by blue-green algae. *Nature*, **142**, 878.

Fujuwara, A. and Okutsu, M. 1960. Cultural and physiological studies on the nitrogen-fixing blue-green algae. *Nostoc spongiaeforme* Ag. (Part 2). *Ref. Soil and Plant Food*, **5** (2), 91.

Gollerbakh, M. M. 1936. Composition and distribution of algae in soils. *Trudy Bot. in-ta Akad. Nauk SSSR*, Ser. 2, (3).

Gollerbakh, M. M. and Syroechkovskii, E. E. 1960. Biogeographical investigations in the eastern Antarctic in the summer season of 1957: in *Sbornik Arkticheskogo i Antarkicheskogo n.-i in-ta. Vtoraya kontinental'naya ekspeditsiya* 1956–1958 *gg.* (*Sovetskaya antarkticheskaya ekspeditsiya*). (Handbook of the Arctic and Antarctic Research Institute. Second continental expedition of 1956–1958. (Soviet Antarctic expedition)), **9**, Morskoi transport, Leningrad, 197–207.

Goryunova, S. V. and Orleanskii, V. K. 1966. Course of development of blue-green algae in the natural conditions of the Krasnodarsk area (Slavyanskii region): in *IX Mezhdunarodnyi Mikrobiologicheskii Kongress. Tezisy dokladov* (Ninth International Microbiology Congress, Summaries of speeches).

Goryunova, S. V. and Rzhanova, G. N. 1964. Vital secretions of nitrogenous organic substances by the blue-green alga *Lyngbya aesbuarii* and their physiological role: in *Sbornik posvyashchennyi pamyati prof. Skadovskogo S.N.* '*Biologiya sinezelenykh vodoroslei*' (Memorial volume for Professor S. N. Skadovskii: The biology of the blue-green algae), Moscow State University.

Gromov, B. V. 1956. Observations on the algae of the soils of certain northern districts of the USSR. *Uch. zapiski LGU*, (216), 170–179.

Heinze, B. 1906. Über die Stickstoffassimilation durch niedere Organismen. *Landwirtsch. Jahrb. Schweiz*, **35**, 889.

Henriksson, E. 1951. Nitrogen fixation by a bacteria-free, symbiotic *Nostoc* strain isolated from Collema. *Physiol. plantarum*, **4**, 542–545.

Hérriset, A. 1946. Fixation de l'azote par le *Nostoc commune* Vauch. *C.R. Acad. Sci.*, **222**.

Hirano, T., Shiraishi, K. and Nakano, K. 1955. *Bull. Schikoku Agrion. Exptl. Stat.*, **2**, 121 (quoted by Watanabe, 1966).

Holm-Hansen, O., Gerloff, G. and Scoog, F. 1954. Cobalt as an essential element for blue-green algae. *Physiol. plantarum*, **7**, 665.

Hosoda, K. and Takata, H. 1955. *Trans. Tottori Soc. Agric. Sci.*, **10** (1), (quoted by Watanabe, 1966).

Iha, K., Alli, M., Singh, R. and Bhattacharya, P. 1965. Increasing rice production through the inoculation of *Tolypothrix tenuis*, a nitrogen-fixing, blue-green alga. *J. Indian Soc. Soil Sci.*, **13** (3), 161–167.

Ishizawa, S. 1964. *J. Sci. Soil and Manure*, **10** (quoted by Watanabe, 1966).

Keller, B. A. 1948. *Osnovy evolyutsii rastenii*. (Essentials of Plant Evolution), Iz-dvo. Akad. Nauk SSSR, Moscow, Leningrad.

Kiyohara, T., Fujita, Y., Hattori, A. and Watanabe, A. 1962. Effect of light on glucose assimilation in *Tolypothrix tenuis*. *J. Gen. Appl. Microbiol.*, **8** (3), 165–168.

Kogan, Sh. I. 1966. Nitrogen-fixing, blue-green algae of the water stretches and soils of South Turkhmenia. *Izv. Akad. Nauk Turk SSR*, (3), 15–23.

Kolesnik, I. I. 1965. Microbiological characterization of the forest soils of the Donets stations. *Lesovodstvo i agrolesomelior. Resp. mezhved. temat. nauchn. sb.*, (7), 7–13.

Kondrat'eva, N. V. 1958. Study of the distribution of blue-green algae in the soils in relation to the agrotechnical methods. *Ukr. bot. zh.*, **15** (4), 61–69.

Kratz, W. and Myers, J. 1955. Nutrition and growth of several blue-green algae. *Amer. J. Bot.*, **42** (3), 282–287.

Kuchkarova, M. A. 1962. The algal flora of rice fields in the Tashkent Region. *Uzb. biol. zh.*, **1**, 35–38.

Kuchkarova, M. A. 1965. Search for and selection of nitrogen fixers from among blue-green algae of the rice fields of Central Asia: in *Materialy Zakovkazkoi konferentsii po sporovym rastenii* (Materials of the Transcaucasion Conference on Spore Plants), Baku, 55–57.

Kulikova, R. M. 1965. Change in the algal flora on cultivation of peat-boggy soil. *Bot. zh.*, **3**, 11–24.

Laloraya, V. and Mitra, A. 1964. Fixation of elementary nitrogen by *Scytonema noffmani*. *Current Sci.*, **33** (20), 619–621.

Land, J. 1942. *Soil Algae. Physiology and Biochemistry of Algae. N.Y.*

Le Van Kan and Sobachkin, A. A. 1963. Use of *Azolla* for fertilization in the Vietnamese Democratic Republic. *Dokl. TSKhA*, (94), 93–97.

Ley, Schang hao, 1959. The effect of nitrogen-fixing blue-green algae on the yields of rice plant. *Acta hydrobiol, sinica*, **4**, 440–444 (in English).

Lhotsky, S. 1946. The assimilation of the free nitrogen in symbiotic Cyanophyceae. *Stud. Bot. Českosl.*, **7** (1), 20–35 (in English).

Maertens, H. 1914. *Beitr. Biol.*, *Pflanzen*, **12**, 439 (quoted by Singh, R. 1961).

Magee, W. and Burris, E. 1954. Fixation of N_2 and utilization of combined nitrogen by *Nostoc muscorum*. *Amer. J. Bot.*, **41**, 777.

Mayland, H. T. and McIntosh, T. H. 1966*a*. Availability of biologically fixed atmospheric nitrogen-15 to higher plants. *Nature*, **209** (5021), 421–422.

Mayland, H. T., McIntosh, T. H. and Fuller, W. H. 1966*b*. Fixation of isotopic nitrogen on a semi-arid soil by algal crust organisms. *Soil Sci. Soc. America Proc.*, **30** (1), 56–60.

Mitra, A. 1961. Some aspects of fixation of elementary nitrogen by blue-green algae in the soils. *Proc. Nat. Acad. Sci. India*, A, **31** (1), 98–99.

Musaev, E. Yu. 1960. *Vodorosli oroshaemykh zemel' i ikh znachenie dlya plodorodiya pochv*. (Algae of irrigated lands and their importance to soil fertility), Ized-vo Akad. Nauk UzbSSR, Tashkent.

Muzafarov, A. M. 1953. Importance of blue-green algae in the fixation of atmospheric nitrogen. *Trudy Bot. in-ta Akad. Nauk UzbSSR*, (2), 3–11.

Nawawy, A. S., Lotei, M. and Fahmy, M. 1958. Studies on the ability of some blue-green algae to fix atmospheric nitrogen and their effect on growth and yield of paddy soils. *Agric. Res. Rev.*, **36** (2), 308–320.

Obukhova, K. M. 1959. Algae of rice fields in the Talda-Kurganskaya and Kzyl-Ordinskaya provinces. in *Sbornik rabotov po ikhtiologii i gidrobiologii in-ta zoologii AN KazSSR*. (Collected works on ichthyology and hydrobiology of the Institute of Zoology of the Academy of Sciences of the Kazakh SSR).

Obukhova, V. M. 1961. Importance of algae in the regime of rice fields. *Iz-v. Akad. Nauk KazSSR, ser. bot. i pochvov*. (1), 91–101.

Odintsova, S. V. 1953. Role of blue-green algae in soil fertility: in *Rol' mikro-organizmov vpitanii rastenii*. (The role of micro-organisms in plant nutrition), Izd-vo Akad. Nauk SSSR, Moscow, 118–122.

Oes, A. 1913. Über die Assimilation des freien Stickstoffs durch *Azolla*. *Z. Bot.*, **5**, 145–164.

Okuda, A. 1962. Nitrogen-fixing micro-organisms in paddy soils. X. Effect of molybdenum on the growth and nitrogen assimilation of *Tolypothrix tenuis*.

Okuda, A. and Yamaguchi, M. 1952. Algae and atmospheric nitrogen fixation in paddy soil. *Mem. Res. Inst. Food, Sci., Kioto Univ.*, **2**, 1–14.

Okuda, A. and Yamaguchi, M. 1960. *Soil and Plant Food*, **6**, 76 (quoted by Watanabe, A. 1965).

Perminova, G. N. 1964. Effect of blue-green algae on the development of micro-organisms in the soil. *Mikrobiologiya*, **33** (3), 472–476.

Pinevich, A. V. and Verzilin, N. N. 1963. Cultivation of protococcal algae in open systems. *Vestn. LGU*, (15), (3), 75–97.

Postolitsa, L. G. 1965. Effect of certain blue-green algae on bacteria. *Nauchn. dokl. vysshei shkoly*, (1), 99–103.

Prasnad, S. 1949. Nitrogen recuperation by green algae in soil of Bihar and their growth on different types. *J. Proc. Inst. Chem.*, **21**, 138.

Pringsheim, E. 1914. *Beitr. Biol. Pflanzen*, 12, 47 (quoted by Singh, R. 1961).

Rao, W. V. B., Coyal, S. K. and Venkataraman, G. S. 1963. *Current Sci.*, **32**, 366.

Rabotnova, I. L. and Konova, I. V. 1950. Effect of the degree of aeration on auto-trophic and heterotrophic nutrition of *Chlorella*. *Mikrobiologiya*, **19** (1).

Relwani, L. and Manna, G. G. 1964. Effect of blue-green algae in combination with urea on rice yield. *Current Sci.*, **233** (222), 687–688.

Relwani, L. L. and Subrahmanyan, R. 1963. Role of blue-green algae chemical nutri-ents and partial soil sterilization on paddy yield. *Current Sci.*, **82** (10), 441–443.

Romanenko, V. V. 1956. Dynamics of the algal film in rice fields. *Trudy Uzb. in-ta malyarii i med. parasitol.*, **2**, 197–200.

Rougieux, R. 1966. Biologie de certain sols du Sersou. Region de Tiaret et de Burgeau. Algérie. *Ann. Inst. Pasteur*, **110** (6), 928–934.

Schneider, R., Bradbeer, C., Singh, R. N., Wang, L.-Chuan, Wilson, P. and Burris, R. 1960. Nitrogen fixation by cell-free preparations from micro-organisms. *Proc. Nat. Acad. Sci.*, **46**, 726.

Shaposhnikov, V. N. and Gusev, M. V. 1964. Role of oxygen in the vital activity of certain blue-green algae: in *Biologiya sinezelenykh vodoroslei* (The biology of blue-green algae), Moscow State University, 119–142.

Shen, C. 1963. A preliminary study of the nitrogen fixation of *Azolla japonica, Turang Tongbao*, **4**, 46–48.

Shioiri, H., Hatsumi, T. and Nishigaki, S. 1943. *J. Sci. Soil and Manure*, **17**, 288; **18**, 59 (quoted by Watanabe, 1966).

Shtina, E. A. 1956. Interaction of soil algae and higher plants. *Vestn MGU*, (6).

Shtina, E. A. 1961. Experience in the cultivation of algae as fertilizer: in *Tezisy dokladov Vsesoyuznogo soveshchaniya po Kul'tivirovannuyu odnokletochnykh vodorosaii* (Proc. All-Union conf. on the cultivation of unicellular algae), Leningrad, 64–65.

Shtina, E. A. 1963. Nitrogen fixation by blue-green algae. *Uspekhi sovr. biologii*, **56** (2–5), 284–299.

Shtina, E. A. 1964. Role of algae in the accumulation of nitrogen in the soil. *Agro-khimiya*, (4), 77–83.

Shtina, E. A. 1964. Participation of algae in the processes of soil formation. *Izv. Akad. Nauk SSSR, ser. biol.*, (1), 72–80.

Shtina, E. A. 1965. Nitrogen fixation by blue-green algae. In: *Ekologiya i fiziologiya sinezelenykh vodoroslei* (The ecology and physiology of blue-green algae), Nauka, Moscow-Leningrad, 160–177.

Shtina, E. A. 1965. Features of the communities of algae in deep chernozems of the Central Chernozem Reserve. *Trudy Tsentral'no–Chernozemnogo zapovednika*, (9), 146–155.

Shtina, E. A. 1965; and Roizin, M. B. 1966. The algae of the podzolic soils of Khibin. *Bot. zh.*, **51** (4), 509–519.

Singh, R. N. 1942. The fixation of elementary nitrogen by some of the commonest blue-green algae from paddy field soils of the United Provinces and Bihar, *Indian J. Agric. Sci.*, **12**, 743.

Singh, R. N. 1961. *Role of blue-green algae in the nitrogen economy of Indian agriculture*. Indian council Agric. Res. New Delhi.

Stewart, D. 1964. Nitrogen fixation by Myxophyceae from marine environments. *J. Gen. Microbiol.*, **36**, 415–422.

Stewart, D. 1965. Nitrogen turnover in marine and brackish habitats. *Ann. Bot. New Ser.*, **29** (114), 229–239.

Stokes, J. L. 1940. The role of algae in the nitrogen cycle of the soil. *Soil Sci.*, **49** (4).

Subrahmanyan, R. 1964. Observations of the role of blue-green algae on rice yield compared with conventional fertilizers. *Current Sci.*, **83** (16), 485–486.

Subrahmanyan, R. and Manna, G. B. 1966. Relative response of the rice plant to blue-green algae and ammonium sulphate in bulk trials. *Current Sci.*, **35** (19), 482.

Subrahmanyan, R., Relwani, L. L. and Manna, G. B. 1965. Nitrogen enrichment of rice soils by blue-green algae and its effect on the yield of paddy. *Proc. Nat. Acad. Sci. India, Sympos. on Land Fertility Improvement by Blue-Green Algae.* Sec. A, **35**, Pt. 3, 382–386.

Subrahmanyan, R. and Sahay, M. W. 1964. Observation on nitrogen fixation by some blue green algae and remarks on its potentiabilities in rice culture. *Proc. Indian Acad. Sci.*, **60**, B, (2), 145–154.

Subrahmanyan, R. and Sahay, M. W. 1965. Observations on nitrogen fixation and organic matter produced by *Anabaena circinalis* Rabh. and their significance in rice culture. *Proc. Indian Acad. Sci.*, B, **61** (3), 164–169.

Sung-Hong Hien. (Zyong-Khong-Khien). 1957. *Ispol'zovanie azolla dlya udobrenii posevov risa v DRV (doklad na soveshchanii po voprosam udobrenii* (Use of *Azolla* in fertilizing rice fields in the Vietnamese Democratic Republic (Report to the meeting on fertilizers)), Moscow.

Takha, M. S. 1963. Effect of the hydrogen ion concentration of the medium (pH) and temperature on the growth of blue-green algae and their fixation of nitrogen. *Mikrobiologiya*, **32**, 6.

Tamiya, M. 1957. Mass culture of algae. *Annual Rev. Plant Physiol.*, **8**, 309–334.

Tikhomirov, B. A. 1957. Dynamic phenomena and the plant cover of the mottled tundra of the Arctic. *Bot. zh.*, **42** (11), 1691–1717.

Tischer, R. G. 1965. *Nature*, **205**, 419 (quoted by Davis, E. B. and Tischer, K. G. 1966).

Tret'yakova, A. N. 1965. Comparative study of the strains of nitrogen-fixing, blue-green algae isolated from different soils of the USSR. *Mikrobiologiya*, (3).

Tret'yakova, A. N. 1966. Effect of light on the growth and fixation of nitrogen by the soil blue-green algae. *Mikrobiologiya*, **35** (4), 721–726.

Trukhin, N. V. 1960. Optimal pH value for the growth of certain blue-green algae. *Byull. in-ta biologii vodokhranilishch Akad. Nauk SSSR*, (6), 3–7.

Trukhin, N. V. 1963. Effect of temperature on the light growth optimum. *Dokl. Akad. Nauk SSSR*, 149 (6).

Tseng Chi-mien. 1959. A method for conservation of the nitrogen-fixing, blue-green algae for inoculations. *Acta hydrobiol. sinica*, **4**, 452–455.

Umarova, Sh. U. 1962. Seasonal variations in the development of algae of cotton fields. *Dokl. Akad. Nauk UzbSSR*, (9).

Uspenskaya, V. I. 1939. Penetration of paints and copper into the cells of algae in connection with the pH and rH within the cells in the medium. *Mikrobiologiya*, **8** (8).

Uspenskii, E. E. 1926. *Zhelezo kak faktor vasprostraneniya vodoroslei.* Iron as a factor in the distribution of algae. (Moscow State University.)

Vaulina, E. N. 1958. Main features of the algal flora of certain soils of Byelorussia. *Izv. Akad. Nauk BSSR, Ser. Biol.*, 5–15.

Venkataraman, G. 1961*a*. The role of blue-green algae in agriculture. *Sci. and Culture*, **27** (1), 9–13.

Venkataraman, G. 1961*b*. Studies on nitrogen fixation by blue-green algae 11. Nitrogen fixation by *Cylindrospermum sphaerica* Prassad under various conditions. *Proc. Nat. Acad. Sci. India*, **A31** (1), 100–104.

Venkataraman, G. 1961c. Nitrogen fixation by *Stigonema dendroideum* Fremy. *Indian J. Agric. Sci.*, **31** (3), 213–215.

Venkataraman, G. 1962. Algae–fertilizers of the future. *Indian Farming*, **6–7**, 15.

Verner, A. R. 1935. Biological *Azotobacter* activators. *Dokl. Akad. Nauk SSSR*, **4**, 55.

Watanabe, A. 1951a. Production in cultural solution of some amino acids by the atmospheric nitrogen-fixing, blue-green algae. *Arch. Biochem. and Biophys.* **34**, 50.

Watanabe, A. 1951b. Effect of nitrogen-fixing, blue-green algae on the growth of rice plants. *Nature*, **168**, 748.

Watanabe, A. 1952. Programs report study of nitrogen-fixing, blue-green algae. *Biol. Lab. Seizyo Univ. Tokyo*, (1) (in English).

Watanabe, A. 1959a. Distribution of nitrogen-fixing, blue-green algae in various areas of South and East Asia. *J. Gen. Appl. Microbiol.*, **5** (1–2), 21–29.

Watanabe, A. 1959b. On the mass-culturing of nitrogen-fixing, blue-green algae *Tolypothrix tenuis*. *J. Gen. Appl. Microbiol.*, **5** (1–2), 85–91.

Watanabe, A. 1959c. Collection and cultivation of nitrogen-fixing, blue-green algae and their effect on the growth and crop yield of rice plants. *Proc. Sympos. on algology*, New Delhi, JCAR.

Watanabe, A. 1959d. Large scale culture of a blue-green algae, *Tolypothrix tenuis*, utilizing hot-spring and natural gas as heated carbon dioxyde sources. *J. Gen. Appl. Microbiol.*, **5** (1–2), 51–57.

Watanabe, A. 1960. Collection and cultivation of nitrogen-fixing, blue-green algae and their effect on the growth and crop yield of rice plants. *Proc. Sympos. algae JCHR*, 1960, 162–166.

Watanabe, A. 1962. Effect on nitrogen-fixing, blue-green algae *Tolypothrix tenuis* on the nitrogenous fertility of paddy soils and on the yield of rice. *J. Gen. Appl. Microbiol.*, **8** (2), 85–91.

Watanabe, A. 1965. Studies on the blue-green algae as green manure in Japan. *Proc. Nat. Acad. Sci. India, Sympos. on land fertility improvement by blue-green algae*. Sect. A35, Pt. 3, 361–369.

Watanabe, A. (Vatanabe). 1966. Blue-green algae as nitrogen fixers: in *IX Mezhdunarodnyi Mikrobiologicheskii Kongress Simpozium V-I* (Ninth International Microbiology Congress, Symposium V-I). Moscow, 58–64.

Watanabe, A. and Kiyohara, T. 1960. Decomposition of blue-green algae as effected by the action of soil bacteria. *J. Gen. Appl. Microbiol.*, **5** (4), 175–179.

Watanabe, A. and Kiyohara, T. 1963. Symbiotic blue-green algae of lichens, liverworts and cycads. Plant and cell physiology: in *Studies on micro-algae and photosynthetic bacteria*, 189–196.

Willis, W. and Green, V. 1948. Movement of nitrogen in flooded soils planted with rice. *Soil Sci. Soc. America Proc.*, **13**, 229–237.

Winter, G. 1935. Über die Assimilation des Luftstickstoff durch endophytische Blaualgen. *Beitr. Biol. Pflanzaen*, **83**, 295.

Wolfe, M. 1954. The effect of molybdenum upon the nitrogen metabolism of *Anabaena cylindrica*. *Ann. Bot.*, **18** (71), 299.

Zhurkina, V. V. 1956. Algae of the rice fields of the (Soviet) Far Eastern Experimental Station: in *Voprosy sel'skogo i lesnogo khozyaistva Dal'nego Vostoka* (Questions of agriculture and forestry in the Soviet Far East), Vladivostok, 71–74.

8 Other Groups of Nitrogen-Fixing Micro-organisms

Before Beijerinck (1901) isolated pure cultures of *Azotobacter* and Vinogradskii (1895) established the role of *Clostridium* in the enrichment of soil nitrogen, the French investigator Berthelot (1885–1892) isolated many diverse soil micro-organisms on meat-peptone broth and gelatine. He said these organisms could fix nitrogen. On media containing different sugars, humic substances and kaolin, some of the organisms fixed nitrogen sufficient to increase the total content in the medium by 80 per cent. However, unidentified cultures and imperfect methods did not encourage belief in Berthelot's claims that nitrogen fixation is widespread among soil micro-organisms.

Many *a priori* assumptions of the widespread ability to fix atmospheric nitrogen among other than 'specialist microbes' (an expression due to Omelyanskii) were made later (for example Lipman, 1903; Keutner, 1905; Keding, 1906; Benecke, 1907; Koch, 1907; Emerson, 1917; Richards, 1917; Omelyanskii, 1923; Turk, 1935). Fixation of atmospheric nitrogen was noted in soils from which *Azotobacter* and *Clostridium* were absent.

Waksman, summarizing existing evidence in 1932, made quite a long list of micro-organisms that can fix atmospheric nitrogen in small quantities. The list included *Bact. lactis viscosum*, *Bact. pneumonia*, *Bact. radiobacter*, *Bact. prodigiosum*, *Bact. aerogenes*, *Bact. pyocyaneum*, *Bact. vulgare*, representatives of the group *Bact. mesentericus* (*Bact. malabarensis* and *Bact. dunicus*), *Bact. astersporus*, *Bact. azophile* and *Planobac nitrifigens*.

Vinogradskii (1945) recorded that in about fifteen years since the discovery of azotobacters many investigators looked successfully for nitrogen-fixing microbes in the soil. "Nearly everywhere they looked, they found the organisms." But several of the properties attributed to these 'nominal' nitrogen assimilators call into question the reliability of the reports. In particular, it is unlikely that the introduction of bound nitrogen into the media, even in considerable doses, would have an inhibitory effect on nitrogen-fixing capacity.

Sensitive isotope methods and other precise procedures have now made it possible to establish beyond doubt the nitrogen-fixing capacity of many soil micro-organisms, and the list of nitrogen fixers has been supplemented by *Azotomonas insolita* (Stapp, 1940), *Azotomonas fluorescens* (Krasil'nikov, 1945), *Pseudomonas azotogensis* (Voets and Debacker, 1956) *Derxia gummosa* (Jensen *et al.*, 1960), *Endosporus azotophagus* (Tchan and Pochon, 1950) and others. Nitrogen-fixing powers have been established in several widely known micro-organisms such as the Actinomycetes (Plotho, 1940), *Rhodospirillum rubrum* (Kamen and Gest, 1949), representatives of the family Thiorhodaceae and Athiorhodaceae (Lindstrom *et al.*, 1950), *Pseudomonas*, *Bacterium*, *Bacillus*, *Mycobacterium* (Mishustina, 1955) and others. Finally it has been established that some species of micro-organisms that do not fix nitrogen in ordinary conditions of cultivation can become vigorous nitrogen fixers in association with other microbial species (Fedorov and Kalininskaya, 1959, 1961; Okuda *et al.*, 1960; Abraham and Subramoney, 1960).

The Family Pseudomonadaceae

All members of the family Pseudomonadaceae are aerobic Gram-negative organisms. They are oblong, rod-shaped (sometimes coccoid), and mono-trichous, lophotrichous or immobile. Several of them can fix nitrogen and these are widespread in soil and water.

Azotomonas insolita was first isolated in 1940 by Stapp. Its cells are small, almost coccoid ($0.6–1.2 \times 0.6–1.8 \mu m$), motile with between one and three polar flagella (according to the strain). Their size and shape varies according to the conditions of cultivation. The pigment which forms—yellow-brown to greenish-brown—usually diffuses into the substrate. A wide range of compounds serves this organism as a carbon source. Out of thirty-two carbon-containing substances that Stapp tested (various alcohols, sugars, polysaccharides, glucosides) *Azotomonas insolita* failed to develop only on erythritol. The metabolic products on carbon-containing substances included acids and gases.

In optimum culture conditions *Azotomonas insolita* fixes as much as 12 mg of nitrogen per gram of sugar in two weeks. Stimulation of nitrogen assimilation is observed in the presence of molybdenum and iron and also when the medium is solidified by introducing agar, for *Azotomonas insolita* is a microaerophil. As a nitrogen source it can use mineral nitrogen compounds, especially nitrates and organic compounds (urea, amino acids, peptone).

Within the cells of *Azotomonas insolita* Macher and Manninger (1961) demonstrated spectrographically, ions of Ag, Al, B, Ca, Cu, Fe, K, Mn, Mg, Na, Ni, Si, Sn and V. These findings are of undoubted importance in the choice of an optimum synthetic medium for *Azotomonas insolita*.

The organism is not rigid in its pH requirements, and can develop in media between pH 3.3 and 9.5. Within the limits 4.5 and 8.0 the rate of growth hardly varies. However, these limits were established by Stapp working with a weakly buffered medium of which the pH changed considerably during the experiment. The acid pH values at the start had become alkaline by the end. It is possible that with a better buffered medium, the limits of optimum pH would have been closer. Optimum temperatures for the development of *Azotomonas insolita* are 25°–30°C. Optimum nitrogen fixation is at 26°C and the lethal temperature is 59°–60°C. According to Khudyakov and Voznyakovskaya (1956) *Azotomonas insolita* is widespread not only in the soil but as an epiphyte on plants.

Azotomonas fluorescens was isolated from compost and described by Krasil'nikov (1945). It was then studied in detail by Kalininskaya (1954) who considered that it had much in common with azotobacters. The cells are 0·5–0·8 × 2·0–5·0 μm in size-shortening with age to 1·3–1·6 μm. The shape and size of the cells vary with the culture conditions. In media supplied with organic forms of nitrogen the cells are relatively small. In media containing amino acids they fluctuate within the limits 0·7 × 0·4–0·5 μm. Even smaller cells are observed in nitrogen-free media containing organic acids. The cells are often paired. They are mobile and have three polar flagella. *Azotomonas fluorescens* characteristically develops a greenish-yellow pigment which diffuses into the substrate. In old cultures, the pigment acquired a brownish tint.

According to Kalininskaya, *Azotomonas fluorescens* develops well on various compounds that contain carbon, sugars, alcohols, some fatty and some aromatic acids. Citric, malic, pyruvic and lactic acids are poorly utilized. Lactose, inulin, and formic, oxalic valeric and salicylic acids are not assimilated at all. On carbohydrates (except for pentose) and some polyhydric alcohols, the productivity of nitrogen fixation by *Azotomonas fluorescens* is approximately equal to 12 mg of nitrogen per gram of substance utilized. For more reduced compounds (mannitol, sorbitol, glycerol) the productivity of nitrogen fixation is 1–2 mg higher. Partial oxidation of the carbon source, which reduces its store of chemical energy, leads to a decrease in the productivity of nitrogen fixation. Thus, for example, productivity is higher when *Azotomonas fluorescens* utilizes lactic rather than pyruvic acid. Such a decrease in productivity with oxidation is observed in several dicarboxylic acids.

The productivity of nitrogen fixation on alcohols decreases with increasing surface activity. Fedorov (1948) was the first to note this phenomenon for *Azotomonas agile* and *Azotomonas chroococcum*, attributing it to the specific influence of surfactants on nitrogen-fixing enzymes. The productivity of nitrogen fixation is very high on benzoic and quinic acids.

Mineral nitrogen compounds are utilized by *Azotomonas fluorescens* as nitrogen sources, but most of them greatly suppress the assimilation of nitrogen from the atmosphere; nitrates are the least inhibitory. The greatest suppression of nitrogen fixation is observed on ammonium salts and urea. *Azotomonas fluorescens* can utilize amino acids and peptone as the sole source of carbon and nitrogen.

A. fluorescens can develop within wide pH limits. Thus, in Kalininskaya's experiments it developed in media of pH 5·1–9·0 and higher, although growth was very weak at 5·6 and above 9·0. The productivity of nitrogen fixation decreased only at pH 5·3 and below. At these pH values both the consumption of sugar and the quantity of nitrogen fixed were insignificant.

The optimum temperature for *Azotomonas fluorescens* is 20°–30°C; development is greatly slowed down below 10°C and above 45°C. A very high concentration of oxygen is not required for active nitrogen fixation.

Pseudomonas leuconitrophilus, isolated from the soil of the Roman Campania in 1906 by Perotti, is a characteristically weak fixer of nitrogen in nitrogen-free media. It has one flagellum, forms a capsule and zoogloea and brings about the liquefaction of gelatine.

Pseudomonas azotocolligans was isolated from acid and alkaline soils of the states of Idaho and Washington in the United States by Anderson (1955) using a synthetic nitrogen-free medium. Cultures grow very slowly, visible colonies appearing usually after a week. At this time they are punctate, transparent, smooth, round and slightly yellow. After two weeks they become less transparent and more yellow, but the pigment formed does not diffuse into the medium. In liquid medium the most vigorous growth is observed around particles of $CaCO_3$.

Cultures use a wide range of carbohydrates, forming small amounts of acid but no gases. In media with a source of bound nitrogen, colonies develop much more rapidly and form pigment more efficiently.

Microscopically, *Pseudomonas azotocolligans* consists of mobile, Gram-negative, thin, straight rods measuring $0.5 \times 2–4.0$ μm. There are between one and four flagella in a polar arrangement.

Pseudomonas azotocolligans is not very specific in its pH requirements. It develops in media of pH 4.0–9.0. It is extremely resistant to sucrose (withstanding up to 40 per cent). It grows at temperatures between 5° and 37°C, optimum development being at 25°C. The pH characteristics of a culture change with temperature: at pH 4.0 and 9.0 there is growth at 28°C but not at 37°C. The culture produces catalase.

After two weeks of incubation at 25°C on a nutrient medium containing 2 per cent sucrose, *Pseudomonas azotocolligans* fixes 0.27–0.36 mg of nitrogen in a Petri dish. (If we take the amount of agar medium in a Petri dish to be 25 ml, then the actual gain in nitrogen is 0.3 mg.) Nitrogen fixation has been confirmed by the isotope method.

Pseudomonas azotogensis was isolated from hot-house soil by Voets and Debacker (1956) on a nitrogen-free medium containing mannitol. This microorganism is an aerobic, Gram-negative rod measuring $0.5 \times 2.0–3.0$ μm. Its motility is given by between one and four polar flagella.

Pseudomonas azotogensis can grow at temperatures of 5°–37°C, with optimum growth at about 28°C. At 50°–55°C it dies. It utilizes a wide range of carbohydrates, without producing any gases or acids. It hydrolyses starch well. In a medium containing carbohydrates and mannitol but no nitrogen it forms smooth, glassy colonies, not pigmenting even on prolonged cultivation. It slowly reduces nitrates to nitrites, but does not liquefy gelatine, nor affect milk. It grows slowly on meat agar. In 7 days of incubation in a liquid, nitrogen-free medium containing 1 per cent glucose *Pseudomonas azotogensis* accumulates up to

1 mg of nitrogen per 100 ml of medium. Proctor and Wilson (1958) used the isotope method to confirm the ability of *Pseudomonas azotogensis* to fix molecular nitrogen.

In 1961 an organism similar to *Pseudomonas azotogensis* was isolated (Paul and Newton) from various Canadian soils (in the provinces of Alberta and Saskatchewan) in which there were no nodule bacteria, and azotobacters only sporadically in irrigated zones (Ivarson, 1953). It had been thought that other forms of nitrogen fixers might be found there (Milne, 1951; Newton, 1954).

In several ways the bacterium isolated by Paul and Newton resembles *Pseudomonas azotocolligans*, but it is more like *Pseudomonas azotogensis*. This bacterium forms colourless colonies on agar media. One strain, isolated from grey forest soils, gave a dark brown, water-soluble pigment in a medium containing sodium benzoate. On meat peptone agar, round, white colonies form, later turning yellow. This micro-organism does not hydrolyse starch, nor liquefy gelatine or form indole or gas, but it does form hydrogen sulphide. It is relatively tolerant of acid conditions, starting to grow at pH 4·9. Nitrogen fixation is comparatively low, 0·1–3·9 mg of nitrogen per gram of mannitol utilized. When incubated for 48 hours on a shaker in a nitrogen-free medium buffered to pH 7·2 with 2 per cent mannitol, the cultures isolated from the soils of Alberta accumulated between 4 and 78 γ of nitrogen/ml.

The DNA composition of *Pseudomonas azotogensis* is markedly different, from that of other members of the genus. Ley and Park (1966) thought it very doubtful whether *Pseudomonas azotogensis* should really belong to the genus *Pseudomonas*.

Pseudomonas methanitrificens first came to notice in 1930 when Schollenberger found that soils with an outlet of natural gas had a higher content of nitrogen. Harper (1939) established the nitrogen content of such soils to be 0·26 per cent, whereas in adjacent ordinary soils it was 0·098 per cent. The increase in nitrogen was erroneously ascribed to the activity of clostridia. In 1952, Davis isolated from this soil a culture of a bacterium that utilized methane and other gaseous hydrocarbons. This bacterium, however, was not investigated for nitrogen-fixing powers, nor was it identified. Davis *et al.* (1964) thought that the increased content of nitrogen in such soils may be due either to the nitrogen-fixing activity of bacteria that utilize hydrocarbons, or to the assimilation of nitrogen by micro-organisms that develop later. The facts incline them towards the first assumption. There is some evidence that bacteria can utilize hydrocarbons as the sole source of carbon (Kataidi *et al.*, 1966).

Pseudomonas methanitrificens was isolated by Davis *et al.* in 1964 from soil taken from the site of a leakage of natural gas from a pipeline (Davis, Coty and Stanley). When cultivated on nitrogen-free solid media, this bacterium forms colonies varying from dull to brilliant, from white to yellow and from smooth to rough. Most often seen are brilliant, light yellow, smooth colonies consisting of weakly mobile rods 1–2 and 2–4 μm long. The cells contain lipid granules usually two per cell. The lipid fraction contains a large amount of poly-β-hydroxybutyric acid. Methane is the only source of carbon used.

To determine the nitrogen fixation, a culture of *Pseudomonas methanitrificans* was placed in a small closed system containing 10 ml of medium. The air was pumped out and methane and air (in a ratio of 3:7) pumped in. The composition of the impure methane introduced was 99–99·2 per cent methane, 0·26–0·32 per cent ethane, 0·12–0·16 per cent CO_2 and 0·26–0·32 per cent nitrogen. In this atmosphere, with the thermostat at 30°C, *Pseudomonas methanitrificans* assimilated 0·66–0·86 mg of nitrogen per 10 ml of medium.

Table 8.1. Distribution of Assimilated Nitrogen in the Cells of *Pseudomonas methanitrificans* and the Medium

Experimental conditions	Specimen	Amount of nitrogen fixed, mg
1 litre of medium, incubation for 2 months	Bacterial cells (dry weight 0·6 g)	42·4
	Culture fluid	3·7
2 litre of medium, incubation for 4 months	Bacterial cells (dry weight 3·23 g)	204·0
	Culture fluid	53·0*

* The high content of nitrogen is explained by autolysis of the cells on prolonged incubation.
This table is taken from Davis, Coty and Stanley (1964).

Later experiments were carried out in larger systems in which the culture developed in 1–2 ml of medium containing no nitrogen or carbon compounds. A purified mixture of methane and air was passed through the medium (in the ratio 1:9) at a rate of about 50 ml per minute.

After two weeks the nutrient solution in the system became turbid as the culture developed. The turbidity gradually increased. After two and four months of incubation the nitrogen was determined in the bacterial cells, the culture fluid and the gas purifier—sulphuric acid. Table 8.1 shows that most of the assimilated molecular nitrogen was found in the cells of the micro-organisms.

Pseudomonas non-liquefaciens was isolated by Novikova (1955) from the surface of various plants. The cultures isolated bound 1·6–4·5 mg of molecular nitrogen per gram of sugar. The cells are short, thin, Gram-negative, immobile rods. They are aerobes and do not liquefy gelatine.

Belyaeva (1954) established that *Pseudomonas pantotropha* assimilates small quantities of molecular nitrogen. This micro-organism can assimilate various organic compounds, but can also live autotrophically, inducing the oxidation of hydrogen. There are also indications that *Pseudomonas herbicola* has weak nitrogen-fixing powers (Novikova, 1955). Unidentified nitrogen-fixers of the genus *Pseudomonas* have been noted by several investigators.

Mishustina (1955) noted the weak nitrogen-fixing capacity of many cultures of *Pseudomonas*, which were able to multiply only in a medium containing small quantities of bound nitrogen. Such cultures are Gram-negative monotrichous or lophotrichous rods. On a nitrogen-free medium or on potato agar, they are mobile in the first hours of growth. In liquid culture 'pseudozoogloeas' often form. On a nitrogen-free solid medium they produce oval, mucilaginous, milk-coloured or transparent colonies, sometimes spreading over the surface of the agar. Most cultures develop poorly on protein media. Some form a yellow pigment.

Proctor and Wilson (1958) isolated from soil and water, six strains of *Pseudomonas* that fix nitrogen in nitrogen-free media containing 2 per cent sucrose. The effectiveness of nitrogen fixation by these strains was within the limits 1·1 to 4·3 γ of nitrogen fixed per mg of carbon utilized. The strains isolated were capable of fixing nitrogen in anaerobic conditions in amounts approximately equal to 5 mg of nitrogen per 100 ml of medium. In an atmosphere containing about 36 atom per cent of $^{15}N_2$ excess by the end of the experiment the content of gas in the culture was greater than 0·1 atom per cent.

Using nitrogen-free media, Ross (1958) isolated three strains of *Pseudomonas* from New Zealand tussock grassland soils and from the leaves of *Festuca*. Two of the strains formed yellowish-green and brown pigments on nitrogen-free media. In many physiological and morphological features, the cultures Ross isolated resembled *Pseudomonas azotocolligans* as described by Anderson (1955).

The cultures isolated by Ross had a low nitrogen-fixing capacity. In liquid media the content of nitrogen did not increase by more than 0·4 mg per 100 ml of medium in 28 days. On solid media the increase doubled but even this value was virtually within the limits of possible experimental error. In 1961 Panosyan *et al.* found that some cultures of *Pseudomonas* isolated from the soils of Armenia could fix nitrogen.

All this is evidence that the family Pseudomonadaceae includes bacteria with a greater or lesser capacity to assimilate molecular nitrogen (Gurfel', 1960; Romanova, 1961; Pshenin, 1965). Missirliu *et al.* (1964) considered that their experiments showed that in acid meadow soils the genus *Pseudomonas* played an essential role in the accumulation of nitrogen.

Family Spirillaceae

The spirillas and vibrions belonging to this family include both heterotrophs and autotrophs which are able to assimilate molecular nitrogen. Beijerinck (1925) isolated the nitrogen-fixing bacterium *Spirillum lipoferum* from garden soil in a medium containing humates. The cells of these bacteria are twisted once. The bacteria are motile and have a characteristic bundle of flagella. They develop in the air and are Gram-negative. The cells contain droplets of fatty inclusions when cultivated in broth with glucose and peptone, and also when in agar medium with calcium malate. On agar-peptone medium the cells of *Spirillum lipoferum* multiply abundantly but lose the characteristic spiral appearance and the ability to synthesize fat. In a medium containing calcium malate they form small, round, transparent dry colonies. Their temperature optimum is about 22°C.

Spirillum azotocolligens, isolated by Rodina (1956) in the Krasnodar Region, is widespread in stretches of water. The cells are rod-shaped, slightly bent and have granular protoplasm; in young cultures they are paired and Gram-positive. Their morphology hardly changes in different nitrogen-free media. The largest cells reach 4·6–6·65 × 1·3–2·0 μm, the smallest 3·9–5·5 × 1·3–2·0 μm. In organic media containing nitrogen, involutive forms often appear; these are flask-shaped, spindle-shaped, inflated and so on. As a rule in media containing nitrogen, growth is sparse, in the form of a thin spreading sheet. When there is bound nitrogen in the medium, small, whitish, slightly convex dry colonies form. On gel plates impregnated with nitrogen-free medium and in Fedorov's medium they fix between 4·0 and 12 mg of nitrogen per gram of mannitol. A bacterium similar to that described by Rodina was discovered by Gulya (1956) in the Alma Ata fish ponds (Kazakhstan).

Spirillum magnum, which was isolated from sea water (Pshenin, 1965), is an aerobic, Gram-negative bacterium. In Fedorov's liquid medium with mannitol the cells of 1–7 day old cultures are thin and weakly convoluted. They are 2·5–20·0 × 0·4–0·7 μm in size. The cells, which have a regular or irregular helical shape, have from a half to eight turns in them. During heteromorphic division, short, rod-shaped or slightly bent daughter cells may branch off. These grow to normal dimensions and assume the characteristic helical shape. In old cultures cells may have as many as seventeen turns. In Fedorov's agar medium with glucose, *Sp. magnum* forms small cells bent in the form of a C or an S. At the end the cells have one (or more) flagella. Microcysts may form; these are either shaped like lemons or have an irregular round shape.

On Fedorov's mannitol-agar, white, brilliant or dull, smooth, slightly protuberant colonies develop. With age they acquire a yellow or greyish-yellow colour. Colonies usually have a mucilaginous or pasty consistency.

Spirillum magnum develops well on media with or without nitrogen. It slowly liquefies gelatine, and uses as its sole source of carbon ethanol, mannitol, glucose, fructose, lactose, maltose, sucrose and acetic, pyruvic, fumaric, malic, lactic and malonic acids. It weakly utilizes glycerol and propionic, butyric, succinic and citric acids. It does not decompose cellulose. When using lactose it produces acids. Acid is formed when *Spirillum magnum* utilizes maltose and mannitol and especially glucose and sucrose. An isolated culture has a considerable nitrogen-fixing capacity (16·1–17·4 mg of nitrogen per gram of glucose).

Spirillum speciosum has seven strains which have been isolated from sea-water (Pshenin, 1965). It is an aerobic, Gram-negative organism. In Fedorov's liquid medium with mannitol the young cells are helical and measure 4·0–23·0 × 0·5–0·8 μm, with up to three, or sometimes four to six turns. In old cultures cells may be 60 μm long and have ten turns. Metachromatin granules can be seen in old cells. There are between one and four flagella. *Sp. speciosum* forms either spore-like bodies or gonidia, depending on the conditions. Microcysts may form in old cultures if the pH of the medium is changed to 4·0–5·0 or 7·8–8·0. In 7-day cultures on Czapek medium, in peptone water with glucose and meat-peptone broth, deformed, bent, ramified and inflated cells are characteristic. In liquid media involutive forms are rarely seen. In Fedorov's agar medium with mannitol or glucose, *Sp. speciosum* forms small, transparent, brilliant convex colonies. The consistency of the colonies is mucilaginous and thread-like. In time, colonies reach 3 mm in diameter and assume a whitish colour. Sometimes colonies spread out over the whole surface of the plate. In agar medium cells often have a straightened, almost rod-shaped form, and start to form cysts more rapidly than in liquid cultures.

Sp. speciosum grows better on mannitol than glucose, and can utilize ethanol and butyric, pyruvic, fumaric, malic, lactic, citric and malonic acids. Glycerol, fructose, and acetic, propionic and succinic acids are less well assimilated. This species does not decompose cellulose, and assimilates ammonium salts, urea, nitrates and asparagine less well. It does not liquefy gelatine. For every gram of glucose utilized, it fixes 4·24 mg of nitrogen.

Spirillum nana was isolated from sea water and from the surface of the thallus of the phyllophore by Pshenin and described, as were *Sp. magnum* and *Sp. speciosum*, as a new species. Unlike them, *Sp. nana* has thin, spiral, S-shaped or vibrion-like cells which are sometimes almost rod-shaped. When immobile they have a half to two and a half turns. When moving, the cells twist in snake-like fashion and have at least one to three complete turns. They are 2·1–4·9 × 0·3–0·4 μm in size. In liquid nitrogen-free media with mannitol or glucose, *Sp. nana* forms a protuberant, folded film. It is characterized by weak assimilation of ammonium salts and nitrates, by the ability to reduce nitrates to ammonia and by its weaker utilization of ethanol and greater utilization of glycerol and succinic acid than *Sp. magnum* and *Sp. speciosum*. It also differs from *Sp. magnum* in its weak assimilation of propionic and butyric acids and better utilization of fructose. It fixes up to 4·8 mg of nitrogen per gram of glucose utilized.

When looking for *Beijerinckia* in soils, Becking (1963) found members of the genus *Spirillum* that fix molecular nitrogen. These are slightly bent mobile rods, with one flagellum. They are 2–4 × 1 μm in size, containing poly-β-hydroxybutyric acid. The bacterium grows on broth and peptone media, but the cells do not accumulate poly-β-hydroxybutyric acid.

Nitrogen fixation by these cells was demonstrated on nitrogen-free media with 1 per cent glucose and 0·005 per cent yeast extract when the nitrogen source was $^{15}N_2$. When the yeast extract was replaced by biotin, pyridoxine or vitamin B_{12} the assimilation of molecular nitrogen ceased. Hydrogenase could be detected in the bacterium only when benzyl viologen or methyl viologen were used as the hydrogen acceptor, instead of methylene blue. The species of *Spirillum* in question was not determined but Becking noted its similarity to *Sp. lipoferum.*

Pshenin (1965) found nitrogen-fixing activity in marine forms of the genus *Vibrio* (discussed later). *Vibrio frequens* is an aerobe but can multiply in micro-aerophilic conditions. It is Gram-negative.

In nitrogen-free liquid and agar media, young (18–24 hour) cells are C or S shaped or coccoid. They have one polar flagellum five to six times the length of the cell, which is itself $1\cdot5$–$3\cdot0 \times 0\cdot5$–$1\cdot5$ μm in size. Polymorphism is characteristic of older cultures with the appearance of ovoid, clostridial, elongated and other forms.

In Fedorov's agar medium containing glucose or mannitol, smooth, brilliant colourless or somewhat whitish round colonies develop after two or three days. They have a smooth or wavy outline. As they age, the colonies become folded and yellow. After 1–1·5 months they reach a diameter of 4–5 mm. The consistency of the colonies is pasty or slightly rope-like. In a similarly constituted liquid medium a greyish-white film develops within 2 to 4 weeks, turning yellow in time. The liquid becomes greyish-white, turbid and very ropey.

As a source of carbon *Vibrio frequens* can utilize glucose, and pyruvic, malic, citric, lactic and malonic acids, and, more weakly, ethanol, glycerol, fructose, and acetic, propionic and butyric acids. It does not utilize cellulose. As a source of nitrogen, it makes good use of asparagine, ammonium salts and nitrates, and poor use of urea. It does not liquefy gelatine, but acidifies peptone water with lactose, maltose, sucrose or glucose and mannitol within 1 week to pH 4·5–5·5. It fixes between 1·8 and 2·2 mg of nitrogen per gram of glucose.

Vibrio hydrosulfureus is a facultative aerobe with a C, S or coccoid shape in nitrogen-free liquid and agar media. Its mobility is determined by the presence of one polar flagellum, two to five times larger than the cell, which is $1\cdot0$–$3\cdot5 \times 0\cdot5$–$0\cdot8$ μm in size. As with *V. frequens*, polymorphism is typical, becoming more accentuated as the culture ages.

On Fedorov's agar medium with mannitol or glucose, two types of colony develop. The first is shaped like a raspberry and its colour varies from light yellow to orange. These colonies are dry, leathery and peel away from the agar in large pieces. After 2–3 weeks they reach 3–5 mm in diameter and 1–5 mm high. The second type of colony is mucilaginous and flat. The colour varies from white to yellow. In analogous liquid media the colony forms a slightly protuberant, yellow or yellowish-orange, brilliant or dull film of leathery and mucilaginous consistency. The fluid of the medium becomes more turbid when the mucilaginous film develops. When colonies develop on meat-peptone broth there is strong evolution of hydrogen sulphide and ammonia; indole does not form.

In peptone water with maltose, sucrose, glucose and mannitol, *V. hydrosulfureus* produces acids, which it does only weakly if at all in lactose. This bacterium liquefies gelatine, and makes good use of ammonium salts and nitrates. It fixes 1·6 mg of molecular nitrogen per gram of glucose.

Vibrio nonhydrosulfureus is a variant of *V. hydrosulfureus* that does not evolve hydrogen sulphide and fixes nitrogen in larger quantities (4·0–8·6 mg per gram of glucose).

The genus *Desulfovibrio* also has members that bind molecular nitrogen (Sisler and Zobell, 1951). Bacteria of this genus are slightly bent rods of differing lengths. Sometimes they are joined in short chains, resembling spirilla. Movement is accomplished by means of a polar flagellum. They are strict anaerobes and reduce sulphates, sulphites, hyposulphates and hyposulphites. *Desulfovibrio* can utilize a considerable range of organic compounds (peptone, asparagine, carbohydrates, a number of organic acids, and so on) which serve as hydrogen donors. Some cultures oxidize hydrogen as the sole source of energy (simultaneously reducing oxidized sulphur compounds). Nitrates are not usually reduced. The temperature optimum is at about 25°–30°C and the maximum between 35° and 40°C. Growth is observed within the pH range 5·0 to 9·0. The bacteria are found in soil, water and other natural substrates.

The nitrogen-fixing abilities of bacteria that reduce sulphates were discovered in a most unusual manner. When a mixture of molecular nitrogen and argon was used as an inert atmosphere for investigations of these bacteria the nitrogen proved not to be inert in the presence of the hydrogenase-forming strain of *Desulfovibrio*. After prolonged incubation the content of nitrogen in the culture decreased. Further investigations, involving changes in the ratio of nitrogen and argon in the gas phase confirmed that *Desulfovibrio* fixes nitrogen. In these experiments a pure culture of *Desulfovibrio* was introduced into 20 litres of inorganic medium prepared in sea water. The space above the medium was filled with a gas mixture consisting of hydrogen, carbon dioxide, nitrogen and oxygen (the two latter in the ratio 83:1). When the bacteria developed there was a decrease in the hydrogen and carbon dioxide taken up by the culture according to the equation $CO_2 + 2H_2 = (HCHO) + H_2O$. ((HCHO) represents the organic matter of the cell of the micro-organism.) As a result of the uptake of hydrogen and carbon dioxide, the concentration of nitrogen and argon in the gas phase increased. But there was a decrease in the ratio of nitrogen to argon, indicating that the bacteria had assimilated nitrogen (Table 8.2).

The experiment was repeated with four pure cultures of *Desulfovibrio* grown in the same conditions at 28°C (Sisler and Zobell, 1951). The change in the ratio of nitrogen to argon was as in Table 8.3.

These observations confirm that molecular nitrogen can be utilized by species of *Desulfovibrio* as the sole source of nitrogen.

In the mud of Bere lake (France) an anaerobic, nitrogen-fixing micro-organism was found and identified as *Desulfovibrio desulfuricans* (Le Gall *et al.*,

1959, 1960). It fixes nitrogen only during heterotrophic nutrition, thus differing from the marine forms of this species that fix nitrogen chemo-autotrophically. Sisler and Zobell thought that bacteria of the genus *Desulfovibrio* could be

Table 8.2. Changes in the Gas Phase caused by *Desulfovibrio*

Component of gas medium	Composition of gas % after			Component of gas medium	Composition of gas % after		
	2 days	42 days	61 days		2 days	42 days	61 days
H_2	82·93	56·04	47·42	H_2S	0	0	0
O_2	0	0	0	N_2	16·63	43·14	51·61
CO_2	0·51	0·27	0·19	Ar	0·20	0·55	0·68
				N_2:Ar	83	78	76

This table is taken from Sisler and Zobell (1951).

important in the fixation of nitrogen in the sea because of their wide distribution in sea deposits.

In a medium used to isolate *Beijerinckia*, Becking (1963) found a vibrion resembling *Desulfovibrio gigas*. This strictly aerobic micro-organism assimilated

Table 8.3. Changes in the Ratio of Nitrogen to Argon caused by *Desulfovibrio*

Culture No.	After 2 days incubation	After 60 days incubation
1	83	67
2	83	76
3	83	79
4	83	55
Control	83	88

molecular nitrogen only in the presence of Difco yeast extract. Vitamin B_{12}, biotin and pyridoxine could not replace the biologically active substances present in the yeast extract. Cells of this micro-organism enriched an atmosphere containing 43 atom per cent $^{15}N_2$ excess by 6·4 atom per cent excess and enriched an atmosphere containing 65 excess atomic per cent by 1·04 excess atomic per cent.

Family Azotobacteraceae

As well as the genus *Azotobacter*, already described in detail, the family Azoto-bacteraceae includes the nitrogen-fixing genus *Derxia*. *Derxia gummosa* assigned by Norris (1960) to an independent genus of the family Azotobacteraceae was described in 1959 by Jensen *et al.* as a new nitrogen-fixing bacterium possessing several features characteristic of this family. A culture was isolated from soils of West Bengal. The three strains isolated were Gram-negative, mobile rods with polar flagella. Mobility is an irregular feature depending on the strain and the medium. In liquid medium containing ammoniacal nitrogen the cells usually have one polar flagellum or two radiating from each pole. Mobile cells are usually shorter than the others.

In a nitrogen-free agar medium with mannitol at 25°C rod-shaped cells begin to develop after 20–24 hours, and the process lasts up to 2 days. The cells are $3 \cdot 0$–$6 \cdot 0 \times 1 \cdot 0$–$1 \cdot 2 \, \mu m$ in size with rounded ends and homogeneous cytoplasm. After 2 days the cells take on an alveolar structure, each surrounded by a mucilaginous capsule. Five day old cultures often contain long thread-like cells, sometimes inflated and bent. Fatty inclusions are found in the large cells. In old cultures the capsules spill out and then the cells are an amorphous mass. The colour of such colonies varies from yellow to rusty brown.

The colonies grow in two ways; they may be in the form of a thin semi-transparent coat or a non-transparent mass. Only colonies of the second type can fix molecular nitrogen. As Tchan and Jensen (1960) showed, the formation of either type of colony of *Derxia gummosa* is determined by the presence in the medium of compounds containing nitrogen. The type of colony that develops is not controlled genetically. Massive mucilaginous colonies will only form in a medium containing bound nitrogen. In a medium devoid of nitrogen *Derxia* develops weakly, forming pellicular semi-transparent colonies.

As a carbon source *Derxia* uses glucose, fructose, ethanol, glycerol, mannitol and sorbitol. On mannose and sodium lactate it grows sparsely. Only traces of the bacterium are detectable on galactose, sucrose, acetate, pyruvate, dulcitol and starch. Butyrates, citrates, benzoates and xylose all suppress the develop-ment of *Derxia*.

Derxia gummosa is a slow-acting nitrogen fixer. In 30 days of incubation at 35°C in a nitrogen-free medium containing 1 per cent glucose or mannitol, it fixes from 9·4–25·0 mg of nitrogen per 100 ml of medium. According to Chakravorty and Das (1965), there is a linear relation between the amounts of sugar (glucose, fructose) consumed and nitrogen fixed. There is no nitrogen fixation if organic acids are the only carbon source. Molybdenum stimulates nitrogen fixation but cannot be replaced by vanadium in the case of *Derxia gummosa*. Sodium azide in a concentration of $10^{-4} \, \mathrm{m \, mol \, l^{-1}}$ suppresses nitrogen fixation by almost 50 per cent. Sodium arsenate, however, is a weak inhibitor of nitrogen fixation even in a concentration of $10^{-2} \, \mathrm{m \, mol \, l^{-1}}$ (Chakravorty and Das, 1965). Various forms of bound nitrogen, particularly glutamic acid, are assimilated more efficiently than molecular nitrogen.

Derxia gummosa will not grow unless the pH is above 5·0; its maximum pH is slightly above 9·0. It actively uses its carbon source and fixes nitrogen at pH 5·9–7·0; nitrogen fixation may be intense in a more acid (with mannitol) or in a more alkaline (with glucose) medium according to the carbon source available. Whenever carbon compounds are utilized the medium becomes acid, sometimes reaching pH 4·5. No gas is evolved.

The optimum temperature for the development of *Derxia gummosa* is between 25° and 35°C; at 15° and 42°C there is weak growth and a temperature of 50°C is lethal. *Derxia gummosa* is an obligate aerobe and does not form catalase.

Derxia indica was isolated by Roy and Subir Sen (1961) from the wetting fluid of jute (*Corchorus olitorius L.*) in a nitrogen-free medium containing mannitol. It is a Gram-negative, aerobic bacterium which does not produce catalase. In solid nitrogen-free medium it forms convex, semi-transparent, oval or round verrucose colonies. In the first days of development the colonies are solid and then become soft and later mucilaginous. On the third and fourth days the colonies have a bright yellow colour, gradually turning to dark brown. Five day old colonies measure about 10 mm across. On nitrogen-free medium with dextrose the colonies are chocolate coloured and slimy. Liquid media used to grow *Derxia indica* acquire a gel-like consistency and a yellow or dark brown colour.

Three distinct morphological stages can be distinguished in the life cycle of the culture that develops on a nitrogen-free mannitol or glucose medium (at 34°–35°C). In the first stage there are characteristic 'striated' cells due to the presence of denser zones in disc-shaped or annular formations. Characteristic of the second stage is the appearance of cells with alveolar structure. This stage usually rapidly follows the first and is especially distinct in 3–4 day old cells. The cells are surrounded by a mucilaginous capsule varying in thickness. In the third stage peculiar to old cultures, the rods are 0·79–4·32 × 0·53–1·16 μm in size and have rounded ends. First and second stage cells have inclusions of fat and volutin. On the basis of these morphological features and developmental peculiarities the new species, *Derxia indica*, was created.

Derxia indica also differs physiologically from *Derxia gummosa*. It develops well on nitrogen-free media containing glucose, ethanol, mannitol, sucrose, laevulose, sorbitol and glycerol, but less well on mannose and maltose. It does not utilize starch, arabinose, lactose, xylose, galactose or rhamnose. On a source of bound nitrogen growth is more vigorous and rapid than on molecular nitrogen. *Derxia indica* makes good use of aspartic and glutamic acids, asparagine, alanine, ammonium acetate, sodium nitrate, peptone and urea. Development is weak on ammonium sulphate and very retarded on glycine. This bacterium can develop in the presence or absence of calcium.

The maximum quantity of nitrogen fixed is 30 mg per 100 ml of nitrogen-free medium containing 2 per cent glucose in 30 days at 31°–32°C. Molybdenum intensifies the process. *Derxia indica* develops at pH 5·0–9·0, but an alkaline medium is optimal. With prolonged incubation the medium is acidified to pH 4·4.

Rhizobiaceae and Achromobacteraceae

In the literature there are indications that *Agrobacterium radiobacter*—a free-living representative of the Rhizobiaceae—can fix molecular nitrogen. This is an aerobic, Gram-negative (sometimes changing with Gram-stain), small motile rod with dimensions of 0·15–0·75 × 0·3–2·3 μm. It usually develops in the form of individual cells, but these are sometimes paired. In some conditions stellate conglomerations form. This bacterium utilizes sugars, glycerol and mannitol, forming carbon dioxide as the end product. No organic acids are formed. The optimum temperature is 28°C, the minimum about 1°C and the maximum 45°C. It tolerates a very alkaline medium (up to pH 12·0).

Skinner (1928) investigating two strains of *Agrobacterium radiobacter* found that one of them had insignificant nitrogen-fixing activity. In 1958 Ross isolated from soil and from leaves of *Festuca* eleven strains of *Agrobacterium radiobacter* that developed well on a nitrogen-free medium. After 28 days of incubation, however, the cultures had accumulated an insignificant amount of nitrogen—1·7 mg/ml on a liquid medium and 0·5 mg per 100 ml on a solid medium. Some strains growing on nitrogen-free media have been found to accumulate nitrogen. For this reason Ross did not consider the Rhizobiaceae important in the nitrogen balance of the soil.

The genus *Achromobacter* is a nitrogen-fixing member of the Achromobacteraceae. The bacteria of this family are small, Gram-negative rods that move by means of peritrichous flagella, although some species are immobile. They are widespread in sea and fresh water and soil. Proctor and Wilson (1959) found their importance as nitrogen fixers hard to evaluate quantitatively, but because representatives of this family are widespread in surface water and soils, their activity is probably significant.

On agar and gelatine media species of *Achromobacter* do not form pigment, but on potato media they may accumulate a yellow carotinoid type of pigment. When sugars are utilized small quantities of acid are formed.

From the rhizosphere of wheat Fedorov and Pantosh (1957) isolated *Achromobacter parvulus* (synonym *Bacterium parvulum*), which fixes 5·9 mg of nitrogen per gram of glucose. It is a very small (0·1–0·2 × 0·3–0·4 μm) immobile, strictly aerobic bacterium forming punctate colonies on agar and gelatine media. It does not liquefy gelatine, and develops slowly in liquid media. When it utilizes glucose, lactose, sucrose, glycerol or ethanol, it does not form acids. It does not hydrolyse starches. It reduces nitrates to nitrites and fixes ammonium vigorously.

Achromobacter hartlebii, which assimilates molecular nitrogen, was isolated by Proctor and Wilson (1958) from soil. It fixes 1·11–1·38 mg of nitrogen per gram of carbon source utilized. Nitrogen fixation has been demonstrated by the isotope method.

The ability to fix nitrogen has been demonstrated in other members of the genus *Achromobacter*. In 1958 Jensen isolated from stagnant, contaminated river water near Copenhagen an accumulation culture of a nitrogen-fixing organism, using a medium containing 0·02 per cent K_2HPO_4 and 1 per cent mannitol. After transfer to a nitrogen-free medium plus glucose, a pure culture of bacteria identified as *Achromobacter* (strain 4) was isolated. The bacterium is rod-shaped, and measures 0·8–0·9 × 1·6–2·2 μm. It is immobile and Gram-negative, and grows on meat-peptone agar and on nitrogen-free media. Colonies on meat-peptone agar are convex, greyish-white, moist, smooth and round. In nitrogen-free media they are grey, semi-transparent and slimy.

Achromobacter is a faculative anaerobe which reduces nitrates to nitrites, does not form catalase, ferments various sugars with the formation of acid, does not form gas or acetone, but assimilates glycerol, sorbitol, mannitol and inositol.

According to Hamilton *et al.* (1965), in anaerobic conditions pyruvic acid is converted by a culture of *Achromobacter* (strain 4B-non-slime variant of strain 4) into acetic and formic acids. In a medium containing 1 per cent glucose or mannitol, small quantities of $CaCO_3$ and yeast extract, it fixes molecular nitrogen—1·5–1·7 mg per gram of glucose consumed and 1·1 mg per gram of mannitol. Nitrogen fixation proceeds more vigorously in anaerobic conditions than in air, but intensified aeration sharply reduces the effectiveness of the process. Nitrogen fixation has been demonstrated in experiments with $^{15}N_2$.

Working with a culture of *Achromobacter* isolated by Jensen, Proctor and Wilson (1958, 1959) confirmed its ability to fix nitrogen in aerobic and anaerobic conditions. About 1·35 mg of nitrogen was fixed per gram of carbon source utilized. The deuterium-hydrogen exchange reaction has been used to establish the presence of weakly active hydrogenase in this culture.

Goerz and Pengra (1961) also worked with a culture of *Achromobacter* obtained by Jensen, but found nitrogen fixation only in anaerobic conditions, when the nitrogen increment was 2·4 mg per 100 ml of medium. The addition of small amounts of ammonium salts to the medium (1·0–3·0 mg per 100 ml) increased the amount of nitrogen fixed to 8·0 mg per 100 ml of medium. The introduction of hydrogen into the gas mixture heavily reduced fixation.

Investigating the development of a culture of *Achromobacter* in different media, Goerz and Pengra found that nitrogen fixation began after variable amounts of the bound nitrogen in the medium had been consumed.

As well as investigating the culture of *Achromobacter* isolated by Jensen, Proctor and Wilson isolated eight cultures of nitrogen fixers from different sources, including the soil, all assigned to the genus *Achromobacter*. Most of the cultures were mobile and reduced nitrates to nitrites. Amounts of molecular nitrogen fixed in a medium containing 2 per cent sucrose in 4 days varied from

1·1 to 1·38 mg per gram of carbon source utilized. Nitrogen fixation was confirmed in experiments with $^{15}N_2$.

All the cultures isolated possessed hydrogenase and fixed nitrogen only in aerobic conditions. Nitrogen fixation began after the nitrogen present in the nutrient medium had been used up.

Mahl et al. (1965) suggested that the facultative anaerobe and nitrogen fixer Achromobacter strain 4 is a non-aerogenic variant of Klebsiella pneumoniae.

Family Enterobacteriaceae

The enteric bacteria include many Gram-negative mobile rods with peritrichous flagella, and also immobile rods. We shall deal only with two of these— Aerobacter aerogenes and Klebsiella, of which individual strains are known to fix nitrogen. Kaufmann (1959) thought that both should be united in one group, Klebsiella, but because we agree in principle with the classification of Bergey, we shall consider them separately and successively.

Aerobacter aerogenes was long considered unlikely as a nitrogen fixer (Bhat and Palacios, 1949) for it was impossible to establish a nitrogen gain in a culture using the Kjeldahl method. The bacterium is a rod measuring 0·5–0·8 × 1·0–2·0 μm, usually in the form of individual cells. These are usually immobile and Gram-negative.

Skinner (1928) established a reliable increase in the nitrogen content of liquid media containing only two out of twenty-five strains of Bact. aerogenes (Aerobacter aerogenes), and this in very insignificant amounts. Hamilton et al. (1953) confirmed this using the isotope method. They obtained a weak but regular nitrogen gain in cultures of two strains of Aerobacter aerogenes (out of sixteen they isolated from faeces and water). Cells of Aerobacter aerogenes in an atmosphere with 32 atom per cent of $^{15}N_2$ excess incorporated 0·16–0·258 atom per cent $^{15}N_2$. In the cells of Azotobacter vinelandii and Saccharomyces cerevisiae, used as positive and negative controls, incorporation was 0·226 1·99 and 0·00–0·013 atom per cent respectively.

Later, in conditions more favourable to development (anaerobiosis and neutral pH), Hamilton and Wilson (1955) demonstrated clear nitrogen increases in cultures of Aerobacter aerogenes by the Kjeldahl as well as by the isotope method. For various strains of Aerobacter aerogenes the increase was between 8·2 and 12·2 mg of nitrogen per 100 ml of medium. Jensen (1956a and b) isolated three strains of Aerobacter aerogenes from river water, of which two, practically identical (N_1 and N_5), had considerable nitrogen-fixing activity. In a nitrogen-free medium containing 1 per cent mannitol or glucose, cultures of these strains grew abundantly, forming grey-white, transparent, slimy colonies consisting of small, non-capsulated rods with heterogeneous protoplasm. Young cells were mobile. On meat-peptone agar greyish colonies formed, which were smooth and convex with even margins, consisting of rod-shaped cells which took up dyes uniformly. These bacteria are Gram-negative, and form catalase and can reduce nitrates to nitrites, but do not liquefy gelatine, nor form H_2S. They utilize arabinose, glucose, xylose, fructose, mannose, galactose, sucrose, maltose, glycerol, sorbitol, mannitol and salicin with the resultant formation of acid and gas. They do not utilize butanol, erythritol or dulcitol.

The cultures develop more vigorously on a source of bound nitrogen, growth being faster with ammoniacal nitrogen than with nitrates. The reason for this is the weak nitrate reductase activity of *Aerobacter aerogenes*, which also determines the slow growth in a medium containing nitrates. The introduction of agar or, better still, yeast extracts, into a synthetic nitrogen-free medium stimulates growth and nitrogen fixation. Because of this Jensen considered that these processes require growth substances that are contained in yeast extracts and apparently also in agar.

Jensen did not find appreciable differences between the amounts of nitrogen fixed by cultures grown in aerobic and anaerobic conditions. In anaerobic conditions the productivity of the cultures was 2·5–4·2 mg of nitrogen per gram of glucose consumed and in aerobic conditions it was 3·6–3·8 mg per gram. Suitable values for nitrogen fixation are 3·5–8·0. He also noted that the culture fixed nitrogen considerably better at 15°–18°C than at 25°–30°C; the minimum temperature for development was 5°C and the maximum was about 40°C. This is apparently associated with the more economic utilization of glucose at low temperatures. The average level of nitrogen fixation was 2·3–4·5 mg of nitrogen per gram of carbon source utilized. To develop on media with or without nitrogen, *Aerobacter aerogenes* requires magnesium (Yoch and Pengra, 1964).

Pengra and Wilson (1958), working with seven strains of *Aerobacter aerogenes*, found that molecular oxygen inhibits nitrogen fixation in culture. A high partial pressure of oxygen completely suppresses nitrogen fixation. These results confirm those of Hamilton and Wilson (1955). Molecular hydrogen is also a competitive inhibitor of nitrogen fixation. (This phenomenon had already been established for aerobic nitrogen fixation (Wilson, 1951).) Pengra (1959) showed that the nitrogen-fixing systems of *Aerobacter aerogenes* are activated in the presence of molecular nitrogen.

Pengra and Wilson found four phases in the development of a culture of *Aerobacter aerogenes* in nitrogen-free media. First is the lag phase, when there is an initial delay in growth that lasts until the bacteria have adjusted to the new conditions of the medium or until these conditions are changed to their requirements. In the second phase, the organism utilizes the insignificant store of nitrogen in the medium in the form of ammonium salts or yeast extract. In the third phase, there is a short delay in growth when the nitrogen stores in the medium are exhausted and the formation of the adaptive enzyme system is just beginning—this has been called the second lag phase. In the fourth phase, the growth of the culture is accompanied by nitrogen fixation. This pattern of growth was confirmed by Procter and Wilson (1953, 1959) for *Aerobacter aerogenes* and by Goerz and Pengra (1961) for *Achromobacter* and *Pseudomonas*.

There are indications that *Klebsiella pneumoniae* may assimilate molecular nitrogen. This micro-organism, which is very close to *Aerobacter aerogenes*, is Gram-negative, immobile, rod-shaped, usually with a capsule. It forms mucilage and measures about 0·3–0·5 × 5·0 µm.

In 1907 and 1908 Löhnis and Pillai obtained appreciable amounts of fixed nitrogen in a culture of *Klebsiella pneumoniae*. The strains investigated produced 1–2 mg of nitrogen per gram of carbon source used. When developing in the presence of algae, their nitrogen-fixing activity was sharply intensified.

Mahl *et al.* (1965) examined thirty-one cultures of *Klebsiella pneumoniae* for nitrogen-fixing powers, and found thirteen capable of assimilating from 1·7 to 6·5 mg of nitrogen per 100 ml of mannitol medium containing a small amount of bound nitrogen.

Klebsiella rubiacearum was isolated by Silver *et al.* (1963) from leaf nodules of *Psychotria bacteriophila* on nitrogen-free mineral agar, and, Centifanto and Silver (1964) made a more detailed study. This bacterium is rod-shaped, with granular inclusions at the ends. It is not resistant to acids. Growth in a liquid, nitrogen-free medium containing glucose is accompanied by the evolution of gas on the fourth to fifth day of culture. When incubated for 24 hours at 28°C in a nitrogen-free medium containing sodium pyruvate, in an atmosphere containing nitrogen-15, a culture incorporates 0·849 atom per cent nitrogen. Nitrogen is assimilated in anaerobic conditions, to the extent of 4·5 mg per gram of glucose utilized (in 3 days).

Ley (1955) considered the genus *Klebsiella* to be genetically close to *Azotobacter* and *Rhizobium* because of the similarity in the ratios of guanine to cytosine in the DNA of the three genera.

Some bacteria of the genus *Arthrobacter*, of the family Corynebacteriaceae, have been reported to assimilate molecular nitrogen. *Arthrobacter* is characterized by the absence of flagella, negative or variable Gram staining, aerobicity, the formation of 'arthrospores' in the cell and unique morphological changes during development.

In 1962, Jensen found a high content of nitrogen in beech trees growing on soil that did not contain *Azotobacter*, but in which he identified as nitrogen accumulators several micro-organisms belonging to different groups, including *Arthrobacter*.

In 1963, Smyk and Ettlinger, using Alexandrov and Zak medium, isolated fifteen strains of *Arthrobacter* from the surface of Karst deposits in the Alps. Twelve of these strains were able to fix nitrogen in nitrogen-free media containing a large amount of calcium carbonate. In a nitrogen-free medium containing 1 per cent glucose, 20–30 mg of nitrogen accumulated per gram of carbon source utilized. Most nitrogen was fixed when large quantities of calcium carbonate and potassium felspar was added.

In 1966 Smyk confirmed nitrogen fixation by these strains using labelled nitrogen. These strains are not identical with other known representatives of the genus *Arthrobacter*. One of their products of metabolism is 2-ketogluconic acid.

Other nitrogen-fixing bacteria that do not produce spores

Many have noted the ability of several non-sporing bacteria to assimilate molecular nitrogen, but have not tried to identify the bacteria. In 1905 Volpino isolated from Italian soils an immobile, non-sporing rod that fixed considerable amounts of nitrogen in a nitrogen-free medium. When, however, traces of ammonia were removed from the atmosphere, the organism almost stopped fixing nitrogen.

From soils of the island of Krakatoa (Sunda Archipelago), Kruyff (1906) isolated an immobile, non-sporulating rod-shaped bacterium closely resembling *Pseudomonas leuconitrophilus* Perotti which could liquefy gelatine and form abundant slimy colonies. It made liquid, nitrogen-free media turn cloudy without forming a film. For every gram of mannitol, it fixed up to 2·5 mg of nitrogen. The bacterium was named *Bacterium krakatau*. In 1910, Kruyff, investigating Javanese soils, isolated a further twelve oligonitrophilic bacteria, of which the most vigorous nitrogen fixers were three very small species: $0·5 \times 2·0, 0·3 \times 2·0$ and $0·4 \times 1·7$ μm. Kruyff did not give a detailed description of the species, but *Azotobacter* is very rarely encountered in these soils. Investigating the heavy clay and terra rosa of the rice fields of southern India, Löhnis and Pillai (1907) isolated several nitrogen-fixing bacteria in nitrogen-free media with added mannitol, dextrose and sodium tartrate with and without chalk. Among them was a micrococcus (0·6 μm across) forming yellow colonies, and able to fix 3 mg of nitrogen per gram of sugar. Nitrogen was also fixed by short non-sporing rods of the type *Bacterium turcosum* and *Bacterium chrysoglocea* which produced a yellow pigment on agar media. Negligible gains in nitrogen were also observed in the accumulation culture of bacterium assigned by Löhnis and Pillai to *Bacterium tartaricum*.

Gray and Smith (1950) isolated from chalk soils of Cambridge (England), pH 7·4–7·6, a mobile, rod-shaped bacterium 6·0–8·0 μm long that took up Gram stain with difficulty. When cells were treated with dyes their inclusions lined up in the form of drops (beads) as in diphtheria organisms. The systematic position of the bacteria is difficult to determine, especially because of the unusual structure of the flagellar apparatus. Although it is peritrichous, the polar flagella are considerably larger than the lateral flagella.

Development is good on the fourth day. On agar containing soil extracts and mannitol at 22°C punctate, dense, white, moist colonies form. On nitrogen-free mineral agar medium containing 1·5 per cent mannitol, the colonies are not transparent, but are jelly-like and amorphous. Growth is abundant when glycerol or glucose is added to the medium.

This organism does not liquefy gelatine even after 6 weeks of incubation. It utilizes lactose, maltose, glucose and sucrose with the production of acids; no gas is evolved. It does not form indole or ammonia, nor does it assimilate salicin. It rapidly reduces nitrates to nitrites. This bacterium is aerobic and psychrophilic, and develops best between 10° and 20°C. The optimum pH is about 7·2–7·4; it does not grow in acid media. In nitrogen-free medium with mannitol at 20°C as much as 0·2 mg of nitrogen accumulates per 100 ml of medium in 3–4 days.

From many soils of the USSR, Mishustina (1955) isolated a rod-shaped bacterium unlike any species previously described. On Ashby's medium it formed convex transparent colonies which were mucilaginous and of an irregular round shape. Growth on meat-peptone agar was very scanty. The cells do not take up Gram stain and in young culture are very mobile, with a mucilaginous capsule. The flagella are peritrichous, and the cells are 1·4 × 0·6 μm in diameter. This micro-organism does not liquefy gelatine but slowly peptonizes milk, producing a brown pigment. It grows in media containing different carbon sources—glucose, sucrose, arabinose, lactose, dextrin and mannitol—without the formation of gas. It develops well on synthetic medium with ammonium nitrate, urea and peptone as sources of nitrogen. The productivity of a culture grown in liquid nitrogen-free medium with purified air blowing through it was 2·0 mg of nitrogen per gram of carbon source utilized.

Novikova (1955) isolated epiphytic bacteria from the surface of lupin, clover, tomatoes, maize and cucumbers and assigned them to the genus *Bacterium*. Some fixed molecular nitrogen, most of them close to *Bact. paracloacae*. There were also cultures identified as *Bact. herbicola* (*Pseudomonas herbicola*) and *Bact. cornea*. The amount of nitrogen fixed was determined after ten days of incubation on calcined sterile sand impregnated with Vinogradskii's nitrogen-free medium (with chalk and sucrose). It varied from 1·5 to 13·0 mg of nitrogen per gram of sugar used. One of the active cultures of *Bact. herbicola* utilized sugar and evolved gas, while the other evolved no gas.

Roy and Mukherjee (1957) isolated from soils in which jute was growing a Gram-negative bacterium with cells which were initially coccoid, in ones, twos or threes. Later the bacterium became rod-shaped with dimensions of 0·75–1·0 × 2·5 μm. There were considerable quantities of fatty cell inclusions. The cells formed a poor suspension in water. On Ashby's medium, colonies were convex and oval measuring 4·6 × 3·3 μm. At first they were transparent and slightly yellow; after 3–4 days the pigment was concentrated in the central part of the colonies which assumed a light chocolate colour. The margins of the colony were a brilliant light colour. Later the margins became grey and the bulk of the colony took on a chocolate colour, and a gelatine-like consistency. In Vinogradskii's medium colonies visible to the eye appeared earlier but developed more slowly. The pigment did not diffuse into the medium nor dissolve in water.

The culture was aerobic but could develop when the pressure of oxygen was low. In this case no pigment formed and the colonies resembled drops of water.

The optimum temperature for them was 28°C. The culture tolerated well an acid medium (to pH 4·0). In Ashby's medium the productivity of nitrogen fixation reached 17 mg per gram of mannitol. The bacterium made good use of carbohydrates, especially glucose and mannitol. But if nitrates or ammonium salts were present in the medium the culture did not develop whatever the carbon source. Roy and Mukherjee (1957) thought this bacterium to be close to the genus *Azotobacter*, although differing in that it could develop in meat-peptone broth.

Hug and Ali (1957) isolated from Indian soils a nitrogen-fixing bacterium that was either a Gram-negative rod, resistant to acid or a cocco-rod measuring 0·6–1·2 × 1·2–1·8 μm. The cells were joined in twos, threes or fours.

Abraham and Subramoney (1960) isolated from soil obtained from coconut retting grounds bacteria that grew well in Ashby's nitrogen-free medium with mannitol and glucose and fixed 6–7 mg of nitrogen per gram of sugar. The bacterium is an oval coccus and cells are often joined in pairs. In young culture they were mobile, in old cultures a thick capsule formed around the cell. On nitrogen-free medium colonies were round with a smooth convex surface. They developed within a fairly wide range of pH values. These bacteria were anaerobes. In 100 ml of Ashby's medium inoculated with 1 g of soil from the retting ground 20·5 mg of nitrogen accumulated in 4 weeks of incubation. Abraham and Subramoney assumed that these bacteria fixed nitrogen in symbiosis with algae.

Two Gram-negative bacteria have been isolated from different soils of the Nigeria types (Moore, 1963) characterized by their preference for different pH values (from 3·8–7·4). In an atmosphere containing labelled nitrogen the bacteria isolated gave an increment of 0·02 to 0·92 atom per cent of the gas.

Balloni *et al.* (1965) established that in Florentine forest soils of pH 5–5·5 nitrogen was fixed by aerobic, nitrogen-fixing bacteria that did not belong to the genera *Azotobacter* or *Beijerinckia*. Florenzano and Balloni (1965) isolated from these soils seven strains of nitrogen-fixers but, in spite of a detailed study, could not assign them to any known species. The bacteria were characterized by the formation of a green pigment that diffused into the medium, especially when growing on mannitol and nitrates. The pigment fluoresced blue in ultra-violet.

When young, these bacteria are very short rods, 1·5 × 0·8–10 μm, Gram-negative, immobile and surrounded by a mucilaginous capsule. They form neither spores nor cysts nor the light-refracting particles characteristic of *Beijerinckia indicum*. As nitrogen sources, they can use urea and nitrates and ammoniacal nitrogen. They can grow at pH 4·6 in nitrogen-free agar medium containing mannitol. On nitrogen-free agar medium containing mannitol, they form mucilaginous, transparent, vitreous colonies. On peptone agar with glucose, they form brilliant, whitish-cream colonies. They cause liquid medium containing peptone and glucose to become turbid and acid. The medium is acidified to a slightly lesser extent when they utilize sucrose, maltose, lactose or mannitol as a carbon source. None of the strains isolated utilized sodium benzoate as a carbon source.

It is very likely that there are many nitrogen-fixing bacteria yet to be discovered. In particular there seem to be many among the non-sporulating bacteria (Kořinek, 1932; Maliyants, 1933; Velankar, 1955; Caumartin, 1957; Fedorov and Pantosh, 1957; Zarma, 1959; Skalon, 1965; Karaguishieva, 1966; Rubenchik *et al.*, 1966).

Family Bacillaceae

The sporulating bacteria of the genus *Bacillus* include species that bind atmospheric nitrogen. This ability was first demonstrated for some strains of *Bacillus asterosporus* identical with *Bac. polymyxa* (Bredemann, 1908).

The nitrogen-fixer *Bacillus polymyxa* was isolated by Hino (1955) from soil of Nagoya (Japan). It was first mistakenly assigned to the genus *Clostridium*. Later its taxonomic position was revised and Hino and Wilson (1958) made a detailed investigation of its physiological and biochemical properties. The bacillus is a short Gram-variable rod, not forming chains and measuring $0.6–1.0 \times 2.0–7.0$ μm. Spore formation is of the clostridial type, and the catalase reaction is positive. A culture develops at temperatures between 28° and 35°C, and death occurs at 45°C.

Bac. polymyxa is a facultative anaerobe, its oxygen requirements changing with the nitrogen source. When atmospheric nitrogen is used, oxygen is not required, and indeed suppresses nitrogen fixation, as experiments with isotopic nitrogen have shown. When oxygen was present in the gas phase (70 per cent helium, 20 per cent oxygen, 10 per cent nitrogen, 33 atom per cent nitrogen-15 excess) cells were enriched with 0.02 atom per cent nitrogen-15. With no oxygen in the gas (90 per cent helium, 10 per cent nitrogen-15 with the same enrichment) cells were enriched with 7.65 atom per cent nitrogen-15. In media containing nitrates, oxygen stimulates the development of *Bac. polymyxa*; in the absence of oxygen the lag phase is lengthened by 40–70 hours (in the presence of oxygen there is almost no lag phase). The utilization of ammoniacal nitrogen does not depend on a supply of oxygen (Katznelson and Lochhead, 1944).

When developing on glucose the principal products of the culture are hydrogen, carbon dioxide, 2-3-butylene glycol, ethanol, glycerol, acetone and some volatile acids. Butyl alcohol and diacetyl are not formed. Biotin but not thiamine is needed. Optimum growth is observed at 30°C, weak growth at 42°C. A culture of *Bac. polymyxa* does not utilize lactose, arabinose or glycerol. During 20 hours growth in anaerobic conditions on a rocker, a culture fixes 10–12 mg of nitrogen per 100 ml of medium.

Fifteen of the seventeen typical strains investigated by Grau and Wilson (1960, 1962, 1963) fixed 3–15 mg of nitrogen per 100 ml of medium, but only in the presence of pyruvic acid. Without iron and molybdenum, growth and nitrogen fixation decreased by 30–50 per cent. When ammoniacal nitrogen was available, lack of molybdenum had no effect, and lack of iron caused only a slight retardation of growth. The effect of calcium was investigated for one strain only, and after seven transfers to a medium without calcium there were no changes in the nitrogen-fixing powers of the culture.

Active hydrogenase was found in suspensions of washed cells of *Bac. polymyxa*, reducing its rate by 50–70 per cent and the final amount of assimilated nitrogen by 50 per cent. Arsenates strongly depress nitrogen fixation, apparently destroying the acetylphosphate that is formed.

In 1962 Jensen observed active nitrogen fixation in forest soils, and considered that *Bac. polymyxa* was involved. These soils are almost anaerobic because of the covering of litter.

Kalininskaya (1966) isolated from chernozems a culture of *Bac. polymyxa* which assimilated 1·5–2·8 mg of molecular nitrogen per gram of carbon source consumed in anaerobic conditions. In aerobic conditions the strains isolated did not fix nitrogen.

According to Jensen (1962) and also Petersen and Esther (1964) there are strains of *Bacillus megaterium* that assimilate molecular nitrogen. The cells of *Bac. megaterium* are large (1·2–1·5 × 2·0–4·0 μm) and sometimes joined in short chains. They are aerobic, mobile and Gram-positive or sometimes Gram-variable. The spores usually form centrally or paracentrally. The bacteria grow at temperatures between 28° and 30°C. The maximum temperature at which growth still occurs is 45°C.

Thermobacillus azotofigens is a nitrogen-fixing bacillus isolated by Rakhno and Tokhver (1955) from manure of soddy-calcareous soil of the Estonian SSR (pH 7·5). The bacillus was isolated on a nitrogen-free medium, and proved to be a thermophil, growing best at 45°–50°C and not above 60°–65°C. It does not grow at 18°–20°C. A pH of 6·5–8·3 is favourable for development, and the optimum temperature decreases somewhat as the pH decreases. In 5 days of incubation at 50°C, *Thermobacillus azotofigens* can fix 2·9–4·5 mg of nitrogen per gram of sucrose utilized.

On a nitrogen-free, salt medium containing 0·5–2 per cent sucrose, 2·5–5·0 per cent soil extract, 1 per cent yeast autolysate and a mixture of trace elements the bacillus forms small mucilaginous colonies. After 1–2 days of cultivation, the colonies are whitish, transparent, mucilaginous, convex and with a round base and even margins. After 2–4 days they spread over the whole plate, a whitish, ropey, viscous, slimy mass with an uneven surface. The culture penetrates into the medium.

The cells usually have the dimensions 5·0–10·0 × 0·5–1·2 μm. Young cells are shorter than old cells, which have blunt rounded tips. Cells are mobile, with peritrichous flagella and are, also inclined to form chains. Terminal, Gram-positive, oval spores are formed. There are no capsules. The cells are obligate aerobes, and do not develop in meat-peptone agar or nitrate medium. They utilize sucrose, glucose, maltose, dextrin, citrates and racemates.

Ryys (1963) noted the wide distribution of *Thermobacillus azotofigens* in cultivated soddy-calcareous and humus-calcareous soils of the Estonian SSR, and their absence, or presence, in insignificant amounts, in uncultivated soils.

Bacillus truffauti, which has the trivial name 'Bacille fixateur d'azote' was first described from soil by Truffaut and Bezssonoff (1922). The cells of this species are peritrichous rods usually measuring $1·5$–$3·0 \times 5·0 \mu$m, sometimes reaching 10–15μm long. The spores are eccentric or oval ($0·9$–$1·2 \times 0·5 \mu$m). The colonies are smooth, colourless, brilliant and sometimes granular or undulating. The bacterium grows well on ordinary synthetic media and on protein media. It liquefies gelatine but does not affect milk. It assimilates various sugars and decomposes starch. When it decomposes carbohydrates it produces alcohol and acetic acid. It is a facultative aerobe. For every gram of glucose utilized it fixes 2–7 mg of molecular nitrogen.

In 1963 Kurdina found many cells of a small nitrogen-fixing ($0·4$–$0·5 \times 0·6$–$0·7 \mu$m) aerobic bacillus closely resembling *Bac. truffauti*, around the roots of wheat, cocksfoot and lucerne in irrigated light chestnut-coloured soils of Kazakhstan. The cells were rods, often paired or in short chains, and completely filled by the spores when they formed. The culture withstood a temperature of 80°C for 10 minutes.

On nitrogen-free medium brilliant, round, transparent mucilaginous colonies with even margins are formed. They do not spread over the medium. In media containing mineral nitrogen, diffuse milky white colonies form. When nitrogen is provided in the form of protein, growth is poorer. The productivity of a culture is 3 mg of nitrogen per gram of glucose utilized. Such a culture differs from *Bac. truffauti* in its smaller cells and inability to liquefy gelatine.

Methanobacterium omelianskii can fix molecular nitrogen, according to Pine and Barker (1954). Cells of this bacterium have dimensions of $0·6$–$0·7 \times 3·0$–$6·0 \mu$m, and are sometimes longer. They produce spherical terminal spores which are not very resistant to increased temperature. The cells are weakly mobile but the type of flagellation has not been established. Gram-staining is variable.

As carbon sources *Methanobacterium omelianskii* utilizes ethyl, propyl and other primary alcohols (butyl, amyl and so on) converting them to the corresponding acids. It oxidizes secondary alcohols to the corresponding ketones. Ethanol is one of the best sources of carbon for this bacterium. Glucose, fatty acids and amino acids are not utilized; ammonium salts may serve as a source of nitrogen but nitrates do not.

This micro-organism is an obligate anaerobe which reduces CO_2 to CH_4. In an atmosphere containing carbon dioxide and 9·5 atom per cent nitrogen-15 excess, the cells of *Methanobacterium omelianskii* accumulate 0·74 atom per cent nitrogen-15 after 4 weeks of incubation. If nitrogen is provided in the form of small concentrations of ammonium chloride (5 γ of nitrogen per ml) the incorporation of labelled nitrogen is increased to 2·11 atom per cent, the lag phase is shortened and growth is induced in the culture. Larger amounts of ammonium chloride (25–30 γ of nitrogen per ml) decrease nitrogen fixation to 0·01 atom per cent.

Buchanan and Rabinowitz (1964) detected ferredoxin, rather different from that of clostridia, as a component of the cells of *Methanobacterium omelianskii* cultivated in an ethanol medium. The optimum temperature for the development of this bacterium is between 37° and 40°C and the maximum is about 46°–48°C. Limits of pH for growth are 6·5 to 8·1.

We shall mention some more nitrogen-fixers of the genus *Bacillus*. In 1895 Caron reported the first field experiments with an aerobic nitrogen-fixing bacillus, *Bac. ellenbachensis*. Morphologically it has been said by some to be related to *Bac. mycoides* and *Bac. megaterium* and by others to *Bac. subtilis*. Polar germination of the spore is characteristic, as in *Bac. anthracis*, unlike *Bac. subtilis*, in which it is equatorial. Stocklasa (1895), Jacobitz (1903), Beijerinck (1904), all confirmed that this micro-organism fixed nitrogen, while simultaneously Krüger (1894) and Stutzer (1904) said that it did not.

Bacillus ellenbachensis is interesting not so much for its nitrogen-fixing powers as for the fact that it was the first micro-organism to be used as an active principle in the bacterial soil fertilizer Alinite. This preparation, a powder containing spores, has sometimes been a failure, but nevertheless played a part in the history of bacterial fertilizers by attracting the attention of agriculturists to the products of applied microbiology. The results of experiments with Alinite have been summarized by Heinze (1902).

Konvalevskii (1898) found that some strains of *Bac. subtilis* and *Bac. prodigiosum* could fix nitrogen. According to Löhnis and Pillai (1908) when *Bac. prodigiosum* fixes nitrogen, dextrose is usually the source of carbon.

Nitrogen fixation was established by Löhnis and Pillai also for the sporulating bacterium they called *Bac. malabarensis* isolated in the summer from soil on the Malabar shore of India. In pure culture, cells of this bacterium are rods of varying length with characteristic constructions at the ends, and transverse striation after staining with different compounds. A similar species was isolated by Omelyanskii (1923) from soils of the Irkutsk region, but in none of the media he tested (containing mannitol, dextrose, calcium malate and so on) did the bacterium fix nitrogen.

In 1909 Pringsheim described a long sporulating rod, growing at 61°C on Vinogradskii's nitrogen-free medium with added soil extract, which fixed nitrogen almost as actively as clostridia.

Isachenko (1914) noted nitrogen fixation by the aerobic, sporulating bacillus he isolated from silt in the northern Arctic Ocean. The bacterium could easily be obtained in pure culture after inoculation into nitrogen-free mineral agar medium with mannitol and 3·5 per cent NaCl. After 24 hours the bacterium began to form abundant spores and within only 48 hours almost all the cells had become spores. Spindle-shaped cells and involutive forms resembling the bacteroids of nodule bacteria are characteristic of this bacterium.

Endosporus azotophagus was isolated by Tchan and Pochon (1950) from French soils and studied by them in detail.

Belyaeva (1954) found that the myxotroph *Bacillus hydrogenes* she studied could oxidize hydrogen and bind small amounts of atmospheric nitrogen. The productivity of nitrogen fixation by cultures of *Bac. hydrogenes* is higher in the presence of glucose (0·8–2·1 mg of nitrogen per gram of glucose utilized).

Mishustina (1955) isolated many sporulating, oligonitriphilic bacteria from soils of the Soviet Union. They included various species distinguished morphologically by the type of spore formation and other signs. All the bacteria were tentatively divided into two groups. First, cultures forming convex, drop-shaped colonies of pasty or ropey consistency on Ashby's medium. They grow poorly in protein media, and ferment sugars, and evolve gas. The cells do not form long threads, rarely form spores, and have well-marked capsules. The most typical representative of this group is strain No. 17, isolated from the solodized soil of north-west Kazakhstan. On Ashby's medium it forms large transparent colonies in the form of droplets that can easily be removed from the agar. The cells are single or paired and surrounded by capsules. In young cultures they are mobile and their dimensions are 4·0 × 0·9 μm. The spores are about 0·9 μm across and terminal. The cells do not affect gelatine or milk, but ferment glucose, sucrose, lactose, arabinose, dextrin and mannitol, evolving gas at the same time. They fix nitrogen in liquid nitrogen-free medium containing yeast autolysate. The productivity of nitrogen fixation is 2·9–3·6 mg per gram of sugar consumed. Similar bacteria have been isolated from soils of the Moscow Region, from the black-earth belt and from the Crimea.

The second group of Mishustina's bacteria (1955) form relatively flat, white, beige or brown colonies in a nitrogen-free medium. Sometimes a turbid halo can be seen around the colony. These bacteria grow equally well on media with or without nitrogen. When utilizing sugars they do not form gas. The cells of young cultures are usually grouped in long threads and have coarse-grained protoplasm. The capsules are small and can be seen only when cells are stained with Indian ink. Bacteria of this type are far more numerous than representatives of Mishustina's first group.

From Nigerian soils with a pH of 3·8–6·8, Moore (1963) isolated in a nitrogen-free, alkaline medium, two cultures of the genus *Bacillus* that fix about 1·4 mg of nitrogen per 100 ml of medium. From the same soils Moore and Becking (1963), using a nitrogen-free medium with 2 per cent glucose at 30°C, isolated many facultative anaerobic bacilli nominally assigned to the genus *Aerobacillus*.

On a nitrogen-free medium these bacilli form round, convex viscous colonies with a diameter up to 5·0 mm. Under the microscope, they appear as large mobile rods, variably staining with Gram stain. They form ellipsoid terminal spores. All the strains isolated developed well on a medium containing sucrose, lactose or dextrin. They made little or no use of glycerol and mannitol. This group may have remained undetected for a long time because most investigators used nitrogen-free media with mannitol when isolating nitrogen fixers.

The cultures isolated by Moore and Becking (1963) fixed more nitrogen in the presence of yeast extract (an increase of 0·0025 per cent); PABA and biotin had no such effect. One of the strains isolated contained hydrogenase.

The amount of nitrogen fixed by the various strains varied from 1·3 to 1·5 mg of nitrogen per 100 ml of the medium. In an atmosphere containing labelled nitrogen, enrichment of the culture amounted to 0·10–0·49 atom per cent of nitrogen-15 excess.

Thus, the sporulating and non-sporulating bacteria include many species that fix nitrogen. It seems reasonable to conclude that these groups play a significant role in the nitrogen balance of the soil (for example, Kaufman et al., 1948; Zhukova, 1956; Aristovskaya, 1957; Parinkina, 1960; Bulgadaeva, 1961; Ilyakhina, 1962; Klevenskaya, 1966; Dudareva, 1966; Khak-Mun Ten, 1966; O. and G. Chesnyak, 1966).

Phostosynthesizing Nitrogen-Fixing Bacteria

Several families of photosynthesizing bacteria include species that can assimilate molecular nitrogen.

According to Burris (1961), the many nitrogen fixers among the photosynthesizing bacteria are not chance occurrences because this is one of the oldest groups of nitrogen fixers on the earth.

The nitrogen fixers among this group of bacteria began to be detected rapidly after the publication by Gest and Kamen (1949) of the results of work with *Rhodospirillum rubrum*. They started from the premise that organisms possessing hydrogenase must also be potential nitrogen fixers, which they confirmed experimentally.

The fixation of atmospheric nitrogen by photosynthesizing bacteria allows them to develop in sites that are poor in nitrogen compounds or which do not contain any at all. The bacteria enrich the surrounding medium with bound forms of nitrogen (Van Niel, 1955; Kamen, 1955).

When nitrogen-15 is fixed briefly in the presence of light and in anaerobic conditions the largest amount of labelled nitrogen is found in glutamic acid and in the ammonium fraction (Wall et al., 1952; Wilson, 1952). This shows that the pathway of nitrogen fixation is similar in photosynthesizing bacteria to that in other nitrogen-fixing bacteria, in particular *Azotobacter* (Fry, 1955; Fogg, 1956).

A detailed description of photosynthesizing bacteria has been given in the monograph by Kondrat'eva (1963). We shall confine ourselves to a description of the properties of the species studied as nitrogen fixers. The properties of the principal families (Athiorhodaceae, Thiorhodaceae and Chlorobacteriaceae) of these bacterial species are given in Table 8.4. One photoheterotrophic micro-organism of the family Hyphomicrobiaceae is also described. Its properties are considered at the end of this section.

Photosynthesizing bacteria that fix nitrogen are chiefly encountered in water. There are indications that they play an essential role in the fertility of the soil of rice-growing areas (Materassi and Balloni, 1965).

Table 8.4. Principal Properties of Purple and Green Photosynthesizing Bacteria

Feature of micro-organism	Purple sulphur bacteria (Thiorhodaceae)	Non-sulphur purple bacteria (Athiorhodaceae)	Green sulphur bacteria (Chlorobacteriaceae)
Morphology	Spherical, rod-shaped bacteria and spirillae; motile and non-motile	Rod-shaped and spirillae; motile	Short, rod-shaped bacteria, motile and non-motile
	Do not form spores; Gram-negative		
Pigments	Bacteriochlorophyll and acyclic carotenoids, lycopine derivatives		Bacterioviridin and monocyclic carotenoids of the V-carotene type
Evolution of oxygen during photosynthesis		No	
Hydrogen donors	H_2S and other sulphur compounds H_2 and organic compounds	Organic compounds H_2 (sulphur compounds)	H_2S and other sulphur compounds, H_2, (organic compounds)
Carbon sources	CO_2 and organic compounds	Organic compounds and CO_2	CO_2 and organic compounds
Vitamin requirements	None (B_{12}?)	Individual vitamins of group B	None
Behaviour in response to oxygen	Strict anaerobes	Strict anaerobes and facultative aerobes	Strict anaerobes
Growth in dark	No	Possible in presence of O_2 for facultative aerobic forms	No
Photoformation of H_2	Yes	Yes	Not studied
Fixation of N_2		Yes	

Properties in parentheses are characteristic only of certain representatives of the family.

This table is taken from Kondrat'eva (1963).

Family Thiorhodaceae, the Purple Sulphur Bacteria

The cells of photosynthesizing purple sulphur bacteria contain bacteriochlorophyll and carotenoid pigments. Their nitrogen-fixing capacity was established in *Chromatium* sp. by Lindstrom *et al.* (1949, 1950) and *Chromatium minus* by Petrova (1959). Representatives of this family are characterized as a rule by strict anaerobiosis and by the ability to grow photoautotrophically, oxidizing hydrogen sulphide and certain other inorganic compounds (sulphites, thiosulphates) during photosynthesis. Oxygen is not evolved in this process. The oxidation of hydrogen sulphide is as follows:

$$CO_2 + 2H_2S \xrightarrow{\text{light}} (CH_2O) + H_2O + 2S$$
$$\text{cell matter}$$

When there is a lack of hydrogen sulphide purple bacteria rapidly oxidize the sulphur present to sulphates. Then the reaction proceeds differently:

$$2CO_2 + H_2S + 2H_2O \xrightarrow{\text{light}} 2(CH_2O) + H_2SO_4$$
$$\text{cell matter}$$

Photosynthesis may also be accomplished in the light when the hydrogen donor is not hydrogen sulphide but an organic substance (organic acids, carbohydrates, alcohols and so on). Certain organic compounds can, it seems, also be used by purple bacteria for anaerobic processes.

Purple bacteria do not grow in strict anaerobic conditions in the dark. In aerobic conditions in the dark some of them can absorb oxygen and oxidize certain organic compounds.

Nitrogen fixers of the genus *Chromatium* are widespread in nature. They are encountered in mineral sources, contaminated pools, estuaries and on the bottom of stretches of water. When growth is most intensive (usually in summer) the numbers of cells of *Chromatium* may reach several hundred thousand to 1–3 million per ml (Kuznetsov, 1952). The anaerobic conditions and hydrogen sulphide necessary for these bacteria are often present only at considerable depths (Lyalikova, 1957) but whenever possible they tend to grow closer to the surface of the water—closer to the light.

Cells of *Chromatium* are ovoid, bean-shaped and cylindrical, with rounded ends, or they are in the form of a comma. They move by means of polar flagella. The size of the cells varies; 1–4 × 2–10 μm, but longer cells are also encountered. They contain the bacteriochlorophyll and carotenoid pigments that give the cells different shades of red coloration. Drops of elemental sulphur (the product of incomplete oxidation of hydrogen sulphide) are deposited in the cell. Some representatives of the genus *Chromatium* can develop when the surrounding pH is between 5·0 and 5·9 (Baas-Becking and Wood, 1955).

When *Chromatium* is cultured in the light (at pH 8·5) in nitrogen-free medium with a suitable hydrogen donor, the amount of nitrogen fixed in 11 days may reach 5·9 mg per 100 ml. When incubation is for 2–3 weeks in such conditions, 7·5–10·0 mg of nitrogen accumulates per 100 ml. In an atmosphere containing 1·6 atom per cent of nitrogen-15 excess *Chromatium* accumulates from 0·298 to 0·442 atomic per cent of nitrogen in the cells (Lindstrom *et al.*, 1950). The level of nitrogen fixation of the strain of *Chromatium* reported earlier by Lindstrom *et al.* (1949) was considerably lower, apparently because incubation was for a shorter time.

Arnon *et al.* (1960) established that in the presence of thiosulphate the fixation of nitrogen by *Chromatium* was stimulated by oxaloacetic acid, which may be an acceptor of amino groups. In 1962, Tagawa and Arnon demonstrated the presence of ferredoxin in *Chromatium* and noted its reactivity in photochemical reactions.

Depending on the conditions of culture of *Chromatium*, the electrons released in the photochemical reaction (from the hydrogen donor) may go on to reduce carbon dioxide or some other carbon compounds (Arnon *et al.*, 1961).

Bennet *et al.* (1964) found that pyruvic acid stimulated the fixation of nitrogen by the D strain of *Chromatium*, and assumed that ferrodoxin was an electron carrier in the reaction.

Nicotinamide adenine dinucleotide may be involved in the reduction of molecular nitrogen by photosynthetic bacteria, as it is in the reduction of carbon dioxide. This is suggested by the ability of extracts of cells of *Chromatium* to fix nitrogen-15 in the dark in the presence of reduced nicotinamide adenine dinucleotide (Kondrat'eva, 1963). Extracts of *Chromatium* can also fix nitrogen-15 in the dark in the presence of hydrogen (Arnon *et al.*, 1961).

Athiorhodaceae, the Non-Sulphur Purple Bacteria

This family consists of purple bacteria which are photoheterotrophs, requiring organic compounds in order to develop—hydrogen donors and material for anabolic processes. Even those representatives that can oxidize only inorganic substrates need not be considered autotrophs, for they still require organic growth factors. The Athiorhodaceae are widespread in soil, pools, ditches and larger stretches of water. They are relatively small, unicellular organisms, made mobile by their polar flagella. They have bacteriochlorophyll and one or more of the carotenoids which makes the cells either yellowish-brown to dark brown or various shades of red. Colour is noticeable only when the bacteria are massed together. They are Gram-negative microaerophils. They develop in strictly anaerobic conditions only in the light, when they photosynthesize. Some may grow in the presence of total atmospheric oxygen tension. For photosynthesis to occur there must be exogenous hydrogen donors such as alcohols, and fatty, hydroxy and keto acids. The process is not accompanied by the evolution of oxygen. In some cases organic compounds are utilized only as hydrogen donors and carbon dioxide is reduced, for example in the conversion of isopropyl alcohol to acetone:

$$2CH_3CHOH\ CH_3 + CO_2 \xrightarrow{\text{light}} 2CH_3COCH_3 + \underset{\substack{\text{cellular} \\ \text{material}}}{(CH_2O)} + H_2O$$

In other cases the same substances are used in the synthesis of cellular material:

$$\text{isopropyl alcohol} \xrightarrow{-2H} \text{acetone} \xrightarrow{+CO_2} \text{acetoacetic acid} \xrightarrow{H_2O}$$

$$\text{acetic acid} \longrightarrow \text{cellular material}$$

In the air the Athiorhodaceae can develop in the dark as well as in the light.

The family has two genera, *Rhodopseudomonas* and *Rhodospirillum*, most species of which can fix nitrogen. Cells of *Rhodopseudomonas* are rod-shaped or spherical, measuring 0·5–1·0 × 1·2–4·0 μm. They are widespread in silt, stagnant water and mud.

In 1951, Lindstrom *et al.* made a detailed investigation of the optimum conditions for nitrogen fixation by representatives of all four species of this genus (*Rhodopseudomonas palustris*, *Rhodopseudomonas capsulatus*, *Rhodopseudomonas gelatinosa* and *Rhodopseudomonas spheroides*, which they received from Hutner and Siegel). They also isolated thirty-two strains of *Rhodopseudomonas* from lake silt close to Madison, Wisconsin. Nineteen of them gave positive response to tests for nitrogen-fixing activity.

All experiments with different cultures of *Rhodopseudomonas* were carried out in both anaerobic and aerobic conditions (in the light and in the dark). Abundant growth and nitrogen fixation were observed in anaerobic conditions with illumination. In the dark, however, the anaerobic cultures did not grow, while the aerobic cultures grew very poorly. The experiments lasted 6–9 days and the temperature was 25°C. Lindstrom *et al.* used nitrogen-free medium (pH 7·2) containing lactic, malic and acetic acids or ethanol as the hydrogen donor. According to the requirements of the strain, they added growth substances. For *Rhodopseudomonas palustris* they added 1 mg/litre of n-amino benzoic acid and for *R. spheroides* and *R. gelatinosa* they also added 50 mg/l of Difco yeast. *Rhodopseudomonas capsulatus* was grown in a medium containing no growth substances. Nitrogen assimilation was most vigorous in media containing lactic and malic acids and their salts.

Rhodopseudomonas palustris is a typical representative of the genus. Its cells are short (0·6–0·8 × 1·2–2 μm), slightly bent rods, very mobile when young. In old cultures irregular, bent and curled long rods (up to 10 μm) appear, sometimes with inflations at one or both ends resembling branching. Old cells are immobile and usually form conglomerations similar to those in cultures of *Corynebacterium* and *Mycobacterium*. They are Gram-negative. When developing in liquid media the cells never produce slime. Their colour changes with the composition of the medium, especially in anaerobic conditions and in the light. Cells developing weakly (in media with malic acid, thiosulphate or glycerol) are pink; in media containing fatty acids they are almost red-brown due to the presence of bacteriochlorophyll and carotenoids. Most strains also produce a water-soluble bluish-red non-carotenoid pigment that diffuses into the medium.

This species uses in particular glutaric acid and ethanol as carbon sources. It can assimilate several amino acids, including leucine, and most of the fatty acids, and hydroxyacids also make good material for metabolism. Cells do not, however, grow in media containing about 0·2 per cent carbohydrates (glucose or mannose or the alcohol sorbitol). Optimum development requires *p*-amino-benzoic acid. All cultures can grow in media containing thiosulphate, which is used as the principal oxidizable substrate.

In media containing yeast autolysate growth is possible if the pH is between 6·0 and 8·5. Development is good in a temperature up to 37°C. Cultures of this species have no odour.

According to Lindstrom *et al.* (1951), if *Rhodopseudomonas palustris* is cultivated in anaerobic conditions in the light for 6 days, it accumulates up to 17·2 mg of nitrogen per 100 ml of medium. In aerobic conditions in both light and dark, and in anaerobic conditions in the dark, nitrogen is hardly fixed at all. The local strains isolated by Lindstrom *et al.*, kept in anaerobic conditions in the light, accumulated from 6·0 to 12·3 mg of nitrogen per 100 ml of medium in 6–9 days. In an atmosphere containing 32 atom per cent excess nitrogen-15, *Rhodopseudomonas palustris* accumulated up to 14·2 atom per cent of nitrogen-15.

In young cultures, cells of *Rhodopseudomonas gelatinosa* are short rods measuring $0·5 \times 1·2 \mu m$. In old cultures, cells reach $15 \mu m$ long. Irregular bent rods are also found, and these have inflations and nodulation. At this stage the cells are similar to *Rhodopseudomonas palustris* but do not form conglomerations that resemble *Mycobacterium*. There is abundant formation of mucilage in all media when individual cells merge. Young cells are very mobile, but because of a strong tendency to conglomerate mobility is hard to detect.

Most of the anaerobic cultures are pale pink. In the presence of high concentrations of yeast autolysate the mass of the mucilaginous cells is dirty brown in colour. The pigments that form in *Rhodopseudomonas gelatinosa* are the same as those of *Rhodopseudomonas palustris*.

In media containing yeast autolysate the cells develop when the pH is between 6·0 and 8·5. Good oxidizable substrates are ethanol, glucose, fructose, mannose and the amino acids alanine, asparagine and aspartic and glutamic acids. This species does not grow in media containing 0·2 per cent propionic acid, glycerol or tartaric acid, nor does it oxidize thiosulphate. It requires biotin and thiamine for development, and liquefies gelatine. It is the most microaerophilic of the species of *Rhodopseudomonas*. Most of the strains do not develop in aerobic conditions, but are strict anaerobes and need illumination so that they can photosynthesize. They have a characteristic odour.

When yeast extract is added to the medium in anaerobic conditions in the light, cells of *Rhodopseudomonas gelatinosa* fix up to 18 mg of nitrogen per 100 ml. In atmospheres containing 32 and 16·8 atom per cent nitrogen-15 excess, they fix 0·810 and 0·292 atom per cent nitrogen-15 respectively.

The cells of *Rhodopseudomonas capsulatus* are spherical or rod-shaped, depending on the pH of the medium. The most common size (in nutrient medium) is from 2·0 to 2·5 μm long. The cells are often immobile. If the pH is above 8·0 the cells grow in the form of threads. A distinguishing feature of the species is the zig-zag arrangement of chains of the Gram-negative cells.

In media with a pH greater than 8·0 the cultures produce mucilage. Anaerobic cultures are brown, with shades ranging from light yellow-brown to dark red-brown. In the presence of oxygen, cultures are dark red. The colour is due to the presence of the bacteriochlorophyll and carotenoid pigments, and rate of change in colour is determined by the intensity of the light.

Rapid and abundant growth is observed in the presence of 0·2 per cent propionic acid and glucose, fructose, alanine and glutamic acid. There is no development on leucine, ethanol, glycerol, mannitol or sorbitol. Thiamine, biotin and nicotinic acid stimulate growth.

Cultures of *Rhodopseudomonas capsulatus* do not liquefy gelatine. Most of them give off a pleasant aroma of peaches. Growth is possible when the surrounding pH is between 6·0 and 8·5. The optimum temperature for growth is 25°C.

In the light, cultures develop well in anaerobic conditions because they can photosynthesize. The strain isolated by Hutner accumulated 7·2–10 mg of nitrogen per 100 ml of medium, while the culture isolated by Lindstrom *et al.* (1951) accumulated 5·1–9·7 mg per 100 ml of medium. In an atmosphere containing 16·8 atom per cent nitrogen-15 excess, the cells accumulated 11·0 atom per cent of the isotope.

From rice soils Okuda *et al.* (1960, 1961) isolated cultures of *Rhodopseudomonas capsulatus* that actively fixed nitrogen in anaerobic conditions in the light. In the dark, nitrogen fixation was weak in aerobic or anaerobic conditions. In natural conditions, when *Rhodopseudomonas capsulatus* was in association with various aerobic micro-organisms, they found that it could fix nitrogen in aerobic conditions in the light. When organic acids were added to the soil the bacteria fixed nitrogen with vigour, increasing in proportion to the amount added.

Using mixed cultures of *Rhodopseudomonas capsulatus* with *Bacillus subtilis* and *Bacillus megaterium* isolated from the same soils, Kobayashi *et al.* (1965) achieved nitrogen fixation in media containing glycerol and starch when pure cultures did not fix nitrogen. In pure cultures the bacilli formed large amounts of pyruvic acid and α-ketoglutaric acids. In mixed culture less acid was produced and nitrogen was fixed; as soon as the acids began to decrease, nitrogen fixation began. For this reason great importance is attached to these acids with regard to the stimulation of nitrogen fixation. Okuda *et al.* (1960) found that nitrogen is fixed more vigorously by *Rhodopseudomonas capsulatus* in a mixed culture with *Proprionibacterium* than when it is in pure culture.

Nitrogen fixation in a mixed culture of *Rhodopseudomonas capsulatus* and *Azotobacter* is about three times as brisk as in pure cultures of either microorganism. In such a mixed culture the growth of *Rhodopseudomonas capsulatus* is yet more vigorous in aerobic conditions in the light. The explanation that the Japanese investigators gave for the increase in nitrogen fixation in mixed cultures of photosynthesizing and heterotrophic bacteria rests primarily on the exchange of metabolic products. For example, a culture of *Azotobacter* accumulates in the medium pyruvic acid, which can be utilized by *Rhodopseudomonas capsulatus* as a hydrogen donor (Katayama *et al.*, 1965; Kobayashi *et al.*, 1965). It is also possible that other microbes that reduce the redox potential of the medium protect the oxygen-sensitive hydrogenase of *Rhodopseudomonas* from inactivation.

Cells of *Rhodopseudomonas spheroides* are usually single and almost spherical. Their diameter (disregarding the mucilaginous capsule) varies between 0·7 and 4 μm. When they are young they are mobile. In old cultures, as a result of intense alkalinization of the medium, or in alkaline media, there are deformed rods which apparently contain spores. Such a situation results from the formation of fatty inclusions. In media containing sugars the cells are ovoid with dimensions of 2·0–2·5 × 2·5–3·5 μm. They are Gram-negative.

Anaerobic cultures are brownish and aerobic cultures red. This microorganism forms bacteriochlorophyll, carotenoid pigments and a water-soluble bluish-red pigment of a non-carotenoid nature which diffuses into the medium. The bacterium does not liquefy gelatine, and all cultures give off an unpleasant odour of decay. Development can be seen between pH 6·0 and 8·5, and thiamine is required. Cultures of *Rhodopseudomonas spheroides* develop in the presence of 0·2 per cent tartaric or gluconic acid, ethanol, glycerol, mannitol, sorbitol, glucose, fructose or mannose. In a medium containing 0·2 per cent propionic acid there is no development. Thiosulphate is not oxidized; the optimum temperature for growth is below 30°C, and oxygen does not prevent growth.

In anaerobic conditions in the light, cultures fix 5·0–11·4 mg of nitrogen per 100 ml of medium. In an atmosphere containing 16·8 or 32·0 atom per cent nitrogen-15 excess, cultures accumulate 2·12 or 5·82 atom per cent of the isotope respectively.

Bacteria of the genus *Rhodospirillum* have Gram-negative cells that have a spiral appearance and move around by means of polar flagella. They contain bacteriochlorophyll and photosynthesize in the presence of exogenous oxidizable substances. They do not produce molecular oxygen, and do not grow in strictly mineral media even when they can utilize hydrogen as oxidizable substrate; they require organic substances. They produce other pigments as well as those involved in photosynthesis.

Nitrogen fixation has been established only in *Rhodospirillum rubrum*, a typical representative of the genus, which includes three other species. The mobile cells of *Rhodospirillum rubrum* have a characteristic spiral form but their size varies, according to the medium in which they are grown, from 0·5 × 2·0 to

$1.5 \times 5.0 \mu$m. In a medium containing alanine, most cells have the form of a ring or half ring. In old cultures involutive, straightened and inflated forms appear, sometimes branching and containing fatty inclusions.

These cells are Gram-negative, and do not liquefy gelatine. The dark red colour is due to the presence of bacteriochlorophyll and carotenoid pigments, with spirillo-xanthine (absorption band 550 nm) predominant. The cells form no water-soluble pigment, and develop when the pH is between 6·0 and 8·5. The cultures give off the odour of slightly decayed yeast. A useful diagnostic sign is good growth in media containing 0·2 per cent ethanol, alanine, asparagine and, aspartic and glutamic acids. The cells of this bacterium cannot utilize thiosulphate and carbohydrates. They are microaerophilic. Immediately after they have been isolated, many strains are unable to develop at atmospheric oxygen tension, but later they may adapt to such conditions. They can develop in strictly anaerobic conditions when illumination makes photosynthesis possible.

Rhodospirillum rubrum was the first of the photosynthetic bacteria in which nitrogen fixation was established. Gest and Kamen (1949) recorded the formation of molecular hydrogen as the principal product of photosynthesis by the strain *Rhodospirillum rubrum* SF, which they had cultured in anaerobic conditions (without nitrogen) in the light in a mineral medium with added malic, fumaric or succinic acids and small amounts of biotin. No hydrogen was evolved when nitrogen was present in the gaseous phase. The idea that in an atmosphere containing nitrogen there is an admixture of oxygen which might suppress photosynthesis was discarded after further experiments.

Hydrogen was not found when the medium contained ammonium ions, and 0·5 g per ml of ammonium chloride completely suppressed the uptake of atmospheric nitrogen. This suggested that the photosynthesizing, non-sulphur purple bacteria contain a nitrogenase system which determines the fixation of molecular nitrogen—a process which is competitive with respect to the evolution of molecular hydrogen. This idea was confirmed by subsequent experiments with isotopes (Kamen and Gest, 1949). A culture kept for 5 days in an atmosphere of one part nitrogen to nine parts helium, and containing 30 per cent nitrogen-15, was enriched with 3·14 atom per cent of nitrogen-15. In the dark enrichment was only 0·189 atom per cent (controls had 0·008 atom per cent enrichment).

Lindstrom *et al.* (1949) confirmed the link between anaerobic nitrogen fixation, photoreduction and the metabolism of hydrogen in *Rhodospirillum rubrum*. They found that this species accumulated 7·7–10·1 mg of nitrogen per 100 ml of medium during 4–6 days of incubation in the light in anaerobic conditions. They obtained this value using a mineral medium with added malic acid, Difco yeast extract, biotin and molybdenum (pH 7·0). In anaerobic conditions in the dark, and in aerobic conditions in the light and the dark, the results were within the limits of experimental error (1·7–2·5 mg of nitrogen per 100 ml and for non-inoculated controls 2·0–2·1 mg of nitrogen per 100 ml).

Ten years later Pratt and Frenkel made a detailed study of nitrogen fixation by a culture of *Rhodospirillum rubrum* and also noted its dependence on light and anaerobiosis. Fixation of nitrogen in the dark was not observed even directly after a period of illumination. The degree of nitrogen fixation varied almost linearly with the intensity of illumination, decreasing sharply in a low luminosity.

Relatively low partial pressures of molecular oxygen completely suppressed nitrogen fixation by *Rhodospirillum rubrum*. Thus, when nitrogen-fixing cultures grown in an atmosphere containing traces of oxygen were transferred to a gaseous mixture containing 4·1 per cent oxygen, nitrogen fixation stopped. When the concentration of oxygen was 1·8 per cent there was partial suppression (by 45 per cent) of nitrogen fixation. The process started up again when the culture was returned to anaerobic conditions.

Kamen *et al.* (1950), and also Lindstrom *et al.* (1949) found an insignificant increase in nitrogen in a culture of *Rhodospirillum* in aerobic conditions in the dark. Pratt and Frenkel thought that this was because the suspensions of *Rhodospirillum rubrum* were incubated for a long time (5–6 days) in closed vessels containing nitrogen enriched with nitrogen-15. In these conditions the pressure of oxygen undoubtedly decreased to an insignificant level, and so the nitrogen fixation which Kamen, Lindstrom *et al.* observed occurred at a partial pressure sufficient for oxidative metabolism but insufficient completely to suppress nitrogen fixation. When they added ammonium chloride to a culture of *Rhodospirillum*, nitrogen fixation was suppressed. Pratt and Frenkel, however, demonstrated a lag period of from 6 to 20 minutes preceding suppression. They found that nitrogen fixation was suppressed totally in the presence of 89 per cent hydrogen, and partially suppressed (by 60 per cent) in the presence of 45 per cent hydrogen. The nature of the suppression by hydrogen (competitive or non-competitive) was not established. Schneider *et al.* (1960) demonstrated considerable fixation of nitrogen (1·0–2·5 atom per cent of nitrogen-15) in experiments with cell-free preparations of *Rhodospirillum rubrum*.

The fixation of molecular nitrogen by this organism requires more nitrogen in the atmosphere than is needed by aerobic nitrogen-fixers. At a partial pressure of nitrogen of about 0·2 atm the rate of nitrogen fixation reaches only half maximum, whereas for *Azotobacter* the same rate of fixation has been observed at a pressure of 0·02 atm (Burris, 1961). Therefore, experiments with *Rhodospirillum* are usually carried out in an atmosphere containing 95 per cent nitrogen and 5 per cent carbon dioxide (Pratt and Frenkel, 1959).

Nitrogen fixation by *Rhodospirillum rubrum* is stimulated by sodium sulphate in media containing nitrogen compounds (Gest *et al.*, 1950).

In experiments carried out by Lindstrom *et al.* (1951), using an unidentified strain of *Rhodospirillum* sp., nitrogen fixation was demonstrated only in anaerobic conditions in the light. During 6–10 days of incubation the culture fixed 6·4–10·0 mg of nitrogen per 100 ml of medium.

In an atmosphere containing 32 atom per cent nitrogen-15 excess, cells of

Rhodospirillum sp. were enriched with up to 12·7 atom per cent of the isotope.

Family Chlorobacteriaceae, Green Sulphur Bacteria

This family includes short, rod-shaped bacteria, both mobile and immobile, which differ from the purple sulphur bacteria in their pigments, which include bacterioviridin and carotenoids of the γ-carotene type. Like purple sulphur bacteria they are strict anaerobes and are photo autotrophs, being able to oxidize during photosynthesis, hydrogen sulphide and other sulphur compounds. They are encountered in the form of both single cells and conglomerations. They develop in media containing high concentrations of hydrogen sulphide when illuminated. Elemental sulphur is deposited in the cells. During photosynthesis they do not evolve oxygen.

At one time, green sulphur bacteria were considered to be unable to utilize organic compounds, and so were designated strict autotrophs. Larsen (1951, 1953) showed this to be wrong; small amounts (about 0·005 mole) of glucose or pyruvic, lactic or acetic acid in the medium cause the growth of a culture of *Chlorobium* and the formation of bacterioviridin to increase somewhat. *Chlorobium* can evidently utilize certain organic compounds (primarily acetic acid) as sources of carbon. But there are some indications that, unlike purple bacteria, *Chlorobium* cannot oxidize acetic acid and use it as a hydrogen donor or a source of carbon dioxide (Sadler and Stanier, 1960).

Schaposchnikov *et al.* (1960) isolated cultures of the mobile green sulphur bacterium *Chloropseudomonas ethylicum* which proved capable of growing in the light in a mineral medium with $NaHCO_3$ and Na_2S, and in media containing different organic compounds when 0·01–0·02 per cent $Na_2S.9H_2O$ was added to ensure anaerobic conditions. This micro-organism oxidizes ethanol and can utilize for its metabolism part of the acetic acid that is formed, but it continues to assimilate carbon dioxide. Thus, organic compounds are not inert substances for green sulphur bacteria but these organisms are mostly autotrophs.

Green sulphur bacteria are widespread in nature, usually encountered in large numbers where purple bacteria develop. The spectra of light absorption differ for green and purple bacteria, and this is important when they develop together. The greens are more sensitive to oxygen than the purples and therefore develop only where oxygen is virtually absent. Mass development can be seen in the summer and autumn.

Cells of *Chlorobium limicola* differ in size and shape, depending on external conditions. In young cultures small cells predominate, with shapes varying from ovoid to short rods, measuring $0·7 \times 0·9–1·5$ μm. Chains often form, making the culture resemble streptococci. Involutive, elongated, irregularly bent and curved rods are also encountered, and these too may be joined in chains.

Club-shaped and spiral forms with sharp twists in them have also been described. But the conditions necessary for the production of such forms have not been clearly established.

In deficient media conglomerates of varying sizes and shapes appear. The cultures are characterized by the presence of immobile mucilaginous strands. They are bright green, except in media which lack iron, when they are yellowish-green.

These are strictly anaerobic photosynthetic bacteria, which utilize hydrogen sulphide, and also elemental sulphur and molecular hydrogen as oxidizable substrates. They form sulphur from the sulphides but do not deposit it within the cells. Sulphur may be the end product of the oxidation of sulphide, but in optimal conditions it is further oxidized to sulphate. The bacteria cannot utilize thiosulphate and tetrathionate as oxidizable substrates. Growth is not observed in organic media that lack sulphide. Mass development is seen when there is a relatively high concentration of sulphide and low pH at sites accessible to the light.

In 1949, Lindstrom et al. found a culture of *Chlorobacterium* that fixed small amounts of molecular nitrogen. Later experiments revealed larger increments of nitrogen, which were determined by the semi-micro method of Kjedahl and the isotope technique (Lindstrom et al., 1950).

Irrespective of the incubation time (3–9 days) and initial content of nitrogen in the medium, accumulation of nitrogen in the bacterial culture reached 3·5–4·0 mg per 100 ml. Development and nitrogen fixation were optimal in the light in a nitrogen atmosphere (0·8 atm). In a hydrogen atmosphere the amount of nitrogen fixed in the same conditions decreased to 1·2–2·2 mg per 100 ml. In an atmosphere containing 1·65 atom per cent of nitrogen-15 excess, the cells of the bacterium studied accumulated 0·203–0·318 atom per cent of the isotope.

Chlorobium thiosulfatophilum has small ovoid or rod-shaped cells measuring 0·7 × 0·9–1·5 μm, often joined in chains. In old cultures club-shaped and spiral forms appear, which have been described as involutive, also usually in chains. The cells always produce mucilage and are bright green, although when iron is scarce they are yellowish-green.

Chlorobium thiosulfatophilum is a strictly anaerobic, photosynthesizing bacterium utilizing sulphites, sulphur, thiosulphate, tetrathionate and molecular hydrogen as oxidizable substrates. It produces sulphates from inorganic sulphur compounds, and does not develop on organic media which lack oxidizable inorganic sulphur compounds. Large numbers of this species are present in the sea and in fresh water, silt and stagnant water.

Pine and Barker (1954) used the isotope method to determine the intensity of nitrogen fixation by *Chlorobium thiosulfatophilum* compared with a strain of *Methanobacterium omelianskii*. In an atmosphere containing 8·1–9·5 atom per cent nitrogen-15 excess, the cell mass of a culture of *Chlorobium thiosulfatophilum* incubated in a mineral medium with thioglycolate for 2 weeks, was enriched with 1·46–2·18 atom per cent of nitrogen-15.

Family Hyphomicrobiaceae

Rhodomicrobium vannielii is a photoheterotrophic bacterium for which Duchow and Douglas (1949) created a new genus. It is closely related to the representatives of the family Athiorhodaceae, but differs morphologically and in the mechanism of its cell division. It was first isolated from silt.

The cells of *Rhodomicrobium vannielii* are oval or round (diameter 1·2–2·8 μm) moving by means of a thin (diameter 0·3 μm) branched flagellum. The cells contain bacteriochlorophyll and carotenoids which give the colonies a colour varying from delicate pink to orange red. Colonies are of irregular shape with uneven surfaces. These Gram-negative bacteria are anaerobic. Hydrogen donors for them may be ethyl, propyl or butyl alcohols and acetic or butyric acids and other compounds, except sugars. They do not utilize sulphides or thiosulphates, nor do they liquefy gelatine. There is catalase in the cells.

Lindstrom *et al.* (1951), working with a culture of *Rhodomicrobium vannielii* given to them by Douglas, found that if kept for 7–8 days on lactic acid in an atmosphere of nitrogen, it would fix 5·6–11·6 mg per 100 ml, and in an atmosphere of hydrogen, it would fix 1·5–4·2 mg per 100 ml. In an atmosphere containing 1·65 atom per cent nitrogen-15 excess, *Rhodomicrobium vannielii* accumulated 0·681–0·921 atom per cent of the isotope, whereas the controls accumulated 0·009–0·014 atom per cent.

Nitrogen-fixing Mycobacteria

Mishustina (1955) found that many soil mycobacteria can assimilate small quantities of molecular nitrogen. On Ashby's nitrogen-free medium, the mycobacteria she studied developed in the form of convex, brilliant, dense, white, pink or yellow colonies with a mucilaginous, ropey or dough-like consistency. The colonies sometimes had turbid haloes. Individual cultures grew well on proteinaceous media. The cells in young culture were characteristically branched, but on ageing they became coccoid.

When cultured in an aerated liquid medium for 6–10 days, a strain of *Mycobacterium* sp. fixed 1·1–1·2 mg of nitrogen per 100 ml of medium, which corresponded to a productivity of 1·3–1·9 mg of nitrogen per gram of sugar consumed.

Fedorov and Kalinskaya (1959, 1960) isolated from soddy podzolic soil several mixed cultures of oligonitrophilic micro-organisms which developed on nitrogen-free media and assimilated considerable quantities of atmospheric nitrogen. At first only those mixtures including a culture assigned to *Mycobacterium flavum* were found to accumulate nitrogen.

Mycobacterium flavum was isolated from the rhizosphere of oats and from the soil of the experimental station of the Timiryazev Moscow Agricultural

Academy. The cells were Gram-negative, non-acid resistant (staining by the Ziel-Neelsen method) straight or slightly bent, immobile rods, which were solitary or joined in pairs, often at an angle to each other and with granules of metachromatin at their ends. The size of the cells was 0.4–0.5×1.5–3.0 μm. When growing in media containing ethanol, the cells were shortened to 0.5–0.7×1.4–1.7 μm or took on a coccoid form with a diameter of about 0.5 μm.

On meat-peptone agar *Mycobacterium flavum* develops slowly, forming semi-mucilaginous, round, smooth, shiny yellow or yellowish-brown colonies. When grown in meat-peptone broth it forms a yellow sediment. It does not liquefy gelatine, but reduces nitrates to nitrites. It forms an orange-yellow coating on slices of potato. The productivity of nitrogen fixation is 14–18 mg per gram of carbon source. This bacterium resembles *Corynebacterium equi* in several features.

When cells of *Mycobacterium flavum* are grown on nitrogen-free media they contain greater dehydrogenase activity than when cultured on ammonium chloride. In the presence of molybdenum cultures pass more rapidly through the development cycle (Il'ina, 1966).

Kalininskaya (1963) isolated more species of *Mycobacterium* that fix nitrogen. She called them *Mycobacterium roseo-album* and *Mycobacterium invisible*.

Mycobacterium roseo-album belongs to the so-called fast-growing myco-bacteria. The cells are immobile, Gram-negative and not resistant to acid. They are 0.5–0.6×1.5–3.0 μm in size. They are colourless, but acquire a pink or red colour together with a characteristic folded structure with age. Meat-peptone is first made cloudy and then a sediment forms. This species does not liquefy gelatine, but it reduces nitrates to nitrites. On potato it forms a brown coating. It fixes as much as 10–12 mg of nitrogen per gram of carbon source used.

Mycobacterium invisible is widespread in soil. The cells are immobile, Gram-negative and not resistant to acid. They measure 0.3–0.4×1.5–3.0 μm. The species does not develop on meat-peptone agar or in broth, on gelatine or potato. The various strains are usually cultured in synthetic medium of the following composition: 0.5 g of K_2HPO_4, 0.3 g of $MgSO_4.7H_2O$, 0.1 g of $CaCl_2.6H_2O$, 0.5 g of NaCl, 0.1 g of $FeCl_3.6H_2O$, 1 g of $CaCO_3$, 1 g of glucose, 1 g of sodium acetate, 1 g of sodium lactate, 1 g of sodium malate, 1 g of peptone and yeast autolysate in a concentration of 100 mg of nitrogen per litre all in 1 litre of water. In this medium *Mycobacterium invisible* develops slowly in the form of small, punctate semi-transparent colonies of a solid consistency.

In pure culture mycobacteria do not fix nitrogen on a glucose medium, and only do so insignificantly on media containing organic acids and ethanol. In the presence of *Pseudomonas radiobacter*, *Pseudomonas fluorescens*, *Bacillus subtilis*, *Flavobacterium fulvum* and so on, the nitrogen-fixing activity of myco-bacteria sharply increases. Kalininskaya (1966a and b; 1967) thought that the stimulating effect of the accompanying microflora on nitrogen-fixing micro-organisms may be due to three properties: first, their ability to decompose compounds inaccessible to nitrogen-fixing micro-organisms, with the formation of products utilized by nitrogen fixers; second, their ability to synthesize biologically active substances such as vitamin B, necessary for the nitrogen fixers, and third, their ability to produce favourable redox conditions in the medium, thus protecting the nitrogen fixers from the inhibitory influence of oxygen.

In all cases investigated the inhibition of nitrogen fixation by oxygen was such that when mycobacteria were cultured on sources of bound nitrogen increased aeration stimulated development. All these cultures possessed hydrogenase. In the presence of molecular hydrogen the amount of nitrogen fixed was halved.

It must be emphasized that the symbiotic relationships which Kalininskaya studied (1967) were not very specific, for various soil micro-organisms may actually be associated in this way with mycobacteria. Because favourable culture conditions, particularly the provision of a mixture of amino acids or a complex of nitrogen-containing and growth substances like yeast autolysate, can ensure vigorous nitrogen fixation by a pure culture of *Mycobacterium* (Kalininskaya and Il'ina, 1963), the symbioses discussed are regarded as being facultative. They are widespread in different types of soil.

In 1963 L'vov isolated from soddy-podzols a new nitrogen-fixing species of *Mycobacterium* which was given the name *M. azot-absorptum* (Plate 13). This is immobile, Gram-negative and not resistant to acid when stained by the Ziehl-Neelsen method. In a 24-hour old culture the cells are thick, sometimes club-shaped, inflated or slightly bent rods, 1·5–3·5 μm long and 0·5–0·7 μm thick. The presence of metachromatin granules is characteristic of the species.

With age (40–65 hours) the cells elongate into thin, unevenly stained rods of different lengths (3–7 μm) and still later (5 days) into long (10–15 μm) and thin bent threads. In still older cultures (10 days) the number of threads increases, some reaching 25 μm long. L'vov considered these long cells to be involutive. The developmental cycle ends with the breakdown of the threads into short rods which grow into thick, club-shaped cells.

On meat-peptone agar *Mycobacterium azot-absorptum* grows rapidly, forming smooth, shiny, round and slightly convex colonies with a rather mucilaginous consistency. With age the colonies turn brown, which distinguishes them from the very similar colonies of *Radiobacter*. On agar slopes the bacterium forms a thick coating with a slightly wavy margin and a moist, rather lumpy surface. This coating is whitish, and in transmitted light it has a pearly appearance, the central part of the surface being more dense than the margin.

This bacterium slowly and weakly liquefies gelatine, and has no effect on milk. It reduces nitrates to nitrites, possesses hydrogenase and active catalase and is sensitive to mycetin.

When incubated for 10 days in a medium containing fumaric acid, *M. azot-absorptum* fixes 3–5 mg of nitrogen per 100 ml. The productivity of nitrogen fixation is 9–11 mg of nitrogen per gram of fumaric acid. The nitrogen-fixing enzyme system in this species of *Mycobacterium* is induced (Lyubimov *et al.*,

1965). In its nitrogen-fixing activity, *M. azot-absorptum* comes halfway between *Azotobacter* and *Clostridium*, fixing nitrogen more slowly than the former but more rapidly than the latter, with a productivity similarly intermediate.

For nitrogen fixation in a pure culture *Mycobacterium azot-absorptum*, which is closely related to *Mycobacterium roseo-album*, requires an accessible source of carbon (certain organic acids, alcohols and carbohydrates) and a small 'starter' dose of bound nitrogen. Maximum nitrogen-fixing activity is found in joint culture with other micro-organisms, of which the strongest stimulator of nitrogen fixation is *Pseudomonas radiobacter*. In mixed culture, *Mycobacterium azot-absorptum* grows well and can tolerate an acid medium. For good growth, aeration of this bacterium should be only moderate, more aeration inhibits growth and nitrogen fixation.

According to Kalininskaya (1966) *Mycobacterium azot-absorptum* develops well on a culture fluid of microbial stimulators, fixing about 4 mg of nitrogen per 100 ml of medium. Sushkina (1965) found nitrogen-fixing activity in most of the mycobacteria in primitive soils of Taimyr and Pamir. Nearly all the pure cultures she isolated fixed atmospheric nitrogen, accumulating in optimum conditions on average up to 5 mg per gram of carbon source. The strains isolated could also destroy primary and secondary minerals, forming chelate-like compounds.

According to Sushkina, mycobacteria, other oligonitrophils, blue-green algae, lichens and mosses enrich the primitive soil with sufficient nitrogen for it to support higher plants. So many mycobacteria bind molecular nitrogen that they probably play a major role in soil fertility.

Proactinomycetes and Actinomycetes which bind molecular nitrogen

In 1957 Metcalfe and Brown isolated *Nocardia cellulans* and *Nocardia calcarea*, both fixing nitrogen. *Nocardia calcarea* is a Gram-positive organism with a characteristic developmental cycle on meat-peptone agar. After 1–2 days at 25°C and pH 7·2 short rods (1·5–2·0 × 1·0 μm in size) and aseptate, unbranched, filamentous cells 10 μm long can be seen, with the branched threads less common. Some threads have lemon-shaped bulges. After four days, short rods appear, cells similar to corynebacteria; unbranched threads (5·0–8·0 × 1·0 μm) are abundant, together with slightly branched threads and cocci (1·0 μm). After seven days, rods and cocci predominate but threads are encountered too. At this stage many rods contain a spore like structure. In the threads these structures, which slightly refract light in phase contrast, look like chains. After 14 days, the colonies consist essentially of short rods, cocci and rods with inflations. Colonies are pink, round, convex and soft without an aerial mycelium.

On agar with glucose or mannitol, the cycle of development is similar to that on meat-peptone agar, but after two days of incubation many long branched threads usually form (10–25 μm). The culture does not liquefy gelatine nor does it hydrolyse starch, but it reduces nitrates to nitrites. When carbohydrates are utilized acids are produced. It develops well on maltose and poorly on lactose.

The culture is similar to *Nostoc salmonicolor*. The productivity of nitrogen fixation varies from 2·0 to 4·5 mg per gram of mannitol, glucose or sucrose. The nitrogen-fixing activity is enhanced by aeration, but does not change when yeast extract is added to the medium.

Nocardia cellulans can fix nitrogen on glucose, mannitol and sucrose as well as on cellulose. This is a Gram-positive bacterium with a characteristic develop-mental cycle. On meat-peptone agar at 25°C and pH 7·0–7·2 after two days, aseptate threads 30–40 μm long appear, often with inflations. During this time short cells are rare. On the fourth day, the threads fragment and short rods appear (1·5–2·0 × 1·0 μm). Branched threads (up to 10 μm long) are en-countered in the medium until the tenth day. After seven days spore-like struc-tures develop in the form of terminal inflations. After 28 days, the colonies consist of very short rods, cocci and spore-like cells. On agar containing glucose the threads fragment less rapidly and usually after 28 days numerous Y-shaped forms are found.

Many branched and unbranched threads (20–30 μm long) appear after 6 days on agar containing cellulose. Many of them have terminal inflations. Fragmenta-tion occurs very rapidly and when cellulose is being actively decomposed short rods and cocci predominate in the culture. In old cultures cocci only are found.

The colonies are convex, soft, and cream in colour. There is no aerial mycelium. The culture does not liquefy gelatine nor hydrolyse starch. It forms nitrites from nitrates, and utilizes paraffin when there is yeast extract in the medium. The culture is similar to those of *Nocardia polychronogenes*, *Nocardia opaca* and *Nostoc erythropolis*.

When a culture of *Nocardia cellulans* in a medium containing mannitol is aerated it fixes between 3·0 and 4·0 mg of nitrogen per gram of carbon source; when cellulose is present in the medium, aeration results in the fixation of from 5·0 to 9·0 mg of nitrogen per gram of carbon source. The culture reacts very strongly to the addition of yeast extract, increasing the productivity of nitrogen fixation to 12·0 mg per gram of cellulose. The component of the yeast extract that is responsible for stimulation has not been identified. The place of yeast extract in cultures of *Nocardia cellulans* cannot be taken by thiamine, biotin or any amino acids. Prolonged subculture leads to loss of nitrogen-fixing activity, but with passage through sterile soil the activity is restored.

Karyagina (1955) has demonstrated some nitrogen-fixing activity in strains of Proactinomycetes. Nitrogen fixation by Actinomycetes was first demonstra-ted by Heinze in 1904. In 1910, Von Peklo, investigating the nodules of *Alnus glutinacea* and *Myrica gale*, concluded that the Actinomycetes in the nodular tissue vigorously fixed nitrogen on nitrogen-free nutrient media. In 1914 Nikolaeva investigated the nitrogen-fixing activity of *Actinomyces elephantis primigenii, Act. denitrificans, Act. griseo-viridis* and *Act. putrificus*, and obtained only negative results.

Actinomyces azotophilus was described (Panosyan, 1945) as a new nitrogen-fixing species of Actinomycetes. It was isolated from alkaline soils of Armenia. It is similar to *Act. albus*, but differs in several properties which served as a basis for its classification as a new species. On agar containing an extract of alkaline soil, Actinomycetes develop in the form of faint, velvety colonies 5–7 mm in diameter with a solid yellow margin 1 mm wide. The centre of the colonies is brown. Branched cells have no septa. In the central portion of the branchings there are dark bodies that take up nuclear stains. Spores form where the cells branch. The organism is an aerobe. When incubated on Ashby's medium it fixes 0·7 mg of nitrogen per 100 ml of medium, and on soil extract it fixes 0·7 mg of nitrogen per 100 ml of medium with a productivity of 8–12 mg of nitrogen per gram of carbon source. As well as molecular nitrogen it utilizes the nitrogen of organic and inorganic compounds, but it does not reduce nitrates.

Novák and Dvořákova (1955) found nitrogen-fixing activity in various species of actinomycetes, including *Actinomyces cylindrosporus* and *Actinomyces griseus*. The productivity of nitrogen fixation on Ashby's medium (with mannitol or glucose) reached 5·8–14·6 mg of nitrogen per gram of carbon source.

Fiuczek (1959), investigating an Actinomycete isolated from alder nodules and assigned to *Streptomyces alni*, established nitrogen-fixing activity. When grown with starch as the carbon source, the culture fixed 2·4 mg of nitrogen per 100 ml of medium. With other sources of carbon no nitrogen was fixed. Plotho (quoted by Fiuczek, 1959) had already established that this strain fixed 5 mg of nitrogen per 100 ml.

Fedorov and Kudryashova (1956–1960) investigated 200 strains of actinomycetes freshly isolated from soddy podzolic soil, and more than fifty collections of strains, and established weak fixation of molecular nitrogen in only 10 per cent of cultures (0·4–5·76 mg of nitrogen per 100 ml of medium). The productivity of fixation was very low—not more than 3–4 mg of nitrogen per gram of glucose. The addition of yeast autolysate to nitrogen-free media increased the growth of Actinomycetes somewhat, and in individual cases increased the productivity of nitrogen fixation. Humic acids did not have a stimulating effect on nitrogen fixation. On the basis of this, Fedorov and Kudryashova (Il'ina) considered that nitrogen-fixing activity depended in the main on the features of the species and the strain of Actinomycete concerned, and not on the presence of activating substances in the media. The most favourable carbon sources for the nitrogen-fixing strains were sucrose, mannitol, succinic acid and starch.

When cultures of Actinomycetes change over to molecular nitrogen oxidative processes always decline and the productivity of synthetic processes increases.

Klevenskaya (1962), investigating the distribution of oligonitrophilic strains of *Actinomyces* in the soils of the Altai mountains in Outer Mongolia, detected nitrogen-fixing activity in *Actinomyces aureofaciens, Act. ruber, Act. albidoflavus, Act. fumosus, Act. grisaus, Act. candidus, Act. flaveolus* and *Act. globisporus*.

Between 0·94 and 1·59 mg of nitrogen accumulated per 100 ml of medium, which corresponded to a productivity of 0·7–2·8 mg of nitrogen fixed per gram of sugar consumed. Nitrogen-fixing Actinomycete cultures were found in various vertical soil zones in the Altai mountains. Their numbers increased with the transition from high to low mountain soils.

In the mountainous and sub-mountainous soils of the Zailiiskii Alatau, Teplyakova (1966) discovered that *Actinomyces globisporus* fixed 2·2–2·4 mg of nitrogen per gram of sugar utilized. The largest numbers of these bacteria were in grey and black earths (up to 8 million per gram of soil), in mountain-meadow, sub-alpine soils they developed only in the summer, chiefly in the upper layers.

In cultivated takyr-like soils the quantity of cells of oligonitrophilic actinomycetes was between 2 and 23 per cent of the total bacterial content. The remaining 77–98 per cent consisted of other oligonitrophilic bacteria (Karaguishieva, 1966).

Naplekova (1966) found that *Act. griseus*, *Act. coelicolor*, *Act. griseoruber* and other species of actinomyctes fixed 0·3–1·2 mg of nitrogen per gram of cellulose utilized. When sucrose was the carbon source, the productivity of nitrogen fixation increased considerably. No connection was demonstrated between dehydrogenase activity and nitrogen-fixing function in the cultures isolated.

Rubenchik *et al.* (1966) established nitrogen fixation by several Actinomycete cultures when insignificant quantities of bound nitrogen were present in the medium.

The material we have outlined suggests that certain representatives of the genera *Nocardia* and *Actinomyces* are able to fix atmospheric nitrogen. In many cases, however, this ability has still to be confirmed with nitrogen-15.

Spirochaetes which fix nitrogen

Pshenin (1965) isolated two strains of nitrogen-fixing spirochaetes from water obtained near the surface of the open sea, and described them as a new species, *Treponema hyponeustonicum*. This is a Gram-negative, saprophytic bacterium living as an aerobe, but which can grow microaerophilically and even anaerobically in the presence of nitrates. It gives a positive reaction for catalase. In young cultures cells of *Treponema* are thin and convoluted, and vary in length, with pointed or rounded ends. They have dimensions of 2·5–18·0 × 0·3–0·6 μm. In Hissberger's solution with sodium citrate the cells take on the form of long rods or threads up to 70 μm long. In favourable conditions spore-like bodies and microcysts may form. In meat-peptone broth prepared with fresh water many branched forms develop. In meat-peptone broth prepared from sea water involutive forms do not form.

After 1–2 days, colonies of *Treponema* on Fedorov's agar with mannitol are transparent or rather dull, with a smooth shiny surface and mucilage of a ropey consistency. After 2–3 weeks on Fedorov's agar with glucose, colonies take on a yellowish-ochre colour. On mannitol and glucose media old colonies are elastic and rubbery.

When *Treponema* is cultured in meat-peptone broth prepared with sea water there is heavy evolution of ammonia and hydrogen sulphide; indole does not form. Nitrates are actively reduced to ammonia by *Treponema*, but ammonium salts and urea are utilized less well. *Treponema* does not decompose cellulose nor hydrolyse starch. In peptone water with lactose, maltose, sucrose, glucose or mannitol, it produces acids. It makes good use of glycerol and glucose and butyric, pyruvic, malic and lactic acids as carbon sources. It utilizes weakly ethanol and fructose and acetic, propionic, succinic and citric acids. Optimum development is at 25°C; 8·8–9·7 mg of nitrogen is fixed per gram of glucose used.

Fungi that fix nitrogen

The first indication of nitrogen-fixing activity in fungi, in particular yeasts, goes back to the last century. Frank (1892, 1893) noted the ability of *Penicillium* to accumulate as much as 4·4 mg of nitrogen in 100 ml of nutrient solution after 10 months of incubation. Remy (1902) found that some strains of *Aspergillus niger* fixed 0·5 mg of nitrogen per gram of dextrose. The ability of species of *Aspergillus* to fix nitrogen was also noted by Senn (1928), Zarma (1959) and Čeruti (1960). Ternetz (1907) found that fungi of the genus *Phoma* could fix up to 22 mg of nitrogen per gram of dextrose, while *Pencillium glaucum* and *Aspergillum niger* could fix up to 2·7 mg. All her fungi were isolated from peat. Lipman (1911–1912) and Omelyanskii (1923) give detailed analyses of these early investigations.

Lipman studied in detail more than twenty species of yeasts, yeast-like organisms and microscopic fungi and found that dextrose, mannitol and lactose were good carbon sources for use in detecting nitrogen-fixation. On the other hand, when he used maltose, his experiments failed.

The productivity of fixation by yeasts (*Saccharomyces apiculatus Saccharomyces ellipsoideus champagne, Saccharomyces ellipsoideus* and *Saccharomyces cerevisiae*) and 'pseudoyeasts' cultivated for a month at 26°–28°C in a medium containing 1·5 per cent of mannitol fluctuated between 0·10 and 1·47 mg per gram of carbon source. Matkovics (1960) and Schanderl (1955) too noted fixation of nitrogen by *Saccharomyces*. In the same conditions the nitrogen-fixing activity of the fungi *Mycoderma vini, Aspergillus niger, Penicillium glaucum* and *Botrytis cinera* fluctuated between 0·04 and 2·38 mg of nitrogen per gram of mannitol. All these findings, of course, need to be confirmed by more refined methods than those so far used.

In 1909 Zikes isolated *Torula wiesneri* from laurel leaves, and found that it fixed 2·5 mg of nitrogen per gram of carbon source. Insignificant quantities of nitrogen have been reported to be fixed by *Oidium lactis, Monilia* and certain species of *Saccharomyces* (Kossowicz, 1915). Kossowicz later changed his mind, deciding that the nitrogen fixation he observed might be explained by the utilization of ammonia in the atmosphere by the micro-organisms. Löhnis and Pillai (1907) emphasized that *Torula* isolated from the soil did not fix nitrogen, and was also characterized by having large mucilaginous capsules and by developing on nitrogen-free media.

In 1954 non-sporulating yeasts were isolated by means of a nitrogen-free medium from prairie podzols (pH 4·5) overgrown with Scottish heather (*Calluna vulgaris*) (Metcalfe *et al.*, 1954). One culture greatly resembled *Saccharomyces*, the other, which formed a red pigment, was *Rhodotorula*. It was not possible to isolate these yeasts from an enriching culture by the usual method because their numbers were so small. Therefore Metcalfe *et al.* perfused columns of soil with an elective nitrogen-free medium containing various carbon sources, and isolated micro-organisms on solid nitrogen-free medium containing mannitol after the cells had multiplied.

When the yeasts thus isolated were cultured on liquid nitrogen-free media which contained mannitol and phosphate and were periodically aerated, an abundant sediment formed on the fourth day, and this contained a considerable amount of nitrogen. The yeast cells contained many fatty globules and retained Gram stain only weakly. The nitrogen-fixing activity of *Saccharomyces* was about 4 mg of nitrogen per gram of manitol used; for *Rhodotorula* it was about 1–3 mg per gram of glucose.

Roberts and Wilson (1954) confirmed, using nitrogen-15, that yeasts could fix nitrogen. In an atmosphere of oxygen, argon and nitrogen enriched with 32 excess atomic per cent nitrogen-15, 0·45–0·48 atomic per cent nitrogen had been incorporated into the cells of *Saccharomyces* after four hours. The cells were actively budding in these conditions. In the same way 1·38–1·41 atomic per cent nitrogen-15 was incorporated into cells of *Rhodotorula*. Roberts and Wilson considered the nitrogen-fixing activity of *Rhodotorula* to be one-tenth and that of *Saccharomyces*, one-fiftieth of that of *Azotobacter*.

Brown and Metcalfe (1957) isolated another yeast-like organism, belonging to the genus *Pullularia*, from the rhizosphere of clover growing on heavy loamy soils at pH 6·8. It developed very well in a nitrogen-free medium containing phosphates and glucose, and after 14 days of incubation accumulated 2·0–2·6 mg of nitrogen in 100 ml of medium, which corresponded to a productivity of 4·0–5·0 mg of nitrogen per gram of glucose. A black pigment is typically formed in this organism.

Metcalfe (1957) isolated a similar organism from chalky soil, prairie soils, the humus of forest soil and the lichen *Cladonia uncialis*. In a nitrogen-containing nutrient medium the organism rapidly lost its power to fix nitrogen. In a nitrogen-free medium it developed in the form of small colonies which, as they aged, came to resemble rhizoids. The organism produced a black pigment vigorously on nitrogen-free agar, but weakly on wort-agar and yeast-agar. Ross (1958) was unable to confirm the nitrogen-fixing activity of a culture of *Pullularia pullulans*. But nitrogen-fixing activity by yeast-like soil organisms has been noted by others too (Verona and Picci, 1961; Jensen, 1962; Moore, 1963).

Moore (1963) found that in an atmosphere of labelled nitrogen various cultures of *Lipomyces* accumulate 0·01–0·6 atomic per cent nitrogen-15 after two days. Other strains accumulated more, however. In Moore's view, the species of *Lipomyces* are micronitrophils rather than nitrogen fixers.

Mazilkin (1957) and Bab'eva and Hasan Meavad (1966) noted that many cultures of *Lipomyces* could grow on Ashby's nitrogen-free medium. When nodule bacteria were isolated from *Lupinus luteus* growing on sandy acid soils in Hungary, Nemeth (1959) also isolated two micro-organisms that formed a red pigment and fixed molecular nitrogen. The lupin plants in question were considerably larger than others around them. The morphology and physiology of the two organisms identified them as *Saccharomyces*. The two species are designated Pr and E3. Their cells are spherical or ovoid and contain two to four round granules. The diameter of the cells is 4–6 μm, increasing in old cultures to 8–9 μm. Many smaller granules appear in the cell protoplasm.

On bean-agar smooth colonies slowly develop, firmly attached to the agar surface. The cultures can utilize ethanol and citric acid as a source of carbon. They also utilize glucose, maltose, sucrose, lactose, mannitol and salicin forming acids but no gas. As the micro-organisms develop, the pH of the medium decreases from 7·2 to 5·2–5·4. Development is optimum at 25°–28°C. These yeasts are obligate aerobes. In the presence of insignificant doses of potassium nitrate, ammonium chloride and ammonium sulphate and peptone, growth and formation of pigment is stimulated. Dispersed light also enhances the formation of pigment.

The pigment of *Saccharomyces* Pr is brighter than that of *Saccharomyces* E3. Nemeth thought that the pigment was involved in nitrogen fixation, particularly since fixation is observed only during the formation of pigment. Spectroscopy showed that the pigment was a haemoprotein of the haemoglobin type. The productivity of nitrogen fixation of these cultures on aerated bean decoction (with 2 per cent glucose) varied from 2·4 to 5·7 mg per gram of glucose.

In conclusion, we feel that much more information is needed about the assimilation of molecular nitrogen by microscopic fungi.

References

Abraham, A. and Subramoney, N. 1960. Nitrogen-fixing organisms in coconut retting grounds in Kerala. *Current Sci.*, 29 (3), 100.

Andersen, G. R. 1955. Nitrogen fixation by *Pseudomonas*-like soil bacteria. *J. Bacteriol.*, 70 (2), 129–133.

Aristovskaya, T. V. 1957. Certain features of the microflora of podzolic soils of the north-western part of the USSR: in *Raboty tsentral'nogo muzeya pochvovedeniya* (Proceedings of the Central Museum of Pedology), (2). Izd-vo Akad. Nauk SSSR, Moscow, 228–249.

Arnon, D. J., Losada, M., Nozaki, M. and Tagawa, K. 1960. Photofixation of nitrogen and photoproduction of hydrogen by thiosulphate during bacterial photosynthesis. *Biochem. J.*, **77** (3), 23.

Arnon, D. J., Losada, M., Nozaki, M. and Tagawa, K. 1961. Photoproduction of hydrogen, photofixation of nitrogen and unified concept of photosynthesis. *Nature*, **190**, 601–610.

Baas-Becking, L. G. and Wood, E. J. 1955. Biological processes in the estuarine environment. *Proc. Koninkl. nederl. akad. wet. ser.*, B, **3**, 160, 173 (in English).

Bab'eva, I. P. and Hasan Meavad. 1966. Ecology of the soil yeasts Lipomyces. *Tezisy Dokladov III Vses. s"ezda pochvovedov.*

Balloni, W., Materassi, R. and Favilli, F. 1965. Observations sur le metabolisme de l'azote dans les sols forestiers. *Ann. Inst. Pasteur*, **109** (3) Suppl., 68–73.

Becking, J. H. 1963. Fixation of molecular nitrogen by aerobic *Vibrio* or *Spirillum*. *Antonie van Leeuwenhoek. J. Microbiol. and Serol.*, **29** (3), 326 (in English).

Beijerinck, M. W. 1901. Über oligonitrophile Mikroben. *Zbl. Bakteriol., Parasitenkunde, Infektionskrankh. und Hyg.*, Abt. 2, **2** (7), 561–582.

Beijerinck, M. W. 1904. L'influence des microbes sur la fertilité du sol et croissance des végétaux supérieurs. *Arch. neerl. extractes et naturetess*, ser. 2.

Beijerinck, M. W. 1925. Über ein Spirillum welches freien Stickstoff binden kann? *Zbl. Bakteriol., Parasitenkunde, Infektionskrankh. und Hyg.*, Abt. 2, **63**, 353–359.

Belyaeva, M. I. 1954. Assimilation of molecular nitrogen by hydrogen bacteria. *Uch. zapiski Kazansk. un-ta.*, **114** (13–18).

Benecke, W. 1907. Über stickstoffbindende Bakterien aus dem Golt von Neapel. *Ber. Dtsch. bot. Ges.*, **25**, 1–7.

Bennet, R., Rigopoulos, N. and Fuller, R. C. 1964. The pyruvate phosphoroclastic reaction and light-dependent nitrogen fixation in bacterial photosynthesis. *Proc. Nat. Acad. Sci.*, **52** (3), 762–768.

Bergey. *Manual of Determinative Bacteriology*. Seventh ed., Williams and Wilkins, Baltimore, 1957.

Bhat, J. and Palacios, G. 1949. Influence of *Aerobacter aerogenes* in the nitrogen status of the soil. *J. Univ. Bombay*, **71**B, 84–87; *Chem. Abstr.*, **44**, 779d.

Bredemann, G. 1908. Untersuchungen uber die Variation und das Stickstoffbindungsvermögen des *Bacillus asterosporus* A.M., ausgeführt an 27 Stämmen verschiedener Herkunft. *Zbl. Bakteriol., Parasitenkunde, Infektionskrankh. und Hyg.*, Abt. 2, **22**, 44–89.

Brown, M. E. and Metcalfe, G. 1957. Nitrogen fixation by a species of *Pullularia*. *Nature*, **180**, 282.

Buchanan, B. B. and Rabinowitz, J. C. 1964. Some properties of *Methanobacterium omelianskii* ferrodoxin. *J. Bacteriol.*, 1964, **88** (3), 806–807.

Bulgadaeva, R. V. 1961. Microflora of soddy calcareous brown soils and black earth of the Angara Region. *Trudy Vost.-Sib. fil. Sib. otdel. Akad. Nauk SSR, ser. biol.*, (27), 114–123.

Burk, D. and Lineweaver, H. 1930. The influence of fixed nitrogen on *Azotobacter*. *J. Bacteriol.*, **19**, 389–414.

Burris, R. G. 1961. Evolution of biological nitrogen fixation: in *V Mezhdunarodnyi Biochimicheskii Kongress. Simpozium III* (Fifth International Biochemical Congress, Symposium III), (5), Moscow, 28–31.

Caron, J. 1895. Landwirtschaftlich-bakteriologische Probleme. *Landwirtsch. Versuchs-Stat.*, **45**, 401.

Caumartin, V. 1957. Recherches sur une bacterie des argiles de cavernes des sediments ferrugineux. *C.R. Acad. Sci.*, **245** (20), 1758–1760.

Centifanto, Y. M. and Silver, W. S. 1964. Leaf-nodule symbiosis I. Endophyte of *Psychotria bacteriophila*. *J. Bacteriol.*, **88** (3), 776–781.

Čeruti, A. 1960. Moderne scoperte sulla fissazione del N_2 *Nuovo giorn. bot. ital.*, (1–2), 288–291.

Chakravorty, S. C. and Das, N. B. 1965. Studies on Nitrogen fixation and respiratory activity of *Derxia gummosa*. *Indian J. Exptl. Biol.*, **3** (4), 234–239.

Chesnyak, O. A. and Chesnyak, G. Ya. 1966. Course of microbiological activity of deep chernozem on prolonged agricultural exploitation: in *Nauchnye osnovy ratsional'nogo ispol'zovaniya pochv chernozemnoi zony SSSR i puti povyssheniya ikh plodorodiya* (Scientific principles of the rational use of the soils of the Black Earth zone of the USSR and ways of improving their fertility), (1), Voronezhsk un-t., Voronezh, 88–89.

Davis, J. B. 1952. Studies on soil samples from a paraffin dirt bed. *Bull. Amer.Assoc. Petrol. Geologists*, **36**, 2186–2188.

Davis, J. B., Coty, V. F. and Stanley, J. P. 1964. Atmospheric nitrogen fixation by methane-oxidizing bacteria. *J. Bacteriol.*, **88** (2), 468–472.

Duchow, E. and Douglas, H. C. 1949. *Rhodomicrobium vannielli* a new photoheterotrophic bacterium. *J. Bacteriol.*, **58** (4), 409–416.

Dudareva, T. E. 1966. The microflora of alkaline soils of Siberia: in *Mikroflora pochv severnoi i srednei chasti SSSR* (The microflora of soils of the northern and central parts of the USSR). Nauka, Moscow, 365–389.

Emerson, P. 1917. Are all the soil bacteria and streptothrices that develop on dextrose agar azofiers? *Soil Sci.*, **3**, 417–421.

Fedorov, M. V. 1948. *Biologicheskaya fiksatsiya azota atmosfery* (Biological fixation of atmospheric nitrogen). Sel'khozgiz, Moscow.

Fedorov, M. V. and Il'ina, T. K. 1959. Stimulating effect of yeast autolysate on the nitrogen-fixing activity of soil Actinomycetes. *Mikrobiologiya*, **28** (4), 541–547.

Fedorov, M. V. and Il'ina, T. K. 1960. Behaviour of individual forms of soil Actinomycetes to different carbon sources when growing on nitrates and molecular nitrogen. *Mikrobiologiya*, **29** (4), 495–500.

Fedorov, M. V. and Kalininskaya, T. A. 1959. Nitrogen-fixing activity of mixed cultures of oligonitrophilic micro-organisms. *Mikrobiologya*, **28** (3), 343–357.

Fedorov, M. V. and Kalininskaya, T. A. 1960. Relationships between individual species of oligonitrophilic bacteria fixing molecular nitrogen in mixed cultures. *Izv. TSKhA*, (2), 125–136.

Fedorov, M. V. and Kalininskaya, T. A. 1961. New forms of nitrogen fixing microorganisms isolated from soddy podzolic soils. *Dokl. TSKhA*, (70), 145–151.

Fedorov, M. V. and Kudryashova, T. K. 1956. Nitrogen-fixing activity of certain soil Actinomycetes. *Dokl. Akad. Nauk SSSR*, **107** (2), 345.

Fedorov, M. V. and Pantosh, D. 1957. Physiological features of the main forms of rhizospheric bacteria of spring wheat. *Dokl. Akad. Nauk SSSR*, **116** (1), 149–152.

Fiuczek, M. 1959. Fixing of atmospheric nitrogen in pure cultures of *Streptomyces alni*. *Acta microbiologica polonica*, **8**, 283–287.

Florenzano, G. and Balloni, W. 1965. Quelques souches de bacteries fixatrices d'azote aerobies acido-tolerantes. *Ann. Inst. Pasteur*, **109** (3) Suppl., 133–135.

Fogg, G. E. 1956. Nitrogen fixation by photosynthetic organisms. *Annual Rev. Plant Physiol.*, **7**, 51.

Frank, B. 1892. Die Assimilation freien Stickstoffs bei den Pflanzen in ihrer Abhängigkeit von Species, von Ernährungsverhaltnissen und von Bodenarten. *Landwirtsch. Jahrb.*, **21**, 1–7.

Frank, B. 1893. Die Assimilation des freien Stickstoffs durch die Pflanzenwelt. *Bot. Ztg.*, **31**, 139–146.

Fry, B. A. 1955. *The Nitrogen Metabolism of Micro-organisms*. London, New York.

Gest, H., Judis, J. and Peck, H. D. 1956. Reduction of molecular nitrogen and relationships with photosynthesis and hydrogen metabolism: in *Inorganic Nitrogen Metabolism*. Baltimore, 298.

Gest, H. and Kamen, M. D. 1949. Photoproduction of molecular hydrogen by *Rhodospirillum rubrum*. *Science*, **109** (2840), 558–559.

Gest, H., Kamen, M. D. and Bregoff, H. M. 1950. Studies on the metabolism of photosynthetic bacteria. 5. Photoproduction of hydrogen and nitrogen fixation by *Rhodospirillum rubrum*. *J. Biol. Chem.*, **182**, 153–170.

Goerz, R. D. and Pengra, R. M. 1961. Physiology of nitrogen fixation by a species of *Achromobacter*. *J. Bacteriol.*, **81** (4), 568–572.

Grau, F. H. and Wilson, P. W. 1960. Nitrogen fixation by *Bacillus polymyxa*. *Bacteriol. Proc.*, **61**, 151.

Grau, F. H. and Wilson, P. W. 1962. Physiology of nitrogen fixation by *Bacillus polymyxa*. *J. Bacteriol.*, **83** (3), 490–496.

Grau, F. H. and Wilson, P. W. 1963. Hydrogenase and nitrogenase in cell-free extracts of *Bac. polymyxa*. *J. Bacteriol.*, **85** (2), 446–450.

Gray, E. and Smith, J. D. 1950. A new aquatic nitrogen-fixing bacterium from three Cambridgeshire chalk streams. *J. Gen. Microbiol.*, **4** (3), 281–285.

Gulya, N. K. (quoted by A. G. Rodina, 1956).

Gurfel', D. 1960. Use of micro-organisms splitting organic phosphorus compounds for improving plant nutrition. *Byull. nauchno-tekhn inf. Estonsk. n.-i. in-ta zemled. i melior.*, (5), 52–56.

Hamilton, I. R. 1962. Nitrogen fixation and pyruvate metabolism in *Achromobacter N. 4-B*, *Bacteriol. Proc.*, **62**, 101.

Hamilton, I. R. 1963. *Nitrogen fixation and pyruvate metabolism by Klebsiella (Achromobacter sp.)*. *Dissert. Abstr.*, **24** (6), 2250.

Hamilton, I. R., Burris, R. H. and Wilson, P. W. 1964. Hydrogenase and nitrogenase in a nitrogen-fixing bacterium. *Proc. Nat. Acad. Sci., USA*, **52** (2), 637–641.

Hamilton, I. R., Burris, R. H. and Wilson, P. W. 1965. Pyruvate metabolism by a nitrogen-fixing bacteria. *Biochem. J.*, **96** (2), 383–389.

Hamilton, P. B., Magee, W. E. and Mortenson, L. E. 1953. Nitrogen fixation by *Aerobacter aerogenes* and cell-free extracts of the *Azotobacter vinelandii*. *Bacteriol. Proc.*, **82**.

Hamilton, P. B. and Wilson, P. W. 1955. Nitrogen fixation by *Aerobacter aerogenes*. *Ann. Acad. Sci. fennicae*, ser. A, **2**, Chemica, 139–150 (in English).

Harper, H. J. 1939. The effect of natural gas on the growth of micro-organisms and accumulation of nitrogen and organic matter in the soil. *Soil Sci.*, **48**, 461–466.

Heinze, B. 1902. Über die Stickstoffassimilation durch niedere Organismen. *Zbl. Bakteriol., Parasitenkunde, Infektionskrankh. und Hyg.*, Abt. 2, **8**.

Heinze, B. 1904. Über die Bildung und Wiederverarbeitung von Glykogen durch niedere pflanzliche Organismen. *Zbl. Bakteriol., Parasitenkunde, Infektionskrankh. und Hyg.*, Abt. 2, **2**, 43–49.

Hino, S. 1955. Studies on the inhibition of carbon monoxide and nitrogen oxide of anaerobic nitrogen fixation. *J. Biochem.*, **42** (6), 775–784.

Hino, S. and Wilson, P. W. 1958. Nitrogen fixation by a facultative bacillus. *J. Bacteriol.*, **75** (4), 403–408.

Hug, F. and Ali, H. 1957. A species of nitrogen-fixing organism. *Pakistan J. Biol. and Agric. Sci.*, 1, 94 (*Biol. Abstr.*, 1961, **36** (23), 81821).

Il'ina, T. K. 1966. Study of the role of molybdenum in the processes of biological nitrogen fixation in *Mycobacteria*. *Mikrobiologiya*, **35** (3), 422–426.

Ilyakhina, Z. V. 1962. Microbiological characterization of the soils of the Krasnoyarsk forest-steppe and the contiguous sub-taiga: in *Trudy I. Sibirskoi konferentsii pochvovedov* (Proceedings of the First Siberian Conference of Pedologists). Krasnoyarsk, 430–438.

Isachenko, B. L. 1914. Investigations of the bacteria of the North Arctic Ocean. in *Trudy Murmanskoi nauchno-prom. ekspeditsii* 1906 (Proceedings of the Murmansk scientific and industrial expedition of 1906), Petrograd.

Ivarson, K. C. 1953. *Studies on non-symbiotic, nitrogen fixation in some Alberta soils.* M. Sci. Thesis, Univ. of Alberta, Edmonton, Alta.

Jacobitz. 1903. Beitrag zur Frage der Stickstoffassimilation durch den *Bac. ellenbachensis. Z. Hyg. und Infektionskrankh.*, **45**, 97.

Jensen, H. L., De, P. K. and Bhattacharya, R. 1959. A new nitrogen-fixing bacterium. *Nature*, **184** (4700) Suppl. N 22, 1743.

Jensen, H. L., Petersen, E. J., De, P. K. and Bhattacharya, R. 1960. A new nitrogen-fixing bacterium: *Derxia gummosa* nov. gen. nov. spec. *Arch. Mikrobiol.*, **36** (2), 182–195 (in English).

Jensen, V. 1956a. A new nitrogen-fixing bacterium from a Danish watercourse. *Arch. Microbiol.*, **29**, 349–353 (in English).

Jensen, V. 1956b. Nitrogen fixation by strains of *Aerobacter aerogenes. Physiol. plantarum*, **9** (1), 130–136 (in English).

Jensen, V. 1958. A new nitrogen-fixing bacterium from a Danish watercourse. *Arch. Microbiol.*, **29**, 349–353 (in English).

Jensen, V. 1962. Activity of various physiological groups of micro-organisms in some forest soils. *Arsskr. Kgl. veterinog landbøhojskole*, København, 180–194 (in English).

Kalininskaya, T. A. 1954. *Fiziologicheskoe osobennosti Azotomonas fluorescens* (Physiological Features of *Azotomonas fluorescens*), Author's Abstract of M.Sc. Thesis, Moscow.

Kalininskaya, T. A. 1963. Nitrogen-fixing *Mycobacteria* isolated from soddy podzolic soils: in *Pochvennaya i sel'skokhozyaistvennaya mikrobiologiya* (Soil and agricultural microbiology). Izd-vo Akad. Nauk Uzb SSR, Tashkent, 73.

Kalininskaya, T. A. 1966a. Importance of symbiotic relationships between micro-organisms in the processes of biological nitrogen fixation in the soil: in *Tezisy dokladov III Vsesoyuznyi Sëzd pochvovedov* (Summary of speeches of the Third All-Union Meeting of Pedologists). Tartu, 87–88.

Kalininskaya, T. A. 1966b. Facultative symbiotrophic nitrogen fixers and their role in the assimilation of molecular nitrogen in soils of various types: in *IX Mezhdunarodnyi Mikrobiologicheskii Kongress, Tezisy dokladov* (Ninth International Microbiology Congress. Summaries of speeches), Moscow, 297.

Kalininskaya, T. A. 1967. Role of microbial symbioses in nitrogen fixation by free-living micro-organisms: in *Biologicheskii azot i ego rol, v zemledelii* (Biological nitrogen and its role in farming). Nauka, 221–220.

Kalininskaya, T. A. and Il'ina, T. K. 1963. Effect of the bound forms of nitrogen and supplementary growth factors on nitrogen fixation by *Mycobacteria*: in *Tezisy Vsesoyuznoi konferenksii po sel'skokhozyaistvennoi i pochvennoi mikrobiologii* (Summaries of the All-Union Conference on agricultural and soil microbiology), Leningrad.

Kamen, M. D. 1955. Nitrogen fixation. in *Physics and Chemistry of Life*. New York, 47.

Kamen, M. D. and Gest, H. 1949. Evidence for a nitrogenase system in the photosynthetic bacterium *Rhodospirillum rubrum. Science*, **109** (2840), 560–562.

Karaguishieva, D. 1966. Free-living nitrogen-fixers of virgin and cultivated takyr-like soils. *Izv. Akad. Nauk KazSSR*, ser. biol., (2), 20–26.

Karyagina, L. A. 1955. *Proaktinomitsety dernovo-podzolistykh i chernozemnykh pochv.*

(Proactinomycetes of soddy-podzolic and black earth soils), Author's abstract of M.Sc. Thesis, Moscow.

Kataidi, M., Gei, T., and Pianaroli, Dzh (Kataidi, M., Gay, T., and Pianaroli, J., 1966. Fixation of atmospheric nitrogen by bacteria utilizing hydrocarbons: in *IX Mezhdunarodnyi Mikrobiologicheskii Kongress, Tezisy dokladov* (Ninth International Microbiology Congress. Summaries of speeches).

Katayama, T., Kobayshi, M. and Okuda, A. 1965. Nitrogen-fixing micro-organisms in paddy soils. 14. *Soil Sci. and Plant Nutr.*, **11** (2), 78–83.

Katznelson, H. and Lochhead, A. G. 1944. Studies with *Bacillus polymyxa* 3. Nutritional requirements. *Canad. J. Res.*, C, **22**, 273–279.

Kaufman, F. 1959. Semeistvo kishechniykh bakterii (Family of enteric bacteria), Medgiz, Moscow (translated from English).

Kaufman, J., Pochon, J. and Tchan, Y. 1948. Recherches sur la microflore oligonitrophile du sol. 1. Nutrition azote minerale. *Ann. Inst. Pasteur*, **75** (1).

Keding, M. 1906. Weitere Untersuchungen über stickstoffbindend Bakterien. *Wiss. Meeresuntersuch. Abt. Kiel.* N. F., **9**, 273–309.

Keutner, J. 1905. Über das Vorkommen und die Verbreitung stickstoffbindender Bakterien im Meere. *Wiss. Meeresuntersuch.*, *Abt. Kiel N.F.*, **8**, 27–55.

Khak-Mun Ten. (Hak Mung t'eng). 1966. Oligonitrophilic micro-organisms of the soils of southern Sakhalin. *Trudy Sakhalinsk kompleksnogo n.-i. in-ta*, (17), 44–51.

Khudyakov, Ya. P. and Voznyakovskaya, Yu. M. 1956. The microflora of the roots of wheat and some properties. *Mikrobiologya*, **25** (2), 184–190.

Klevenskaya, I. L. 1962. Distribution of oligonitrophilic Actinomycetes in the soils of the Altai Mountains and their nitrogen-fixing activity: in *Genezis pochv Zapadnoi Sibiri* (Genesis of the soils of West Siberia), 93–99.

Klevenskaya, I. L. 1966. The microflora of black earth soils of Siberia. in *Mikroflora pochv severnoi i srednei chasti SSSR* (The microflora of the northern and central parts of the USSR), Nauka, Moscow, 250–273.

Kobayashi, M., Katayama, T. and Okuda, A. 1965. Nitrogen-fixing micro-organisms in paddy soils. 13. *Soil Sci. and Plant Nutr.*, **11** (2), 74–77.

Koch, A. 1907. Ernährund der Pflanzen durch frei im Boden lebende stickstoffsammelnde Bakterien. *Mitt. Landwirtsch. Ges.*, **12**, 110–126.

Kondrat'eva, E. N. 1963. *Fotosinteziruyushchie bakterii* (Photosynthesizing Bacteria), Izd-vo Akad. Nauk SSSR.

Konvalevskii, S. L. 1898. Assimilation of free atmospheric nitrogen by microbes. *Russkii arkhiv patologii klin. med. i bakteriologii*, **6**, 251.

Kořinek, I. 1932. Über oligonitrophile Mikroben in Meere. *Zbl. Bakteriol., Parasitenkunde Infektionskrankh und Hyg.*, Abt. **2**, 86, 201–206.

Kossowicz, A. 1915. Die Bindung des elementaren Stickstoffs durch Saccharomyceten (Hefen und Schimmelpilze). *Z. Garungsphysiol.*, **5**, 26.

Krasil'nikov, N. A. 1945. New species of heterotrophic nitrogen fixers *Azotomonas fluorescens*: in *Referaty rabot OBN Akad. Nauk SSSR za 1945*, (Reviews of works of the Department of Biological Sciences of the Academy of Sciences of the USSR for 1945). Izd-vo Akad. Nauk SSSR, Moscow.

Krüger. 1894. *Beitrage zur Physiologie und Morphologie niederer Organismen*, H, 4.

Kruyff, E. de. 1906. Sur une bacterie aérobe fixant l'azote libre l'atmosphere: *Bacterium krakatau. Bull. Dept. Agric. Indes. néerl.*, **4**, 9.

Kruyff, E. de. 1910. Quelques remarques sur des bacteries aérobes fixant l'azote libre de l'atmosphere dans les tropiques. *Zbl. Bakteriol.*, **26**, 54.

Kurdina, R. M. 1963. A sporulating aerobic nitrogen fixer: in *Fiziologiya i biokhimiya mikroorganizmov* (The physiology and biochemistry of micro-organisms), Izd-vo Akad. Nauk KazSSR, Alma-Ata, 58–59.

Kuznetsov, S. I. 1952. *Rol'. mikroorganizmov v krugovorote veshchestv v ozerakh* (The role of micro-organisms in metabolic turnover in lakes), Izd-vo Akad. Nauk SSSR, Moscow.

Larsen, H. 1951. Photosynthesis of succinic acid by *Chlorobium thiosulphatophilum*. *J. Biol. Chem.*, **193**, 167.

Larsen, H. 1953. On the microbiology and biochemistry of the photosynthetic green sulfur bacteria. *Kgl. norske vid. selskabsskr.*, (1), 1 (in English).

Le Gall, J., Senez, J-C. and Pichinoty, F. 1959. Fixation de l'azote par les bactéries sulfato-reductrices. *Ann. Inst. Pasteur*, **92** (3215), 223–230.

Le Gall, J. and Senez, J-C. 1960. Biochimie cellulaire. Influence de la fixation de l'azote sur la croissance de *Desulforibrio desulfuricans*. *C.R. Acad. sci.*, **250**, 404–406.

Lewis, S. B. 1950. *Nitrogen fixation by the Athiorhodaceae*. M.Sc. Thesis, Univ. Wisconsin, Madison.

Ley, J. de. 1955. DNA base composition of *Klebsiella rublacearum. Antonie van Leeuwenhoek. J. Microbiol. and Serol.*, **31** (2), 203–204.

Lindstrom, E. S., Burris, R. H. and Wilson, P. W. 1949. Nitrogen fixation by photosynthetic bacteria. *J. Bacteriol.*, **58** (3), 313–316.

Lindstrom, E. S., Lewis, S. M. and Pinsky, M. J. 1951. Nitrogen fixation and hydrogenase in various bacterial species. *J. Bacteriol.*, **61** (4), 481–487.

Lindstrom, E. S., Tove, Sh. R. and Wilson, P. W. 1950. Nitrogen fixation by the green and purple sulfur bacteria. *Science*, **112** (2903), 197–198.

Lipman, J. G. 1903. Experiments on the transformation and fixation of nitrogen by bacteria. *New Jersey State Exptl. Stat. Rept.*, 217–235.

Lipman, Ch. B. 1911–1912. Nitrogen fixation by yeasts and other fungi. *J. Biol. Chem.*, **10**, 169–182.

Löhnis, F. and Pillai, N. K. 1907. Über stickstoff-fixirende Bakterien. *Zbl. Bakteriol., Parasitenkunde. Infektionskrankh. und Hyg.*, Abt. 2, **19**, 87.

Löhnis, F. and Pillai, N. K. 1908. Über stickstoff-fixirende Bakterien. *Zbl. Bakteriol., Parasitenkunde. Infektionskrankh. und Hyg.*, Abt. 2, **20**, 781.

L'vov, N. P. 1963. Nitrogen-fixing species of the genus *Mycobacterium. Dokl. TSKhA*, (94), 265–269.

L'vov, N. P. 1963. New free-living nitrogen-fixing micro-organisms. *Izv. Akad. Nauk SSSR*, ser. biol., (2), 271–282.

L'vov, N. P. 1964. *Svobodnozhivuyushchie azotfiksiruyushchie mikroorganizmov dernovo-podzolisto pochv* (Free-living nitrogen-fixing micro-organisms of soddy-podzolic soils). Author's abstract of M.Sc. Thesis, Moscow.

L'vov, N. P. and Lyubimov, V. I. 1965. Study of the physiology of a new nitrogen-fixing mycobacterium *Mycobacterium azotoabsorptum* sp. n. *Izv. Akad. Nauk SSSR, Ser. Biol.*, (2), 250–256.

Lyalikova, N. N. 1957. Study of the process of assimilation of free carbon dioxide by purple sulphur bacteria in lake Belovody. *Mikrobiologiya*, **26** (1).

Lyubimov, V. I., L'vov, N. P. and Loseva, L. P. 1965. Induced character of the enzymes of nitrogen fixation in *Mycobacterium azoto-absorptum* N. sp. *Izv. Akad. Nauk SSSR, Ser. Biol.*, (3), 392–394.

Macher, F. and Manninger, E. 1961. Die Verwendung der Spektrochemie zur Bestimmung der Zusammensetzung von Bakterien. *Arch. Mikrobiol.*, **38** (3), 201–208.

Mahl, M. C., Wilson, P. W., Fife, M. A. and Ewing, W. H. 1965. Nitrogen fixation of members of the tribe Klebsielleae. *J. Bacteriol.*, **89** (6), 1482–1487.

Maliyants, A. A. 1933. Microbiological study of the bottom of the Caspian Sea. *Trudy Azerb. neft. issl. in-ta*, (18), 1–87.

Materassi, P. and Balloni, W. 1965. Quelques observations sur la présence des micro-organismes autotrophes fixateurs d'azote dans les rizières. *Ann. Inst. Pasteur*, (109), (3) Suppl., 218–223.

Matkovics, B. 1960. Extreme Werte der atmosphärischen N_2 Bindung bei den aus den Wurzelknöllchen von *Lupinus lutes* isolierten Hefen. *Naturwissenschaften*, 47 (3), 92.

Mazilkin, I. A. 1957. Soil fungus *Lipomyces* and its distribution in the soils of Yakutia. *Mikrobiologya*, 26 (4), 477–480.

Metcalfe, G. 1957. (Quoted by Brown, M. E. and Metcalfe, G. 1957).

Metcalfe, G. and Brown, M. E. 1957. Nitrogen fixation by new species of *Nocardia*. *J. Gen. Microbiol.*, 17 (3), 567–572.

Metcalfe, G., Chayen, S., Roberts, E. R. and Wilson, T. G. G. 1954. Nitrogen fixation by soil yeasts. *Nature*, 174, 841–842.

Milne, R. A. 1951. *Studies on the non-symbiotic nitrogen fixation of some Saskatchewan soils*. M. Sci. Thesis, Univ. Alberta, Edmonton.

Mishustina, E. E. 1955. Oligonitrophilic micro-organisms of the soil. *Trudy in-ta mikrobiol. Akad. Nauk SSR*, (4), 110–129.

Missirliu, E., Papacostea, P., Preda, C., Manolescu, V., Popa, E. and Cămirzen, G. 1964. Contributions to knowledge about alpine acoperite soil microflora in association with *hardetum* strictae from the Bucegi massif. *Studii tehn. şi econ.*, C, (12), 67–93 (in Rumanian).

Moore, A. W. 1963. Occurrence of non-symbiotic nitrogen-fixing micro-organisms in Nigerian soils. *Plant and Soil*, 19 (3), 385–395.

Moore, A. W. and Becking, J. H. 1963. Nitrogen fixation by *Bacillus* strains isolated from Nigerian soils. *Nature*, 198, 915–916.

Nemeth, G. 1959. A new nitrogen-fixing micro-organism producing a red pigment. *Nature*, 183, 1460–1461.

Nemeth, G., Uresch, F., Foder, G. and Lang, L. 1961. Chemical character of the pigments in a new nitrogen-fixing micro-organism. *Nature*, 191, 1413–1414.

Newton, J. D. 1954. Microbial maintenance of nitrogen in western Canada's grey wooded, black earth and brown prairie soils. in *Trans. Fifth Internat. Congr. Soil Sci.*, 3, 76–87.

Nikolaeva. 1914. Characterization of some Actinomycetes. *Arkhiv biol. nauk*, 18 (3).

Norris, D. O. 1960. *Notes on the classification of the family Azotobacteriazeae*. Glasgow, Dept. Bacteriol.

Novák, B. and Dvořákova, H. 1955. Fixation of atmospheric nitrogen by actinomycetes. *Shor. Českosl. acad. zeměd. věd., Rostl. výroba*, 28 (3–4), 257–259.

Novikova, N. S. 1955. Ability to fix atmospheric nitrogen in bacterial representatives of the epiphytic microflora. *Mikrobiologiya*, 24 (6), 705–709.

Okuda, A. and Yamaguchi, M. 1960. Nitrogen-fixing micro-organisms in paddy soils. 6. Vitamin B_{12} activity in nitrogen-fixing blue-green algae. *Soil and Plant Food*, 6 (2), 76–85.

Okuda, A., Yamaguchi, M. and Kobayashi, M. 1961. Nitrogen-fixing micro-organisms in paddy soils. 7. *Soil Sci. and Plant Nutr.*, 7 (3), 115–118.

Omelyanskii, V. L. 1923. *Svyazyvanie atmosfernogo azota pochvennymi mikroorganizmov* (Fixing of atmospheric nitrogen by soil micro-organisms). Petrograd.

Panosyan, A. K. 1945. A new nitrogen-fixing species of *Actinomyces. Dokl. Akad. Nauk ArmSSR*, (2), (3), 89–93.

Panosyan, A. K., Arutyunyan, R. Sh. and Tarayan, Sh. S. 1961. Effect of the relationships of certain soil bacteria on nitrogen assimilation in the cultivation of various farm crops. 1: in *Mikrobiologicheskii sbornik AN Arm SSR* (Microbiology handbook of the Academy of Sciences of the Armenian SSR), (2), 219–222.

Parinkina, O. M. 1960. Some information on the oligonitrophilic micro-organisms of podzols: in *Sb. Rabot Tsentr. muzeya pochvovedeniya*, 3.

Paul, E. A. and Newton, J. D. 1961. Studies of aerobic non-symbiotic nitrogen-fixing bacteria. *Canad. J. Microbiol.*, 7 (1), 7–13.

Peklo, M. 1910. Die pflanzlichen Aktinomykosen. *Zbl. Bakteriol., Parasitenkunde. Infektionskrankh und Hyg.*, Abt. 2, **27**, 451.

Pengra, R. M. 1959. Nitrogen fixation and hydrogen metabolism of *Aerobacter aerogenes. Dissert. Abstr.*, **19** (11), 2719–2720.

Pengra, R. M. and Wilson, P. W. 1958. Physiology of nitrogen fixations by *Aerobacter aerogenes. J. Bacteriol.*, **75**, 21–25.

Perotti, K. 1906. Su una nova specie di bacterii oligonitrofili. *Ann. Bot.*, **4**, 213.

Petersen, E. J. and Esther, H. 1964. On nitrogen fixation in Danish deciduous forests. *Kgl. veterin. og landbohojskole ausskr.*, 209–226 (in English).

Petrova, E. A. 1959. Sources of nitrogen nutrition for purple sulphur bacteria. *Dokl. Akad. Nauk SSSR*, **126** (5), 1100–1102.

Pichinoty, F. 1960. Reduction assimilative du nitrate par les cultures aerobies d'*Aerobacter aerogenes. Folia microbiol. ČSR*, **5** (3), 165–170.

Pine, M. J. and Barker, H. A. 1954. Studies on the methanobacteria XI. Fixation of atmospheric nitrogen by *Methanobacterium omelianski. J. Bacteriol.*, **68** (5), 589–591.

Plotho, O. von. 1940. Beitrage zur Kenntnis der Morphologie und Physiologie Actinomyceten. *Arch. Microbiol.*, **11B** (1), 33–72.

Pratt, D. C. and Frenkel, A. W. 1959. Studies on nitrogen fixation and photosynthesis of *Rhodospirillum rubrum. Plant Physiol.*, **34** (3), 333–337.

Pringsheim, H. 1909. Über die Verwendung von Cellulose als Energie quelle zur Assimilation des Luftstickstoffs. *Zbl. Bakteriol.*, Abt. 2, **24**, 488.

Proctor, M. H. and Wilson, P. W. 1958. Nitrogen fixation by Gram-negative bacteria. *Nature.*, **182**, 891.

Proctor, M. H. and Wilson, P. W. 1959. Nitrogen fixation by *Achromobacter* spp. *Arch. Mikrobiologie*, **32** (3), 254–260.

Pshenin, L. N. 1964. Nitrogen-fixing bacteria of the near-surface layer of the Black Sea. *Trudy Sevastopol'sk. biol. stantsii, Akad. Nauk UkrSSR*, **15**, 3–7.

Pshenin, L. N. 1965. *Vidovoi sostav i raspredelenie azotfiksivuyushchikh mikroorganizmov v vode Chernogo morya. Plankton Chernogo i Azovskogo morei* (Species composition and distribution of nitrogen-fixing micro-organisms in the waters of the Black Sea. Plankton of the Black and Azov seas). Naukova dumka, Kiev, 3–9.

Pshenin, L. I. 1966. *Biologiya morskikh azotfiksatorov* (Biology of marine nitrogen fixers). Naukova dumka, Kiev.

Rakhno, P. Kh. and Tokhver, V. I. 1957. Possibility of assimilation of molecular nitrogen at a temperature of 50°C by individual soil bacteria. *Dokl. Akad. Nauk SSSR*, **112** (1), 144–145.

Remy, C. 1902. Stickstoffbindung durch Leguminosen. *Verhandl. ges. Dtsch. Naturf. und Aerzte*, **74** (1), 221.

Richards, E. H. 1917. The fixation of nitrogen in faeces. *J. Agric. Sci.*, **8**, 299–311.

Rodina, A. G. 1956. Aquatic spirillae fixing molecular nitrogen. *Mikrobiologiya*, **25** (2), 145–149.

Rodina, A. G. 1964. Nitrogen-fixing bacteria in the grounds of mangrove thickets in the Gulf of Tonkin. *Dokl. Akad. Nauk SSSR*, **155** (6), 1437–1439.

Romanova, A. P. 1961. *K mikrobiologii ozera Baikal: sezonnaya dinamika chislennosti bakterii i protsessov krugovorota azota v vodnoi tolshche i gruntakh yuzhnogo Baikala* (Microbiology of Lake Baikal, seasonal changes in the numbers of bacteria and the processes of nitrogen turnover deep in the water and in the soils of South Baikal) Author's abstract of M.Sc. Thesis, Irkutsk.

Ross, D. J. 1958. Biological studies of some tussock grassland soils. 5. Non-symbiotic nitrogen-fixing bacteria. *N.Z. Agric. Res.*, **1**, 958–967.

Ross, D. J. 1960a. A note on the occurrence of non-symbiotic nitrogen-fixing bacteria in some introduced-pasture soils. *N.Z. J. Agric. Res.*, **3** (2), 245–249.

Ross, D. J. 1960b. Biological studies of some tussock grassland soils XVI. Non-symbiotic nitrogen-fixing bacteria of two cultivated soils. *N.Z. J. Agric. Res.*, 3 (2), 224–229.

Roy, A. B. and Mukherjee, M. K. 1957. A new type of nitrogen-fixing bacterium. *Nature*, 180, 236.

Roy, A. B. and Subir Sen. 1961. A new species of *Derxia*. *Nature*, 194, 604–605.

Rubenchik, L. I., Bershova, O. I., Smalii, V. T., Zinov'eva, Kh. G., Andreyuk, E. I. Mal'tseva, N. N., Knizhnik, Zh. P., Kozlova, I. A., Smirnova, E. N. and Kogan S. B. 1966. Nitrogen-fixing micro-organisms of the soils of the Ukrainian SSR: in *IX Mezhdunarodyi Mikrobiologicheskii Kongress. Tezisy dokladov* (Ninth International Microbiology Congress. Summaries of speeches), Moscow, 299–300.

Ryys, O. 1963. Distribution of free-living nitrogen-fixing bacteria in soddy podzolic soils of the Esthonian SSR. *Izv. Akad. Nauk EstSSR, Ser. Biol.*, 12 (4), 274–281.

Sadler, W. R. and Stanier, R. Y. 1960. The function of acetate in photosynthesis by green bacteria. *Proc. Nat. Acad. Sci., USA*, 46 (10), 1328.

Schanderl, H. 1955. Zur Frage der Bindung des atmosphärischen Stickstoffs durch Hefen. *Brauwissenschaft*, (7), 157–159.

Schaposchnikov, W. N., Kondrateyva, E. N. and Fédoróv, W. D. (Shaposhnikov, U. N. Kondrat'eva, E. N. and Federov, V. D.). 1960 A new species of green sulphur bacteria. *Nature*, 187, 167.

Schneider, K. C., Bradbeer, C., Singh, R. N., Wang Li Chuan, Wilson, P. W. and Burris, R. H. 1960. Nitrogen fixation by cell-free preparations from micro-organisms. *Proc. Nat. Acad. Sci. USA*, 46 (5), 726–733.

Schollenberger, C. J. 1930. Effect of leaking natural gas upon the soil. *Soil Sci.*, 29, 261–266.

Senn, G. 1928. The assimilation of the molecular nitrogen of the air by lower plants especially by fungi. *Biol. Rev., Cambridge Philos. Soc.*, 3 (1), 77–91.

Silver, W. S., Centifanto, J. M. and Nicholas, D. J. D. 1963. Nitrogen fixation by the leaf-nodule endophyte of *Psychotria bacteriophila*. *Nature*, 199, 396–397.

Sisler, D. and Zobell, C. E. 1951. Nitrogen fixation by sulfate reducing bacteria indicated by nitrogen/argon ratio. *Science*, 113, 511–512.

Skalon, I. S. 1965. Microbiological characterization of the rhizosphere of plants in natural communities of dry steppe and desert-steppe sub-zones of Central Kazakhstan. *Trudy Bot. in-ta Akad. Nauk SSSR*, Ser. 3, 17, 151–169.

Skinner, C. E. 1928. The fixation of nitrogen by *Bacterium aerogenes* and related species. *Soil Sci.*, 25, 195–205; *J. Bacteriol.*, 15, 9–17.

Smyk, B. 1966. Fixation of atmospheric nitrogen by strains of *Arthrobacter*: in *IX Mezhdunarodnyi Mikrobiologicheskii Kongress, Tezisy dokladov* (Ninth International Microbiology Congress. Summaries of speeches), Moscow, 288–289.

Smyk, B. and Ettlinger, L. 1963. Recherches sur quelques espèces d'*Arthrobacter* fixatrices d'azote isolées des roches karstiques Alpines. *Ann. Inst. Pasteur*, 105 (2), 341–348.

Stapp, C. 1940. *Azotomonas insolita* ein neuer aerober stickstoffbinderder Mikroorganismus. *Zbl. Bakteriol., Parasitenkunde, Infektionskrankh und Hyg.*, Abt. 2, 102, 1–19.

Stoklasa, J. 1895. Studien über die Assimilation elementaren Stickstoffes durch die Pflanzen. *Landwirtsch. Jahrb.*, 24, 827.

Stutzer, A. 1904. Die Nutzbarmachung des Stickstoffs der Luft fur die Pflanze. *Landw. Presse*, (10–12), 17, 19.

Sushkina, N. N. 1965. Role of micro-organisms in the primary soil-forming process. Preliminary communication. *Vestn. Moskovsk. un-ta, ser. biol., pochvoved.* (3) 72–75.

Tagawa, K. and Arnon, D. J. 1962. Ferrodoxins as electron carriers in photosynthesis and in the biological production and consumption of hydrogen gas. *Nature*, **195**, 537–543.

Tchan, Y. T. and Jensen, H. L. 1954–1963. Studies of nitrogen-fixing bacteria. 8. Influence of N-content of the media on the N-fixation capacity and the colony variation of *Derxia gummosa* Jensen *et al.* (1960). *Proc. Linnean Soc. N.S. Wales*, **88**, 379–385.

Tchan, Y. T. and Pochon, J. 1950. Une espèce nouvelle de bactérie fixatrice d'azote moleculaire isolée du sol *Endosporus azotophagus* n. sp. *C.R. Acad. sci.*, **30**, 422–430.

Teplyakova, Z. F. 1966. The microbial population of the soils of Kazakhstan: in *Mikroflora pochv yuzhnoi chasti SSSR* (The microflora of the soils of the southern part of the USSR), Moscow, Nauka, 189–225.

Ternetz, An. 1904. Assimilation des freien Stickstoffs durch einen torfbewohnenden Pilze. *Ber. Dtsch. bot. Ges.*, **22**, 267.

Ternetz, An. 1907. Über die Assimilation des atmospharischen Stickstoffs durch Pilze. *Jahresb. wiss.*, *Bot.*, **44**, 353.

Tesić, Z. P. and Todorović, M. S. 1958. Investigation of specific properties of silicate bacteria. *Zemljiste Biljka*, (1–3), 233–239.

Truffaut, G. 1923. Une nouvelle bacterie du sol: le bacillus Truffaut, fixateur aerobe d'azote atmospherique. *Sci. sol.*, **2**, 3.

Truffaut, G. and Bezssonoff, N. 1922. Un nouveau bacille fixateur d'azote. *C.R. Acad. sci.*, **175**, 544.

Turk, L. M. 1935. Studies of nitrogen fixation in some Michigan soils. *Michigan State Coll. Agric. Stat. Bull.*, **143**, 1–36.

Van Niel, C. B. 1955. Natural selection in the microbial world. *J. Gen. Microbiol.*, **13** (1), 201.

Velankar, N. K. 1955. Bacteria in the inshore environment at Madapam. *Indian J. Fish.*, **2** (1), 96–112.

Verona, O. and Picci, G. 1961. Sul potere azoto-fissatiore de *Candida pulcherrima*. *Ann. Fac. agrar. Univ. Pisa*, **22**, 233–234.

Vinogradskii, S. N. 1945. Presumed saphrophytic nitrogen fixers. in *Mikrobiologiya pochvy* (Soil microbiology), Izd-vo Akad. Nauk SSSR, 1952, 535–548.

Voets, J. B. and Debacker, J. 1956. *Pseudomonas azotogensis* nov. sp. a new free living nitrogen-fixing bacterium. *Naturwissenschaften*, **43** (2), 40–41.

Volpino. 1905. Sopra un interessante microorganismo radunatore d'azote isolato dal terreno. *Riv. igiene e sanita publica*, **16**.

Voznyakovskaya, Yu. M. 1959. Importance of nitrogen fixation in the soil by non-symbiotic micro-organisms. *Agrobiologiya*, (1), (115), 37–48.

Waksman, S. A. 1932. *Principles of soil microbiology*, 2nd ed. Baltimore, Williams Co.

Wall, J. B., Wagenknecht, A. C., Newton, J. W. and Burris, R. 1952. Comparison of the metabolism of ammonia and molecular nitrogen in photosynthesizing bacteria. *J. Bacteriol.*, **63** (5), 563.

Wilson, P. W. 1951. Biological nitrogen fixation: in *Bacterial physiology*, chap. 14. Ed. by Werkman, C. H. and Wilson, P. W., Academic Press, New York.

Wilson, P. W. 1952. The comparative biochemistry of nitrogen fixation. *Advances Enzymol.*, **13**, 345.

Wilson, T. G. and Roberts, E. R. 1954. Attempts at nitrogen fixation *in vitro*. *Nature*, **174**, 795.

Yoch, D. C. and Pengra, R. M. 1964. Magnesium requirements of *Aerobacter aerogenes* for assimilation of molecular and combined nitrogen. *J. Bacteriol.*, **88** (3), 808–809.

Yocum, C. S. 1960. Nitrogen fixation. *Annual Rev. Plant Physiol.*, **11**, 25.

Zarma, M. 1959. Molecular nitrogen-fixing micro-organisms in the Black Sea. II.

Some micro-organisms not belonging to the genus *Azotobacter* in the Black Sea. *Lucrările Sesiunii ştiinţe A. vol. festiv.*, 567–573.

Zhukova, G. A. 1956. Microbiological study of the virgin soils of the Kol'skii peninsula. *Mikrobiologiya*, **24** (5), 569–576.

Zikes. 1909. Über eine den Luftstickstoff assimilierende Hefe: *Torula wiesneri. Sitzungsber. Akad. Wiss. Wien. Math. naturwiss. Kl.*, **108**, 1091.

9 The Chemistry of the Fixation of Molecular Nitrogen by Micro-organisms

Possible Pathways

The unravelling of the chemistry which underlies the activation and subsequent conversion of molecular nitrogen might lead to new methods of manufacturing nitrogenous compounds and, above all, to cheaper mineral nitrogenous fertilizers. It is therefore not surprising that the chemistry of nitrogen fixation is studied widely in research institutes and industrial laboratories.

'Just as the theoretical study of the mechanism of the flight of birds led to the construction of a flying machine heavier than air,' wrote A. N. Bakh (1950), 'we hope that a theoretical study of the concerted action of biological redox catalysts responsible for the binding of atmospheric nitrogen by bacteria will indicate the most favourable conditions for the commercial synthesis of ammonia.'

The atoms in the nitrogen molecule (N_2) are joined by three bonds, with an energy of 225 kcalories/mole. The first bond has the highest energy of 127 kcalories/mole, the second has an energy of 60 kcalories/mole and the third 38 kcalories/mole. The nitrogen molecule is inert and does not form compounds with elements or other substances, with the exception of metallic lithium, with which it forms lithium nitride (Li_3N), at room temperature and under atmospheric pressure.

The inertness of nitrogen accounts for the need to activate it and to increase the chemical activity of the compound or elements that it is reacting with. Ways of activating nitrogen include ionization (Zacharias, 1962; Peive, 1965; Ivanov et al., 1965) and dissociation (Peive, 1965).

The synthesis of ammonia from atmospheric nitrogen by the Haber process requires a temperature of about 500°C and high pressure (about 350 atm). The cells of micro-organisms, however, carry out this process in ordinary conditions.

Ammonia can be synthesized in ordinary conditions by prolonged (up to 2 weeks) cathode reduction of nitrogen (or reduction under the action of $SnCl_2$) in the presence of aqueous solutions of molybdate and tungstate ions (Haight and Scott, 1964). But no more than 0·001 mole of ammonia per litre is formed.

Considerable amounts of energy are expended when nitrogen is fixed and it is natural to assume that the energy exchange of nitrogen fixers somehow differs from that of non-nitrogen-fixing saprophytic organisms. The available evidence is in complete agreement with this, and shows that the intensity of the energy exchange in nitrogen fixers, particularly with regard to the degree of consumption of molecular oxygen, is very high—appreciably higher than for other micro-organisms. Thus, in one hour *Azotobacter* cells consume 4,000–5,000 μl. of oxygen per mg dry weight as against only 194–467 μl for *Pseudomonas aeruginosa*, 5–147 μl for *E. coli* and 3–58 μl per mg for *Brucella abortus*. Oxygen is also taken up briskly by nodule bacteria—1,000–2,000 μl in 1 hour per mg dry weight (Kretovich and Lyubimov, 1964). The uptake of oxygen by the cells must in no way be confused with the concept of aerobicity of the medium. As we shall show, the higher the degree of aerobicity of the medium, the higher the degree of inhibition of the process of nitrogen fixation.

Kretovich and Lyubimov also pointed out that the enzymes of nitrogen fixers are more active than the corresponding enzymes of saprophytic micro-organisms. They draw this conclusion from the considerably higher resistance of enzymes of nitrogen fixers to inhibitors compared with the corresponding enzymes of saprophytic micro-organisms.

Of course the energy exchange of nitrogen-fixers is closely connected with the fixation of nitrogen but, unfortunately, the relationship has been insufficiently studied so far. Grin (1961) showed that the energy released in respiration is stored by *Azotobacter* in the form of ATP in structures that are identical to mitochondria. Future investigations must establish how this energy is used during nitrogen fixation.

Since the discovery of nitrogen-fixing micro-organisms, there have been various ideas about the pathways involved in the fixation of molecular nitrogen. But the true mechanism is still not known. The current hypotheses, like earlier ones, attempt to a greater or lesser extent to explain the course of nitrogen fixation on the basis of knowledge of the intermediate products, and the catalytic mechanisms acting at each stage in the process. We shall outline current ideas by making a brief analysis of the widespread views. We shall consider the various hypotheses in the order that they were put forward, paying particular attention to the characterization of intermediates. Then we shall deal with the most realistic catalytic mechanism for the possible pathways of nitrogen fixation. Here too we shall analyse a series of hypotheses. Inevitably there will be some over-lapping of material in the two discussions.

In critically summing up all the hypotheses, Kretovich and Lyubimov (1964) produced a hypothetical scheme for nitrogen fixation, which they thought, in the first stages, could proceed in one of two directions—by a reductive pathway or an oxidative one. Each pathway has many stages and is catalysed by a special enzyme system. The first pathway is the more likely, and we shall consider the arguments that support it.

It is known that in a medium containing developing nitrogen fixers, both anaerobic and aerobic, quite a low redox potential is set up. This suggests that nitrogen fixation is a reductive process. It is also known with some degree of

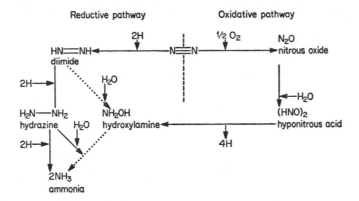

reliability that oxygen above a certain concentration sharply inhibits nitrogen fixation. This is understandable, for oxygen is a vigorous hydrogen acceptor and depresses the formation of the reduced products containing nitrogen. According to Meyerhof and Burk (1928), *Azotobacter* stops assimilating molecular nitrogen if the content of oxygen in the atmosphere reaches 60 per cent. Nitrogen is most vigorously bound by *Azotobacter* when the concentration of oxygen is 1 per cent. A calculation shows that nitrogen fixation ceases if the ratio of the soluble forms of nitrogen and oxygen reaches 1:3.

According to Bergersen (1966) nitrogen fixation in crushed nodules occurs only in the presence of oxygen. A similar phenomenon has been noted for other species of nitrogen fixers, but the critical level of oxygen obviously differs. Thus, for example, *Bacillus polymyxa* ceases to assimilate nitrogen in 10 per cent oxygen. In such conditions this bacterium can multiply on ammonium salts. *Aerobacter aerogenes* actively fixes nitrogen only in anaerobic conditions, but can grow even in the presence of oxygen if it is on ammonium salts (Hamilton and Wilson, 1955).

Oxygen has often been observed to suppress nitrogen fixation (for example, Wilson, 1952; Fedorov, 1952; Hamilton and Wilson, 1955). The degree of depression of nitrogen fixation by oxygen depends largely on the nature of the carbon source (Dilworth and Parker, 1961).

It is most probable that the chemistry of nitrogen fixation, if not identical, is similar in aerobic and anaerobic micro-organisms. The recognition of the possibility of primary oxidation of molecular nitrogen by aerobic micro-organisms would lead to the inevitable conclusion that nitrogen fixation by aerobes and anaerobes proceeds along different pathways.

There is convincing indirect evidence that nitrogen fixation is accomplished by reduction in the high productivity of the process in anaerobic bacteria. As a rule, for every gram of carbon source consumed *Clostridium pasteurianum* assimilates four to seven times less nitrogen than *Azotobacter* and other aerobic micro-organisms. But when carbohydrates are oxidized the yield of energy is forty-five times greater than in butyric acid fermentation, and so the anaerobes

13

fix six to ten times more nitrogen per unit energy received than do the aerobes.

An important argument in support of a reductive pathway for nitrogen fixation is the detection of reduced compounds in cultures and cell-free preparations of many nitrogen fixers. Thus, labelled ammonia is produced by cell-free preparations of nitrogen fixers containing nitrogen-13 and nitrogen-15 (for example Allison and Burris, 1957; Nicholas and Fisher, 1960a, 1960b; Carnahan and Castle, 1963). In certain conditions, in particular in a young *Azotobacter* culture, the presence of ammonia can be established (for example, Kostychev *et al.*, 1926; Burris and Wilson, 1946; Zelitch *et al.*, 1951; Kurdina, 1957).

Virtanen (1947) detected hydroxylamine in the first stage of development of cultures of nitrogen fixers, both on free and on bound nitrogen. Later, when there was a good growth of culture, hydroxylamine disappeared. But when growth is weak it can be observed for a long time. This compound is very mobile, but all the same may be formed during nitrogen fixation.

Another important fact is that several presumed intermediates of the reduction of molecular nitrogen, such as hydroxylamine (in the form of the oxime of α-ketoglutaric acid) and hydrazine, are assimilated by cultures of nitrogen fixers (Kretovich and Lyubimov, 1964; Bundel' *et al.*, 1966). The incorporation of hydroxylamine into proteins by *Azotobacter vinelandii* has been established by the isotope method (Bundel' *et al.*, 1966). This is very important, for many investigators have denied the possibility of assimilation by nitrogen fixers either of hydroxylamine (Burk and Horner, 1935; Fedorov, 1945; Rosenblum and Wilson, 1951) or hydrazine (for example, Blom, 1931; Burk and Horner, 1935), considering them to be highly toxic to micro-organisms.

The possibility that hydroxylamine is one of the intermediate metabolites of nitrogen fixers is also indicated by the detection of hydroxylamine reductase in cell-free enzyme preparations of *Azotobacter* that catalyse the reduction of hydroxylamine to ammonia (Kretovich *et al.*, 1966).

But hydroxylamine is a strong inhibitor of electron transfer in the respiratory chain and of the hydrogen donor system involving $DNPH_2$, as experiments with *Azotobacter* and *Chromatium minutissima* have shown (Ivanov *et al.*, 1965; Ivanov and Demina, 1966).

Bach (1957) established that *Azotobacter* synthesizes cyclic derivatives of hydrazine of the azine type as a result of reaction with α-ketoglutaric acid.

Hydrazine and hydroxylamine may be intermediates in the formation of ammonia but they may also be connected with keto acids and conversion, through several stages, to amino acids. Reacting with α-ketoglutaric acid, hydrazine (Nicholas, 1963) is converted to glutamine through dihydropyridazinon-5-carboxylic acid.

At the same time the primary oxidation of a molecule of nitrogen is unlikely in the first stages of nitrogen fixation (oxidation pathway) for several reasons. Nitrous oxide (N_2O), which in this case must be the first product of the oxidation of nitrogen, is poorly assimilated by free-living and symbiotic nitrogen fixers.

Even in low concentrations it is relatively toxic for micro-organisms. When hydrated N_2O gives hyponitrous acid (HON — NOH) or nitramide (H_2N — NO_2) which is not utilized at all by nitrogen fixers (Rosenblum and Wilson, 1950, 1951). Attempts to find oxidized nitrogen compounds (N_2O, N_2O_3, NO_2, NO_3) were also unsuccessful (Beijerinck and van Delden, 1902; Kellermann and Smith, 1914; Kostychev et al., 1926). And so we can understand why most investigators consider nitrogen fixation to be a reductive process.

But this conclusion was the last of a chain of different hypotheses. At the turn of the century hypotheses were put forward about the primary oxidation of the nitrogen molecule (for example Gautier and Drouin, 1888; Remy and Rösing, 1911). The crux of these hypotheses was the initial oxidation of molecular nitrogen, subsequent reduction of the oxidized compounds to ammonia and utilization of the latter for amino acid synthesis. The catalysts of oxidation were presumed to be iron compounds. A distinctive point of view was put forward in 1909 by Loew and Aso, who believed that the nitrogen molecule was initially hydrolysed. They regarded fixation of nitrogen as the reverse of the decomposition of ammonium nitrite:

$$N \equiv N + 2H_2O \rightleftarrows NH_4NO_2$$

The NH_4^+ cation may enter directly into the synthesis of amino acids while the NO_2^- anion is utilized in synthetic processes after being reduced. Calculations show that such a reaction requires a large expenditure of energy and very high pressures—about 10^{+51} (L'yuis and Rendall, 1936). Loew and Aso's theory cannot therefore be substantiated, which is also true for the view of Angeli (1908), who assumed that the hydrolysis of nitrogen results in the formation of dihydroxyhydrazine which is transformed to hydroxylamine and dihydroxyammonia. These assumptions involving oxidation or hydrolysis of nitrogen have no experimental foundation.

Vinogradskii (1894) was the first to propose that molecular nitrogen is reduced when it is fixed. He assumed that the product was ammonia, believing that in the bacterial cell activated nitrogen interacts with activated hydrogen. According to Vinogradskii, activation of nitrogen is accomplished by the enzyme nitrogenase. He concluded (1894, p. 340) with the interesting sentence: 'To sum up in a few words I would say that in our case the phenomenon appears to be the result of the collision of gaseous nitrogen with hydrogen generated within the protoplasm and maybe the direct result of this would be the synthesis of ammonia'.

Like earlier hypotheses, this is marred by the absence of experimental proof. But it fits the more reduced conditions present in the nitrogen-fixing microorganisms. It is also upheld by several of the points we have already discussed.

Vinogradskii's hypothesis was supported by several scientists (for example Stoklasa, 1908; Kostychev et al., 1926a and b; Burris, 1956). Most later hypotheses developed Vinogradskii's assumption, describing in more detail the scheme of the process. Most hypotheses newly advanced were supported by experimental evidence obtained for individual stages of the process.

At present there is a whole group of hypotheses on the reductive pathway of assimilation of atmospheric nitrogen. In turn, this group can be divided into several subgroups depending on the nature of the intermediate products assimilated directly by the cell.

Some investigators (for example Vinogradskii, 1894; Kostychev et al., 1926a and b) consider that before molecular nitrogen is utilized by the cell it must have been bound in ammonia. These are the advocates of the 'ammonia' hypotheses (first subgroup). But there is no reason why ammonia should necessarily be the primary product. Its formation may be preceded by several other compounds.

Blom (1931), Virtanen (1947) and others are the authors of the 'hydroxylamine' hypotheses (second subgroup). The third subgroup comprises the 'hydrazine' reductive hypotheses (for example, Fedorov, 1945; Bach, 1957; Yocum, 1960). The authors of these hypotheses do not consider that hydroxylamine or hydrazine are necessarily the primary product, but believe that the assimilation of hydroxylamine (or hydrazine) by the cells occurs directly, bypassing the ammonia stage.

Variants of the 'ammonia' hypotheses have been proposed by Wieland (1922), Wilson and Burris (1947, 1953) and Gapon (1947). Wieland (1922) considered molecular nitrogen to be initially reduced to the diimide, after which it becomes hydrazine, which on reduction gives ammonia.

$$N \equiv N + 2H \rightarrow HN = NH + 2H \rightarrow H_2N - NH_2 + 2H = 2NH_3$$

Wilson and Burris (1947) experimenting with nitrogen-15, also concluded that diimide was the initial product of nitrogen fixation and that hydrazine was formed afterwards. The reduction of nitrogen is brought about, in their view, by the hydrogen of the radicals.

The diimide is very unstable and rapidly destroyed, forming hydrazine and nitrogen. When nitrogen is fixed the diimide can hardly exist in the free form, and evidently links up with organic substances.

Blom (1931) considered it unlikely that either the diimide or hydrazine were formed. According to his hypothesis, nitrogen is first hydrolyzed with consequent formation of dihydroxyhydrazine. Its subsequent reduction leads to the formation of hydroxylamine.

$$N\equiv N \ + \ 2H_2O \rightleftharpoons HO-\overset{\overset{\displaystyle H}{|}}{N}-\overset{\overset{\displaystyle H}{|}}{N}-OH \ + \ 2H \rightleftharpoons HO-\underset{\underset{\displaystyle H}{|}}{\overset{\overset{\displaystyle H}{|}}{N}}-H \ + \ H-\underset{\underset{\displaystyle H}{|}}{\overset{\overset{\displaystyle H}{|}}{N}}-OH$$

Hydroxylamine, considered to be one of the first products during the assimilation of molecular nitrogen, is a chemically active compound which is readily reduced to ammonia and oxidized to give several compounds (nitrous oxide, nitrogen dioxide, nitrogen trioxide). Hydroxylamine forms oximes with organic compounds containing the $C = O$ group (ketones, aldehydes). Amines and amino acids readily form from oximes.

Blom suggested that the assimilation of molecular nitrogen was a process similar to the utilization of molecular oxygen in respiration and probably due to the action of an iron-containing catalyst. This must be in the form of the more reactive oxide. Such a catalyst would be related to the haemes and would form unstable compounds with nitrogen. But because molecular nitrogen is inert, substances such as oxygen, ammonia, and nitrates would couple more readily with the catalyst and would impede nitrogen fixation.

Virtanen (1947), who worked mostly on symbiotic nitrogen fixation, Gapon (1947) and others thought, like Blom, that the bivalent iron of the haeme enzymes was very important in the activation of molecular nitrogen. Blom and Virtanen considered that iron atoms took part in the synthesis of hydroxylamine, while Gapon thought they were involved in the synthesis of ammonia. The principal tenets of the Gapon hypothesis are as follows: first, molecular nitrogen is adsorbed by an enzyme of the haeme type, that contains iron ($Fe^{++} \rightleftharpoons Fe^{+++}$) and is activated. The rate of this adsorption determines the rate of formation of ammonia from active nitrogen, which occurs practically instantaneously. Reduction of nitrogen is accomplished by molecular hydrogen. The adsorption of nitrogen by the enzyme (E) leads to the formation of an active nitride containing surplus energy equal to the sum of the activation and adsorption energies and the heat of adsorption.

$$E + N_2 + E \rightarrow EN{=}NE$$

At this stage Gapon postulated the simultaneous participation of two molecules of the haem enzyme located at a definite distance from each other. Gapon did not explain how the enzymes in the cell get into such a 'position'. Next in the presence of hydrogen, the nitride is reduced either directly to ammonia, or through the diimide, and the enzyme is regenerated.

$$E—N\!\!=\!\!N—E + H_2 \rightarrow E—NH—NH—E$$
$$E—NH—NH—E + H_2 \rightarrow 2E—NH_2$$
$$E—NH_2—1/2H_2 \rightarrow E + NH_3$$

When there is a hydrogen deficit the nitride may be hydrolysed with consequent formation of diimide: the latter is hydrated to hydroxylamine.

$$E—N—N—E + H_2O \rightarrow EOH + HN\!\!=\!\!NH$$
$$HN\!\!=\!\!NH + 2H_2O \rightarrow 2NH_2OH$$

Thus, according to this theory, hydroxylamine is an abnormal product in nitrogen fixation.

The weak point of Gapon's theory is that it ignores the fact that the cell is alive. He considers all the stages of nitrogen fixation on the basis of physico-chemical laws.

The 'hydrazine' hypotheses include those of Fedorov (1952) and Bach (1957). Fedorov considered that in nitrogen-fixing bacteria assimilation of molecular nitrogen is catalysed by a special two-component enzyme. Specific proto-plasmic proteins act as a colloidal carrier, and the active group provided by a compound which contains carboxyl, amino and two adjacent carbonyl groups. This compound is apparently a derivative of, and similar to, diketoglutaric acid.

Different colloidal carriers are assumed to take part in the two different stages of the process: proteins with first a high and then a low oxidative potential. The interaction of molecular nitrogen with the carbonyl groups is a result of the breaking of the double bonds between carbon and oxygen and the formation of others between the carbon atoms and between the nitrogen atoms.

After the attachment of nitrogen it is assumed that there is a change in the colloidal carrier, which is then reduced by hydrogen released by dehydrogenases from the oxidizable organic substance. During this reduction hydrazine deriva-tives form.

The nitrogen from these compounds is accepted by the keto acids and reduced to amino acids.

(a) R—C=C—R   R—C=O   R—C=C—R$_1$ R—C=N—NH$_2$
       | | | | | |
      OHO   + COOH + OHOH   +   COOH   + H$_2$O
      |
      HN—NH$_2$

(b) R—C=N—NH$_2$ R—OHNH$_2$
       | + 2 (2H) | + NH$_3$
      COOH   COOH

As this scheme shows, ammonia and hydroxylamine form only as secondary reaction products.

More precise thermodynamic calculations (Karnaukhov 1965) have shown that the primary reactions of Fedorov's scheme are energetically detrimental, so that they are unlikely to exist. Fedorov's theory was also criticized by Wilson (1952). In any case, it is unlikely that fixation of molecular nitrogen is catalysed by a single enzyme.

In Bach's scheme (1957) hydrazine is also an intermediate product. A major role in the activation of molecular nitrogen is assigned to polyvalent metals, as in the schemes of Blom, Virtanen and Gapon. Bach's scheme is possible both for free-living and symbiotic nitrogen fixers.

The hypotheses we have just considered have been critically evaluated in thermodynamic and kinetic terms by Karnaukhov (1965), who decided that any hypothesis is untenable that depends on the activation of nitrogen by metals (bivalent, trivalent and tetravalent states) forming part of the active groups of enzymes. This is because the activation of nitrogen by metals cannot in normal conditions ensure breakage of even one bond in the nitrogen molecule. This requires high temperature (or other conditions) incompatible with life.

Karnaukhov also found kinetically unrealistic the mechanism, proposed by Wilson and Burris, of reduction of nitrogen through radicals because the formation of NH=NH requires the simultaneous collision of a molecule of nitrogen and two radicals. The probability of such collisions occurring in ordinary conditions is remote.

Karnaukhov advanced his own scheme of nitrogen fixation, postulating that nitrogen-fixing micro-organisms have catalytic systems that activate molecular nitrogen (Fig. 36). In his view nitrogen fixation begins with the chemisorption of the nitrogen molecule at special, chemically-active centres. This is followed by the activation of nitrogen. Winfild (1951) and Mortenson (1962) also considered that the chemisorption of nitrogen on the surface of a catalyst is necessary for activation.

According to Karnaukhov, the active centres of the catalyst are oxaloacetic and, in some cases, α-ketoglutaric acid fixed on the protein structure by means of two of their own carboxyl groups and two carboxyl groups of the protein molecules and also two atoms of polyvalent metals (Fig. 36, scheme 1). This attachment may also be accomplished through the sulphydryl and amino groups. The metal of the catalyst is apparently molybdenum, which is capable of forming strong covalent bonds with carboxyl and sulphydryl groups in the active centre. Iron and cobalt may also be important in the catalytic scheme.

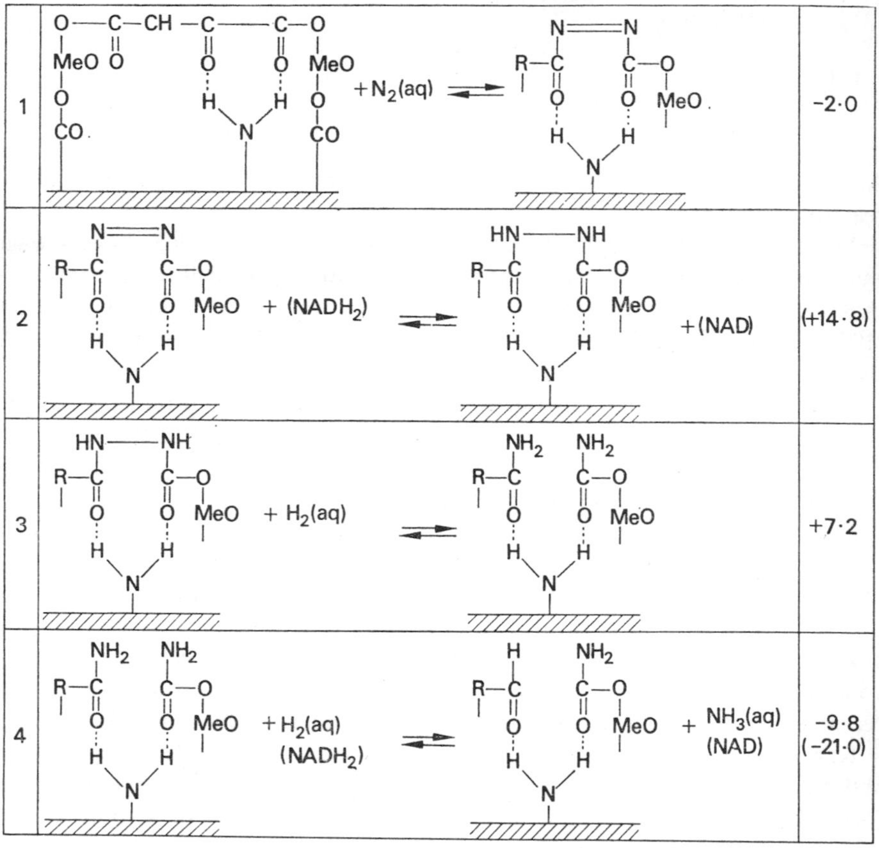

Fig. 36. Variants of active centres and their interaction with nitrogen and hydrogen. $\Delta H°$, kcalories/mole. The values of $\Delta H°$ are given on the right. The shaded portion denotes the protein moiety. The broken line denotes hydrogen bonds; R, radical.

Chemisorption of nitrogen occurs as a result of breakage of a bond in the nitrogen molecule and a bond between the carbonyl and carboxyl carbon atoms of the α-keto acid, with the subsequent formation of an azo compound. The change in the enthalpy amounts to 2 kcal/mole, which means that the process is possible.

When the azo compound is reduced by specific $NADH_2$ (reduced nicotinamide adenine dinucleotide) a hydroazo compound forms (Fig. 36, scheme 2).

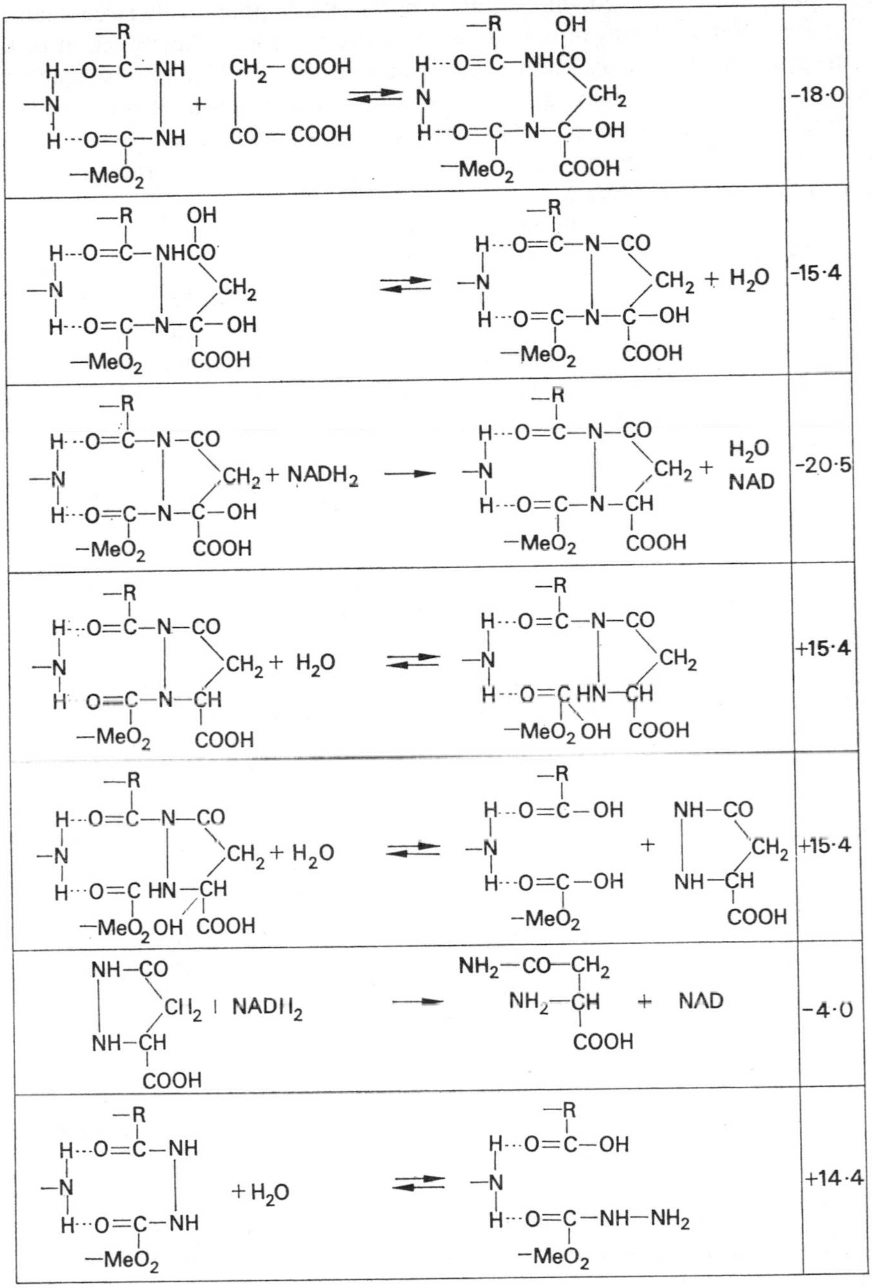

Reaction	ΔH°
$\begin{array}{c} -R \\ H\cdots O{=}C{-}NH \\ -N \qquad + \\ H\cdots O{=}C{-}NH \\ -MeO_2 \end{array} \quad \begin{array}{c} CH_2{-}COOH \\ \\ CO{-}COOH \end{array} \quad \rightleftharpoons \quad \begin{array}{c} -R \quad OH \\ H\cdots O{=}C{-}NHCO \\ N \qquad CH_2 \\ H\cdots O{=}C{-}N{-}C{-}OH \\ -MeO_2 \quad COOH \end{array}$	$-18\cdot0$
$\begin{array}{c} -R \quad OH \\ H\cdots O{=}C{-}NHCO \\ -N \qquad CH_2 \\ H\cdots O{=}C{-}N{-}C{-}OH \\ -MeO_2 \quad COOH \end{array} \quad \rightleftharpoons \quad \begin{array}{c} -R \\ H\cdots O{=}C{-}N{-}CO \\ -N \qquad CH_2 + H_2O \\ H\cdots O{=}C{-}N{-}C{-}OH \\ -MeO_2 \quad COOH \end{array}$	$-15\cdot4$
$\begin{array}{c} -R \\ H\cdots O{=}C{-}N{-}CO \\ -N \qquad CH_2 + NADH_2 \\ H\cdots O{=}C{-}N{-}C{-}OH \\ -MeO_2 \quad COOH \end{array} \quad \rightarrow \quad \begin{array}{c} -R \\ H\cdots O{=}C{-}N{-}CO \\ -N \qquad CH_2 + \begin{array}{c}H_2O\\NAD\end{array} \\ H\cdots O{=}C{-}N{-}CH \\ -MeO_2 \quad COOH \end{array}$	$-20\cdot5$
$\begin{array}{c} -R \\ H\cdots O{=}C{-}N{-}CO \\ -N \qquad CH_2 + H_2O \\ H \quad O{=}C{-}N{-}CH \\ -MeO_2 \quad COOH \end{array} \quad \rightleftharpoons \quad \begin{array}{c} -R \\ H\cdots O{=}C{-}N{-}CO \\ -N \qquad CH_2 \\ H\cdots O{=}C \; HN{-}CH \\ -MeO_2 \; OH \; COOH \end{array}$	$+15\cdot4$
$\begin{array}{c} -R \\ H\cdots O{=}C{-}N{-}CO \\ -N \qquad CH_2 + H_2O \\ H\cdots O{=}C \; HN{-}CH \\ -MeO_2 \; OH \; COOH \end{array} \quad \rightleftharpoons \quad \begin{array}{c} -R \\ H\cdots O{=}C{-}OH \qquad NH{-}CO \\ -N \qquad + \qquad CH_2 \\ H\cdots O{=}C{-}OH \qquad NH{-}CH \\ -MeO_2 \qquad\qquad COOH \end{array}$	$+15\cdot4$
$\begin{array}{c} NH{-}CO \\ \qquad CH_2 + NADH_2 \\ NH{-}CH \\ COOH \end{array} \quad \rightarrow \quad \begin{array}{c} NH_2{-}CO{-}CH_2 \\ NH_2{-}CH \quad + \; NAD \\ COOH \end{array}$	$-4\cdot0$
$\begin{array}{c} -R \\ H\cdots O{=}C{-}NH \\ -N \qquad\qquad + H_2O \\ H\cdots O{=}C{-}NH \\ -MeO_2 \end{array} \quad \rightleftharpoons \quad \begin{array}{c} -R \\ H\cdots O{=}C{-}OH \\ -N \\ H\cdots O{=}C{-}NH{-}NH_2 \\ -MeO_2 \end{array}$	$+14\cdot4$

Fig. 37. Formation of asparagine at the active centres of oxaloacetic acid
(ΔH°, kcalories/mole).

This process is associated with the absorption of a comparatively large quantity of energy (14·8 kcalories/mole) and is therefore impeded. The reduction of the hydroazo compound may lead to the formation of two amide groups which are then converted to ammonia (Fig. 36, schemes 3 and 4). Monosubstituted hydrazine may also form and this is reduced to ammonia and hydroxylamine. When the bound nitrogen is split off, the catalyst may be regenerated by oxidation with oxygen and NAD.

Amino acids might also be produced in the forward reaction of the hydroazo compounds with keto acids. Heterocyclic compounds would form in such a case. Figure 37 is an example of the numerous possible pathways of formation of aspartic acid from oxaloacetic acid by this means. These reactions for oxaloacetic acid can be applied to other keto acids (α-ketoglutaric, pyruvic and so on). Thus, from pyruvic acid is formed α-alanine and from α-ketoglutaric acid, glutamic acid is formed.

In Karnaukhov's view, the catalyst combines nitrogenase and hydrogenase activity in one structural centre. This is unlikely, however, especially since Mortenson et al. (1962) obtained separate fractions with these activities from cell-free preparations of Cl. pasteurianum.

Reviewing available material, Nicholas (1963) proposed the scheme for molecular fixation, shown in Fig. 38. In this scheme the conversion of nitrogen is principally reductive.

Turning to consider nitrogen fixation in nodule bacteria, we recall that they do not fix molecular nitrogen in pure culture. However, some investigators have been able to achieve an increase in the content of nitrogen by cultivating Rhizobium in bacteriological media, but it was insignificant (Fedorov, 1952). This work involved symbionts of non-leguminous plants. For this reason nitrogen fixation in symbiotic micro-organisms is usually studied in the nodules and not in pure culture (Krasheninnikov, 1916; Aprison et al., 1954; Magee and Burris, 1956).

The mechanism of nitrogen fixation in the nodules undoubtedly has certain special features that distinguish it from the process in free-living micro-organisms. First, two partners with quite different properties function in unison. Second, leghaemoglobin is always present, undoubtedly taking some part in the process of fixation. Third, in the symbiotic system bacteroids form, apparently associated with nitrogen fixation. Nevertheless, the development of ideas about the mechanisms of nitrogen fixation has been largely similar for both free-living and symbiotic micro-organisms, often without reference to these special features.

Many investigators (Vinogradskii, 1895; Blom, 1931; Fedorov, 1952) considered that their schemes for free-living nitrogen fixers were also applicable in essence to symbiotic nitrogen fixers. We shall consider the principal propositions of these hypotheses. Others, reflecting to differing degrees the features of a symbiotic system, will be presented later.

Finding that the nodules of leguminous plants released amino acids, especially aspartic acid and alanine, Virtanen and Laine (1936) assumed that leghaemoglobin was converted to less oxidized forms in the nodules and itself converted molecular nitrogen to hydroxylamine. This hypothetical first phase, criticized

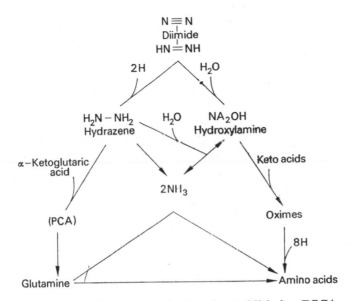

Fig. 38. Scheme for nitrogen fixation due to Nicholas, DPCA, dihydropyridazinone-5-carboxylic acid:

Dihydropyridazinone–
5–carboxy acid

by Fedorov (1952, p. 455), can be expressed as N_2 + leghaemoglobin (Fe^{+++}) $\rightleftharpoons NH_2OH$ + leghaemoglobin (Fe^{++}).

Next, in Virtanen and Laine's scheme, hydroxylamine condenses with the oxaloacetic acid formed in the plant by the metabolism of carbohydrates to give

oximes (> C=NCH) and hydroxamic acids $\left(-C \diagup^{NHCH)}_{\diagdown O} \right)$, which readily

form aspartic acid.

This scheme can be expressed as:

$$N_2 \rightarrow N_2{}^+ \xrightarrow[\substack{\text{transfer of} \\ \text{electrons by} \\ \text{Fe-containing} \\ \text{catalytic} \\ \text{system}}]{+ O_2 + H_2} \text{oxides } N_2 \longrightarrow NH_2OH \underset{\substack{\text{oxalo-} \\ \text{acetic} \\ \text{acid}}}{\overline{\qquad\qquad}} (\ldots NH_3)$$

$$\longrightarrow \text{oximes} \xrightarrow{\text{reduction}} \text{aspartic acid} \xrightarrow[\longleftarrow]{\text{reamination}} \text{glutamic acid}$$

In this scheme, the function of the plant host in the symbiotic system is principally to provide an acceptor for hydroxylamine and then to produce oxaloacetic acid.

There is no unanimity of opinion as to the nature of the primary compounds formed in the nodules as a result of the reaction with atmospheric nitrogen. Even if investigators recognize that these are amino compounds they disagree about their composition.

Thus Virtanen and Laine consider that aspartic acid predominates in the nodules with alanine, which they consider to be a product of the conversion of aspartic acid, taking second place.

Turchin et al. (1963) used nitrogen-15 to show that labelled nitrogen in the nodules of leguminous plants is incorporated mostly into asparagine and to a lesser extent into aspartic and glutamic acids. Asparagine was found in the nodules of leguminous plants by Sen and Burma (1953). Aprison et al. (1954) using soya and Bond (1963) using *Myrica* found that nitrogen-15 was incorporated mostly into glutamine.

Recently, Kennedy (1966) found that nodules of seradella kept in an atmosphere of nitrogen-15 for 45 seconds incorporated the isotope mostly into ammonia, and then glutamic acid, and then glutamine. Considerably less nitrogen-15 went into aspartic acid, alanine and asparagine. According to Kennedy, these amino acids are of secondary origin. When nodules were kept for 5 minutes in the same conditions, the amount of isotope incorporated into amino acids increased in direct proportion to the increased time. Accumulation of ammonia ceased after 5 minutes. In Kennedy's view this gas is utilized in the synthesis of amino compounds.

The presence of glutamic acid, alanine, aspartic acid and other amino compounds among the first products of nitrogen fixation in the nodules of leguminous plants has been noted by Aprison et al. (1954) and Aseeva et al. (1966) among others.

In 1963 Virtanen and Miettinen produced a more developed scheme for nitrogen fixation, but it has the same major features: hydroxylamine reacts with α-keto acids leading to the formation of oximes which are then converted to amino acids. Reamination ultimately gives the full range of amino acids needed by the plant.

For a long time Demolon (1951) worked on the chemistry of nitrogen fixation by the symbiotic system, principally developing Vinogradskii's ammonia hypotheses in relation to the symbiotic situation. In his view the dehydrogenases of the nodule bacteria activate hydrogen, after which the activated hydrogen is attached to atmospheric nitrogen to give ammonia. At this stage the host plant supplies keto and unsaturated acids, and aspartic acid and its amide, and other amino acids are formed.

The essence of Turchin's hypothesis (1959) is that the nodule bacteria penetrate the roots of leguminous plants and secrete some unknown substance—factor B (Fig. 39). This induces the formation of nodular (bacteroid) tissue on

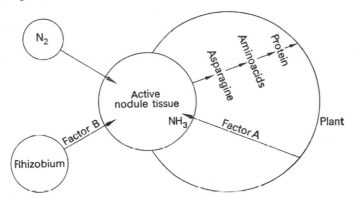

Fig. 39. Turchin's scheme of nitrogen fixation.

the surface of which atmospheric nitrogen is fixed. Nitrogen fixation, however, is feasible only if the nodular tissue is supplied with a special substance—factor A—synthesized by the host leguminous plant. The nature of this factor is unknown. It may consist of carbohydrates or some other substances. One of the first products of fixation is ammonia, which is rapidly transformed to the amide group of asparagine. This amino acid, the form in which ammonia is transported through the plant, is used in the synthesis of other amino acids which are subsequently formed into protein, of both the host and the nodule bacteria.

Bergersen's hypothesis (1966a and b) is much more developed and refined than Turchin's. He started (1966) from the premise that plant and bacteria function as one unit and that the electron transport link between them is leghaemoglobin. Bergersen thought that nitrogen was fixed in the cells of the host which contain small groups of bacteroids surrounded by membranes. A group of bacteroids surrounded by a membrane was considered to be a 'nitrogen fixation unit'. A model of such a nitrogen-fixing system is given in Fig. 40, which shows the primary reactions in the activation of molecular nitrogen and its reduction to ammonia occurring in a sphere with the membrane as its surface.

The host plant is a source of carbon-containing compounds which are broken down to provide energy-rich compounds needed for the activation and reduction of nitrogen. Activated nitrogen is the end acceptor in an electron transport chain which begins with the bacteroids and includes leghaemoglobin. The products of incomplete oxidation of the carbon-containing compounds serve as acceptors of ammonia, and in the bacteroids, amino acids are formed which become accessible to the higher plant.

Bergersen's scheme satisfies the principal requirements of symbiotic nitrogen

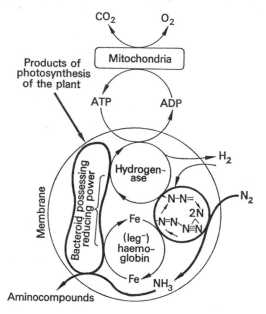

Fig. 40. Bergersen's scheme of nitrogen fixation.

fixation. The plant acts as the donor of carbon-containing compounds—products of photosynthesis—and the source of energy in the form of ATP from the mitochondria. The nodule bacteria in the bacteroid stage reduce leghaemoglobin and the ATP-dependent hydrogenase system. The path from nitrogen to ammonia is regarded as reductive, involving leghaemoglobin and ATP-dependent hydrogenase at different stages.

From all the schemes of nitrogen fixation we have considered, the following conclusion can be drawn. Nitrogen fixation carried out by micro-organisms is a reductive process consisting of several sequential reactions. Each stage must be catalysed by a special enzyme or group of enzymes. Ammonia seems to be the first stable product, but it is possible that compounds yet undetected fill this role, or that the partially reduced forms of nitrogen are not free but fixed at their active centres. Nor should we exclude the possibility that the formation of nitrogen-containing organic substances bypasses ammonia altogether. If that is so ammonia may be a secondary product of the reactions.

Keto acids undoubtedly play a major role during nitrogen fixation. Reacting with hydroxylamine, they form oximes, and are then transformed into amino acids. Keto acids may also link up with ammonia to give amino acids.

Some keto acids, for example pyruvic and α-ketoglutaric acids, are the components of the phosphoroclastic reaction coupled with nitrogen fixation, and also take part in the generation of ATP for nitrogen fixation and are good sources of energy for bacteria. According to Kretovich and Lyubimov the keto-acids are the 'crossroads' at which the two most important pathways of metabolism meet—the exchange of energy and the exchange of nitrogen. It is also possible that certain intermediate compounds of the metabolism of these acids influence nitrogen fixation (Lyubimov, 1963).

In the cells of nitrogen-fixing micro-organisms (*Azotobacter vinelandii*, *Mycobacterium azot-absorptum*) Kretovich *et al.* (1964, 1966) found several keto acids besides pyruvic—hydroxypyruvic, oxaloacetic, α-ketoglutaric, β-methylbutyric, α-keto-β-methylvaleric (in *Azotobacter vinelandii*) the hemi-aldehyde of succinic acid and glyoxylic acid (in *Mycobacterium azot-absorptum*). Some of these substances had not been detected in nitrogen fixers before. Nitrogen fixers produce considerably more of some keto acids (for example α-ketoglutaric) when developing on molecular nitrogen. In 1936, Virtanen and Laine established the presence of oxaloacetic, α-ketoglutaric and pyruvic acids in pea nodules.

Kretovich *et al.* (1964, 1966) found glutamic dehydrogenase in cells of *Azotobacter vinelandii*. When $NADH_2$ is available this enzyme promotes the formation of glutamic acid from α-ketoglutaric acid and ammonia. In cell homogenates, glutamic dehydrogenase is localized in the subcellular particle fraction which sediments when centrifuged at 105,000 g. *Mycobacterium azot-absorptum* also seems to contain alanine and aspartic dehydrogenases, which explains why ammonia reacts with pyruvic and oxaloacetic acids. Similar enzymes have also been found in the nodules of leguminous plants (Grimes and Fottrell, 1966; Fottrell, 1966). All the amino acids necessary for the micro-organisms can form from glutamic acid and other amino compounds by reamination.

Certain metallic ions are necessary for nitrogen fixation. These are chiefly molybdenum, iron and in part cobalt (Mo^{6+}, Fe^{2+}, Co^{2+}). The requirements for them are considerably higher when the bacteria develop on molecular nitrogen (for example, Shug *et al.*, 1954; Carnahan and Castle, 1958; Peive and Zhiznevskaya, 1961; Il'ina, 1966).

The metals seem to be essential in the chemisorption of molecular nitrogen and possibly also in the subsequent conversions; in particular, they are needed for the functioning of several systems. Electron paramagnetic resonance (Nicholas, 1963) showed that the subcellular particles of *Azotobacter* contain molybdenum, protein-bound iron, flavine and also cytochromes b, c and a.

Protein-bound iron has been found in extracts of *Clostridium pasteurianum* and iron and manganese in cell-free extracts of *Klebsiella*.

Yakovlev *et al.* (1965) pointed out that when cell-free preparations of nitrogen-fixing micro-organisms are fractionated, a parallel is revealed in the contents of Mo and Fe, which confirms the idea that these organisms have enzyme systems in which both metals have the same function.

Cobalt is probably necessary for both free-living nitrogen fixers (in particular *Azotobacter*) and symbiotic nitrogen fixers of leguminous and other plants. This metal forms part of vitamin B_{12} and nitrate reductase. Its role in nitrogen fixation has not been identified.

Nitrogen fixation can be inhibited in certain conditions. Molecular hydrogen is a specific suppressor, although it does not affect the development of micro-organisms on bound nitrogen.

The effect of hydrogen on free-living and symbiotic nitrogen fixers has been much investigated, always with similar results. Except with photosynthesizing nitrogen fixers, inhibition of nitrogen fixation has always been observed (Wilson and Umbreitt, 1937; Wyss *et al.*, 1941; Wilson, 1951). Hydrogen has less effect on anaerobic than aerobic micro-organisms. This may be a consequence of adaptation to hydrogen, for it is released by them (Rosenblum and Wilson, 1951).

The inhibitory effect of hydrogen can be attributed either to inhibition of the formation of $NADH_2$ (hydrogen donor), or to disturbance of the chemisorption of nitrogen, by the adsorption of hydrogen on to the iron atom in the hypothetical enzyme 'nitrogenase' (Ivanov *et al.*, 1965). We shall discuss the role of $NADH_2$ and nitrogenase in the next section.

Nitrogen fixation in *Rhodospirillum* and other photosynthesizing micro-organisms is not suppressed by molecular hydrogen. In this case it may itself act as a hydrogen donor.

Inhibitors of nitrogen fixation include oxygen, which apparently blocks the iron part of 'nitrogenase'. Carbon monoxide is also a suppressant. The supression of cytochrome oxidase requires 10–100 times as much carbon monoxide as the suppression of nitrogen fixation.

The presence in the medium of bound forms of nitrogen, in particular ammonium salts, depresses fixation of molecular nitrogen by micro-organisms. According to Burk (1934), 5–10 mg of nitrogen per 100 ml. of medium arrests the process of nitrogen fixation. Only in an aerobic nitrogen fixer isolated from the soils of Java was nitrogen fixation not depressed in the presence of ammonium salts (Roy and Mukherjee, 1957).

The inhibitory effect of ammonium on nitrogen fixation by *Chromatium minitissimum* was attributed by Ivanov and Demina (1966) to the incorporation of ammonia in the reductive amination of keto acids and also, possibly, to competitive interaction between molecular nitrogen and the ammonium ion, for the iron atom of the cytochromes.

It is also necessary to consider the pressure of molecular nitrogen in nitrogen fixation. The significance of this factor was first indicated by Burk (1934). Later the subject was taken up by several microbiologists (for example, Wyss *et al,*. 1941;

Wilson, 1952; Burris, 1956). The results of their work suggested that nitrogen fixation is controlled by the relation established by Michaelis for enzyme reactions (Michaelis and Rona, 1930; Bresler, 1966). *Azotobacter vinelandii, Nostoc muscorum* and nodule slices fix half the maximum amount of nitrogen when the pressure of the gas is 0·02 atmospheres. This applies to anaerobic micro-organisms (*Clostridium pasteurianum* and *Rhodospirillum rubrum*) when the pressure of nitrogen is 0·2 atmospheres.

Catalytic Bases of Nitrogen Fixation

Until the 1930s the participation of particular catalytic systems in nitrogen fixation was judged either solely on the basis of *a priori* suppositions or, in part, on the basis of individual experiments with cells of nitrogen-fixing micro-organisms. The investigations of Bakh, Ermol'eva and Stepanian (1934) on nitrogen fixation by extracts of cells of *Azotobacter* prompted a new enzymological study of the process.

Investigation of the catalytic systems of cell-free preparations of nitrogen-fixing bacteria have opened up great possibilities for the elucidation of the chemical reactions of nitrogen fixation (for example, Wilson and Burris, 1947; 1953; Virtanen, 1947, 1952; E. Allen and O. Allen, 1958; Pratt, 1962; Carnahan and Castle, 1963; Virtanen and Miettinen, 1963; Lyubimov, 1963).

Cell-free preparations of many micro-organisms have been used in these investigations: they are *Clostridium pasteurianum* (Carnahan *et al.*, 1960), *Azotobacter vinelandii* (Nicholas and Fisher, 1960a and b; Nimeck *et al.*, 1963; Yakovlev *et al.*, 1965), *Rhodospirillum rubrum* (Schneider *et al.*, 1960) *Bacillus polymyxa* (Grau and Wilson, 1961), *Chromatium sp.* (Arnon *et al.*, 1960, 1963) and blue-green algae (Schneider *et al.*, 1960).

As we shall show later, in certain conditions, cell-free preparations of free-living nitrogen fixers fix different amounts of atmospheric nitrogen. It is much more difficult to demonstrate nitrogen fixation directly by nodule bacteria, as in pure culture they do not bind molecular nitrogen, and crushed nodules rapidly lose their nitrogen-fixing powers (Aprison *et al.*, 1954).

Without giving details of the technique for preparing cell-free extracts of nitrogen-fixers we note that it may differ somewhat from investigator to investigator. For example, Carnahan (1960) crushed cells of *Clostridium pasteurianum* in a Hug press or subjected them to autolysis, which was also used by Schneider *et al.* (1960) and Nicholas and Fisher (1960), who exposed cultures of *Azotobacter vinelandii* to ultrasonics at low temperature. Lysozyme can be used to destroy the cells (Grau and Wilson, 1961; Nicholas and Fisher, 1960; Yakovlev and Levchenko, 1964) and so can phage (Nimeck *et al.*, 1963).

Because cell-free preparations are only slightly stable, long experiments are not possible, nor is the fixation of large amounts of nitrogen. And so it became necessary to use labelled nitrogen. The stable isotope nitrogen-15 was first used by Burris and Wilson (1957) and later by many others. Acute experiments using nitrogen-13 (half life 10·05 min) were carried out by Nicholas (1963), studying the cell-free extracts of *Azotobacter vinelandii.*

Reviewing several studies of cell-free preparations, we have tried to assemble for the different species of micro-organisms the current picture of the activation of nitrogen and its further conversions from the point of view of enzyme systems that are involved. We consider first the anaerobic nitrogen fixer *Clostridium pasteurianum.*

Cell-free preparations of *Clostridium pasteurianum* undoubtedly bind molecular nitrogen. Short term experiments with nitrogen-15 have shown that the fixed nitrogen is in the form of ammonia (Carnahan *et al.*, 1960). Attempts to establish the presence of hydrazine, hydroxylamine, nitrite and dihydropyridazinone 5-carboxylic acid was unsuccessful (Lyubimov, 1957, 1963).

Mortenson *et al.* (1962) were able to isolate two fractions from cell-free extracts, one containing a hydrogen donor system (HDS) and the other a nitrogen-activating system (NAS). The HDS had hydrogenase and phosphoroclastic activity and the NAS had nitrogenase. Thus, the HDS incorporates hydrogenase and components (and also cofactors) of the phosphoroclastic system while the NAS incorporates an enzyme (or group of enzymes) which plays a specific role in nitrogen fixation. We shall briefly characterize the components of the HDS and NAS.

Hydrogenase has a prosthetic group consisting of flavine adenine dinucleotide bound to molybdenum and non-haeme iron. The bivalent cobalt probably links the enzyme to hydrogen (for example, Winfild, 1955; Peive, 1965). Hydrogenase activates hydrogen and increases its ability to reduce other substances.

$$H_2 \rightleftharpoons 2H^+ + 2e''$$

When nitrogen fixers develop on bound nitrogen, hydrogenase does not form. It should be remembered that certain species of bacteria, for example *Bacterium coli*, possess hydrogenase but do not fix molecular nitrogen.

One of the components of the phosphoroclastic reaction of the hydrogen donor system is pyruvic (possibly α-ketoglutaric) acid, H_3PO_4, and also pyruvic dehydrogenase. The phosphoroclastic reaction proceeds with the formation of acetyl phosphate, hydrogen and carbon dioxide:

$$CH_3COCOOH + H_3PO_4 \rightleftharpoons CH_3COO (H_2PO_3) + CO_2 + H_2$$

Nitrogen fixation by cell-free preparations of *Clostridium pasteurianum* is coupled with the phosphoroclastic reaction, as Carnahan *et al.* (1960) showed by measuring the effect of different concentrations of sodium pyruvate on this system. Figure 41 shows that optimum nitrogen fixation is closely linked to the uptake of pyruvic acid by cell-free preparations (Fig. 42). It has been established that for 100 molecules of pyruvic acid expended, one molecule of nitrogen is

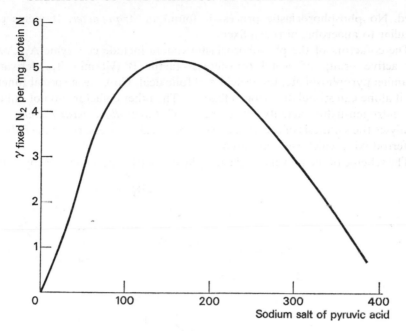

Fig. 41. The effect of sodium pyruvate on nitrogen fixation by cell-free extract of *Cl. pasteurianum* (after Carnahan et al.).

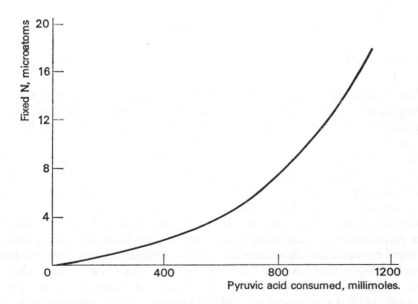

Fig. 42. Relation between nitrogen fixation and uptake of sodium pyruvate in cell-free preparations of *Clostridium pasteurianum* (after Carnahan et al.).

fixed. No phosphoroclastic process is found in *Azotobacter*. It seems to be peculiar to anaerobic nitrogen fixers.

The cofactors of the phosphoroclastic system include coenzyme A (CoA)—the active group of acetyl transferase—factor B (vitamin B_{12} derivative), thiamine pyrophosphate, lipoic acid and folic acid. CoA has a special function, for it alone can stimulate nitrogen fixation. The other cofactors do not influence the nitrogen-fixing activity of extracts of *Clostridium pasteurianum*. CoA catalyses the synthesis of acetyl phosphate and is also necessary for the reduction of ferrodoxin, which we shall discuss later.

The scheme of the phosphoroclastic splitting of pyruvic acid by nitrogen-fixing

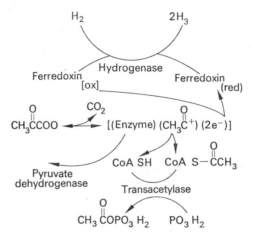

Fig. 43. Phosphoroclastic splitting of pyruvic acid (see text).

extracts of *Clostridium pasteurianum* (Mortenson, 1964, 1966) is shown in Fig. 43.

The NAS includes an enzyme (or group of enzymes)—so called hypothetical nitrogenase—activating the nitrogen molecule. The existence of nitrogenase in the cells of nitrogen fixers was first postulated by Vinogradskii (1894). The NAS has neither hydrogenase nor phosphoroclastic activity.

Winfild's view (1955) was that the active centre of nitrogenase represents two hydrogenases 'turned towards each other' by iron atoms, the distance between which corresponds to the length of the molecule of nitrogen ($\sim$0·3–0·4 nm).

Nitrogen can be fixed only when the HDS and NAS act together. Table 9.1 shows the effects of catalysts on nitrogen fixation by different fractions of cell-free preparations of *Clostridium pasteurianum* (after Mortenson *et al.*, 1962).

When cells of *Clostridium pasteurianum* are cultivated in a medium containing bound nitrogen, they have the HDS but no NAS, which is active only when the cells are grown on molecular nitrogen. Thus 'nitrogenase' can be induced (Wilson, 1952; Lyubimov *et al.*, 1965; Lyubimov, 1957). Purification can increase the activity of the NAS 121 times (Dua *et al.*, 1964, 1965).

Table 9.1. The Different Factors as Catalysts of Nitrogen Fixation.

Test specimen	Content of protein nitrogen (mg)	Nitrogen fixed (mg/hour)
Cell-free preparation	18·6	233
NAS (nitrogen activating system)	14·8	0
HDS (hydrogen donor system)	10·8	0
NAS + HDS	14·8 + 10·8	106

Electron carriers, in particular nicotinamide-adenine dinucleotide (NAD)—a hydrogen acceptor activated by hydrogenase—are very important for nitrogen fixation by *Clostridium pasteurianum*. NAD may accept hydrogen through an unknown carrier X. Reduced NAD (DPNH$_2$) is the principal donor of protons and electrons when activated nitrogen is reduced. Consequently NAD occupies an intermediate position between the HDS and the NAS. But it is quite in order to consider NAD as a component of the HDS. The formation of NADH$_2$ may occur also through the reduction of NAD by hydrogen transferred by dehydrogenases from the tricarboxylic cycle (Ivanov *et al.*, 1966).

The link between pyruvic dehydrogenase and hydrogenase is ferrodoxin (F$_D$), an electron-transporting, non-haeme, iron-containing protein.

Pyruvic dehydrogenase $\rightarrow$ F$_D$ $\rightarrow$ hydrogenase $\rightarrow$ H$_2$.

Here F$_D$ acts as a cofactor of the phosphoroclastic system. The HDS catalyses the conversion of pyruvic acid only in the presence of F$_D$. The introduction of ferredoxin into the medium increases the amount of acetyl phosphate synthesized (Fig. 44).

Ferredoxin also catalyses the reduction of nitrates in enzyme preparations of *Clostridium pasteurianum* and takes part in the reversible conversion of hypoxanthine to xanthine and in the reduction of formic acid, etc.

Ferredoxin was discovered and isolated from cell-free extracts of *Clostridium pasteurianum* by Mortenson *et al.* (1962). Although it stimulates the release of hydrogen during the phosphoroclastic reaction, its direct link with nitrogen fixation still cannot be considered proven (Ivanov *et al.*, 1965). The molecular weight of ferredoxin is 5,600–6,000 (Lovenberg *et al.*, 1963; Mortenson, 1962). Its molecule contains seven iron atoms and seven molecules of inorganic labile sulphide which form part of the active centre (Blomstrom *et al.*, 1964). The protein part of the molecule of ferredoxin consists of fifty-five amino acid residues.

Ferredoxin has been obtained in crystalline form (Lovenberg *et al.*, 1963; Valentine, 1964). Its redox potential (E$_0^1$) at pH 7·1 is very close to that of the hydrogen electrode (-420 mV). Ferredoxin has oxidized and reduced forms

(Tagawa and Arnon, 1962); solutions of the oxidized form have an absorption spectrum with maxima at 390, 300 and 280 nm.

It has been suggested that the fixation of molecular nitrogen by extracts of *Clostridium pasteurianum* requires high energy phosphate (McNary and Burris, 1962). This would mean that adenosine triphosphate (ATP) was involved, together with ferredoxin and hydrogen (Dilworth *et al.*, 1965; Hardy *et al.*, 1964, 1965).

The generation of ATP requires acetyl phosphate; reductants such as $KBrH_4$

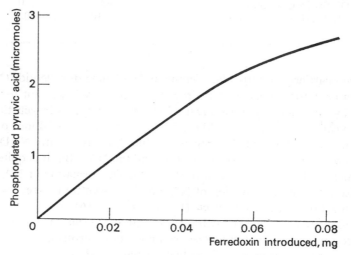

Fig. 44. Catalysis of the phosphoroclastic reaction in extracts of *Clostridium pasteurianum* by ferredoxin (after Mortenson *et al.*).

and $Na_2S_2O_4$ will not do. The idea is that ATP first activates nitrogenase, which then links up with nitrogen.

The need for ATP and ferredoxin to initiate nitrogen fixation by extracts of *Clostridium pasteurianum* has been confirmed by several experiments (d'Eustachio and Hardy, 1964; Mortenson *et al.*, 1965). ATP can be replaced by an ATP-generating system consisting of creatine phosphate, creatine phosphokinase and ADP.

Mortenson (1964) proposed the scheme shown in Fig. 45 for the participation of ferredoxin and ATP in nitrogen fixation.

This scheme reflects the dual role of pyruvic acid as precursor of acetyl phosphate, which is a supplier of phosphate for the formation of ATP, and as a reductant of ferredoxin. According to this scheme, reduced ferredoxin and ATP take part in the formation of active 'nitrogenase', ATP activating the reduced and not the oxidized form of 'nitrogenase'. This view of the function of ferredoxin was shared by d'Eustachio and Hardy (1964).

On the basis of their experimental results, Hardy *et al.* (1965) proposed a scheme for the part of nitrogen fixation which ensures the flux of electrons for the reduction of nitrogen (Fig. 46). According to this scheme hydrogen reduces the presumed electron carrier X to X_{red}, through hydrogenase and ferredoxin. X may also be reduced by $Na_2S_2O_4$ without the participation of hydrogenase

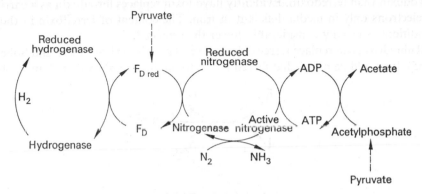

Fig. 45. Mortenson's Scheme (see text).

and ferredoxin. Next in the scheme of Hardy *et al.*, ATP is involved in the conversion of $X_{red *}$ to the activated form X_{red}. Subsequent oxidation of the reduced compound is accompanied either by release of hydrogen or reduction of nitrogen (which is chemisorbed on 'nitrogenase'). The hypothetical electron carrier (X) would be an iron-containing protein possessing paramagnetic properties (Hardy *et al.*, 1965).

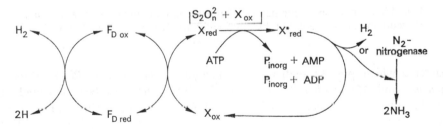

$X^* =$ activated form of X

Fig. 46. Scheme of Hardy *et al.* (see text).

Ferredoxin is assigned a different role by Hardy *et al.* from that in Mortenson's scheme (1964). In the later scheme it does not take part in the activation of 'nitrogenase', for with $Na_2S_2O_4$ as a source of electrons, nitrogen fixation is catalysed by extracts devoid of ferredoxin.

From the cells of *Clostridium pasteurianum* grown in the presence of ammonia or nitrogen but with very little iron, Knight and Hardy (1966) isolated a protein compound—flavodoxin—which is involved in electron transport.

Flavodoxin can replace ferredoxin in extracts of *Clostridium pasteuranium*, taking on the function of primary acceptor of the acetyl group of pyruvic acid, ensuring thus the formation of acetyl phosphate, the evolution of hydrogen and the fixation of nitrogen. But these processes are less vigorous in the presence of flavodoxin than ferredoxin. Evidently flavodoxin replaces ferredoxin as a carrier of electrons only in media deficient in iron. The content of ferredoxin in these conditions is usually considerably lower than normal.

Rubredoxin can replace ferredoxin in several reactions (Lovenberg and Sobell, 1965). It should be noted that ferredoxin is found in several species of anaerobic

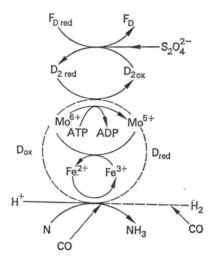

Fig. 47. Mortenson's scheme for the reduction of nitrogen to ammonia. Ferredoxin can be replaced by dithionite; hydrogen forms in the absence of nitrogen.

bacteria which possess hydrogenase (including *Chromatium*, *Rhodospirillum rubrum*, and *Methanobacillus omelianskii*.) But it has not been detected in aerobic micro-organisms. Yakovlev *et al.* (1966) reported that they had detected non-haeme, iron-containing protein in *Azotobacter vinelandii*. They called this ferridin. These substances must be regarded as representatives of a new group of electron carriers.

In 1966, Mortenson considerably advanced our knowledge of the structure of the hydrogenase of *Clostridium pasteurianum*. There are apparently at least two, possibly three protein components (in Fig. 47 D_1 and D_2) associated with the reduction of molecular nitrogen to ammonia. They form only in cells which are utilizing gaseous nitrogen. D_1 has a molecular weight of about 100,000, and contains in its molecules six atoms of iron and one of molybdenum. D_2 has a molecular weight of about 40,000, and has one or two iron atoms and one or two sulphide groups. The cycle of reductive processes in which these compounds participated was represented by Mortenson in the form shown in Fig. 47.

Many investigations have involved cell-free preparations of aerobic, nitrogen-fixing micro-organisms, in particular *Azotobacter*. After Bakh *et al.* (1934) had demonstrated nitrogen fixation by extracts of *Azotobacter*, he repeated the experiments, and his results were negative. Other results were not at first very convincing (Nason *et al.*, 1957). According to Nicholas (1963) nitrogen-fixing preparations of *Azotobacter* can be obtained by destroying the cells in the culture fluid. Nicholas considered that the culture fluid contained a certain unknown component of the HDS of *Azotobacter*, without which nitrogen fixation cannot proceed.

Attempts to isolate the HDS and NAS from *Azotobacter* cells have not been successful. We can, however, take it that the NAS and HDS also exist in aerobes, although quite strongly modified. The HDS of *Azotobacter* possesses hydrogenase activity but no phosphoroclastic activity. Ivanov *et al.* (1965) considered that there was a direct pathway for the reduction of DPN in which the primary dehydrogenases participated. But hydrogenases too could be involved.

According to Bulen *et al.* (1965) when cells of *Azotobacter* are in medium containing ATP and sodium dithionite ($Na_2S_2O_4$) two hydrogenases can be found, one of which forms only in the absence of bound nitrogen and the other which forms irrespective of its absence. The first hydrogenase catalyses the ATP—dependent release of hydrogen from dithionite and the second the uptake of hydrogen by different acceptors, for example certain dyes, and does not take part in nitrogen fixation. Yakovlev and Levchenko (1966), however, considered this second hydrogenase to be involved in nitrogen fixation too. It is possible that this is correct. In their experiments they used intact *Azotobacter* cells, while Bulen *et al.* used fractions of cell-free preparations, which, as they admitted, might have lost some factor that is necessary to couple the activity of the hydrogenase with the reduction of activated nitrogen.

Hydrogenase is localized in the cytoplasmic membrane and the mitochondroid structures of the cell (Yakovlev and Levchenko, 1966). This is not surprising, for on the one hand the principal components of the electron transport chain and the enzymes of oxidative phosphorylation are localized in the mitochondroid structures and, on the other hand, the reduction of nitrogen requires electrons and a source of energy in the form of ATP.

Yakovlev *et al.* (1964, 1965) established that the mitochondroid fraction of *Azotobacter* contains large deformed 'mitochondria' ($0.7 \times 0.9 \mu m$), smaller 'mitochondria' ($0.2 \times 0.4 \mu m$) and miniature particles with a membranous structure ($0.05 \times 0.02 \mu m$) characterized by high tetrazolium reductase activity.

According to Yakovlev (1966) the greatest nitrogen-fixing activity is in the particles which sediment at $187,000 g$. The fact that such particles are never revealed by cytochemical techniques in undamaged cells strongly suggests that they are enzymes of the 'mitochondria' which appear when the cells are destroyed and fractionated.

A connection between the enzymes which catalyse nitrogen fixation and these very small particles has been noted by several other investigators (for example,

Nicholas and Fisher, 1960; Bulen et al., 1965; 1966). Table 9.2 helps to confirm this connection.

Table 9.2. Nitrogen-Fixing Activity of Various Fractions of
Cell-Free Preparations of Azotobacter vinelandii

Fraction	Amount of nitrogen fixed (%)
Supernatant	
25,000 g	100
144,000 g	40
Sediment	70

Figures represent the percentages of the nitrogen fixed by the supernatant obtained by centrifugation at 25,000 g (from Nicholas and Fisher, 1960).

Nicholas and Fisher found the highest nitrogen-fixing activity in the supernatant of Az. vinelandii obtained by centrifugation at 25,000 g, and the lowest at 144,000 g. In this fraction the number of very small particles was also lowest.

Ivanov et al. (1965) have found active alcohol dehydrogenase in Azotobacter. In their view alcohol dehydrogenase and other primary dehydrogenases, for example aldehyde dehydrogenase (also present in Azotobacter), mediate the direct pathway of reduction of DPN, for no phosphoroclastic activity is found in Azotobacter. Ivanov (1966) considered nitrogen fixation to be a variant of oxygen respiration (oxybiosis) and by analogy with this process, he introduced the terms nitrobiosis—respiration of nitrogen.

This hypothesis assumes that molecular nitrogen, like molecular oxygen, is activated and reduced by the terminal cytochromes of the electron transport chain. The activation of nitrogen and oxygen occurs synchronously at the surface of the cytochromes. It is possible that one of the cytochromes of Azotobacter has the dual function of activating both nitrogen and oxygen, which distinguishes Azotobacter from other aerobic micro-organisms which cannot fix nitrogen. Compared with activated oxygen, activated nitrogen is a less active hydrogen acceptor. Ivanov's hypothesis is shown schematically in Fig. 48.

Briefly, in this scheme, $DPNH_2$ reduces the cytochrome system which then transfers H^+ and e^- to the nitrogen reducing system.

Chemisorption ensures the breaking of two bonds in the nitrogen molecule, with the expenditure of 323 kcalories/mole. The breaking of the third bond ($=N$——$N=$) requires 38 kcalories/mole, and can be affected by penetration of H^+ into the molecule of activated nitrogen. Quarternary nitrogen is assumed to be formed in the last stages of the scheme. The first four stages were borrowed from Mortenson *et al.* (1962, 1963). Ivanov considered this scheme to be applicable to other species of nitrogen fixers, including anaerobic forms.

The enzyme systems of nitrogen fixation in facultative, anaerobic and photosynthetic micro-organisms have been studied in much less detail than those in other micro-organisms. We shall note briefly some of the most interesting aspects of these groups.

Pyruvic acid is essential for nitrogen fixation by facultative, anaerobic bacteria of the *Bacillus polymyxa* type. Its conversion to acetyl phosphate, carbon dioxide and hydrogen shows that these bacteria have a phosphoroclastic system. A detailed study of hydrogenase and nitrogenase activity in *Bacillus polymyxa* was made by Grau and Wilson (1961, 1963).

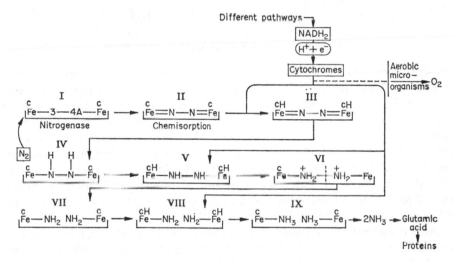

Fig. 48. The structure of nitrogenase is identified with that proposed by Winfild. c = hydrogenase centre.

The HDS of photosynthetic nitrogen fixers comprises hydrogenases, dehydrogenases, the electron carriers DPN and ferredoxin, and also the photosynthetic apparatus (Arnon *et al.*, 1960; Ivanov and Demina, 1966). The product of nitrogen fixation in photosynthetic bacteria, as in *Azotobacter* and *Clostridium*, is ammonia (Wall *et al.*, 1952).

Nitrogen fixation by the photosynthetic *Rhodospirillum rubrum* is stimulated in the presence of α-ketoglutaric acid, especially in the light (Table 9.3).

Pyruvic acid was added to a cell-free preparation of this bacterium and had virtually no effect on the course of the reaction either in the light or in the dark (Table 9.3). The relatively high level of nitrogen fixation in the absence of keto acids is presumably due to endogenous energy stores. To fix nitrogen *Rhodospirillum rubrum* requires ATP or an ATP generating system (Bulen *et al.*, 1965).

There are several conclusions to be drawn from a review of the work with

Table 9.3. Effect of Pyruvic and α-Ketoglutaric acids on Nitrogen
Fixation by Cell-Free Extracts of *Rhodospirillum rubrum*

Substance added	Experimental conditions	Atom % excess $^{15}N_2$
—	Light	0·80
—	Dark	1·04
Pyruvic acid	Light	1·17
—	Dark	1·23
α-Ketoglutaric acid	Light	3·08
—	Dark	1·58

This table is taken from Schneider *et al*. (1960).

cell-free preparations of various species of free-living, nitrogen-fixing micro-organisms. First, the electron flux generated during the exchange of pyruvic acid or the activation of molecular hydrogen is used to reduce molecular nitrogen. Second, iron-containing non-haeme proteins of the ferredoxin type serve as electron carriers at certain stages in the process. Third, nitrogen is fixed only in the presence of ATP or an ATP generating system, but there is still not unanimity as to the mode of utilization of ATP in the reduction of nitrogen. Fourth, it can be assumed that the main pathway of nitrogen fixation is identical in different species of micro-organisms, although the stages seem to differ. Fifth, all the enzymes of nitrogen fixers are inducible.

After considering the enzyme system of nitrogen-fixing bacteria, it is interesting to compare the binding of nitrogen by chemical catalysts. In 1960, Syrkin remarked that chemisorption of several diatomic molecules including nitrogen, occurring with low activation energy on transition metals can be explained by the formation of three-centre bonds: on the one hand the acceptor bonds through the pair of electrons of the paramagnetic bond (π-bond) and the empty p-orbitals of the metal, and, on the other hand, the donor bonds through the pairs of d-electrons of the metal and diffuse orbitals of the non-boundary adsorbed molecule. Such complexing with low activation energy may be of crucial importance in heterogeneous catalysis.

Nitrogen can be activated and made to react at room temperatures and atmospheric pressure if it is complexed with the coordination-unsaturated derivatives of the transition metals. With this in mind, Vol'pin and Shur (1964, 1965) investigated systems formed by the exposure of the salt or complex of a transition metal to the donors of the carbanion (a magnesium or aluminium organic compound) or the hydride ion (lithium alumohydride and sodium borohydride). Such systems readily form paramagnetic complexes with various unsaturated compounds (carbon monoxide, olefines, acetylenes and so on), activate molecular hydrogen and thereby, in several cases, catalyse the most diverse reactions, such as ring-closure, polymerization and hydrogenation (Sloan et al., 1963; Vol'pin and Shur, 1965). The activity of many of these systems is connected with the intermediate formation of more or less stable compounds of the transition metals with metal-carbon or metal-hydrogen bonds. Vol'pin and Shur (1964, 1965) established that several systems formed from derivatives of transition metals also form complexes with the nitrogen molecule, which is simultaneously activated while ammonia is formed. Thus, when nitrogen was passed at a pressure of 1 atmosphere through a mixture of anhydrous $CrCl_3$ and excess $LiAlH_4$ in diethyl ether, after 8–10 hours ammonia formed to the extent of 2 mole per cent of the initial $CrCl_3$ (the yield of ammonia was expressed by Vol'pin and Shur in mole per cent of the initial salt of the transition metal). Chlorides of copper, titanium, iron, cobalt and palladium did not catalyse the formation of ammonia when nitrogen was passed over them, at 1 atmosphere pressure.

Activation of nitrogen has also been achieved using other catalysts, such as a mixture of anhydrous $CrCl_3$ and excess ethereal solution of C_2H_5MgBr (yield of ammonia 17 mole per cent). The greatest yield of ammonia (67 mole per cent) at room temperature and atmospheric pressure was obtained with dicyclopentadienyltitanium dichloride—$(\pi—C_5H_5)_2TiCl_2$ with ethylmagnesium bromide—C_2H_5MgBr in ether (Vol'pin et al., 1964). When the reaction was extended for 10 hours or more and the pressure was increased the yield of ammonia increased considerably. The ability of molecular nitrogen to react with several systems based on compounds of the transition metals has been confirmed using nitrogen-15 (Vol'pin and Shur, 1965).

It is not known how the nitrogen molecule is activated in these systems. The formation of π-complexes of the nitrogen molecule with alkyl derivatives of the transition metal or its hydride probably has something to do with it. Hydrogen may then pass to nitrogen from the alkyl group.

According to Vol'pin and Shur (1964) ammonia is formed in the system when hydrogen from the molecules of the reagents (or solvent) combines with nitrogen, forming a complex incorporating the atom of the transition element. The fixed nitrogen may be released only on hydrolysis in the form of ammonia, and is held meanwhile in the reaction products as the nitride of one of the metals involved in the reaction (Vol'pin and Shur, 1965; Nechiporenko et al., 1965).

Eischens and Jacnow (1964) established the formation of the compound Ni—N≡N as a result of the chemisorption of molecular nitrogen on the surface of nickel. Borod'ko et al. (1966) observed the formation of a similar complex on the surface of cobalt.

It is interesting that the formation of surface compounds of cobalt and nickel with nitrogen is reversible and virtually without expenditure of activation energy. Unlike the paramagnetic complexes $(M \ldots \overset{\text{N}}{\underset{\text{N}}{\|\|}})$[1] these complex compounds have, like diazo compounds, a linear configuration $(M \ldots N{=}N)$ (M is the active particle or active centre on the surface). According to Eischens and Jacnow, hydrogenation of nitrogen occurs on the surface of the metal, with the formation of an unstable compound of the type $Ni\diagup\diagdown\begin{smallmatrix}N\equiv N\\ H\end{smallmatrix}$, which in turn undergoes further regrouping. In this connection Borod'ko *et al.* (1966) assumed that when molecular nitrogen was fixed the active particles could be complex hydrides of the transition metals which were able to coordinate molecular nitrogen.

Work on such schemes, though they are far removed from biological systems, will in the long run help to explain the mechanism of activation of the nitrogen molecule and solve the problem of nitrogen fixation.

Evolution of Nitrogen-fixing Micro-organisms

According to Oparin (1957) the primitive conditions of the earth were characterized by the absence of oxygen and the presence of organic compounds and nitrogen in the bound state. In this setting, the 'pioneers' of life might have become heterotrophic micro-organisms assimilating organic substances and ammonium or nitric acid compounds.

The store of bound nitrogen necessary for the maintenance and development of life was restricted and in time the lack began to limit the growth of organisms (Burris, 1961). At this time the development of the ability to assimilate molecular nitrogen assumed great importance for the survival of micro-organisms. Organisms with this ability were better able to compete than those without it.

It can be supposed that the ability to fix molecular nitrogen appeared as a result of adaptation to particular ecological conditions in representatives of diverse groups of micro-organisms during a long period of evolution (Imshenetskii, 1961). Of course, any assumption about the evolution of nitrogen fixation is clearly *a priori*; nevertheless, certain hypotheses are possible on the basis of general ideas.

According to Imshenetskii the most primitive and apparently the earliest nitrogen fixers were anaerobic organisms—*Clostridium pasteuranium* and closely related species of sporulating anaerobes.

Photosynthetic anaerobic bacteria—nitrogen fixers—came into existence much later. It is logical to assume that the primitive photoautotroph capable of fixing nitrogen was similar in its properties to today's green sulphur bacteria. The requirements of such organisms for hydrogen, hydrogen sulphate or similar hydrogen donors could have been satisfied readily in the conditions then prevailing.

The loss of certain properties by green bacteria could have led to the appearance of the purple sulphur bacteria and from them the purple non-sulphur bacteria like *Rhodomicrobium* and *Rhodopseudomonas*. These have retained the ability to fix nitrogen and to photosynthesize.

Aerobic micro-organisms which fixed nitrogen could have appeared only after the earth's atmosphere had been enriched with oxygen. These were primarily blue-green algae which could have evolved from photosynthetic bacteria. The most remarkable evolutionary change was the appearance of the mechanism for the production of oxygen. The primary process of photosynthesis is the photolysis of water. In the bacteria, hydrogen sulphide and similar donors serve to reduce oxidants. The blue-green algae, however, developed the ability to do without such oxidants as oxygen. In terms of energy this process had enormous advantages over the use of reductants and made possible the development of aerobic forms of life. Nitrogen-fixing, blue-green algae, which at first lived only in water, later moved to dry land and became inhabitants of the soil.

The aerobic, nitrogen-fixing bacteria must also have appeared at this time. The forms transitional between anaerobes and typical aerobes are bacteria such as *Bacillus polymyxa*, which develops in very low concentrations of oxygen, and *Aerobacter aerogenes*, which is a facultative organism. It actively fixes nitrogen in anaerobic conditions and does so considerably more weakly in aerobic ones.

Species of *Aerobacter* can be regarded as possible ancestors of aerobic, free-living nitrogen fixers such as *Azotobacter* and *Beijerinckia*. These forms of free-living, nitrogen-fixing bacteria evidently evolved relatively late.

Imshenetskii (1961) believes that the primitive nitrogen fixers must have used as a carbon source various hydrocarbons or products of their degradations, in particular organic acids. The earth's primary mantle was rich in such compounds, and several chemical substances close to hydrocarbons are utilized by nitrogen fixers; for example, *Azotobacter* grows well in media containing benzoic and hydroxybenzoic acids.

The ability of nitrogen fixers to utilize carbohydrates apparently appeared later when, as a result of photosynthesis, these compounds began to accumulate in the cells of higher plants.

The final stage in the evolution of nitrogen fixation was the appearance of symbiotic relationships between micro-organisms and higher plants. This category contains the symbiosis of various groups of nitrogen fixers with ferns, cycads, leguminous and non-leguminous plants. In some cases, for example in nodule bacteria, nitrogen fixation became possible only with symbiosis because certain obligatory components of the nitrogen-fixing system were formed by the higher plants.

The development of the ability to fix nitrogen in saprophytic micro-organisms is confirmed by the ability of the diverse groups of nitrogen fixers to make use of mineral nitrogen when it is available.

In spite of the possible differences in the origin of various nitrogen-fixing micro-organisms there is considerable similarity between the enzyme systems concerned with the assimilation of nitrogen. Thus, various inhibitors have the same effects on different species of nitrogen-fixing micro-organism: the organisms contain hydrogenase, which is usually characteristic of primitive forms of life; ammonia is formed when nitrogen is fixed by cell-free preparations of different nitrogen fixers. It would be wrong to suppose that there are no significant differences in the structure of nitrogenase (to be more exact, the system of nitrogen-fixing enzymes) and in the reductive system of individual micro-organisms. They must exist and we have endeavoured to show this in outlining the material in the preceding sections.

In conclusion, we recall that many investigators are inclined to regard nitrogen fixation as a distinctive respiratory act identical with oxygen respiration. This view was first advanced by Blom (1931), who postulated that nitrogen is assimilated by catalysts of a haeme nature. Later, a similar view was put forward by others (Parker, 1954; Parker and Scutt, 1960). Ivanov et al. (1965, 1966) suggested that nitrogen is activated on the surface of the cytochromes of aerobic and anaerobic (*Chromatium*) bacteria.

During adaptation to existence in an atmosphere containing oxygen, a certain part of the enzyme complex may have remained unchanged while the rest was modified. Thus, for example, cytochrome is absent from *Clostridium pasteurianum* but present in nitrogen-fixing photosynthesizers. Apparently cytochromes can be involved in nitrogen fixation. Later these organisms may have gone over to respiration. Future investigations will show whether these ideas are correct.

References

Allen, E. K. and Allen, O. N. 1958. Biological aspects of symbiotic nitrogen fixation. in *Handbuch der Pflanzenphysiologie—Encyclopedia of Plant Physiology*. W. Ruhlaud (ed.), **8**, 48–118, Springer, Berlin.

Allison, R. M. and Burris, R. H. 1957. Kinetics of fixation of nitrogen by *Azotobacter vinelandii*, *J. Biol. Chem.*, **224**, 351–364.

Angeli, 1908. *Sammlung chemischer und chemisch-technischer Vortrage*. Bd. 13. *Zusammenfassung der Arbeiten von Angeli von Kurt Arndt.*, 31.

Aprison, T. H., Magee, W. E. and Burris, R. H. 1954. Nitrogen fixations by excised root nodules. *J. Biol. Chem.*, **280**, 22.

Arnon, D. J., Losada, M., Nozaki, M. and Tagawa, K. 1960. Photofixation of nitrogen and photoproduction of hydrogen by thiosulfate during bacterial photosynthesis. *Biochem. J.*, **77**, 23–24.

Arnon, D. L., Losada, M., Nozaki, M. and Tagawa, K. 1961. Photoproduction of hydrogen, photofixation of nitrogen and a unified concept of photosynthesis. *Nature*, **190**, 601–605.

Arnon, D., Tagawa, K. and Tsujimoto, H. 1963. Role of ferredoxin in the energy conversion process of photosynthesis. *Science*, **140**, 378.

Aseeva, K. B., Evstigneeva, Z. G., Kretovich, V. L. 1966. Amino-acid composition of the nodules of the alder, *Phaseolus* and the lupin. *Dokl. Akad. Nauk SSSR*, **169** (2), 463–465.

Bach, M. K. 1957. Hydrazine and biological nitrogen fixation. *Biochim. et Biophys. Acta*, **26**, 104–113.

Bakh, A. N., 1950. *Sobranie trudov po khimii i biokhimii*. (Collected Works on Chemistry and Biochemistry), Izd-vo Akad. Nauk SSSR, Moscow.

Bakh, A. N., Ermol'eva, Z. V. and Stepanian, M. P. 1934. The binding of atmospheric nitrogen at ordinary temperature and pressure via enzymes extracted from nitrogen bacteria. *Dokl. Akad. Nauk SSSR*, **1** (1), 22–24.

Beijerinck, M. W. and Denden van. 1902. Über die Assimilation des freien Stickstoffes durch Bakterien. *Zbl. Bakteriol.*, Abt. 2, **9**, 3–42.

Bergersen, F. 1960. Biochemical pathways in legume root nodule nitrogen fixation. *Bacteriol. Revs*, **24** (2).

Bergersen, F. 1965. Ammonia—an early stable product of nitrogen fixation by soybean root nodules. *Austral. J. Biol. Sci.*, **18** (1), 1–9.

Bergersen, F. 1966a. Some properties of nitrogen-fixing breis prepared from soybean root nodules. *Biochim. et Biophys. Acta*, **130** (2), 304–312.

Bergersen, F. 1966b. Nitrogen fixation in the root nodules of legumes: in *IX Mezhdunarodnyi Mikrobiologicheskii Kongress, Simpozium V-1*. (Ninth International Microbiology Congress, Symposium V1), 69–72.

Blom, J. 1931. Ein Versuch die chemischen Vorgange bei der Assimilation des molekularen Stickstoffs durch Mikro-organismen zu erklären. *Zbl Bakteriol., Parasitenkunde. Infektionskrankh und Hyg.*, Abt. 2, **84**, 60–86.

Blomstrom, D. C., Knight, E., Phillips, W. and Weiner, J. 1964. The nature of iron in ferredoxin. *Proc. Nat. Acad. Sci. USA*, **51** (6), 1085–1092.

Bond, G. 1963. The root nodules of non-leguminous Angiosperms. Symbiotic associations: in 13 *Sympos. Soc. Gen. Microbiol. Cambridge*, 72–92.

Borod'ko, Yu. G., Shilov, A. E. and Shteinman, A. A., 1966. Pathways of activation of molecular nitrogen. *Dokl. Akad. Nauk SSSR*, **168** (3), 581–584.

Bresler, S. E., 1966. *Vvedenie v molekulyarnuyu biologiyu*. (Introduction to molecular biology), Nauka, Moscow-Leningrad.

Bulen, W. A., Burns, R. C. and Le Comte, I. R. 1964. Nitrogen fixation in cell-free, system with extracts of *Azotobacter*. *Biochem and Biophys. Res. Communs*, **17** (3) 265–271.

Bulen, W. A., Burns, R. C. and Le Comte, I. R. 1965. Nitrogen fixation: hydrosulfite as electron donor with cell-free preparations of *Azotobacter vinelandii* and *Rhodospirillum rubrum*. *Proc. Nat. Acad. Sci., USA*, **53** (3), 532–539.

Bundel', A. A., Kretovich, V. L. and Borovikova, I. V., 1966. Assimilation by azotobacters of ^{15}N-hydroxylamine in the form of the oxime. *Mikrobiologiya*, **35** (4), 573–580.

Burk, D., 1934. Azotase and nitrogenase in *Azotobacter*. *Ergeb. Enzümforsch.*, **3**, 23–56.

Burk, D. and Horner, C. K. 1935. Über Hydroxylamin, Hydrazin und Amide als Intermediarprodukte bei der N_2 Fixation durch *Azotobacter*. *Naturwissenschaften*, **23**, 259–260.

Burns, R. C. and Bulen, W. A. 1965. ATP-dependent hydrogen evolution by cell-free preparation of *Azotobacter vinelandii*: *Biochim. et Biophys. Acta*, **105** (3), 537–445.

Burris, R. H. 1956. Nitrogen fixation: in *Atomic Energy Commission. Report No. 7512*. US Govt. Print. Office, Washington.

Burris, R. N. 1961. Evolution of biological nitrogen fixation: in *V Mezhdunarodnyi Biokhemicheskii Kongress—Simpozium III*. (Fifth International Biochemical Congress. Symposium III), Moscow, 28–31.

Burris, R. H. and Wilson, P. W. 1946. Ammonia as an intermediate in nitrogen fixation by *Azotobacter*. *J. Bacteriol.*, **52** (5), 505.

Burris, R. H. and Wilson, P. W. 1957. Methods for measurement of nitrogen fixation:

in *Methods in Enzymology*. **4**, Kaplan N. O. and Colowick, S. (Eds), Academic Press, New York, 355.

Butkevich, V. S. and Kolesnikova, N. A., 1941. Formation of ammonia on fixation of molecular nitrogen by azotobacters. *Dokl. Akad. Nauk SSSR*, **23** (1), 66–69.

Carnahan, J. E. and Castle, J. C. 1958. Some requirements on biological nitrogen fixation. *J. Bacteriol.*, **75** (2), 121–124.

Carnahan, J. E. and Castle, J. C. 1963. Nitrogen fixation. *Annual Rev. Plant Physiol.*, **14**, 105–197.

Carnahan, J. E., Mortenson, K. E., Mower, H. and Castle, J. C. 1960. Nitrogen fixation in cell-free extracts of *Clostridium pasteurianum*. *Biochim. et Biophys. Acta*, **44** (3), 520–535.

Demolon, A. 1951. *Revue gen. bot.*, **58**, 110 (quoted by Pochon, G. and de Barjac, 1960. Izd-vo. Inostr. Lit., Moscow).

Dilworth, M. and Parker, C. 1961. Oxygen inhibition of respiration in *Azotobacter*. *Nature*, **191**, 520–521.

Dilworth, M. Y., Subramanian, D., Munson, T. O. and Burris, R. H. 1965. The adenosine triphosphate requirement for nitrogen fixation in cell-free extracts of *Clostridium pasteurianum*. *Biochim et Biophys. Acta*, **99** (3), 486–503.

Dua, R. D. 1964. Stability of the nitrogen-fixing enzyme complex and purification of nitrogenase and hydrogenase. *Diss. Abstrs.*, **24** (9), 3525.

Dua, R. D., Burris, R. H. 1965. Studies of cold lability and purification of a nitrogen-activating enzyme. *Biochim. et Biophys. Acta*, 504–510.

D'Eustachio, A. J. and Hardy, R. W. F. 1964. Reductants and electron transport in nitrogen fixation. *Biochem. and Biophys. Res. Communs.*, **75** (4), 319–323.

Eischens, R. P. and Jacnow, T. 1964. *Proc. Third Internat. Congr. on Catalysis, Amsterdam* (quoted by Borod'ko *et al.*, 1966).

Fedorov, M. V. 1945. Effect of surfactants on the rate of fixation of atmospheric nitrogen by azotobacters. *Dokl. Akad. Nauk SSSR*, **49** (8), 629–632; (9), 702–705; (1), 501–504; **51** (1), 61–64; **58** (1), 81–84.

Fedorov, M. V. 1952. *Biologicheskaya fiksatsiya azota atmosfery*. (Biological fixation of atmospheric nitrogen), Sel'khozgiz, Moscow.

Fottrell, P. F. 1966. Dehydrogenase isoenzymes from legume root nodules. *Nature*, **210**, 198–199.

Gapon, E. N. 1947. Theory of fixation of atmospheric nitrogen by micro-organisms. *Dokl. Akad. Nauk SSSR*, **LVIII** (2), 249–252.

Gautier, A. and Drouin, R. 1888. Recherches sur la fixation de l'azote par le sol et les végétaux. *C.R. Acad. Sci.*, **106**, janvier-juin. 944–947.

Gautier, A. and Drouin, R. 1891. Sur la fixation de l'azote par le sol arable. *C.R. Acad. Sci.*, **113**, juillet-decembre, 820–825.

Geiko, N. S., L'vov, N. L. and Lyubimov, S. I. 1965. The keto acids of *Mycobacterium azot-absorptum Dokl. Akad. Nauk SSSR*, **165** (3), 699–700.

Grau, F. H. and Wilson, P. W. 1961. Cell-free nitrogen fixation by *Bacillus polymyxa*. *Bacteriol. Proc.*, **193**.

Grau, F. H. and Wilson, P. W. 1963. Hydrogenase and nitrogenase in cell-free extracts of *Bacillus polymyxa*. *J. Bacteriol.*, **85** (2), 446–450.

Grimes, H. and Fottrell, P. F. 1966. Enzymes involved in glutamate metabolism in legume root nodules. *Nature*, **212**, 295–296.

Grin, D. (Green, D.) 1961. Structure and function of subcellular particles: in *IX Mezhdunarodnyi Mikrobiologicheskii Kongress. Plenarnaya lektsiya* (Ninth International Microbiology Congress, Plenary Lecture). Moscow.

Haight, G. P. and Scott, R. J. 1964. Molybdate and tungstate-catalysed fixation of nitrogen. *J. Amer. Chem. Soc.*, **86** (3), 7430744.

Hamilton, P. B., Shug, A. L. and Wilson, P. W. 1957. Spectrophotometric examination of hydrogenase and nitrogenase in soybean nodules and *Azotobacter*. *Proc. Nat. Acad. Sci. USA*, **43**, 297–304.

Hamilton, P. and Wilson, P. 1955. Nitrogen fixation by *Aerobacter aerogenes*. *Ann. Acad. Sci. fennicae*, ser. A, **11** (60), 139–150 (in English).

Hardy, R. W. F. and D'Eustachio, A. J. 1964. The dual role of pyruvate and energy requirements in nitrogen fixation. *Biochem. and Biophys. Res Communs*, **15** (4), 314–318.

Hardy, R. W. F., Knight, E. and D'Eustachio, A. J. 1965. An energy dependent hydrogen evolution from dithionite in nitrogen-fixing extracts of *Clostridium pasteurianum*. *Biochem. and Biophys. Res. Communs.*, **20** (5), 539–544.

Hardy, R. W. F., Knight, E., Donald, M. and D'Eustachio, A. J. 1965. *Paramagnetic protein from nitrogen-fixing extracts of Clostridium* Pan Pietro, A, (Ed.), Antioch. Press., Ohio.

Hoch, C. E., Little, H. N. and Burris, R. H. 1957. Hydrogen evolution from soybean root nodule. *Nature*, **179**, 430 431.

Il'ina, T. K. 1966. Effect of copper, cobalt and other trace elements on fixation of nitrogen by soil *Mycobacteria*. *Mikrobiologiya*, **35** (2), 323–327.

Imshenetskii, A. A. 1961. Evolution of biological fixation of nitrogen: in *V Mezhdunarodnyi Biokhemicheskii Kongress. Simpozium II* (Fifth International Biochemical Congress, Symposium II), 10–18.

Ivanov, I. D. 1966. Nitrobiosis-nitrogen respiration: in *IX Mechdunarodnyi Mikrobiologicheskii Kongress. Tezisy dokladov* (Ninth International Microbiology Congress. Summaries of speeches), Moscow, 284.

Ivanov, I. D. and Demina, N. A. 1966. Effect of inhibitors on the electron transport chain in purple bacteria in connection with the fixation of molecular nitrogen. *Mikrobiologiya*, **35** (5), 780–784.

Ivanov, I. D., Sitonite, Yu. P. and Belov, Yu. N. 1965. Nitrogen fixation as a hydrogen-acceptor process. *Mikrobiologiya*, **35** (2), 193–199.

Karnaukhov, Yu. I. 1965. Mechanism of biological fixation of molecular nitrogen. *Izv. Akad. Nauk SSSR, Ser. Biol.*, (5), 714–730.

Kellermann, K. F. and Smith, N. R. 1914. The absence of nitrate formation in cultures of *Azotobacter*, *Zbl. Bakteriol., Parasitenkunde, Infektionskrankh und Hyg.*, Abt. 2, **40** (1–8), 479–482 (in English).

Kennedy, J. R. 1966. Primary products of symbiotic nitrogen fixation. I. Short-term exposures of seradella nodules to $^{15}N_2$. II. Pulse-labelling of seradella nodules with $^{15}N_2$. *Biochim et Biophys. Acta*, **130** (2), 285–303.

Knight, E. J. and Hardy, R. W. F. 1966. Isolation and characteristics of flavodoxin from nitrogen-fixing *Clostridium pasteurianum*. *J. Biol. Chem.*, **241** (12), 2752–2756.

Kostychev, S. P. Ryskal'chuk, A. T., Shvetsova, O. I. 1926a. Chemical investigations on the binding of molecular nitrogen by the microbe *Azotobacter agile*. *Trudy Otd. s-kh. mikrobiol, GIOA*, 1, 91–98.

Kostychev, S. P. Ryskal'chuk, A. T. and Shvetsova, O. I. 1926b. Biochemische Untersuchungen über *Azotobacter agile*, *Z. physiol. Chem. Hoppe-Seyler's*, **154**, 11–17.

Krasheninnikov, F. 1916. Assimilation of gaseous nitrogen by the root nodules of legumes: in *Sbornik posvyashchennyi K. A. Timiryazevu ego uchenikami* (Festschrift for K. A. Timiryazev by his pupils), Moscow, 307–324.

Kretovich, V. L., Bundel', A. A., Geiko, N. S., Losera, L. P., Lyubimov, V. I., L'vov, N. I. Yakovleva, V. I. 1966. Keto acids in the metabolism of nitrogen-fixing microorganisms: in *IX Mezhdunarodnyi Mikrobiologicheskii Kongress. Simpozium V-1, Tezisy dokladov* (Ninth International Microbiology Congress Symposium V-1. Summaries of speeches), Moscow, 90–71.

Kretovich, V. L. and Lyubimov, 1964. The biochemistry of nitrogen fixation. *Priroda*, (12), 14–21.

Kurdina, R. M. 1957. Fixation of atmospheric nitrogen by azotobacters. *Izv. Akad. Nauk KazSSR, ser. biol.*, (12), 56–69.

Leaf, G., Gardner, J. and Bond, C. 1958. Observation on the composition and metabolism of the nitrogen-fixing root nodules of *Alnus*, *J. Exp. Bot.*, **9**, 320.

Loew, and Aso. 1909. Changes of availability of nitrogen in soils. II. *Zbl. Bakteriol. Parasitenkunde, Infektionskrankh. und Hyg.*, **22** (14–17), 452.

Lovenberg, W., Buchanan, B. B. and Rabinowitz, J. C. 1963. Studies on the chemical nature of clostridial ferredoxin. *J. Biol. Chem.*, **238** (12), 3899–3913.

Lovenberg, W., Sobel, B. E., 1965. Rubredoxin: a new electron transfer protein from *Clostridium pasteurianum. Proc. Nat. Acad. Sci. USA*, **54**, 1.

Lyubimov, V. I. 1957. Biochemical aspects of biological fixation of molecular nitrogen. *S-kh. biologiya* (1), 3–25.

Lyubimov, V. I. 1963. Fixation of molecular nitrogen by cell-free preparations of micro-organisms. *Izv. Akad. Nauk SSSR, ser. biol.*, 681–692.

Lyubimov, V. I. *et al.*, 1965. Induced character of the enzymes of nitrogen fixation in *Mycobacterium azot-absorptum* n. sp. *Izv. Akad. Nauk SSSR, ser. biol*, (3).

L'yuis and Rendall (Lewis and Randall) 1936. *Khimicheskaya termodinamika* (Chemical thermodynamics), Goslitizdat, Moscow.

Magee, W. E. and Burris, R. H. 1956. Oxidative activity and nitrogen fixation in cell-free preparations from *Azotobacter vinelandii J. Bacteriol.*, **71** (6), 635.

McNary, J. E. and Burris, R. H. 1962. Energy requirements for nitrogen fixation by cell-free preparations from *Clostridium pasteurianum. J. Bacteriol.*, **84** (3), 588.

Meyerhof, O. and Burk, D. 1928. Über die Fixation des Luftstickstoffs durch *Azotobacter. Z. phys. Chem.*, A, **139**, 117–142.

Michaelis, L. and Rona, P. 1930. *Practicum der physicalische Chemie.*, 4 Aufl.

Mortenson, L. 1961. Inorganic nitrogen assimilation and ammonia incorporation. in Gunsular, J. C., Stanier, R. J. *The Bacteria*, 8 *Biosynthesis*, 119.

Mortenson, L. E. 1962. A simple method for measuring nitrogen fixation by cell-free enzyme preparations of *Clostridium pasteurianum. Analyt. Biochem.*, **2** (3), 216–220.

Mortenson, L. E. 1964. Purification and analysis of ferredoxin from *Clostridium pasteurianum. Biochim. et Biophys. Acta*, **81** (1), 71–77.

Mortenson, L. E. 1964. Ferredoxin and ATP requirements for nitrogen fixation in cell-free extracts of *Clostridium pasteurianum. Proc. Nat. Acad. Sci.*, *USA*, **52** (2), 272–279.

Mortenson, L. E. 1966. Components of cell-free extracts of *Clostridium pasteurianum* required for ATP—dependent H_2 evolution from dithionite and for N_2 fixation. *Biochim, et Biophys, Acta*, **127** (1), 18–25.

Mortenson, L. E., Mower, H. and Carnahan, J. 1962. Nitrogen fixation by enzyme preparation. *Bacteriol. Revs.*, **26**, 42–50.

Mortenson, L. E., Valentine, R. C. and Carnahan, J. B. 1962. An electron transport factor from *Clostridium pasteurianum. Biochem and Biophys. Res. Communs*, 7 (6), 448–452.

Mortenson, L. E., Valentine, R. and Carnahan, J. 1963. Ferredoxin in the phosphoroclastic reaction of pyruvic acid and its relation to nitrogen fixation in *Clostridium pasteurianum. J. Biol. Chem.*, **238** (2), 794–799.

Naik, M. S. and Nicholas, D. J. D. 1966. $NADH_2$-benzyl viologen reductase from *Azotobacter vinelandii. Biochim. et Biophys. Acta*, **118** (1), 195–197.

Nason, A., Takahashi, H., Hoch, G. and Burris, R. 1957. Nitrogen fixation in sonicates of *Azotobacter. Federat. Proc.*, **16**, 224.

Nechiporenko, G. N. *et al.*, 1965. Mechanism of nitrogen fixation in the reacting system $(C_2H_5)_2TiCl_2-C_2H_5MgBr$. *Dokl. Akad. Nauk SSSR*, **164** (5), 1062–1064.

Nicholas, D. I. D. 1963. The biochemistry of nitrogen fixation. Symbiotic associations. *13 Sympos. Soc. Gen. Microbiol.* Cambridge, Univ. Press.

Nicholas, D. I. D. and Fisher, D. J. 1960a. Nitrogen fixation in extracts of *Azotobacter vinelandii. Nature*, **184**, 26, 735–736.

Nicholas, D. I. D. and Fisher, F. J. 1960b. Nitrogen fixation in extracts of *Azotobacter vinelandii. J. Sci. Food and Agric.*, **77** (10), 603–609.

Nicholas, D. I. D., Fisher, D. J., Redmond, W. J. and Wright, M. A. 1960. Some aspects of hydrogenase activity and nitrogen fixation in *Azotobacter* spp. and in *Clostridium pasteurianum. J. Cen. Microbiol.*, **22**, 191–205.

Nicholas, D. I. D., Silvester, D. I. and Fowler, J. F. 1961. Use of radioactive nitrogen in studying nitrogen fixation in bacterial cells and their extracts. *Nature*, **189**, 634–636.

Nimeck, M. W., Wilson, P. W. and Nicholas, D. I. D. 1963. Nitrogen fixation in cell-free extracts of *Azotobacter vinelandii* prepared by lysis with phage A$_{22}$. *Nature*, **200**, 709.

Oparin, A. I. 1957. *Vozniknovenie zhizni na Zemle* (Origin of life on earth), Izd-vo, Akad, Nauk SSSR, Moscow.

Parker, C. A. 1954. Effect of oxygen on the fixation of nitrogen by *Azotobacter. Nature*, **173**, 780.

Parker, C. A. and Scutt, P. B. 1960. The effect of oxygen on nitrogen fixation by *Azotobacter. Biochem et Biophys. Acta*, **38** (2), 230.

Peive, Ya. V. 1965. The biochemistry of trace elements and the problems of nitrogen nutrition of plants. *Vestn. Akad. Nauk SSSR*, (4), 42–50.

Peive, Ya. V. and Zhiznevskaya, G. Ya. 1961. Effect of molybdenum and copper on the activity of nitrate reductase in plants: in *Mikroelementy i urozhai* (Trace elements and the harvest), Riga.

Peive, Ya. V. Yagodin, B. A. and Popazova, A. D. 1957. Role of cobalt, copper and molybdenum in raising the activity of hydrogenase in the nodules of broad beans. *Agrokhimiya*, (1), 94–99.

Pratt, J. M. 1962. Fixation of nitrogen. *J. Theoret. Biol.*, **2**, 251–258.

Remy, Th. and Rosing, C. 1911. Über die biologische Reiswirkung natürlicher Humusstoffe. *Zbl. Bakteriol, Parasitenkunde*: in *fektionskrankh. und Hyg.*, Abt. 2, **30** (1–3), 349–383.

Roberg, M. 1936. Beitrage zur Biologie von *Azotobacter*. III. Zur Frage eines aussernall der Zelle den stickstoffbindenden Enzymes. *Jahresb. wiss. Bot.*, **83**, 567.

Rosenblum, R. D. and Wilson, P. W. 1950. Molecular hydrogen and nitrogen fixation by *Clostridium. J. Bacteriol.*, **59**, 83–91.

Rosenblum, E. D. and Wilson, P. W. 1951. The utilization of nitrogen in various compounds by *Clostridium pasteurianum. J. Bacteriol.*, **61**, 475–480.

Roy, A. B. and Mukherjee, M. K. 1957. A new type of nitrogen-fixing bacterium. *Nature*, **180**, 236.

Schneider, K. C., Braubeer, C. Singh, R. N., Wang, L. C., Wilson, P. W. and Burris, R. H. 1960. Nitrogen fixation by cell-free preparations from micro-organisms. *Proc. Nat. Acad. Sci. USA*, **46** (5), 726–733.

Sen, S. P. and Burma, D. P. 1953. A study with paper chromatography of the amino-acids in legume nodules. *Bot. Gaz.*, **115**, 190.

Shug, A. L., Hamilton, P. B. and Wilson, P. W. 1956. Hydrogenase and nitrogen fixation: in *Inorganic nitrogen metabolism*. Elroy, W. D., Glase B. (Eds.). John Hopkins Press, Baltimore, 344–360.

Shug, A. L., Wilson, P. W., Green, D. R. and Mahler, H. R. 1954. The role of molybdenum and flavin in hydrogenase. *J. Amer. Chem. Soc.*, **79** (12), 3355–3356.

Sloan, M. F., Matlack, A. S. and Breslow, D. S. 1963. Soluble catalysts for the hydrogenation of olefins. *J. Amer, Chem. Soc.*, **85** (24), 4014–4018.

Spencer, D., Takahashi, H. and Nason, A. 1957. Relationship of nitrite and hydroxylamine reductases to nitrate assimilation and nitrogen fixation in *Azotobacter agile*. *J. Bacteriol.*, **73**, 553.

Stoklasa, J. von. 1908. Beitrag zur Kenntnis der chemischen Vorgënge bei der Assimilation des élementaren Stickstoffs durch *Azotobacter* und *Radiobacter*. *Zbl. Bakteriol., Parasitenkunde, Infektionskrankh. und Hyg.*, Abt. 2, **21** (1–3), 484–509, 620–632.

Syrkin, Ya. K. 1960. Formation of three-centre bonds on chemosorption. *Zh. strukturnoi khimii*, **7** (2), 189–199.

Tagawa, K. and Arnon, D. 1962. Ferredoxins as electron carriers in photosynthesis and in the biological production and consumption of hydrogen gas. *Nature*, **195**, 537–543.

Turchin, F. V. 1959. New information on the mechanism of fixation of nitrogen (atmospheric) in the nodules of leguminous plants. *Pochvovedenie*, (10), 14–17.

Turchin, F. V. Berseneva, Z. N. and Shidkikh, G. G. 1963. Fixation of atmospheric nitrogen *in vitro* by enzyme preparations from nodulated legumes and higher plants not infected with bacteria. *Dokl. Akad. Nauk SSSR*, **149** (3), 731.

Turchin, F. V. *et al.*, 1958. Study of the biological fixation of atmospheric nitrogen in the nodules of leguminous plants using ^{15}N: in *Fiziologiya rastenii, agrokhimiiya, pochvovedenie*. (Plant physiology, agrochemistry and pedology), Moscow.

Valentine, R. C. 1964. Bacterial ferredoxin. *Bacteriol. Revs*, **28** (4), 497–517.

Vinogradskii, S. N. 1894. Assimilation by microbes of gaseous atmospheric nitrogen. *C.R. Acad. Sci.*, **118** (in Russian).

Vinogradskii, S. N. 1895. Assimilation of free atmospheric nitrogen by microbes. *Arkhiv. biol. nauk*, **3**, 293.

Virtanen, A. I. 1947. The biology and chemistry of nitrogen fixation by legume bacteria. *Biol. Revs*, **22**, 239.

Virtanen, A. I. 1952. Some aspects of biological nitrogen fixation. *Ann. Acad. Sci. fennicae*. ser. A, II, (43) (in English).

Virtanen, A. I. and Hakala, M. 1949. Anaerobic nitrogen fixation and formation of oxime nitrogen. *Acta chem-scand.*, **3**, 1044–1049 (in English).

Virtanen, A. I. and Laine, T. 1936. Fixation of nitrogen in the root nodules. *Suomen kem* **9**, 5 (in English).

Virtanen, A. I. and Miettinen, J. K. 1963. Biological nitrogen fixation: in *Plant Physiol.*, Edit. Steward, F. III. 5390669.

Vol'pin, M. E. and Shur, V. B. 1964. Fixation of nitrogen on complex catalysts. *Dokl. Akad. Nauk SSSR*, **156** (5), 1102–1104.

Vol'pin, M. E., Shur, V. B. and Ilatovskaya, M. A. 1964. Fixation of nitrogen by systems based on dicyclopentadientyltitanium dichloride. *Izv. Akad. Nauk SSSR, ser. Khim.*, (9), 1728–1729.

Vol'pin, N. E. and Shur, V. B. 1965. Problems of chemical fixation of molecular nitrogen. *Vestn. Akad. Nauk SSSR*, (1), 51–58.

Wall, J. S., Wagenknecht, A. C., Newton, J. W. and Burris, R. H. 1952. Comparison on the metabolism of ammonia and molecular nitrogen in photosynthesing bacteria. *Bacteriol, Revs*, **63**, 563.

Wilson, P. (Vil'son, P.) 1951. Biological fixation of nitrogen: in *Fiziologiya bakterii* (Physiology of bacteria), Inostr. Lit., Moscow, 1954, 363.

Wilson, P. W. 1952. The comparative biochemistry of nitrogen fixation. *Advances Enzymol.*, **13**, 345–375.

Wilson, P. W. and Burris, R. H. 1947. The mechanism of biological nitrogen fixation. *Bacteriol. Revs.*, **11**, 41.

Wilson, P. W. and Burris, R. H. 1953. Biological nitrogen fixation, a reappraisal. *Annual Rev. Microbiol.*, **7**, 415–432.

Wilson, T. G. G. and Roberts, E. R. 1954. Attempts at nitrogen fixation *in vitro*. *Nature*, **174**, 795.

Wilson, P. W. and Umbreit, W. V. 1937. Mechanism of symbiotic nitrogen fixation. III. Hydrogen as a specific inhibitor. *Arch. Mikrobiol.*, **8**, 440–457.

Winfild, M. E. 1951. Adsorption and hydrogenation of gases on transition metals. *Austral. J. Sci. Res.*, **4**, A, 385.

Winfild, M. E. 1955. Reactions of hydrogen gas in solution. *Rev. Pure and Appl. Chem.*, **5**, 217.

Wyss, O., Lind, C. G., Wilson, J. B. and Wilson, P. W. 1941. Mechanism of biological nitrogen fixation. 7. Molecular H_2 and the pN_2 function of Azotobacter. *Biochem. J.*, **35**, 845.

Yakovlev, V. A. 1966. Investigation of nitrogen fixation in *Azotobacter vinelandii*: in *IX Mezhdunarodnyi Mikrobiologicheskii Kongress, Simpozium V-1* (Ninth International Microbiology Congress, Symposium V-1).

Yakovlev, V. A., Vorob'ev, L. V., Levchenko, L. A., Linde, V. R., Slepko, G. I., Syrtsova, L. A. 1965: Investigation of the biological fixation of molecular nitrogen. *Biokhimiya*, **30** (6), 1167–1179.

Yakovlev, V. A. and Kevchenko, L. A. 1964. Localization of dehydrogenases associated with nitrogen fixation in *Azotobacter vinelandii*. *Dokl. Akad. Nauk SSSR*, **159** (5), 1173.

Yakovlev, V. A. and Levchenko, L. A. 1966. Hydrogenase and succinic dehydrogenase of *Azotobacter vinelandii* and their link with nitrogen fixation. *Dokl. Akad. Nauk SSSR*, **171** (5), 1224–1227.

Yakovlev, V. A. Syrtsova, L. A. and Vorob'ev, L. V. 1966. Isolation and study of iron-containing protein of a non-haeme nature from *Azotobacter vinelandii*. *Dokl. Akad. Nauk SSR*, **171** (2), 477–481.

Yakovlev, V. A. Lyubimov, V. I., Loseva, L. P., and Kretovich, V. L. 1964. Glutamic dehydrogenase of *Azotobacter vinelandii*. *Dokl. Akad. Nauk SSSR*, **158** (6), 1427–1429.

Yocum, C. S. 1960. Nitrogen fixation. *Annual Rev. Plant Physiol.*, **11**, 25 (quoted by T. A. Kalininskaya, *Vestn. Akad. Nauk SSSR*, (7), 44–50 1962).

Zacharias, B. 1962. Effect of UV-irradiated nitrogen gas on biological nitrogen fixation in batch cultures of *Azotobacter vinelandii*. *Biotechnol and Bioengineering*, **4**, 87–97.

Zelitch, J., Rosenblum, E. D., Burris, R. H., Wilson, P. W. 1951. Isolation of the key intermediate in biological nitrogen fixation by *Clostridium*. *J. Biol. Chem.*, **191** (1), 295–298.

Index